HUMAN GROWTH
AND DEVELOPMENT
THROUGHOUT LIFE

HUMAN GROWTH AND DEVELOPMENT THROUGHOUT LIFE

A NURSING PERSPECTIVE

Edited by

Patty Maynard Hill, R.N., B.S.N., M.Ed.
Assistant Professor
School of Nursing
University of North Carolina
Chapel Hill, North Carolina

Patricia Humphrey, R.N., B.S.N., M.P.H., F.N.P.
Assistant Professor
School of Nursing
Duke University
Durham, North Carolina

175 YEARS OF PUBLISHING
1807 JW 1982

A WILEY MEDICAL PUBLICATION
John Wiley & Sons / New York • Chichester • Brisbane • Toronto • Singapore

Cover design: Wanda Lubelska
Production Editor: Cheryl Howell

Library of Congress Cataloging in Publication Data:

Main entry under title:

Human growth and development throughout life.

  (A Wiley medical publication)
  Includes index.
  1. Life cycle, Human.   2. Human growth.
3. Developmental psychology.   4. Nursing.
I. Hill, Patty Maynard.   II. Humphrey, Patricia.
III. Series.
RT69.H85   612′.6   81-16472
ISBN 0-471-05814-9   AACR2

Printed in the United States of America

10 9 8 7 6 5 4 3 2 1

To my parents, Mr. and Mrs. R. W. Maynard; my husband, Gary; and our children, Gary, Jr., Christopher, and Caroline

P. M. H.

To, Mom, Dad, Jim, Maggie, and P. H. C.

P. H.

Contributors

Glenda F. Butnarescu, R.N., M.S.
Associate Professor
School of Nursing
University of Texas Health Science Center at
  San Antonio
San Antonio, Texas

Jacki Delecki, C.R.N., M.S.N.
Instructor
Family Nurse Practitioner Program
School of Nursing
University of Washington
Seattle, Washington

Patty Maynard Hill
Assistant Professor
School of Nursing
University of North Carolina
Chapel Hill, North Carolina

Patricia Humphrey
Assistant Professor
School of Nursing
Duke University
Durham, North Carolina

Betty H. Landsberger, Ph.D.
Assistant Professor
School of Nursing
University of North Carolina
Chapel Hill, North Carolina

Amie Modigh, R.N., M.S.N., G.N.P.
Clinical Assistant Professor
Duke University

Durham, North Carolina
University of North Carolina
Chapel Hill, North Carolina
East Carolina University
Greenville, North Carolina
Health Care Director
The Sailors' Snug Harbor
Sea Level, North Carolina

Bernice D. Owen, R.N., M.S.
Lecturer
Doctoral Candidate
School of Nursing
University of Wisconsin, Madison
Madison, Wisconsin

Eleanor Reiff-Ross, M.A., M.Ed., Ph.D.
Assistant Professor
Center for Research
School of Nursing
University of Maryland
Baltimore, Maryland

Patricia Short-Tomlinson, R.N., M.N.
Associate Professor
School of Nursing
University of Oregon Health Sciences Center
Portland, Oregon

Elsie Williams-Wilson, R.N., M.S.
Associate Professor
School of Nursing
The Ohio State University
Columbus, Ohio

Preface

The sequence of human growth and development follows an orderly and predictable course, yet each human being is unique in many respects. This apparent paradox comes to the attention of students of growth and development almost as soon as the topic is addressed. This text examines the physiological, psychosocial, and cognitive aspects of human growth and development throughout the life span, with special reference to applicability to nursing and allied health. It provides the student with a sound theoretical base. It additionally fosters an appreciation of the need for individualized health care.

The first four chapters are concerned with theories of, influences on, and principles of growth and development, and the reproductive process as seen from a developmental perspective. Chapters 5 through 13, the "core" chapters, treat each phase of the life span, from infancy (including the neonatal period) through advanced old age, and the final chapter deals with death and the grief process. Each chapter concludes with a summary, and the text is further supported by tables, over 90 illustrations, a glossary, and appendixes.

Although the emphasis is on normal growth and development, each core chapter includes sections on anticipatory guidance and health and developmental problems common to that phase of the life span. The core chapters also feature suggestions on how to approach clients of each age group in such a way as to promote effective communication and the establishment of rapport.

Throughout the text, the nurse is referred to as *she*, and, in most cases, the client is identified as *he*. This has been done solely for the sake of clarity; no discrimination against male nursing students and nurses or female clients is intended.

We believe that *Human Growth and Development Throughout Life* fulfills the need for a concise text on human growth and development that is especially concerned with applicability to the delivery of holistic health care in all settings.

P. M. H.
P. H.

Acknowledgments

The authors gratefully acknowledge the contributions of the following people: our contributors, for investing the time required to write for the book; Bud Bynum, John Elkins, and Sam Gray for their photographs; Gay Blackwell, Kathy Christy, Holly Cook, Rosanna Gaskill, Louise Gilmore, and Janice Walker for assistance with typing. Elizabeth Tournquist for editorial assistance; the many educators (including our faculty colleagues) who encouraged our efforts and provided support when we got bogged down in the manuscript; those who have preceded us in writing on the subject of growth and development, for it is they who inspired and challenged us, first as students, later as practitioners, and later still as educators; our students, who have taught us much and inspired us to continue to grow; and our families and special friends, who supported and encouraged our endeavors, understood when we were busy, and assured us that we would indeed finish the book.

P. M. H.
P. H.

Contents

1

Psychosocial Theories Relating to Human Growth and Development

ELEANOR REIFF-ROSS

Systematic observation and confirmation are the hallmarks of scientific inquiry and are means of building the data base of a science. However, in studying complex and diverse phenomena such as human growth and development, one must also study the classical thinkers in human development even though some findings are not derived from experimental designs or statistical analyses. Nursing research places great emphasis on appropriate research methods and professionals are expected to incorporate research findings into practice. Nursing practice is in fact built on a foundation that is both theoretical and empirical.

PREVENTIVE CARE VERSUS THERAPEUTIC CARE

One of the most influential American social theorists of the twentieth century, Talcott Parsons, wrote:

> . . . it is clear that there is a functional interest of the society . . . in the minimization of illness. As one special aspect of this, attention may be called to premature death. From a variety of points of view, the birth and rearing of a child constitute a "cost" to the society through pregnancy, child care, socialization, formal training and many other channels. Premature death, before the individual has had the opportunity to play out his full quota of social roles, means that only a partial return for this cost has been received. (Parsons, 1964)

Thus, with regard to the optimal functioning of the social system, preventive care is preferable to rehabitative care. Since the early 1950s there has been a proliferation of health maintenance organizations; however, it would be an exaggeration to say that they are typical of the health care system. Regarding nursing practice, it is true that the preventive approach is increasingly more common.

SECURITY AND ATTACHMENT AS FACTORS IN CHILD AND ADULT BEHAVIOR

Anxiety and fear are present from the moment the human being takes his first breath until the day he dies. Human attitudes, morale, decisions, and mental and physical health status are shaped by perceived personal security. Fear, which is a concomitant of the absence of security, appears to be largely acquired after birth rather than innate. Probably the neonate arrives with at least one fear, that of falling. Fears differ from anxieties in that fears tend to be object-directed, specific, and gounded in reality, whereas anxieties tend to be diffuse and irrational. For example, the client who refuses to be assured of his or her good health even though all physical signs indicate its presence is displaying anxiety. Naturally, some fears—such as fear of being mugged in a high-crime neighborhood—are useful and life-preserving, but groundless fears can be described as anxiety. Sometimes fears or anxieties become incapacitating in that they seriously interfere with one's daily functioning. An intervention by a psychiatrist, psychologist, or psychiatric nurse might be required.

It is apparent that people differ widely in the scope and depth of their fears and anxieties. How can this variability be accounted for? This chapter summarizes the viewpoints of several leading psychologists whose various orientations focus on different aspects of human life.

THE TRANSITION TO EXTRAUTERINE EXISTENCE

Otto Rank proposed that the newborn experiences a "birth trauma," the effects of which may remain for life (Sander and Rank, 1925). "Separation anxiety" is a well-documented construct in psychological and nursing literature, yet it is used to refer to the anxiety experienced by the young child when he is separated from his mother; little is remarked about the *immediate* separation of neonate and mother at birth, an experience shared by all living creatures. Could it not be that therein is the germ of some lifetime anxiety? In nursing practice, partial remedies exist; it would seem that the best interest of mother and child would be served by housing them together in one unit from the moment of birth, and such "rooming-in" is indeed an accepted practice in many hospitals.

ISOLATION OF THE CHILD

Separation of the mother from the neonate, infant, or child has been shown to have permanent deleterious effects. The gravity of the effect depends on the length of separation and the stage of development of the child at the time. Even enforced absences because of illness of child or mother, routine "propping" of nursing bottles rather than bottle-feeding with personal contact, or job-induced absences of the mother at critical stages may have irrevocable negative effects.

The British child psychiatrist John Bowlby (1951) reported depressive reactions in infants as young as 12 months when separated from their mothers. He observed three stages in the depressive reaction: the first was characterized by strong protest and crying, the second stage was one of despair, anxiety, and depression in which the infant can be characterized as withdrawn, inactive, and making no demands on the environment. The third phase is that which seemed to portend long-term effects; in it, the infant appears detached, interacting in a shallow manner with caretakers, aloof and withdrawn when the mother finally appears. While Bowlby reported no follow-up data, he did describe the effects of long separation on British children during World War II; the children displayed some psychopathological symptoms—detachment, inability to relate to others or to form lasting relationships, avoidance and rejection of help from another, poor achievement in school, lying, stealing, cheating, and high-level pervasive anxiety.

By far the most widely known work on the effects of maternal deprivation is that of Harry Harlow with infant rhesus monkeys at the University of Wisconsin's Primate Laboratory. Harlow constructed two "surrogate mothers." One was carved from a block of wood, layered with sponge rubber, and covered with terry cloth. It had a light bulb in its back to radiate heat. The second surrogate mother was constructed of wire mesh and was warmed by radiant heat. The main difference between the two surrogate mothers was what Harlow termed the "contact comfort" of the terry cloth mother;

physical contact with the wire mother created no sensation of softness. Harlow conducted experiments in which both the wire and cloth surrogate mothers "lactated" by means of attached bottles. The infant monkeys fed by the cloth mother preferred her 15 to 20 times as much as the wire mother. The key variable was the "contact comfort" experienced with the soft, yielding, surrogate mother. However, Harlow's monkeys grew up to be monster monkeys. They did not relate to other adult monkeys; the females would not willingly copulate with the males, and if raped, the females would not nurture their own newborns. Harlow's work also supports the hypothesis that mother love is not innate; the odor, appearance, and responsiveness of the offspring appear to be key variables in the development of a bond between mother and child (Harlow and Harlow, 1966).

Increasing attention is being given to the distinctive life experiences of children. Corollary to the recognition of children is recognition of the rights of children. As far back as 1944, Margaret Ribble vividly described the plight of an infant left by his mother in an understaffed hospital. The infant could not retain food; he lost weight progressively; his muscular growth and development were reversed; and he appeared to be in a life-threatening situation. Subsequent mothering did help to restore some normal physiological growth and development, but the child suffered physiological retardation and permanent emotional damage (Ribble, 1944).

Rene Spitz (1945) is associated with *anaclitic depression*, a term used to characterize infants who show signs of emotional withdrawal, arrested development, lack of appetite, inability to sleep, and weight loss. Whereas Bowlby's and Ribble's work lacked the observation of a contrast group, Spitz encountered two groups of children living in institutional settings. One group, whose mothers were unable to support them, lived in a foundling home; the other group, whose mothers were convicts, lived in a nursery attached to the women's prison. Both groups of children were admitted shortly after birth, and both groups were given the best care. However, the foundling home children lived in a monotonous environment. No one played with them, cuddled them, or otherwise interacted with them. They were cared for by overworked personnel who had no time for playing with the children. The prison mothers, however, could spend a few hours a day with their babies. Spitz found the convicts' children to be better developed.

Nineteen out of 23 of the foundling home children were identified by Spitz as suffering from anaclitic depression. Eventually these 19 children worsened into extreme lethargy and physical deterioration, which Spitz termed "hospitalism." Finally, over a two-year period, 37 percent of these children died; none of the nursery prison children died! Infants, whether humans or monkeys, appear to require "contact comfort," not only for optimal functioning but for survival. Although the research methods used by Spitz and Ribble were not controlled in the scientific sense, their conclusions have been supported time and time again by later observers. Even adults appear to react to a lack of human contact with insecurity and anxiety.

SMOTHERING THE CHILD

Popular wisdom has it that "you can't get too much of a good thing," but this does not apply to mothering; when the child receives such excessive mothering that independence, a sense of competence and self-esteem do not emerge, he may suffer from arrested personality development and insecurity, which may be manifested in selfish, demanding, undisciplined behavior, difficulty in befriending other children, and a tendency toward isolation. These are the findings of David Levy, a child psychoanalyst and distinguished experimental and clinical researcher.

Describing one of his cases, Levy profiles an 11-year-old boy whose mother "has never gone anywhere without" him, and who continued to sleep with the boy even after he attained his eleventh year. Colicky as a baby, he received overprotective care from his mother. At the age of $3\frac{1}{2}$, his life was jeopardized by double lobar pneumonia to which empyema was a sequel. Despite a poor prognosis, the boy recovered, gaining weight as he reached his next birthday. An examination at that time found the boy's genitals to be very small; Fröhlich's syndrome (adiposogenital dystrophy) was diagnosed, and several years of hormone therapy were begun. The mother took it upon herself to make a daily examination of the boy's genitals, a routine she was continuing at the time of Levy's examination. At around the age of 7, the boy was being demonstrated in a clinic; the mother claimed that at that time she overheard a doctor tell his students that the boy would not live past 17. This was the second occasion on which the boy's death was predicted (the first being during his pneumonia); Levy remarks, "these experiences would act presumably as factors keenly intensifying apprehension about the child's death and would reinforce the mother's general oversolicitude about her son's health" (Levy, 1966).

It should be obvious that too much mothering can be as detrimental as not enough mothering. It is incontrovertible that a normal mother–child relationship requires consistent and balanced loving care by the mother or by another devoted and loving person.

COGNITIVE AND PERSONALITY DEVELOPMENT

Why is it that primates are so greatly reinforced by close bodily contact, warmth, and relationships with others? Why do human beings throughout life seek the contact of other human beings? And most important, why does "free-floating anxiety" result when the social system fails to support an individual?

The fetus is *attached* to the mother during uterine life. As far as is known, the fetus experiences no pain or discomfort, provided that the mother is healthy. During the birth process, the attachment is abruptly and permanently ended. The infant throughout its extrauterine existence seeks bodily contact and warmth. The failure to regain these essential attributes of the prebirth life can result in the transformation of literally "free-floating" ease to free-floating anxiety.

Free-floating anxiety can never be permanently removed, only abated. How one reduces free-floating anxiety varies with age, experience, level of development, intelligence and socioeconomic position.

Differences in Cognitive Styles

What one experiences at any given moment is a function of three factors: level of sensory and motor development, emotional state, and one's perception of "reality." The pioneering work of Edward Sapir, Roger Brown, William Sechrest, Jerome Bruner, and others confirms that individual perceptions of reality conform to a preexisting network of categories within the mind.

Another theorist, Jerome Kagan, investigated cognitive styles of middle-class first-graders who were assigned to match a previously identified figure in the Matching Familiar Figures test. Kagan found that the longer the child took to make his choice of comparable or matching figure, the fewer were the child's errors; Kagan called this variable *cognitive tempo*. He characterized the slower, more accurate children as "reflective" and the faster respondents as "impulsive." One year later, a follow-up study showed that the "reflective" youngsters made fewer errors in an oral reading than did the "impulsive" youngsters (Kagan and Lewis,

1965). This series of studies raises an interesting question regarding the relationship between cognition and personality: Is I.Q. a function of personality?

PIAGET'S THEORIES OF DEVELOPMENT

The theories of cognitive development formulated by Jean Piaget (1896–1980) have had a great impact on the American educational system. Initially trained as a zoologist, his entrance into the field of cognitive development was serendipitous. He started observing the behavior of his own three children and carefully recorded each change in their development. Piaget's methods differed from those of many other theorists, clinicians, and researchers in that he studied a few children intensively in their natural environment rather than a large sample in a contrived context or laboratory.

Piaget's best-known contribution is his theory of the child's cognitive development. Piaget saw development as occurring in four stages: sensorimotor, preoperational thought, concrete operations, and formal operations.

The *sensorimotor stage* lasts from birth to age 2. In it the child uses sucking and grasping to learn what the real world is like. The child may shake his rattle and discern between objects that produce a sound and those that do not. Piaget theorized that the child "organizes reality" by arranging experiences in mental categories, or *schemata*, as Piaget called them.

Children are in the *preoperational stage* from 2 to 7 years of age. A significant achievement of this cognitive stage of development is the elaboration of an "inner map" of the external world. The principal medium for this elaboration of cognitive representation is language.

The child makes great strides in increasing his vocabulary after the age of 2. Children of this age are typically egocentric, experiencing some difficulty in differentiating between themselves and the external world. Children at this age often attribute emotions to inanimate objects and attribute reality to the content of their dreams. The child's thinking is rigid and irreversible. In a famous experiment, Piaget asked children to pour beads from a short wide beaker into a taller thinner one. He then asked if one of the beakers contained more beads; the children uniformly selected the taller thinner beaker as containing more beads even though they had performed the operation themselves and knew that no additional beads had been used to fill the thinner beaker.

In another well-known study of preoperational thinking, a 4-year-old child was asked the name of his brother; he responded with "Jim." Asked if Jim had a brother, the child gave an emphatic "no." The child was exhibiting the *egocentrism* and *irreversibility* of thinking that typify this stage.

Concrete operations mark the next stage of development, which generally lasts from 7 to 11 years of age. The child loses the rigidity characterized by preoperational thought. Piaget used another bead game to test development during this stage. A group of children of different ages were given a box of 20 wooden beads in which 2 were white and 18 brown. The children were asked whether the wooden beads or the brown beads would make a longer necklace. The children under 7 said brown, oblivious of the fact that all the beads were wooden. The children over 7 were amused by the "foolish" question; they knew that all the beads were wooden. At this stage children have learned to retrace steps in thinking when confronted by error, to correct themselves, and start over again. At this stage, children can regard more than one aspect of an object at a time. They have acquired the logical operations of combining, reversing, and forming associations. However, the children can think only in terms of objects they can touch or imagine touching; they lack the ability of abstract thinking which is manifested in the next stage, formal operations.

At ages 11 to 15 years children enter the *formal operations* stage in which they make comparisons, perceive relationships, and derive generalizations from logical analyses of real events. To illustrate the characteristics of this stage, Piaget presented a group of children in the formal and concrete operations stages with a group of objects. The children were asked to sort them into two piles, things that would and would not float. Piaget allowed them to use the objects in a pail of water. Few younger children could correctly classify the objects, while the older children had a high success rate. They made comparisons between the objects and gradually concluded that the relationship between size and weight determined whether an object would float. Through this method the older children "rediscovered" Archimedes's law of floating bodies. The children were able to generalize without having to test each object (Piaget, 1957).

FREUD

Possibly the best-known contribution of Sigmund Freud (1856–1939) to modern thinking is his revelation of the role of the unconscious in the everyday life of the adult, but also especially striking were his thoughts on the sexual life of children. Before the late nineteenth century, it was commonly held that children were innocent of sexual thoughts. (Ironically, at the time this thinking prevailed, children were nonetheless clothed in miniature adult garments, made to look like foreshortened adults, and put to work to earn their daily bread, as were adults.) In Freud's view, sexual function begins at the beginning of life, and the foundation of adult personality is established at birth.

Freud postulated that each of the three primary erogenous zones—mouth, anus, and genitals—is the focus of stimulation in each of the corresponding stages of psychosexual development. The force that impels the individual to seek satisfaction via the erogenous zones is termed the *libido*. An excess of either satisfaction or frustration at any stage results in fixation at that stage and failure to advance to the next stage. How the child interacts with the environment is determined by the stage and character of the fixation. Oral fixation is regarded as being correlated with drug addiction, compulsive eating and obesity, compulsive shopping, sarcasm, and verbal fluency. Anal fixation is regarded as responsible for miserly behavior, obsession with neatness and orderliness, obstinacy, and an "obsessive-compulsive" personality. One can readily think of problems with such a facile classification of the complexities of human behavior. How is the obese miser classified? What is a good, constructive approach?

The first stage of psychosexual development is the *oral stage*. The mouth is the neonate's connection with the external world. It serves as a vehicle for contact, stimulation, nourishment, and pacification. The second stage is the *anal stage*, in which gratification is derived from both the elimination and retention of feces. The third stage is the *phallic stage,* in which self-stimulation of the genitals is prominent. By this time the child is about three years old, may masturbate frequently, and is developing a strong attachment to the opposite-sexed parent. In boys, this attachment is called the *Oedipus complex*, and in girls, the *Electra complex*. Resolution of the oedipal conflict occurs in the *latency stage*. The last is the *genital stage*, which marks the onset of puberty; in the genital stage, gratification is derived from lovemaking with an opposite-sexed partner.

In Freudian terms, the *id* is the cornerstone of the personality, the source of psychic energy from which develop the *ego* and *superego*. All the basic instincts of the individual are contained in the id; it is totally unconscious. Knowledge of the id is indirectly acquired through dreams, unconscious slips of the tongue ("Freudian slips"), and other inexplicable behaviors of the individual. The ego is the controller of the id; whereas the id responds to the pleasure principle, the ego is governed by the reality principle. The young child is governed by the id and gradually develops the ego and superego. The superego functions as the child's conscience, inhibiting amoral impulses. When id, ego, and superego work in harmony, one can love and hate without guilt. Unbridled emotions indicate the dominance of the id, while excessive anxiety and inhibition indicate an overactive superego.

ADLER AND JUNG

Among the most prominent of Freud's disciples are Alfred Adler (1870–1937) and Carl Gustav Jung (1875–1961) both of whom broke with Freud and established their own schools of psychoanalysis. Adler agreed with Freud that the individual is molded by unconscious and innate forces, but in Adler's view, the person's social makeup is determined from the moment of birth and is adjusted throughout the life span. Adler saw the individual as having control in choosing a life-style, not as being at the mercy of instinctive forces as Freud maintained. Adler viewed the child as being overcome by feelings of inferiority in size, power, and knowledge in comparison with older siblings and adults. To compensate for this perception of inferiority, the child endeavors to attain a superiority in status, competence, or power. It is the need for superiority that dominates adult life.

Jung's school of analytical psychology also recognized the importance of the unconscious, .

but Jung dichotomized the unconscious into *personal unconscious* and *collective unconscious*. It is in the personal unconscious that repressed thoughts and forgotten experiences are housed; they may become accessible to the conscious mind when released by an appropriate stimulus. The collective unconscious represents the treasury of all human memory, the experiences and behaviors of antecedent generations. Just as the body is the product of centuries of evolution, so is the mind the repository of centuries of *archetypes*, as Jung called the thought forms of prehistoric eras. Each person has the archetypes of "mother," "father," and "hero" in the collective unconscious, which explains basic human similarities across time and space.

ERIKSON

Erik Erikson (1902–   ) studied with Freud in Vienna but is now regarded as an ego psychologist with psychoanalytical leanings. He supports the psychosexual theory of development but stresses the need for compatibility between the individual and society. For such compatibility to exist, the child must experience a sense of trust in himself and others, and a sense of personal worth. Freud focused on the childhood years as being crucial to adult personality development but for Erikson, the development of personality is a lifelong process. Erikson defined the healthy personality in positive terms and not solely as the absence of pathology. Erikson delineated eight paired opposites, each requiring a resolution in favor of one or the other and corresponding to eight stages of life.

Trust versus Mistrust

The development of trust or mistrust in the first year of life is the first stage. The child who receives a large quantity of nurturance, fon-

dling, and reasonably prompt attention to his or her needs tends to see the world as a safe place, i.e., the child places trust in himself, people, and the environment.

It should be noted that the fundamental developmental task of each stage is not totally resolved at the appropriate stage, but reappears subsequently. However, should it not be resolved at all at an earlier stage, it can be worked on afterwards. This feature of Erikson's theory is at variance with the fatalistic quality of much Freudian theory, which postulates that resolution of a developmental conflict must occur when the conflict first appears.

Autonomy versus Shame and Doubt

In the second and third years of life, the child's developmental task is to attain a sense of autonomy through walking, climbing, and simple motor activities. The child needs to be allowed to do things at his own speed and in his own way in order to develop a sense of mastery over the environment. If the child is denied opportunities for this mastery, a sense of shame and doubt emerges and may be a long-term handicap.

Initiative versus Guilt

In the third and fourth years of life, the child who is allowed to explore appropriate motor activities such as running, bike riding, swimming, and skating will develop a strong sense of initiative. When the child is told that the activity is poorly done or that the play is "childish" a sense of guilt may transfer to the later stages of life.

Industry versus Inferiority

In the sixth to eleventh years of life, the child must learn to master selected tasks. As he does so, he attains a feeling of competence and a sense of industry. If the child fails—as may happen if inappropriate goals are set for him, or if others are overly critical—the child develops a sense of inferiority.

Identity versus Role Confusion

From the onset of adolescence to the eighteenth year, the individual develops a firm sense of personal distinction and identity through explorations of the environment, and the "trying on" of different roles in different social contexts. Failure to do so may result in confusion as to one's purpose and role in life, or in the development of a negative role such as failure, coward, or clown.

Intimacy versus Isolation

In young adulthood the individual tries to reach out to make a personal commitment to another. Failure to do so may result in lifelong emotional problems and isolation.

Generativity versus Self-Absorption

In middle age the individual looks beyond himself, to the family, community, and nation. Failure to do so may result in extreme self-absorption and a sense of deepening depression as the years advance.

Integrity versus Despair

In old age, one looks back on one's life as being fulfilling or futile. If it is judged as futile, the life has been misspent and despair ensues, (Erikson, 1964).

THE SOCIAL APPROACH

Other social theorists such as Erich Fromm, Albert Bandura, and Frieda Fromm-Reichmann are less chronological in their approaches. Fromm sees human personality as resulting from social needs. For him, the fear of loneliness is the organizing factor of human behavior. The human being is alone in the world and seeks to alleviate his or her aloneness by bonding with others in marriage, clubs, and other forms of relatedness to people (Fromm, 1955).

Bandura is a proponent of social learning theory, which states that we learn from vicarious experiences, observing what consequences follow the behavior of others. Social learning theory also claims that people reinforce themselves by approving or disapproving their own acts. For Bandura, the worst punishment an individual can experience is self-contempt (Bandura and Walters, 1963).

Fromm-Reichmann was a psychotherapist who believed that certain types of schizophrenia are socially precipitated. One of her psychiatric clients was the subject of the novel *I Never Promised You a Rose Garden*.

HORNEY

Trained as a psychoanalyst, Karen Horney (1885–1952) was an iconoclastic neo-Freudian who vigorously rejected Freud's emphasis on sexual instincts and his views of women. Horney regarded environmental, not biological, factors as the key variables in shaping behavior. Of the environmental factors, the quality of the child's relationships with others has the most impact on growth and development. Horney regarded anxiety, not sexual desires, as the driving force in life. She defined anxiety as the need for safety. Aggression may appear in behavior that masks the source of aggression. For example, a neglected child may act aggressively to elicit parental attention. The child perceives parental neglect as a threat to survival and safety. For Horney, there are three types of neurotic behavior; all concern the purchase of safety at the expense of independence: submission, or moving toward people; aggression, or moving against people; and detachment, or moving away from people.

Freud characterized women as perceiving themselves as incomplete since they lack a penis; Horney's reply is that women desire the male privileges of independence and sexual freedom, not the penis. She criticizes the social system for teaching women to be weak, dependent, and reliant on the male, and for relegating women to the home in the "3-K" tradition (i.e., *Kirche, Kinder, Kuchen,* the German for "church, children, cooking") (Horney, 1967).

THE BEHAVIORAL SCHOOL

In the 1960s and 1970s, the philosophy of the behavioral school tended to dominate psychological thinking, but this school no longer enjoys its former prestige. However, its principles of behavior modification, largely derived from experimentation with animals such as rats, monkeys, and pigeons, continue to be used to help people modify smoking patterns, reduce weight, and remove phobias. Behaviorists ignore all mental processes that are not overt, measurable, or visible. This, then, excludes thoughts, feelings, and dreams and other unconscious processes from the behaviorists' concern. Behaviorist thinking has garnered some disrepute as a by-product of the popular awareness of the use of psychosurgery, heightened by such works as Ken Kesey's *One Flew Over the Cuckoo's Nest* and the enormously popular movie version of that novel.

The work of a contemporary of Freud, the Russian physiologist Ivan Pavlov (1849–1936), provided the impetus for the behavioral school of experimental psychologists.

Classical Conditioning

Pavlov is justifiably regarded as the pioneer researcher of the behavioral school. Pavlov was studying the digestive processes of dogs when he inadvertently discovered what came to be called *classical conditioning*. He had inserted tubes into the salivary glands of dogs to measure how much saliva dogs produce when they ingest food. He observed that the dogs salivated even before being fed. He concluded that they salivated in anticipation of being fed at the sound of the approaching footsteps of the experimenter. Pavlov devised an experiment to "condition" the dogs, that is, to link their salivation to the sound of a bell. He rang a bell repeatedly immediately before feeding the dogs. The dogs began to salivate as soon as they heard the bell and before receiving food. Finally, they salivated at the sound of the bell even when food was not forthcoming. By discontinuing the pairing of the food with the ringing of the bell, Pavlov eventually elicited a cessation of the conditioned response of salivation; this is called *extinction*.

Even 5- to 10-day-old babies can be conditioned to blink when they hear a tone. Babies will naturally blink in response to a puff of air blown into their eyes. A tone can be paired with the puff of air just as Pavlov paired the bell with the dogs' food, and the same classical conditioning results.

Operant Conditioning

In *operant conditioning* the subject performs a task that is rewarded, or *reinforced*, after the performance. Reinforcement is defined as a stimulus that increases the probability that an event will recur on the next occasion. Unlike classical conditioning, the reinforcement is not uniformly paired with each trial. Reinforcement may be either *fixed* or *variable*. For example, in garment piecework, workers may be paid for every 5 dresses completed; this is fixed reinforcement. Graduate students, for example, may experience variable reinforcement in that their major advisors may approve their research proposals but the doctoral committee may later present objections to them. Variable reinforcement cannot be predicted.

Reinforced behavior tends to persist. For example, a child in a pediatrics unit may throw a temper tantrum to get attention; if attention is given, it acts as a reinforcer to the behavior. A more constructive response would be to ignore the tantrum whenever possible and to be routinely attentive to those children who are not disruptive.

B. F. Skinner (1904–   ) is considered the best-known living American behavioral scientist. He believes society can be more effectively organized and managed through the application of the principles of behavior modification. Skinner devised programmed instruction that uses behavioral principles. A topic is broken down into numerous units or frames and the learner proceeds through the material step by step. The answer accompanies each frame so that the learner is provided with immediate feedback, which is the reinforcement. There are no teacher-imposed deadlines or tests; the learner learns at his or her own pace.

Even language learning, according to Skinner, is governed by behavioral principles. At birth, the infant possesses no language; gradually the infant acquires a vocabulary through reinforced imitation of models. These individual language units are pieced together into larger units. Those larger units when properly used by the developing child are reinforced and thus become incorporated into the child's linguistic structure (Skinner, 1957).

In the view of Skinner and those who follow his approach, motivation is an irrelevant aspect of behavior; only those behaviors that are reinforced persist.

Any technique has the potential to be misused. J. B. Watson, a precursor of modern behavioral theorists, reported in 1920 an experiment on an infant that vividly illustrates the misuse of behavior modification. The infant, known as "Little Albert," was a healthy baby who enjoyed playing with a dog, a rabbit, a ball of cotton, a white rat, and a fur coat. At 11 months of age, Watson and his assoicates involved Albert in an experiment. When the baby reached for the white rat, Watson struck a steel bar just behind him, making an unexpected and loud noise. After experiencing two such pairings, the baby began to whimper. A week later the baby would not touch his old playmate. Systematic operant conditioning was performed for seven trials. Thereafter Little Albert cried and fearfully avoided the white rat. One week afterward the fear response had generalized to the friendly rabbit, the fur coat, the ball of cotton, and the dog. Anything furry frightened the child. Before the experimenter could undo the conditioning response, Little Albert was removed from the hospital. One can only pity the poor child, who was apparently destined to go through life in dread of furry objects (Watson and Rayner, 1920).

The salutary use of behavioral technique is exemplified in the work of Joseph Wolpe (1958), who developed the technique of *reciprocal inhibition*. One type of reciprocal inhibition is called *desensitization*. It is used to mitigate phobic anxiety induced by such stimuli as snakes. The therapist may use drugs or hypnosis to promote relaxation. Slowly and gradually the client is told to imagine the least anxiety-producing stimulus of a list of stimuli. When the client is comfortable doing this, the therapist proceeds to the next most anxiety-provoking stimulus until the client is comfortable with every item on the list. As with Little Albert, the client will generalize the feeling of relaxation to related stimuli. This technique of desensitization has been successfully applied to such problems as stage fright, test anxiety, various kinds of phobia, impotence, and frigidity.

Reinforcing the Sick Role

Health professionals, particularly nurses, are expected to be nurturant, warm, comforting, and friendly toward their clients, yet these very behaviors in some instances may be harmful. For example, imagine the fictional case of Mrs. John Elder. She is 74, and has five adult children. She has been hospitalized for a urinary tract infection which has cleared up; she is ready to be released from the hospital, but she won't go. Why should she? Her bills at the hospital are paid for. No one visits her at home. Her meals are brought to her room. She regularly interacts with compassionate nurses and other health professionals. She feels warm and secure when she is in her semiprivate hospital room, where she can talk to her roommate. At home she is all alone.

In such a scenario it would appear that the nurse and others may be reinforcing the sick role. What could they do to promote the client's health? The Office on Aging could be contacted; most states provide such an agency. The agency can suggest state resources to assist the elderly such as Meals on Wheels, gerontological social workers, and housekeepers. If it is appropriate, religious institutions may also be contacted. Once the client is visited by a representative of a service agency or humanitarian group, and the at-home program is instituted, the nurse can start to extinguish the sick-role behavior. The aim is to disengage the client from a sick role that is no longer medically justified. Firmness, courtesy, and a declining schedule of nursing reinforcements would be indicated. Fewer client visits, reduced interaction time with the client, and a gradual interpersonal withdrawal should be contemplated at the same time that other professionals are entering the client's life.

Children too may experience reinforcement in the sick role. Both over-protective and depriving mothers may encourage this phenomenon. For the emotionally deprived youngster, being sick in bed often brings an unusual degree of attention from the mother. To reverse the inappropriate sick role of the child, family therapy may be indicated.

DIFFERENTIAL BEHAVIORS OF THE SEXES

The Freudian dictum that anatomy is destiny has a hoary past, although some deemphasis on biology as the root cause of differential human behavior has appeared since the early 1970s. In the 1940s the French writer Simone de Beauvoir wrote a polemic against the treatment of women called *The Second Sex*; published in the United States a decade later, and still later throughout the world, it was at first largely ignored, but has subsequently come to be regarded as heralding the dawn of the women's rights movement in the United States.

Are women behaviorally or intellectually different from men? If women believe they are incapable of success in the quantitative and spatial sciences—to take but one example of a sex-role sterotype that is already rejected, as of this writing, by many parents, educators, and the more responsible producers of mass media—it would make sense to avoid the fields of medicine, engineering, chemistry, and physics. Self-fulfilling prophecy would dictate that even when they explore these areas, failure is the probable result. But is this prediction supported by research data?

Eleanor Maccoby reviewed differential behavioral findings and found negligible sex differences in the most widely used tests of general intelligence. In verbal ability, females typically outperformed males by a wide margin after the age of 10. Up to age 10, there is no clear superiority of either sex in verbal fluency. In adolescence, girls outperform boys in standard tests of verbal ability by a quarter of a standard deviation, an advantage they maintain throughout life.

In quantitative ability the profile is reversed. Up to puberty, there is again no sex difference in quantitative (mathematical) ability, but after puberty, females consistently rank below males. The same is true of spatial ability; until puberty neither sex outperforms the other. At puberty, males show superior spatial acuity in problem solving, a distinction they maintain throughout life. This characterization does not hold for all females, however; about 25 percent of female adolescents and adults score above the male median score on various standardized tests of spatial ability (Maccoby, 1966).

Other researchers have found some small sex differences in various perceptual-motor skills such as discrimination reaction time, color perception, finger dexterity, card sorting, and the like. On the whole, neither sex is consistently superior in performance.

While Maccoby found no solid research to justify adolescent sex role socialization practices, she found that much behavior is culturally determined. Indeed, she found that early reinforcement of the aggressive response in male children is linked to the aggressive personality of the adult male.

Large sex differences do appear at the extremes of I.Q. distribution. Both in the very high and very low ends of the scale, males predominate. Males are also more likely to die in infancy, and are more likely to be hyperactive, to have reading problems in school, to commit suicide, to develop childhood schizophrenia, to commit aggressive acts, and to be hospitalized at some point in their lives for mental or physical illness. Maccoby (1966) reported that after the atomic bombing of Hiroshima, more male than female fetuses were born defective or dead.

Clearly, neither biology nor environment can independently explain all these variances. The evidence on differential behaviors of the sexes is not conclusive.

THE DEVELOPMENT OF MORALITY

Philosophical interest in the development of moral thought, feeling, and behavior can be traced back to the story of Cain and Abel. Experimental research in moral behavior dates from the mid-1920s with the classical work of Hartshorne and May (1928). They investigated the behavior of young school children in classroom situations in order to assess the prevalence of cheating on tests. They concluded that: (1) it was impossible to divide the children into cheaters and noncheaters because almost everyone cheated at some time; (2) it was not pos-

sible to predict from a past cheating experience whether that child would cheat in another similar situation; (3) the child's stated values about honesty had little relevance to cheating behavior; (4) children tended to assess the risk of detection, and noncheaters tended to be merely more cautious, not more honest, than cheaters; (5) honest behavior is largely determined by "immediate situational factors of group approval and example" and not by internal moral values; and (6) in those cases where honesty was a convention of group membership, the moral values were specific to the child's social class and group, and were not characteristic of all groups or classes.

Piaget's observation of the moral behavior of children confirmed the findings of Hartshorne and May as being applicable to the first of two levels of moral development that Piaget conceptualized. Piaget saw the first level as a *morality of constraint* in which children comply with rules and submit to authority out of fear of punishment. They measure wrongdoing by the amount of harm or damage done, not by the intent of the wrongdoer. At this level of morality, children believe all wrongdoings are ultimately punished either by parents or by nature; illness, accidents, or bad fortune are seen as just punishment for bad behavior. Piaget theorized that older children are governed by a *morality of cooperation*, in which wrongdoing is evaluated in terms of the intent of the person, and rules of behavior are consensually established by the child and the agent of authority (parent, teacher, etc.).

STRESS AND ADAPTATION

There are varying explanations of the relationship between stress and emotion. The earliest theory, known as the James-Lange theory of emotion, is attributed to the American psychologist William James (1842–1910) and the Danish psychologist Carl Lange (1834–1900), who independently arrived at the same formulation in the 1880s. Before their theory, it was generally held that the fear response occurs thus: (1) You see a snake (for example), (2) You feel fear, and (3) You flee. James and Lange dissented. They explained that emotion follows physiological response, so that the sequence becomes: (1) You see a snake, (2) You flee, and (3) You feel fear. The experience of fear is caused not by the sight of the snake, but in reaction to the flight from the perceived threat (James, 1884).

Psychologist W. B. Cannon (1927) described emotional and physical responses as occurring simultaneously, not in sequence. Upon sighting a threatening object such as a snake, one's nerve impulses travel to the thalamus, at which point some impulses travel on to the cortex and others travel to the muscles and viscera. Those impulses that travel to the cortex cause an emotional response to be felt; those impulses that travel to the muscles and viscera cause a physiological reaction, which may include an increase in respiration rate, an increase in pulse rate, a rise in blood pressure, the secretion of adrenalin and additional sugar into the bloodstream, the release of additional cells into circulation by contraction of the spleen, an increase in the tension of striated muscles, dilation of the pupils, an increase in perspiration, and a decrease in gastrointestinal activity. The individual may appear flushed, tense, even trembling. These imbalances resulting from an emotional reaction to stress are the mechanisms whereby the body ultimately reestablishes the state of balance, or *homeostasis* (as Cannon called it), that is temporarily lost in such "fight or flight" situations.

Hans Selye (1907–   ) is a prominent biologist and endocrinologist who is regarded as the "father" of stress theory. Selye distinguishes between *stress*, which he sees as a nonspecific response by the body to any demand made upon it, and *distress*, which he regards as harmful and negative stress.

As Cannon explained, an individual faced with stress responds with "fight or flight," i.e., an active or passive response. Selye equates different internal substances with the two responses; he calls these "syntoxic" and "catatoxic." "Syntoxic substances act as tissue tranquilizers, creating a state of passive tolerance which permits peaceful coexistence with aggressors . . ." (Selye, 1974, p. 152). Catatoxic substances destroy damaging agents such as toxic compounds.

John Calhoun has done much laboratory research on population density as an environmental stressor in rats. He provided an adequate supply of food, water, and nest-building materials. As he increased the population density of the pens, nest-building declined and then stopped entirely. Females were besieged by packs of males and were unable to escape their unwanted attentions. The females experienced a high rate of mortality caused by disorders in pregnancy and parturition. Calhoun reports increased abnormalities in sexual behavior such as hypersexuality, bisexuality, and homosexuality. Infant care was ignored, and there were occurences of cannibalism. No infant survived the experiment (Calhoun, 1962).

Environmental stress is established as a contributing factor to psychotic behavior. R. D. Laing, the eminent British existential psychologist, feels that we need to reassess our view of psychosis as abnormal behavior. Laing feels that the environment is abnormal, and that psychosis is a reasonable response to a crazy world (Laing, 1967).

THE HUMANISTIC SCHOOL

How does one measure success in life? How does one define success? Success may be defined as the favorable result of effort expended or the achievement of a goal, for which there may be some recognition. Abundant data in longitudinal studies of human beings show a relationship between achievement and longevity;

achieving people tend to live longer. Achievement also seems to be an important component in human growth and development at every stage of life. What are the basic needs of people? Are they solely acquisitive? Where do success and achievement fit?

Abraham Maslow (1908–1970), who is credited with the development of a *humanistic* school of psychology, asked himself these questions and others. Maslow felt that psychology functioned as a "sickology" in which psychopathology and evil or aggressive behavior had been the focus, rather than the good, cooperative, altruistic side of human behavior. Maslow espoused a psychology of health based on adherence to one's own naturally good self, but cautioned against a hasty distinction between good and evil, health and sickness.

> Does sickness mean having symptoms? I maintain now that sickness might consist of not having symptoms when you should. Does health mean being symptom-free? I deny it. Which of the Nazis at Auschwith or Dachau were healthy? Those with striken conscience or those with a nice, clear, happy conscience? Was it possible for a profoundly human person not to feel conflict, suffering, depression, rage, etc.? (Maslow, 1962 p. 7)

In Maslow's view, the human being has a basically good nature. However, the tendency toward the fullest development of that innate goodness is easily swayed by social pressures. Maslow postulated two fundamental human motivations: *deficiency*, which motivates the individual to restore physical and psychological equilibrium that has been upset, and *growth motivation*, in which the individual strives toward maximal fulfillment of potential. Maslow is widely quoted in nursing literature because of the special relevance of his needs theory to nursing practice.

Maslow's hierarchy of motives assumes that lower, basic, more simple needs must be satisfied before the individual can satisfy higher needs. At the bottom of the hierarchy are phys-

iological needs—food, water, and hygiene. The next level of needs is safety needs, followed by belongingness needs. They are followed by esteem needs, and finally by self-actualization needs. Maslow's theory of the self-actualized person is based upon his study of living self-actualized people as well as of famous dead people such as Beethoven, Lincoln, Eleanor Roosevelt, and Einstein. Maslow formulated a list of 15 attributes of self-actualized persons:

1. More efficient perception of reality. Self-actualized people are more effective judges of events and people, and can tolerate ambiguity more readily than others.
2. Acceptance of self and others. The self-actualized come to terms with themselves and others as they are.
3. Spontaneity. They are more idiosyncratic in thinking than others but are seldom unconventional in behavior.
4. Problem-centered. They focus their energies on problems outside of themselves, frequently involving themselves in broad social issues.
5. Detachment. They require privacy and are not concerned about being alone.
6. Autonomy. Self-actualized people are relatively independent of society but are not unconventional for the sake of being different.
7. Deep appreciation. Self-actualized people retain a deep appreciation of life's basic experiences even though they have been repeated often.
8. Mystic experiences. They have had a feeling of wonder, awe, a deep sense of ecstasy, having a sense of limitless horizons available to them.
9. *Gemeinschaftgefuhl*, that is, a sense of oneness and identity with all people, regardless of race, nationality, religion, or any other attribute.
10. Interpersonal Relations. Self-actualized people experience very deep personal relationships with a chosen few.

11. Democracy. They show respect for all people.
12. Discrimination between means and ends. Self-actualized people often enjoy the means or process leading to the ends more than others do.
13. Sense of humor. They have a gentle, philosophical sense of humor.
14. Creativeness. Self-actualized people are richly creative.
15. Resistance to enculturation. The self-actualized tend to resist blind conformity to a cultural mold, but they do fit into their own culture. (Maslow 1970)

Psychologist Gordon Allport (1897–1967) did seminal work on group interactions and the formation of prejudices. He argued, along with Maslow, that an ethical standard must be used in judging abnormality (Allport, 1937). The writings of Allport, Maslow, and other humanists stress the ethical and moral element in human behavior as necessary for maximal individual development.

Carl Rogers (1902–   ) is similarly oriented. Rogers is best identified with nondirective client-centered therapy. He believes that a motivated client can, with skillful guidance, specify problems and discover personal solutions. As Rogers sees it, the purpose of therapy is to provide the means for personal growth through individual responsibility for decision-making. The therapist does not "know best"; he accepts, reflects on, and supports the feelings of the client in a nonjudgemental and nondirective fashion. Roger's strategies reflect respect for the client as a valuable, intelligent, and resourceful person who has within himself or herself the potential for growth and health (Rogers, 1946).

FAMILY THERAPY

Theorists have been reporting the imminent dissolution of the family since the 1920s, but, in the vein of Mark Twain's alleged retort to reports of his demise, the report is "grossly exaggerated." Nonetheless, family life is undergoing change. Philippe Ariès documented the history of changes in the family and in the life cycles of the individual. The postponement of adult duties for children and the recognition of adolescence as a transitional period are attributed by Ariès to the nuclear family's response to the industrial revolution (Ariès, 1962).

From the time of Freud until the 1950s, most psychotherapy focused on the individual. Nathan Ackerman (1908–1971) was one of the pioneers in treating the emotional troubles of the whole family as a unit, rather than treating the individual alone. In family therapy the therapist is emotionally linked to the pervasive anxiety of family members, and each family member is similarly involved. Ackerman contends that family therapy brings to light the "transmission of . . . conflict and coping from person to person and from one generation to the next" and enables the family members to end "the vicious cycle of blame and accusation for things past" (Ackerman, 1966 p. 43).

Therapists have long believed that the experiences they had in their own families helped shape them as professionals. Murray Bowen (1913–   ) believed all families are pretty much alike. He decided that the study of his own family of origin would provide as much data as he needed. The formulation of his family systems theory is largely an outgrowth of this study. Bowen's contributions involve many concepts, of which two are paramount: the triangle, and the differentiation of self.

Bowen's "triangle" comprises a "comfortably close" couple and an outsider whose position is less comfortable in periods of calm but more comfortable in periods of stress; the classic example is the father–mother–child triangle. One of the most common patterns in such a triangle is one of basic tension between the parents; to escape the tension, the father (often characterized as "passive," "weak,"; or "distant") gains the outside position, so that the conflict is between the mother and child. The mother, who is often typified as "domineering" or "castrating," enlists the support of the child, whose ability to function as an individual is hampered by his pawnlike position in these maneuvers. Bowen comments "families replay the same triangular game over and over for years, as though the winner were in doubt. . . . Over the years the child accepts the always-lose oucome more easily . . ." (Bowen, 1978).

Bowen's differentiation-of-self scale ranges from 0 to 100 and is, essentially, a scale of emotional maturity. The lowest level (0) represents the least possible degree of differentiation, and the highest level represents a theoretical "Perfect Self" which no one has achieved. In Bowen's view, lower-level people are more vulnerable to stress and illness, and "their dysfunction is more likely to become chronic when it does occur"; higher-scale people are more emotionally resilient (Bowen, 1978 p. 472).

Bowen seems to have borrowed heavily from Cannon and Maslow; the self-actualized individual is the equivalent of Bowen's highly differentiated self. Unless basic family conflicts are resolved, they tend to recur throughout the lives of the individuals. Bowen believes these dysfunctional individuals have "inherited" the dysfunction from previous generations; an important component of Bowen's family systems theory is the genogram, diagram of the individual's genealogical tree with description of interpersonal relationships of each person.

SEXUAL NEEDS

Sex occupies the thoughts of most people throughout the lifespan. The modern scientific investigation of sexual behavior is best known to the general public through the work of William Masters (1915–   ) and Virginia Johnson (1925–   ). This team of physiological researchers, the most significant investigators since Alfred Kinsey, directly observed and recorded physiological responses of adults during sexual intercourse, and also studied sexual dysfunctions. Masters and Johnson and others refute the myth of the sexless aged. Sexual behavior can continue from purberty to near death. There is, however, a waning of sexual activity with advancing age, especially after age 60. Males take longer to achieve erection, and also take longer to ejaculate. The amount of time needed to resume sexual activity after ejaculation also increases with age. For the aging male, orgasm may take the form of a seepage of semen rather than the explosive ejaculation characteristic of the young male. For women, there is a relatively slight decline in capacity with advancing years.

Masters and Johnson (1966) found a high correlation between early expression of sexuality and sexuality in older men. They reported that over the years depressed male sexuality may be attributed to mental fatigue (not physical fatigue), overindulgence in food, and, above all, overindulgence in alcohol and general poor health.

THE EMERGENCE OF THANATOLOGY

In some cultures—for example, many oriental cultures and traditional Jewish culture—treatment of the elderly is deferential and respectful. This deference is not arbitrary, but quite practical; by dint of their accumulation of years and property, the elderly are able to dispense wisdom and distribute wealth. Also, a high regard for ancestors, if not ancestor worship, is integral to some Eastern and Western religions. For example, the Jews regularly invoke their biblical ancestors Abraham, Isaac, and Jacob in their litanies. In many cultures, survival to a great age is regarded as a mark of divine esteem.

Elisabeth Kübler-Ross (1926–   ) is a psychiatrist who has become preeminent in the emerging field of thanatology. She has personally interviewed over 200 people in order to understand the stages of dying. Kübler-Ross finds that the central problem in handling imminent death is the dread of dying, which is as prevalent in American culture as the above-mentioned deference toward the aged is rare. People tend to regard death as a punishment, not as a natural event. The friends and family of a dying person may stay away from him, since they are reminded of their own mortality, and the dying one is thus denied company, compassion, and love (Kubler-Ross 1969).

A solution as Kübler-Ross sees it, is the establishment of hospices for the dying. The important feature of the hospice concept is not the segregation of terminally ill people, but a "hand-picked" staff selected especially on the basis of positive attitudes toward the dying. Such people can provide an atmosphere of hope, love, and caring, and they ultimately go on to become specialists in terminal care (Kübler-Ross, 1969).

Kübler-Ross delineated five psychological stages that dying people typically pass through before death. Not every person experiences the sequence in order, and some never reach the final stage, that of acceptance. The first stage is denial; the client refuses to accept the diagnosis of terminal illness and may consult other sources—physicians, faith healers, and occultists. The second stage is anger. In this stage, the client finally accepts the prognosis but is resentful and angry. A typical reaction is "Why me? I don't deserve to die."

Bargaining is the third stage. The client may bargain with God, with the illness, or with the doctor in a frantic attempt to gain time. This stage is not as visible as others, but where it exists it represents a healthy coping strategy in the face of approachjing death. In the fourth stage, depression, the client fully accepts the death verdict. "He mourns past losses first and then begins to lose interest in the outside world. He reduces his interest in people and affairs, wishes to see fewer and fewer people and silently passes through preparatory grief. If he is allowed to grieve, if his family has learned to let go, he will be able to die with peace and in a final stage of acceptance" (Kübler-Ross, 1969).

SUMMARY

Anxiety and fear, or the absence of security, characterize human life from the beginning. People differ greatly in the scope of their fears and how they cope with them. Several major theorists, whose viewpoints are summarized in this chapter, have advanced explanations for these differences, as well as ideas about many other aspects of human experience.

The protracted separation of an infant from his mother can have permanent deleterious effects. John Bowlby and Rene Spitz both studied this phenomenon and confirmed that maternal–child separation created psychological and developmental disorders in children. Harry Harlow's classic experiments with rhesus monkeys underscored the importance of "contact comfort," which is no less important to human infants. The other side of the coin, overprotectiveness, can be just as harmful, as is illustrated in studies by David Levy.

Piaget's investigations of cognitive and moral development are outstanding contributions to the understanding of human behavior. His work (which will be referred to repeatedly throughout this book) is of the utmost significance in terms of the content itself, but is also noteworthy as an example of flexible methodology; Piaget's observations are highly *unscientific* in that they derive largely from his study of a "population" of three—his own three children. Nonetheless, his theories have been corroborated time and again by independent (and more objective and orthodox) researchers.

Freud's exploration of the role of the unconscious is a cornerstone not only of psychology but of modern thinking. In the Freudian view, the id is the source of psychic energy and is mediated by the ego and superego. Among Freud's most prominent disciples are Alfred Adler and Carl Gustav Jung. Adler theorized that the need for superiority dominates adult life, but he ascribed to the individual more control over instinct than Freud postulated. Jung's best-known contribution to psychological thought is the dichotomy of personal and collective unconscious.

Erik Erikson also studied with Freud. His work reflects a psychosexual viewpoint but stresses the need for a sense of harmony between the individual and society. Erikson (whose theories, like those of Piaget, are of prime importance and are discussed throughout this text) views development as a progression of tasks, expressed as paired opposites; the first of these is the development of a sense of trust (versus mistrust) and is prerequisite to further development.

Karen Horney is regarded as "neo-Freudian" but she vigorously rejected Freud's emphasis on sexuality and his views on women. Horney saw anxiety, not libidinous impulses, as the driving force in life. She criticized the social system for teaching women to be weak and dependent.

The behavioral school of psychological thought no longer enjoys the prestige it did in the 1960s and 1970s, but behavior modification techniques continue to be used to help people change or eliminate habits and to overcome phobias. B. F. Skinner, the best-known proponent of the school, has also applied its principles to programmed instruction.

Abraham Maslow is regarded as the founder of a humanistic school of psychology. Maslow is perhaps best known for his concept of the self-actualized person, which relates to another of his theoretical contributions, the hierarchy of needs. Others identified with humanistic psychology are Gordon Allport and Carl Rogers.

The 1950s saw a major development in psychotherapy, the advent of family therapy. Two of its pioneers, Nathan Ackerman and Murray Bowen, share the view that families tend to perpetuate certain destructive patterns among their members, and that family therapy, rather than individual "client-centered" therapy, is needed to break these cycles.

William Masters and Virginia Johnson are, to the general public, the best-known researchers of sexuality since Kinsey. Their physiolog- ical research has led to a new understanding of human sexuality, especially as regards sexuality in the elderly. Contrary to widely held beliefs, depressed sexuality is not an inevitable consequence of normal aging, but most often the result of general poor health, mental (not physical) fatigue, and overindulgence in food and alcohol. Masters and Johnson also found that men whose sexual expression began early in life were more likely to be sexually active in old age.

Elisabeth Kübler-Ross is preeminent in the field of thanatology. She is an advocate of the establishment of hospices for the dying, which offer the care and compassion that friends and families of the terminally ill are so often unable to give, largely because of the negative attitudes about death that are deeply ingrained in American culture.

REFERENCES

Ackerman, N. *Treating the troubled family.* New York: Basic Books, 1966.

Adams, H. B., Carrera, R. N., Cooper, C. D., Gibby, R. G., and Tobey, H. R. Personality and intellectual changes in psychiatric patients following brief partial sensory deprivation. *American Psychologist,* 1960, *15,* 448.

Adler, A. *The practice and theory of individual psychology.* New York: Humanities Press, 1971.

Allport, G. W. *Personality: A psychological interpretation.* New York: Holt, 1937.

Anastasi, A. and Foley, J. P., Jr. *Differential psychology: individual and group differences in behavior.* New York: Macmillan Co., 1949.

Ariès, Philippe. *Centuries of childhood.* New York: Vintage Books, 1962.

Aronson, E., Turner, J., and Carlsmith, J. M. Communicator credibility and communication discrepancy. *Journal of Abnormal and Social Psychology,* 1963, *67,* 31–36.

Aronson, E. *The social animal.* San Francisco: W. H. Freeman, 1972.

Asch, S. E. Opinions and social pressure, *Scientific American.* 1955, *193*(5), 31–35.

Bandura, A., and Walters, R. H. *Social learning and personality development.* New York: Holt, Rinehart and Winston, 1963.

———. *Adolescent aggression.* New York: Ronald Press, 1959.

Bayley, N. Behavioral correlates of mental growth: birth to thirty-six years. *American Psychologist,* 1968, *23,* 1–17.

———. Consistency of maternal and child behaviors in the Berkeley growth study. *Vita Humana,* 1964, 7, 73–95.

Bettetheim, B. Individual and mass behavior in extreme situations. *Journal of Abnormal and Social Psychology,* 1943, *38,* 417–452.

Bexton, W. H., Heron, W., and Scott, T. H. Effects of decreased variation in the sensory environment. *Canadian Journal of Psychology,* 1954, *8,* 70–76.

Bowlby, J. *Child care and the growth of love.* London: Penguin Books, 1951.

Bowen, M. *Family therapy in clinical practice.* New York: Jason Aronson. 1978.

Brill, A. A. (ed.) *The basic writings of Sigmund Freud.* New York: Random House, 1928.

Brownfield, C. *The brain benders: a study of the effects of isolation.* New York: Exposition Press, 1972.

Bruner, J. *A study of thinking.* New York: John Wily & Sons, 1967.

Butler, R. N., and Lewis, M. *Aging and mental health.* St. Louis: C. V. Mosby, 1973.

Calhoun, J. B. Population density and social pathology. *Scientific American,* 1962, *206,* 139–148.

Cannon, W. B. The James-Lange theory of emotion: a critical examination and an alternative theory. *American Journal of Psychology,* 1927, *39,* 106–124.

de Beauvoir, S. *The second sex,* New York: Bantam, 1961.

Deutsch, H. *The psychology of women.* New York: Grune and Stratton, 1945.

Dewey, J. *Democracy and education.* New York: Macmillan, 1916.

Ellis, A. Rational psychotherapy. *Journal of General Psychology,* 1958, *59,* 35–49.

Erikson, E. H. *Childhood and society.* New York: Norton, 1964.

Festinger, L. *A theory of cognitive dissonance.* Evanston, Ill.: Row, Peterson, 1957.

Fromm, E. *The sane society.* New York: Fawcett, 1955.

Gibson, E. J., and Walk, R. D. The visual cliff. *Scientific American,* 1960, *202*(4), 67–71.

Harlow, H. F., and Harlow, M. K. Learning to love. *American Scientist,* 1966, *54,* 244–272.

Hartshorne, H., and May, M. A. *Studies in deceit.* New York: Macmillan, 1928.

Horney, K. *Feminine psychology.* New York: Norton, 1967.

James, W. What is an emotion? *Mind,* 1884, *9,* 188–205.

Jung, C. G. *Two essays on analytical psychology.* New York: Dodd, Mead, 1928.

Kagan, J., and Lewis, M. Studies of attention in the human infant. *Merrill-Palmer Quarterly of Behavior and Development,* 1965, *11,* 95–127.

Kohlberg, L. The cognitive-developmental approach to socialization. In D. Goslin, (Ed.) *Handbook of Socialization.* Chicago: Rand McNally, 1969.

Kübler-Ross, E. *On death and dying.* New York: Macmillan, 1969.

Laing, R. D. *The politics of experience.* New York: Pantheon, 1967.

Levy, D. *Maternal overprotection.* New York: Norton, 1966.

Lorenz, K. *On aggression.* New York: Harcourt Brace Jovanovich, 1966.

Maccoby, E. *Development of sex differences.* Stanford, Calif.: Stanford University Press, 1966.

Maslow, A. H. *Motivation and Personality.* New York: Harper & Row, Publishers, 1970.

Maslow, A. H. *Toward a psychology of being.* Princeton, N.J.: Van Nostrand, 1962.

Masters, W. H., and Johnson, V. E. *Human sexual response.* Boston: Little, Brown, 1966.

Milgram, S. Some conditions of obedience and disobedience to authority. *Human Relations,* 1965, *18,* 57–75.

Parsons, T. *The social system.* New York: Free Press, 1964.

Piaget, J. The child and modern physics. *Scientific American,* 1954, *196*(3), 46–57.

Ribble, M. A. Infantile experience in relation to personality development. In J. McHart (Ed.), *Personality and the behavior disorders,* (Vol. 2), New York: Ronald Press, 1944.

Rogers, C. Significant aspects of client-centered therapy. *American Psychologist,* 1946, *1,* 415–422.

Sander, S., and Rank, O. *The development of psychoanalysis.* New York and Washington, D.C.: Nervous and Mental Disease Publishing Co., 1925.

Schacter, S., and Singer, J. Cognitive, social and physiological determinants of emotional state. *Psychological Review,* 1962, *69,* 379–399.

Segall, M. H., Campbell, D. J., and Herskowits, M. J. *The influence of culture on perception.* New York: Bobbs-Merrill, 1966.

Selye, H. *The stress of life.* New York: McGraw-Hill, 1956.

———. *Stress with distress.* Philadelphia: Lippincott, 1974.

Skinner, B. F. *Verbal behavior.* New York: Appleton-Century-Crofts, 1957.

Spitz, R. Hospitalism: an inquiry into the genesis of psychiatric conditions in early childhood. *Psychoanalytic Study of the Child,* 1945, *1.*

Watson, J. B., and Rayner, R. Conditioned emotional reactions. *Journal of Experimental Psychology,* 1920, *3,* 1–14.

Wolpe, J. *Psychotherapy by reciprocal inhibition.* Stanford, Calif.: Stanford University Press, 1958.

2

Influences on Growth and Development

BETTY H. LANDSBERGER

THE HISTORY OF DEVELOPMENTAL STUDIES

The study of development began with a search for answers to *problems* of development, rather than with a description of the normal process. Freud's insights (1930) were developed from his treatment of the mentally ill. Binet studied intellectual development in order to sort out academically incompetent children, and Bowlby (1952) and Spitz (1945) observed the effects of maternal deprivation when they studied institutionalized infants who were failing to thrive. However, the causative factors in extreme cases of failure to develop normally have not often

been shown to account for differences among people in the normal range. Escalona examined studies in child development and child psychiatry in an attempt to find those factors that can be expected to produce illness or abnormal development on a population-wide basis; she concluded that "severe and chronic poverty" was the only such factor (Escalona, 1974).

Large studies of normal development in normal populations are not frequently undertaken. The Health Examination Survey (National Center for Health Statistics, 1969) and the National Child Development Study in Britain (Davie, Butler, and Goldstein, 1972) are among the only really large-scale efforts, and they do not reach beyond adolescence. No matter how thorough, any study of thousands of subjects obviously cannot be as intensive as those of smaller populations, or of individuals. Thus, as of this writing, less is known about normal human development than about certain developmental disorders.

When the study of human development began in the years between the world wars it focused almost entirely on early childhood, from about 2 to 6 years of age. Early childhood was not just at center stage; between 1900 and 1950, that period of development was practically the whole show. Not until midcentury were infants studied in newborn nurseries, and substantial research on adolescence did not begin until the late 1950s. The first studies of preadolescence came still later. The late 1960s saw the first studies of old age, and middle age became a popular field of study in the 1970s.

The study of early childhood brought with it an emphasis on some processes that are particularly dramatic in that period. One of these was maturation, both biological maturation, examined in research on fine-muscle coordination, for example, and psychological maturation, as exemplified in language usage, memory, and general problem-solving ability. Interest in psychological maturation led inexorably toward the conceptualization of "intelligence," the testing of which has dominated much of the study of human development and related fields since Alfred Binet's work in France early in the twentieth century.

The Influence of Specialization

The fact that the study of life-span development has had a late start is largely attributable to a trend present in research generally, and especially in such fields as biology, psychology, and medicine—that is, specialization. Anyone entering these fields in recent years has had to learn "more and more about less and less." Because the study of human development is in-

trinsically multidisciplinary, it is inevitable that specialization creates a less-than-optimal perspective on that study.

There is no question that specialized research has yielded much knowledge and the answers to many research questions, but regarding knowledge of human growth and development, it is necessary to ask whether they have been the appropriate questions. It has been suggested that while scientists have concentrated on increasingly specific topics, the broader and more central questions of human development have been addressed by artists, novelists, and dramatists. Arthur Miller conveyed the plight of the middle-aged man in *Death of a Salesman* before the researchers took up the topic for study; *Medea* was with us for centuries before requests for proposals on the topics of child abuse and family violence appeared in the *Federal Register*.

Some Paradoxes in the Study of Human Development

The study of development is faced by paradoxes other than the one posed by the need for and danger of specialized study. One is that of humanity versus individuality: each person has much in common with other human beings, and is yet unique. Equally paradoxical is the fact of continuity versus change: through the changes brought by development runs such continuity that it is frequently possible to predict a person's behavior in new situations. At the social level, there is continuity in problems such as poverty and war despite radical changes in politics and technology.

HEREDITARY FACTORS

Hereditary factors result in the nature of the *genotype*, the genetic makeup present in the germ cells established at conception. The combination of environmental factors and the genotypical design yields the *phenotype*, the total living, thinking, acting person. An additional set of influences is the level of physiological functioning of the parents, particularly the mother at conception and during the course of pregnancy and the perinatal period, the level of physiological functioning of the developing child throughout gestation and labor and delivery.

It is generally agreed that the programming of the genetic code is entirely responsible for such hereditary factors as body type, skin and eye color, acuity of senses, and so on. Barring extraordinary environmental circumstances, the process of maturation results in the gradual development of the characteristics called for by the genotype. These include the fundamental facts of sex and race, and also one's potential longevity.

Abnormalties caused by an irregularity in the structure or number of chromosomes occur in about 25 percent of conceptions (Hetherington and Parke, 1979). The vast majority of these abnormlities terminate at the embryonic stage as spontaneous abortions.

Among the more common developmental anomalies caused by chromosomal aberrations is *Down's syndrome,* which is characterized by the presence of a *trisomy*, i.e., an abnormal extra chromosome, of chromosome 21; affected individuals are marked by varying degrees of mental retardation and by various physical traits, typically including a small, flattened skull, epicanthal folds, and a flat-bridged nose.

Turner's syndrome results from an absence of the second sex chromosome; instead of the normal XY (male) or XX (female) arrangements, the affected person has an XO karyotype, resulting in an apparently female phenotype, but with rudimentary or absent gonads that do not mature at puberty. In contrast, *Klinefelter's syndrome* is characterized by the presence of an extra sex chromosome (XXY karyotype), affects males, and may cause mental retardation and sterility.

Research conducted during the 1970s has led scientists to conclude that exposure to radiation and high temperatures, and the ingestion of various chemical substances can cause genetic damage (National Academy of Sciences, 1972; Meselson, 1971; Neel, 1972). It has been estimated that toxic chemicals are responsible for about 45 percent of spontaneous abortions (Norwood, 1980, p. 7).

MATERNAL FACTORS

As was mentioned earlier, several factors pertaining to the biological parents, especially the mother, constitute a separate set of influences, aside from heredity and environment per se, on the development of the individual.

Age

The optimal maternal age at pregnancy is between 20 and 34 years. Certain deviations from normal in the newborn are associated with maternal age below 20, and the risks are especially great for mothers below 15; the most common of these is low birth weight. Birth weights below 2,500 g, or 5.5 lb, place the neonate at higher risk for a number of deviations in health and development, from mortality in infancy to decreased psychomotor functioning lasting for many years (Newberger, Newberger, and Richmond, 1976). When maternal age is 35 or over, there is a somewhat greater risk of low birth weight, but it is birth defects that are particularly likely to result from these late pregnancies (DHEW, 1979).

Medical History

Various aspects of the mother's medical history may be of significance as regards the development of the fetus. A history of prior fetal loss indicates a risk for the outcome of a present pregnancy. Furthermore, the number of previous pregnancies, even those with healthy outcomes, is a matter of some concern; a slower rate of development throughout childhood is associated with multiparous mothers who have given birth to as many as four babies (Davie, Butler, and Goldstein, 1972).

Preexisting chronic diseases such as hypertension and diabetes may threaten the quality of the intrauterine environment. Neonatal respiratory problems as well as stillbirths have been linked in some cases to maternal diabetes. Regarding stillbirth, risk is highest in the last few weeks of pregnancy. The majority of babies of diabetic mothers are larger than average, the largeness being partly attributable to maternal hyperglycemia. The large size contributes sometimes to difficult delivery and to birth trauma. Problems that commonly appear after birth include respiratory distress syndrome, and approximately half of diabetic offspring have low blood sugar levels.

Still another factor to be explored in the maternal medical history is the existence of incompatibility between blood types of mother and child. ABO incompatibility is usually mild and amenable to therapy. A more serious kind is Rh blood incompatibility between an Rh-positive baby and an Rh-negative mother. An inadequate oxygen supply to the fetus results from this incompatibility, and this is frequently followed by miscarriage, stillbirth, or neonatal death. Fortunately, various advances have been made in the management of this condition (Scipien, 1975).

Nutritional Status

Lifelong maternal deficiencies in proteins, vitamins, minerals, or carbohydrates are associated with poor infant health throughout the first year and beyond. Postneonatal mortality rates are higher when this nutritional condition exists in the mother. Nutritional deficiencies during the pregnancy itself are associated particularly

with poor condition during the first 28 days (Kessner, et al., 1973; Murphy and Landsberger, 1980). Dietary deficiencies during pregnancy are particularly damaging during the last trimester, when fetal growth is very rapid; less growth in brain tissue has been documented in studies by Dobbing (1973).

One consequence of greater affluence in industrialized nations has been a general improvement in the nutritional status of their populations. White male 1-year-olds born in North America were almost 2 in. taller on the average in 1960 than their counterparts born in 1880 (Meredith, 1978).

Endocrine Functioning

The physiology of the embryo, fetus, and newborn is influenced by the manner in which the mother's endocrine functioning accommodates to the physiological demands of pregnancy. The critical nature of the hormonal balance in the prenatal period for the development of masculinity and femininity in later stages has been indicated by findings of various studies (Young, Goy, and Phoenix, 1964).

Results from animal research have shown that the normal presence of the sex hormones during pregnancy influences the development of sex-linked physical characteristics and behavior that appear much later in the individual's development (Young, Goy, and Phoenix, 1964). It has been suggested that during gestation the hormones organize the psychological and biological tendency toward masculinity or femininity (Hetherington and Parke, 1979). Money and Ehrhardt (1978) studying human beings, found that both fetal and maternal hormones may alter the course of sexual development; they reported that in the absence of sufficient androgen after the seventh or eighth week of gestation, embryos with XY karyotypes developed female characteristics despite their chromosomal patterns.

Exposure to Teratogens

Although the teratogenic potential of such factors as high noise levels, radiation exposure, air and water pollution, and food additives has received much attention since the 1970s, the following discussion focuses on teratogens such as are voluntarily consumed by the mother: cigarettes, alcohol, and narcotics. The dangers of these substances are those posed by teratogenic agents in general: no protection is afforded by the placental barrier, and the substances that enter the mother's system also enter the fetal system (see Table 2-1). Wide variations in the effects of teratogens exist because of differences in the timing of the ingestion of the teratogen with respect to fetal age, differences in general physical status, and in dosage, absorption, metabolism, and excretory function of both mother and fetus.

The existence of teratogens among pharmaceutical drugs is significant not only because of documented effects of particular drugs upon fetal development, but because the prescribed and elective use of drugs by pregnant women is very common. The Collaborative Perinatal Project found that the percentage of women who used as many as five to ten drugs during pregnancy rose from 14 percent to 33 percent during the 7-year span of the research (Collaborative Perinatal Project, 1977); the proportion of those who used no drugs at all declined from 9 to 4 percent (Norwood, 1980, p. 57).

The implications of such a trend become clear when one considers that teratogens are not necessarily known as such when they are first made available. A tragic case in point is the Thalidomide disaster of the 1950s and 1960s. That drug had been prescribed for mothers as a sedative or as an antinausea measure. Most mothers had no adverse effects themselves from the drug; indeed, most of their pregnancies ran a perfectly normal course. However, in many instances the offspring were born with various

Table 2-1. Common Drugs and Their Teratogenic Effects

Drug	Effect
Alcohol	Craniofacial, limb, and cardiac defects; intrauterine growth retardation, developmental delay, and mental retardation. More than two drinks per day can be hazardous.
Androgens	Long-term high doses can cause masculinization of female fetuses.
Anticonvulsants	Craniofacial and limb abnormalities; mental deficiency.
Diethylstilbestrol (DES)	Female offspring of women who receive DES during first trimester have increased risk of developing adenosis of vagina and adenocarcinoma of vagina and cervix in adolescence or young adulthood, and of infertility.
Radioactive Iodine	May cause transient or permanent hypothyroidism when given after 14th week of pregnancy.
Tetracycline	When administered to mother during second or third trimesters, may cause staining, caries, and enamel hypoplasia of deciduous teeth.
Narcotics (heroin, morphine, etc.)	Maternal addiction leads to fetal addiction, manifested in withdrawal signs during neonatal period.
Nicotine	Women who smoke one pack or more of cigarettes per day place their offspring at increased risk of prematurity, low birth weight, stillbirth, and neonatal death.

malformations, the most common being pho-
comelia, an anomaly characterized by missing
limbs and the feet and/or hands being attached,
like flippers, directly to the torso.

More recently, the synthetic hormone diethylstilbestrol (DES) has proven to have long-
delayed teratogenic effects on some offspring.
DES was prescribed between 1947 and 1964 as
a means of preventing miscarriage. Two million
women may have taken the drug. Not until the
1960s, when the first generation of affected offspring reached adolescence, was it found that
many daughters of women who had taken DES
were developing vaginal abnormalities, including vaginal cancer. It is therefore recommended
that women born after 1946 and before 1965 be
asked whether their mothers were among the
many for whom DES was prescribed. Sterility
in male offspring of DES users may also be a
long-term result (*Good Housekeeping,* 1980).
The potential problem is so widespread that a
DES Registry has been established in Washington, D.C., and the National Cancer Institute
has a department of DES at its headquarters in
Bethesda, Maryland.

A number of commonly used pharmaceutical
drugs have been found in some instances to have
teratogenic effects. These drugs include qui-
nine, reserpine, and tetracycline. One study has
proposed that when pregnant women take even
the common aspirin in high doses, blood disorders may result in their offspring (Eriksson,
Catz, and Yaffe, 1973).

Illicit drugs also pose threats to normal fetal
development. In some studies, heroin and morphine have been linked to low birth weight and
prematurity (Eriksson, Catz, and Yaffe, 1973).
Infants born to heroin and morphine addicts
have been observed to go through the same
withdrawal symptoms as adults, including severe hyperirritability, rapid respiration, and
hyperactivity. Death can result from these
physiological reactions if the infant is not taken
off the drug gradually (Brazelton, 1970). Eriksson, Catz, and Yaffe reported that the sus-

tained use of LSD has led to chromosomal
breakages (1973).

By far the most common teratogenic agents,
however, are nicotine and alcohol. Eighty percent of pregnant women are reported to drink
alcohol, and over half smoke (Hetherington and
Parke, 1979). Effects that have been linked to
nicotine include prematurity and low birth
weight, and abnormalities such as dwarfism
have been attributed to alcohol use. Higher dosages of both have been found to be especially
dangerous, and the *combination* of smoking and
drinking increases the chances of developmental
deviations (Little, 1975). As with the narcotic
drugs, infants of alcoholic women display withdrawal signs during the first 24 to 48 hours after
delivery. Unless there is careful treatment the
reaction may be fatal.

Infectious Diseases during Pregnancy

Various infections that may attack the pregnant
woman may constitute threats to the fetus, depending, again, on their timing as well as on
the severity of the attack. Rubella during the
first 3 months of pregnancy is known to cause
anomalies of the eyes, heart, and brain, and
sometimes death. Syphilis in the mother is
known to cause serious damage or death to the
fetus. The same is true of gonorrhea, herpes,
and a number of other viral diseases. A herpes
II infection during the last trimester is an au-
tomatic indication for cesarian section; babies
born vaginally contract the disease, and 90 percent die (Kessner, et al, 1973).

Emotional States

Many researchers have suggested that stress experienced by the pregnant woman can be damaging to the developing fetus as well as to the
physiological functioning of the gravida (Fer-
riera, 1969; Jaffe, 1969; McDonald, 1968; Sameroff and Zax, 1973). For example, the heart
rate and activity of the fetus have been shown

to increase when the material cortisone level rises because of anxiety (Murray and Zentner, 1979).

Perinatal Factors

The conditions surrounding the labor and delivery affect the physiological condition of the newborn, although it must be made clear that most births—at least 90 percent of them—do proceed normally. Among the conditions that affect the remaining 10 percent of births are maternal toxemia or hemorrhage; premature separation of the placenta, blockage of the uterine opening caused by the position of the placenta in the lower portion of the uterus, or prolapse of the umbilical cord. Obstetrical procedures such as the administration of anesthetics and medication to the mother and the use of forceps during delivery can also have adverse effects upon the newborn.

Cases of prolongd labor or of breech birth, in which the buttocks presents first rather than the head, increase the chances of anoxia, a lack of oxygen and excess of carbon dioxide in the brain of the newborn. A labor lasting 7 to 10 hours is least likely to be detrimental; it has been found that labors lasting under 2 hours and those terminated by cesarean section sometimes lead to anoxia resulting from exposure to oxygen too suddenly (Schwartz, 1956). The effects of some anoxia at birth usually disappear with age. However, anoxia is the leading cause of perinatal death, and extreme cases in which the child survives sometimes show later mental retardation or cerebral palsy.

Combinations and Interactions of Maternal Factors

The most serious threat to development comes from the all-too-common instance of two or more of the foregoing factors occurring together. Picture, for instance, a situation in which a very young pregnant woman with nu-

tritional deficiencies resulting from lifelong poverty also has a debilitating disease, and lives in a stress-producing inner-city location. Such a woman is bound to provide very poorly for the developmental needs of a fetus she carries. Added to all of the other problems is the fact that pregnant women living in low-income areas are likely to receive prenatal care later on in the pregnancy, if at all (Gortmaker, 1979). The increasing frequency of pregnancies in situations containing a number of problems was focused upon in a report prepared under the auspices of the National Academy of Sciences (1976). That report recommended a broad array of measures; the one most emphasized was to increase the economic resources of women in such circumstances to counteract as soon as possible some of the effects of poverty.

INTRINSIC FACTORS

Regarding the developing organism itself during the prenatal and perinatal stages, the functioning of the body systems—especially the central nervous, endocrine, cardiovascular, and digestive systems—is crucial for a good developmental start. Although protected from bacterial infections and from some trauma by the intrauterine environment, the organism's functioning may be placed at risk by the various maternal factors described above. Some fetal reactions prove fatal during the prenatal period. When this is not the case, the trauma of birth itself may damage a neonate with less-than-perfect development and with poorly functioning body systems. This is especially true when there are problems with respiratory and circulatory systems (Tanner, 1978).

The survival and healthy functioning of preterm infants—that is, those born at or before 37 weeks' conception—or of those whose birth weight is less than 5.5 lb (2,500 g) is imperiled. Some problems facing preterm infants and very small infants are inherent in the treatments needed for survival. For example, if the child must be placed in an incubator, there will be an absence of normal stimuli in the environment, and the child must endure long separations from the mother and other family members, which in turn delays the bonding process. The very small infant's erratic sleep patterns are also costly to health and development, since secretions of the growth hormone increase during sleep.

ENVIRONMENTAL FACTORS

Since the early 1950s, research in obstetrics, pediatrics, epidemiology, nutrition, economics, psychology, anthropology, sociology, and gerontology has established the influence of environment on health and development. Such a great number of scientific fields is involved because of the multitude of different kinds of factors in the environment that influence growth and development. To begin, there is the quantity and quality of the child's nutrition. Then, consider the condition of the air he breathes and the quality of the water he drinks. Type of housing, school and work environments, and aspects of one's neighborhood and community can increase or decrease the likelihood of exposure to such environmental factors as accidents, toxic substances, noise, and crowding. The term "poverty" generally connotes a combination of such factors which typically accompanies low-income family life.

The social environment is known to be important to the individual's health and development from the first hour until the last. The immediate relationship to the family or surrogate family, which provides for the helpless infant's survival, soon takes on emotional significance and shapes the infant's well-being, provides stimulation for cognitive development, and ultimately influences the child's sense of self-esteem (Brazelton, 1976).

The development of individuals in their society is also influenced very much by the availability of natural resources, and by agricultural productivity, technology, management skills, educational opportunities, and political responsiveness to oppression or repression. Finally, environmental quality is influenced by the cultural attitudes and beliefs that dictate the kinds of family structure deemed acceptable, the relationship of human beings to nature, their relationship to music and other aesthetic factors, and to the supernatural, as well as to other human beings (Boulding, 1979). Not only is the nature of these cultural factors important; the *rapidity of cultural change* perhaps exerts even more influence than any one pattern.

SUMMARY

The focus of early studies of human development was on deviations from normal. Large studies of normal development are infrequently undertaken, thus more is known about developmental disorders than about normal growth and development, and not until the middle of the twentieth century were periods other than early childhood studied in depth.

Broadly speaking, human development is influenced by hereditary and environmental factors, but also by parental (and especially maternal) factors, and intrinsic factors. Among the most important aspects of maternal influence on growth and development is exposure to teratogens, including licit and illicit drugs, and such environmental teratogens as noise, air and water pollution, and so forth. Other important facets of maternal influence are nutritional status, history, and exposure to infectious diseases. The most significant intrinsic factors include the neonate's weight, functioning of body systems, and proximity to the mother.

Environmental factors include not only purely physical surroundings but also such social features as housing, school environment, and nature of neighborhood and community. These have an impact not only on such more obvious environmental influences as likelihood of exposure to toxic materials or accidents, but also on the developing child's perception of the world and, ultimately, of himself.

REFERENCES

Ainsworth, M. D., and Bell, S. M. Some contemporary patterns of infant-mother interaction. In J. Ambrose (Ed.), *Stimulation in early infancy.* London: Academic Press, 1969.

Anthony, J. E., and Koupernik, C. (Eds.). *The child in his family,* (Vol. III). New York: John Wiley & Sons, 1974.

Baldwin, A. L. *Theories of child development.* New York: John Wiley & Sons, 1967.

Baltes, P., and Schaie, K. W. (Eds.) *Life span developmental psychology.* New York: Academic Press, 1978.

Benedict, R. Continuities and discontinuities in cultural conditioning. *Psychiatry,* 1938, *1,* 161–167.

Biller, H. B., *Father, child and sex role.* Lexington, Mass.: Lexington Books, 1971.

Birch, H. G., and Gussow, J. D. *Disadvantaged children: health nutrition and school failure.* New York: Harcourt, Brace & World, 1970.

Boulding, E. *Children's rights and the wheel of life.* New Brunswick, N.J.: Transaction Books, 1979.

Bowlby, J. *Maternal care and mental health.* Geneva: World Health Organization, 1952.

Brazelton, T. B. *Infants and mothers: differences in development.* New York: Dell, 1969.

———. Effects of prenatal drugs on the behavior of the neonate. *American Journal of Psychiatry,* 1970, *126,* 1261–1266.

———. Early parent-infant reciprocity. In Vaughan, V. C., and Brazelton, T. B. (Eds.), *The family: Can it be saved?* Chicago: Year Book Medical Publishers, 1976.

Bronfenbrenner, U. *Is early intervention effective?* Washington, D.C.: Department of Health, Education and Welfare, Office of Child Development, 1974.

———. Who cares for America's children? In Vaughan, V. C., and Brazelton, T. B. (Eds.), *The family: Can it be saved?* Chicago: Year Book Medical Publishers, 1976.

———. *The ecology of human development.* Cambridge, Mass.: Harvard University Press, 1979.

Coles, R. *Erik Erikson, the growth of his work.* Boston: Little, Brown, 1970.

Collaborative Perinatal Project. *Birth defects and drugs in pregnancy.* Littleton, Mass.: Publishing Sciences Group, 1977.

Craig, G. J. *Human development.* Englewood Cliffs, N.J.: Prentice-Hall, 1980.

Davie, R., Butler, N., and Goldstein, H. *From birth to seven: a report of the National Child Development Study.* London: Longmans, 1972.

Department of Health, Education and Welfare. *Healthy people.* Washington, D.C.: Office of Health Promotion and Disease Prevention, 1979.

Department of Health and Human Services. *Promoting Health, Preventing Disease: Objectives for the Nation.* Washington, D.C.: Office of Disease Prevention and Health Promotion, 1980.

Dobbing, J. Nutrition and the developing brain. *Lancet,* 1973, *1,* 48.

Elkind, D. Egocentrism in adolescence. *Child Development* 1967, *38,* 1025–1034.

Erikson, E. *Childhood and society.* New York: Norton, 1950.

———. *Identity: youth and crisis.* New York: Norton, 1968.

Eriksson, M., Catz, C., and Yaffe, S. Drugs and pregnancy. In H. Osofsky (Ed.), *Clinical obstetrics and gynecology,* (Vol. 16).New York: Harper and Row, 1973.

Escalona, S. Intervention programs for children at psychiatric risk. In Anthony, J. E., and Koupernik, C. (Eds.), *The child in his family,* (Vol. 3). New York: John Wiley & Sons, 1974.

Evans, R. *Dialogue with Erik Erikson.* New York: Harper and Row, 1967.

Ferreira, A. *Prenatal environment.* Springfield, Ill.: Charles C Thomas, 1969.

Freud, S. *Civilization and its discontents.* London: Hogarth Press, 1930.

Goy, R. W. Organizing effects of androgen on the behavior of rhesus monkeys. In R. P. Michael (Ed.), *Endocrinology and human behavior.* New York: Oxford University Press, 1968.

Good Housekeeping. DES: the new problems. October, 1980, 7.

Goleman, D., 1,528 little geniuses and how they grew. *Psychology Today,* February 1980, *13*(9), 28–53.

Gortmaker, S. L. Poverty and infant mortality in the United States. *American Sociological Review,* 1979, *44,* 280–297.

Greenberg, M., and Morris, N. Engrossment: the newborn's impact upon the father. *American Journal of Orthopsychiatry,* 1974, *44,* 520–531.

Haggerty, R. J., Roghmann, K. J., and Pless, I. B. *Child health and the community.* New York: John Wiley & Sons, 1975.

Harrington, M., *The other America.* New York: Macmillan, 1962.

Harvard Child Health Project, (Vol. 2). *Children's medical care, needs and treatments*. Cambridge, Mass.: Bellinger, 1977.

Heinonen, O. P., Sloane, D., and Shapiro, S. (Eds.) *Birth defects and drugs in pregnancy*. Littleton, Mass.: Public Sciences Group, 1977.

Hetherington, E. M., and Parke, R. D. *Child Psychology*. New York: McGraw-Hill, 1979.

Hobbs, N. *Issues in the classification of children*. San Francisco: Jossey-Bass, 1975,

Illych, I. *Deschooling society*. New York: Harper and Row, 1971.

Jaffe, J. M. *Prenatal determinants of behavior*. Oxford: Pergamon, 1969.

Kamin, L. J. Jensen's last stand (A book review of Jensen, A. R., *Bias in mental testing*). *Psychology Today*, February 1980, 117–123.

Karnofsky, D. A. Drugs as teratogens in animals and man. *Annual Review of Pharmacology*, 1965, *5*, 477–482.

Kenniston, K. *All our children*. New York: Harcourt Brace Jovanovich, 1977.

Kessner, D. M. et al. *Infant death: an analysis by maternal risk and health care*. Washington, D.C.: National Academy of Sciences, 1973.

Klaus, M. H., and Kennell, J. H. *Maternal-infant bonding*. St. Louis: Mosby, 1976.

Korner, A. F., Individual differences at birth: implications for early experience and later development. In J. Westman (Ed.), *Individual differences in children*. New York: John Wiley & Sons, 1973.

Lamb, M. E. *The role of the father in child development*. New York: John Wiley & Sons, 1976.

Levinson, D. *The seasons of a man's life*. New York: Knopf, 1978.

Liederman, P. H. Mothers at risk: a potential consequence of the hospital care of the premature infant. In E. J. Anthony and C. Koupernik (Eds.), *The child in his family*, (Vol. III). New York: John Wiley & Sons, 1974.

Little, R. *Maternal alcohol use and resultant birth weight*. Unpublished doctoral dissertation, Johns Hopkins University, 1975.

Macfarlane, J. W. From infancy to adulthood. In P. Mussen, Conger, J. and Kagan, J. (Eds.), *Basic and contemporary issues in developmental psychology*. New York: Harper and Row, 1975.

McDonald, R. L. The role of emotional factors in obstetrics complications. *Psychosomatic Medicine*, 1968, *30*, 222–237.

Mead, M., and Wolfenstein, M. (Eds.) *Childhood in contemporary cultures*. Chicago: University of Chicago Press, 1955.

Meredith, H. V. *Human body growth in the first ten years of life*. Columbia, S.C.: State Printing Office, 1978.

Meselson, M. Foreword. In A. Hollander, (Ed.), *Chemical mutagens*. New York: Plenum Press, 1971.

Milio, N. *The care of health in communities*. New York: Macmillan, 1975.

Miller, C. A., Health care of children and youth in America. *American Journal of Public Health*, 1975, *65*, 353–358.

Miller, D. R. et al. Fatal disseminated Herpes Simplex virus infection and hemorrhage in the neonate. *Journal of Pediatrics*, 1970, *76*, 405.

Money, J., and Ehrhardt, A. *Man and woman, boy and girl*. Baltimore: Johns Hopkins, 1972.

Montessori, M. *The absorbent mind*. New York: Holt, Rinehart and Winston, 1967.

Murphy, L. B., and Moriarty, A. E. *Vulnerability, coping, and growth*. New Haven: Yale University Press, 1976.

Murphy, N., and Landsberger, B. A data-based approach to reducing infant mortality in three rural counties of a southern H.S.A. *Advances in Nursing Science*, 1980, *2*, 97–109.

Murray, R. B., and Zentner, J. *Nursing assessment and health promotion through the life span*. Englewood Cliffs, N.J.: Prentice-Hall, 1979.

National Academy of Sciences. *Biological effects of ionizing radiation* (BEIR Report). Washington, D.C.: National Academy of Sciences, 1972.

National Academy of Sciences. *Toward a national policy for children and families*. Washington, D.C.: National Academy of Sciences, 1976.

National Center for Health Statistics. Plan, operation and response results of a program of children's examinations. *Vital and Health Statistics*, Series 1, No. 5, PHS Publ. No. 1000, Public Health Service. Washington, D.C.: U.S. Government Printing Office, 1967.

National Center for Health Statistics. Plan and operation of a health examination survey of U.S. youths 12–17 years of age. *Vital and Health Statistics*, Series 1, No. 8, Public Health Service. Washington, D.C.: U.S. Government Printing Office, 1969.

National Center for Health Statistics. Factors associated with low birth weight. *Vital and Health Statistics*, Series 21, No. 37, Public Health Service. Washington, D.C.: U.S. Government Printing Office. 1980.

Neel, J. V. The detection of increased mutagen rates in human populations. In *Mutagenic effects of environmental contaminants*. New York: Academic Press, 1972, pp. 99–119.

Newberger, E., Newberger, C., and Richmond, J. Child health in America: toward a rational public policy. *Milbank Memorial Fund Quarterly*, 1976, *54*, 249.

Norwood, C. *At highest risk*. New York: McGraw-Hill, 1980.

Piaget, J. Piaget's theory. In Neuberger, P. (Ed.), *The process of child development*. New York: New American Library, 1976.

Riley, M. W. (Ed.). *Aging from birth to death*. Boulder, Colorado: Westview Press, 1979.

Sameroff, A. J., and Zax, M. Perinatal characteristics of the offspring of schizophrenic women. *Journal of Nervous and Mental Diseases,* 1973, *46,* 178–185.

Schwartz, P. Birth injuries of the newborn. *Pediatric Archives,* 1956, *73,* 429–450.

Scipien, G. M. et al. *Comprehensive pediatric nursing.* New York: McGraw-Hill, 1975.

Sears, R. R., Your ancients revisited: a history of child development. In *Review of child development research,* (Vol. 5), Chicago: University of Chicago Press. 1975, 1–73.

Spitz, R. A., Hospitalism: an injury into the genesis of psychiatric conditions in early childhood. *Psychoanalytic Study of the Child* 1945, *1,* 53–74.

Tanner, J. M. *Foetus into man.* Cambridge, Mass.: Harvard University Press, 1978.

Thomas, A., Chess, S., and Birch, H. G. *Temperament and behavior disorders in children.* New York: New York University Press, 1968.

Thomas, A., Chess, S. *Temperament and development.* New York: Brunner/Mazel, 1977.

Urban, H. B., The concept of development from a systems perspective. In Baltes, P. (Ed.), *Life-span development and behavior.* New York: Academic Press, 1978.

U.S. General Accounting Office. Early childhood and family development programs improve the quality of life for low-income families: a report to the Congress by the Comptroller General. Report #HRD-79-40, Washington, D.C.: General Accounting Office. February 6, 1979, 96.

Wallace, H. M. *Health care of mothers and children in national health services.* Cambridge, Mass.: Ballinger Publishing Company, 1975.

White, S. Human research and human affairs. In White, S., (Ed.), *Human development in today's world.* Boston: Little-Brown, 1976.

Young, W. C., Goy, R. and Phoenix, C. Hormones and sexual behavior. *Science,* 1964, *143,* 212–218.

3

Principles of Growth and Development

JACKI DELECKI

Life is a dynamic, changing process. Human life proceeds in stages from infancy to adulthood to senescence. Humans spend one-third of their lives in preparation for the latter two-thirds. Following a comparatively lengthy prenatal period, the human is born, still dependent, as during fetal life. An extended stage of immaturity follows as the human passes through infancy, childhood, and adolescence, before achieving the capacity for independent function. Unlike other animals, humans do not inherit a repertoire of instinctive behaviors. The prolonged dependent stage permits the individual to incorporate learned experience.

Although growth and development may be defined as separate physiological processes, they cannot be separated in nature; cells increase in size and number as they differentiate. Growth and development are complex interactive processes by which humans mature. All human beings share the same mechanisms of growth and development and proceed through the same stages of life. The rate of maturation and the final outcome are widely variable because of the extensive and diverse influences on each individual. The mature person is the product of the interaction of genetic potential and the molding influences of the internal and external environment.

The human organism undergoes dramatic

changes in the progression from fertilized ovum to purposeful adult. The processes of growth and development orchestrate this progression through direct effects at the cellular level. *Growth* may be defined as a physiological process by which the organism assimilates or transforms essential, nonliving nutrients into living protoplasm (Timiras, 1972). Growth is not merely the addition of materials to achieve an increase in size; for example, abnormal water retention may cause an increase in size but cannot be called growth. Growth involves the incorporation of new materials. This active process may then increase the size and/or number of cells.

Development is defined as a physiological process by which the individual progresses from an undifferentiated state to a highly organized and functional capacity. The cells of the human organism not only grow and divide but also become differentiated. Differentiation means the cells become committed to performing specialized functions. Development implies an increase in skill and in complexity of function. As growth is measured in pounds and inches, development is measured in abilities.

The processes of growth and development are under the direction of the individual's genetic code, which is determined by the mother's contribution in the ovum and the father's contribution in the sperm. The genetic code is recorded in the deoxyribonucleic acid (DNA) of the fertilized ovum. DNA directs the synthesis of proteins, which in turn direct growth and differentiation. All humans share the same mechanisms of genetic coding; therefore, they follow the same sequence of growth and development.

The sequence of growth and development can be divided into four morphological stages. The first stage, occurring in the late ovum and early embryonic period, is a rapid period of growth with an increase in cell number and cell size; little cell differentiation takes place during this stage. The second stage is characterized by simultaneous cell growth and cell differentiation; this stage ends at maturity. The third stage is adulthood, which is characterized by functional stability and competency. During this stage, growth maintains the body at an optimal level by repair and renewal of cells. The fourth stage occurs during old age, when growth is unable to keep the body in balance. During this stage, cells are lost from various systems without repair or replacement. The loss of cells results in inefficient and deteriorating functioning of the body systems.

PATTERNS OF GROWTH AND DEVELOPMENT

The genetic code determines the rate of the growth and development of the various body parts, which occur in synchrony with overall growth and development. As the rate of growth varies in one area of the body, body proportions change, (see Fig. 3-1). For example, at birth the lower limbs of the infant are less well developed than the upper limbs. The ratio of lower limb length to total height is $1:3$. In the adult the ratio of lower limb length to total height is $1:2$. The child is relatively "top-heavy" as he begins to walk. There will be a downward shift in center of gravity as the child grows; improved stability follows.

During the prenatal and postnatal periods, growth and development follow a *cephalocaudal* pattern, that is, development proceeds from head (cephalo) to toe (caudal). The head matures before the neck, the neck before the chest, the chest before the pelvis, and so on.

Growth and development also follow a *proximodistal* pattern, i.e., they proceed from near the center axis of the body outward toward the extremities. For example, the upper arm grows before the lower arm, which grows before the hand.

The cephalocaudal and proximodistal pat-

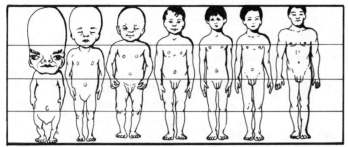

2 mo.(fetal) **5** mo. Newborn **2** yr. **6** yr. **12** yr.   **25** yr.

Figure 3-1. *Changes in body proportion from second fetal month to adulthood. (Redrawn from Robbins, et al.: Growth, New Haven, Yale University Press, 1928)*

terns of growth and development are evident in the motor development of the infant. He rolls over before he can sit; his arm and chest development enable this maneuver. As his trunk and legs develop, he will be able to sit unsupported. He will walk when his trunk and legs develop further. This progression from rolling over to sitting to walking follows the cephalocaudal pattern.

The proximodistal pattern is also seen in the infant's motor development. The infant uses his hand as a unit before he can control the movement of his fingers. He will use his palm and thumb to manipulate objects before he is able to use a pincer grasp with thumb and index fingers; thus, development progresses from hand to fingers. The principles of sequential order of maturation and proportional growth provide a theoretical basis on which to order the phenomena observed in clinical practice.

STAGES OF GROWTH AND DEVELOPMENT

The human life span can be divided into two distinct biological periods, prenatal and postnatal. These stages can be further subdivided into several periods, each distinguished by identifiable morphological, physiological, biochemical, and psychological factors (see Table 3-1). The prenatal period includes three stages: the ovum, embryo, and fetus. Postnatal life begins with birth and proceeds through the neonatal period, infancy, childhood, adolescence, adulthood, and senescence. The progression from one stage to the next is dependent upon completion of the previous stage. Attained growth is both an indicator of past growth processes and a predictor of future growth potential.

Prenatal Period

The most rapid growth and development in the human life cycle occurs in the prenatal period. During the embryonic period, the formation of organs occurs rapidly. Inadequacies or teratogens in the uterine environment, even those that exist only briefly, may cause profound malformations during this vulnerable period. In the early fetal period, the fetus's length increases rapidly. During the last trimester, the fetus makes rapid gains in weight (see Figure 3-2). By the conclusion of the fetal period, organogenesis is completed, and the fetus is a "complete" human being.

Neonatal Life

The neonatal period is defined as the first month of life. The first 24 hours of life comprise the most significant time of this period, as the neonate must make the transition from an intrauterine to atmospheric environment. Infant mortality is highest during this transition.

Table 3-1. Stages in the Life Cycle

Period	Approximate Age or Stage	Some Characteristics
Embryonic	First trimester of prenatal life	Rapid differentiation; establishment of systems and organs
Early fetal	Second trimester of prenatal life	Accelerated growth; elaboration of structures; early functional activities
Late fetal	Third trimester of prenatal life	Rapid increase in body mass; completion of preparation for postnatal life
Parturient	Period of labor and delivery	Risk of trauma and anoxia; cessation of placental function
Neonatal and early infancy	First month of postnatal life	Postnatal adjustments in circulation; initiation of respiration and other functions
Middle infancy	1 mo to 1 yr	Rapid growth and maturation; maturation of functions, especially of nervous system
Late infancy	1 to 2 yr	Decelerating growth; progress in walking and other voluntary motor activities, and in control of excretory functions
Preschool	2 to 6 yr	Slow growth; increased physical activity; further coordination of functions and motor mechanisms; rapid learning
School	Girls: 6 to 10 yr Boys: 6 to 12 yr	Steady growth; developing skills and intellectual processes
Prepubertal (late school or early adolescent)	Girls: 10 to 12 yr Boys: 12 to 14 yr	Accelerating growth; rapid weight gain; early adolescent endocrine and sex organ changes
Pubertal (adolescence proper)	Girls: 12 to 14 yr Boys: 14 to 16 yr	Maturation of secondary sex characteristics;
Postpubertal	Girls: 14 to 18 yr Boys: 16 to 20 yr	Maximum postnatal growth increase; decelerating and terminal growth; rapid muscle growth and increased skills; rapid growth and maturing functions of sex organs; need for self-reliance and independence

Reproduced with permission from Timiras, P. S. *Developmental Physiology and Aging.* New York: Macmillan, 1972. Copyright © 1972 by Paola S. Timiras.

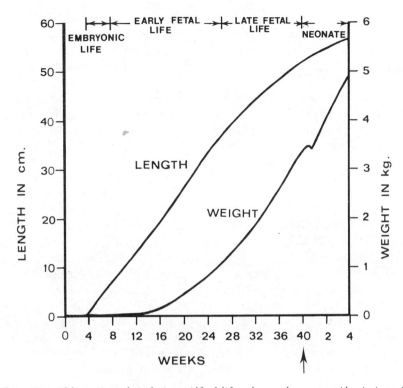

Figure 3-2. *The most rapid linear growth is during midfetal life, whereas the most rapid gain in weight is during the last trimester. (Redrawn from Smith, Bierman: The biological ages of man, Philadelphia, W. B. Saunders Co., 1973.)*

Infancy

Infancy is a period of rapid growth and development. During the first three months of life, the infant gains an average of two pounds per month; by five months of age, the birth weight has doubled. Birth weight is tripled by the end of the first year and quadrupled by the end of the second year. The infant also grows rapidly in height. By the end of the first year the infant's lenth is 50 percent greater than at birth. During this period, the infant brain grows more rapidly than the whole body and attains nearly adult size by two years of age. This rapid growth and development is demonstrated by the remarkable gains in abilities that the infant makes. The infant becomes a mobile individual with the capacity to communicate.

Childhood

The preschool years (age 2 to 6) and the school years (age 6 to 12) represents a period of relatively slow growth. The annual increment in weight averages approximately 5 lb until the ninth or tenth birthday. The height increment during the school years is approximately 2 in. each year. During this period, the child's social world expands as he enters school.

Adolescence

The adolescent period comprises the years in which the child "grows up." It is a period of profound changes in physical, mental, and emotional development. The period is marked by an accelerated gain in weight and height.

Sexual maturity is brought about by changes in the endocrine and central nervous systems.

Adulthood and Aging

The adult years are a period of competency and ability. Unlike the development stages, the adult years do not have a sequence of precisely timed physiological events. This is also true of the aging years from 65 years until death.

HORMONAL INFLUENCE ON GROWTH AND DEVELOPMENT

The growth and development of an individual is influenced by four major sources of hormones: the central nervous system, and the thyroid, adrenal, and gonadal glands. Although the hormones have different chemical structures and are produced by different organs, they all influence growth and development by stimulating protein synthesis.

It is hypothesized that each hormone has its major effect on growth and development at a particular period of life. The thyroid hormones exert their major effects during the prenatal period and childhood; the adrenal and gonadal androgens have their major effect in the adolescent period, and the central nervous system is influential during both childhood and adolescence (Timiras, 1972).

Growth Hormone

Growth hormone is secreted by the anterior portion of the pituitary gland in response to growth hormone releasing factor, which is produced by the hypothalamus; under the influence of various stimuli, the hypothalamus controls the anterior pituitary and, thereby, the release of growth hormone. Thus, the pituitary is an intermediary gland. The hypothalamus also acts as a control center for the release of thyroid

hormones from the thyroid gland, and androgens from the adrenal cortex and gonads. Sensitive to the levels of circulating hormones, it emits releasing factors to stimulate or suppress the thyroid and adrenal glands. The hypothalamus, pituitary gland, and the target organs (i.e., the thyroid and adrenal glands) function together in a classic feedback mechanism (Lowrey, 1973) (see Figure 3-3).

Growth hormone is chiefly influential in protein synthesis. At the cellular level, growth hormone stimulates increased transport of amino acids into cells and increased incorporation of amino acids into protein.

The role of growth hormone in the prenatal period has not been clearly delineated. It is probable that growth hormone plays an early role in the growth and development of the fetus. Growth hormone is essential for growth in childhood. If the pituitary gland is destroyed, the child's linear growth will cease.

The influence of growth hormone brings about an increase in the child's height and weight, changes in skeletal proportions and head and face contours, dental development, and ossification of the long bones. However, growth hormone is not primarily responsible for the dramatic changes of adolescence; the androgens are the main source of adolescent growth. It is thought that normal levels of growth hormone are necessary in order for the adrogens to exert their effect in adolescence.

Thyroid Hormones

Thyroxine (T4) and triiodothyrodine (T3) are the major thyroid hormones and are secreted by the thyroid gland. Their release is regulated by the hypothalamus and the pituitary gland. The thyroid hormones exert their influence by stimulating protein synthesis.

In contrast to growth hormone, thyroid hormones have a great influence on fetal and neonatal growth. In the absence of thyroid hormones during this period, there is a lack of brain

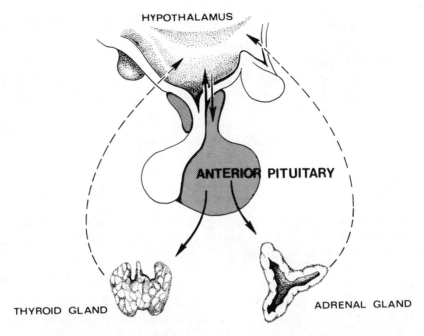

Figure 3-3. *Hypothalamic–pituitary–thyroid/adrenal mechanism.*

growth, probably in both cell number and cell size. Thyroid hormones are necessary for overall growth throughout childhood and adolescence; more specifically, they promote growth and development of the brain and skeleton, sexual maturation, and tooth eruption.

Androgens

An increase in androgens is felt to be responsible for most of the growth and development that takes place during adolescence. The androgens exert their influence like growth hormone and thyroid hormone, i.e., by stimulating protein synthesis. They are responsible for overall growth and skeletal and muscular maturation.

Testosterone is the principal androgen produced by the testes. Dehydroepiandrosterone is the principal androgen produced by the adrenal cortex. The release of the androgens is mediated by the pituitary-hypothalamus feedback mechanism. The exact mechanism reponsible for the rise in the androgens in adolescence is not clearly understood; a change in the sensitivity of the hypothalamus-pituitary system to the androgens has been implicated (Falkner and Tanner, 1978). The estrogen hormones produced by the ovaries in the female do not appear to have any growth-promoting properties in adolescence.

The androgens and estrogens are responsible for sexual maturation during adolescence. Androgens promote penile and testicular growth and development along with facial and pubic hair growth, the estrogens stimulate breast, ovarian, and uterine growth and development in the adolescent female.

HEREDITY AND ENVIRONMENT

The human organism experiences a lengthy period of growth and development, influenced by internal factors (those that are intrinsic to the individual, i.e., genetic potential) and external factors (those that are outside of the individual, i.e., environmental influences). As has been said, the final outcome of growth and development is a result of the dynamic interplay of genetic potential and environmental factors. The respective influences of the two forces are not easily differentiated from one another in such a dynamic, complex interaction.

Genetic Potential

The information stored in the genes exerts a tremendous influence on the individual's maturation. Genetic factors affect the individual's rate of growth, the time of the onset of the adolescent growth spurt, and the final size and shape of the body. This genetic influence is evident in familial patterns of growth. Certain families demonstrate patterns of accelerated growth and early maturation while other families demonstrate a retarded growth pattern that can be described as "late blooming," characterized by slow growth and delayed maturation. Studies of twins confirm the genetic influence on growth; monozygous twins (i.e., those who arise from one zygote and share the same genetic information) have similar body shape, size, and fat deposition even if they are raised in different environments.

Genetic factors play a leading part in the growth difference between males and females. Boys and girls have different patterns of growth and development, and an obvious difference in final outcome. Studies indicate that some of the genes that play a role in the determination of size and pace of maturation are on the X and Y chromosomes (sex chromosomes).

Girls mature more rapidly than boys throughout the developmental years. Adolescence begins sooner for girls, and they attain sexual maturity and maximum height approximately two years earlier than boys. Girls begin their growth spurt at approximately age 10 to

11 years and will have reached adult height by about 16 years. Boys begin their growth spurt at age 12 to 13 years and do not reach adult stature until 18 years of age.

Genetic factors are also influential in the differences seen in different ethnic groups. Data collected during World War II showed individuals of Italian and Jewish descent to be the shortest; people of Scottish, English, and German descent were the tallest (Lowrey, 1973). Further studies have shown that the differences in stature in various ethnic groups are caused by the difference in *rate* of growth; there is little significant difference in body dimensions between ethnic groups at birth (Lowrey, 1973).

Environmental Factors

The environment supplies the external factors that make growth and development possible. Environmental factors include temperature, nutrition, drugs, infections, radiation, and so on. These external factors are influential throughout the life span.

The impact of an environmental factor on growth and development is dependent on the characteristics, intensity, duration, and timing of the factors. The organism as a whole, and specific organs, is particularly susceptible to environmental factors during specific stages called *critical periods*. Changes induced by environmental factors at critical periods can alter normal growth and development, and may cause long-lasting functional abnormalities. Critical periods are characterized by rapid cell growth and differentiation. It is evident that the fetal and infant periods are the most vulnerable stages in the life span because of the rapid growth and development that take place during these stages.

The central nervous system is very susceptible to environmental factors during the fetal period. For example, the exposure of the rapidly developing fetal central nervous system to a dose of radiation could produce permanent abnormalities. Depending on the quantity and duration of the dose, the individual could sustain profound central nervous system impairment, from which recuperation would not be possible.

If an environmental insult occurs during a noncritical period of growth, recovery, through what is known as *catch-up growth,* is possible. For example, temporary deprivation of adequate nutrition in an older child may retard the child's growth in height and weight. When adequate nutrition is reestablished, growth accelerates to "make up" the loss in weight and height. This accelerated growth period is defined as catch-up growth. Catch-up growth may restore the expected weight and height, depending on the duration of the malnutrition.

SKELETAL GROWTH

Skeletal growth is commonly taken as an index of overall growth and development. Height, a reflection of an individual's skeletal growth, is a common clinical measure. Bone age may also be used as a clinical measurement of physiological maturity.

Bone is a specialized tissue that supports the body and protects vital organs. The human skeleton exists first in the form of connective tissue, then cartilage, and finally bone. The adult skeleton is composed of bone with no remnants of the original membranous skeleton. The closure of the fontanelles in infancy demonstrates this progression from connective tissue to bone in skeletal development.

During childhood, cartilage is found throughout the skeleton. The length of the skeleton grows by dint of the proliferation of cartilage. The cartilage changes to bone by the process of ossification. It is through the two processes of cartilage proliferation and ossification that the bones grow.

Cartilage is produced in the metaphyses of bones and in the cartilage plate. Ossification proceeds from the epiphysis (end) of the bone, which contains the ossification centers.

Skeletal growth and development involves the continual growth of ossification centers; skeletal maturation is initiated by the appearance of ossification centers and is terminated when these centers are fused. In humans, the first ossification center appears in the two-month-old fetus. The individual has approximately 400 ossification centers at birth, and twice as many by maturation. There are approximately four ossification centers for each mature bone. New ossification centers appear according to a regular schedule. The growth of new ossification centers is under hormonal control; growth hormone seems to have the most influence on the growth of new ossification centers.

The stages of bone growth occur at regular intervals that vary little between individuals. This relative uniformity permits the use of bone age measurement as a reliable index of maturation. (The diagnostic use of bone age measurement will be further discussed in the assessment section of this chapter.) Bone growth ends when the cartilage is ossified and the epiphysis unites with the diaphysis (center) of the bone. Epiphyseal fusions occurs after puberty when adult stature has been attained.

Skull Growth

The skull is large at birth because of the relatively large size of the brain. After birth, the skull continues to grow in order to accommodate the very rapid further growth of the brain. In the first two years of life, the skull's capacity grows from 400 ml to 950 ml and its circumference increases from about 33 cm to 47 cm.

At birth, the bones that comprise the cranial vault are separated by gaps filled with fibrous connective tissue; the gaps allow the bones to slide over each other to some extent as the neon-

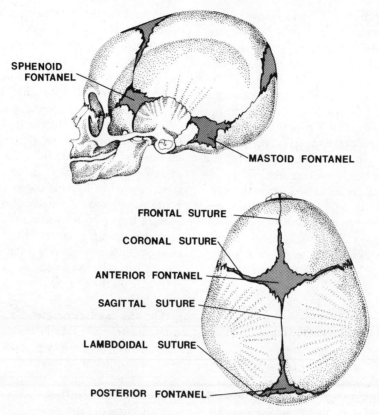

SPHENOID
FONTANEL

MASTOID FONTANEL

FRONTAL SUTURE

CORONAL SUTURE

ANTERIOR FONTANEL

SAGITTAL SUTURE

LAMBDOIDAL SUTURE

POSTERIOR FONTANEL

Figure 3-4. *The skull at birth, showing the fontanelles and sutures.*

ate passes through the mother's narrow birth canal. The largest gap, the anterior fontanelle, can be felt at the top of the skull, between the frontal and parietal bones. The anterior fontanelle is approximately 3 cm across and is not filled in by ossification until the second year of life. Other smaller fontanelles lie between the bones of the skull; they are ossified more quickly than the anterior fontanelle (see Figure 3-4).

As ossification of the fibrous connective tissue proceeds, the individual bones come into contact with each other along a series of fibrous joints known as sutures. It is at the sutures that skull growth takes place. Skull growth will continue until after puberty, accommodating the growing brain. After skull growth stops, the sutures will be obliterated by their ossification and fusing together.

ASSESSMENT OF GROWTH AND DEVELOPMENT

The assessment of growth and development is an integral part of pediatric health care. The child's weight, height, and developmental milestones, which serve as indicators of overall health, are evaluated at regular intervals at well-child visits.

Weight and height are useful clinical measurements because they reflect overall body system functioning. Height and weight measuring is a nonintrusive, quick, and easy-to-learn procedure. Since the individual's measurements are plotted on a standard growth grid for interpretation, weight and height must be measured in a standard fashion. All measurements should be made in a room with adequate lighting, comfortable temperature, and a firm, level floor. Children should wear only minimal light-weight clothing or none at all. Shoes should not be worn.

Length

During the first two years of life, the child's length must be measured while the child is in the recumbent position, since values on the growth grid were all obtained with the infants in a recumbent position. This is particularly important for children of small stature since supine length will be greater than standing height.

A recumbent-length measuring table is required to obtain a measurement (see Figure 3-5). Two persons are needed, one to hold the infant's head in contact with the headboard, and one to bring the movable footboard firmly against the child's heels. To position the feet so that the soles are directed vertically against the footboard, gentle traction may be applied to the legs. Recumbent measurement should be recorded to the nearest 0.1 cm.

Length should be measured as standing height for children over 2 years old. Either a Frankenburg plane (see Figure 3-6) or a wooden or metal tape attached to the wall can be used to accurately measure stature. The child should face forward, with heels together, back as straight as possible, and with heels, buttocks, and the upper part of the back touching the wall. The measurement should be recorded to the nearest 0.1 cm.

Weight

The child should be weighed while wearing a minimal amount of clothing, if any. Infants should be weighed on an infant scale having a capacity of approximately 15 kg. The older children's scale should be of the beam-balance type, with a capacity of about 150 kg. The accuracy of the scales should be checked and adjusted approximately three times a month, and each time the scales are moved. The infant should be placed centrally on the scale and the older child should stand centrally on the platform.

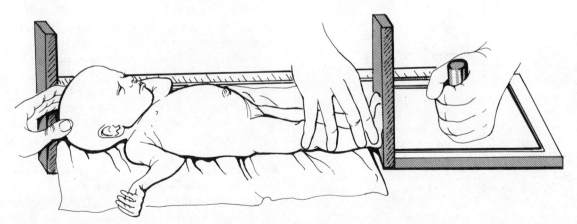

Figure 3-5. *Measuring length with a recumbent-length table. (Redrawn from Evaluation of body size and physical growth of children, The Maternal and Child Health Program, U.S. Department of Health, Education and Welfare, 1976.)*

Figure 3-6. *Measurement of height using a Frankenburg plane. (Redrawn from Evaluation of body size and physical growth of children, The Maternal and Child Health Program, U.S. Department of Health, Education and Welfare, 1976.)*

Weight should be recorded to the nearest 0.1 kg.

Head Circumference

Head circumference is measured until the child is 2 years old. It should be measured with a flexible narrow-width tape. The tape should be placed on the infant's head so as to obtain the maximal frontal-occipital circumference. The head circumference should be recorded to the nearest 0.1 cm.

The pattern of the growth of the cranium is very different from that of the body; its growth is nearly completed by 6 years of age, so growth evaluation is most relevant in infancy and early childhood. It has been determined from studies of large populations that cranial growth is not significantly different in various ethnic and socioeconomic groups; the same standards may therefore be used in any population (Nellhaus, 1968).

The method for interpretation of head circumference on the growth grid is identical to that for interpretation of height and weight. A measurement outside the normal range does not absolutely indicate cranial abnormality; rather, it is the *rate* of growth that must be evaluated. "A child whose head size increases either more rapidly or more slowly than the group pattern is highly suspect of intracranial pathology" (Frankenburg and Campbell, 1975, p. 62).

Bone Age

Another clinical tool that assists the clinician in evaluating the child's progress toward maturity is bone age assessment. Chronological age is an unreliable guide because of the large differences in maturity rates, and height and weight measurements are useful only in gauging the growth of children who follow expected norms. Bone age can be helpful in interpreting the atypical child's growth.

The progress of bone growth can be recorded by x-ray. The appearance of the centers of ossification can be noted because their calcium content makes them radiopaque. The sequence of bone growth is similar in every person. The timing of ossification may differ quite widely depending on whether skeletal development is advanced, average, or delayed. Because of the potential dangers of radiation, only the hand

and wrist are x-rayed; these parts are chosen because they have a large number of centers of ossification.

The x-ray of the child's hand and wrist is then compared to an atlas of skeletal growth. From large numbers of radiographs of children of a given chronological age, a norm is established for bone maturity. In accordance with the appropriate norm, the child is given a score for level of maturation. The atlas has the same drawback as growth grids and developmental assessment tools, i.e., that individual variations caused by genetic differences may account for atypical maturation, that cannot be reconciled with a norm.

The clinical purpose of bone age assessment is to identify the level of skeletal maturation. This information assists the examiner in making a decision concerning the child with abnormal growth by allowing a comparison with norms for the same chronological age.

Growth Grids

The National Center for Health Statistics (NCHS) has developed a growth grid that is commonly used in clinical practice. It is based on a distance curve. The distance curve shows the height, weight, and head circumference of a child at various ages. It enables one to record the measurements attained by the child as he grows toward adult size. The standards for each age are derived from measurements of a large number of children of the specific age. An average (mean) and a range (fifth through ninety-fifth percentiles) for each height, weight, and head circumference have been determined for each age. Ideally, every child's growth should be evaluated in relation to norms established for children of the same sex, ethnic group, and socioeconomic status. NCHS growth grids are based on measurements of American children and distinguish male and female patterns. These standards are applicable to white and black American populations but are not reliable for children of Asian and Native American descent,

since such children are typically of smaller stature (see Appendix A).

Once the child's measurements are accurately taken and recorded on the appropriate growth grid, the child's relationship to the patterns shown on the growth grids needs to be interpreted. The rationale for the use of growth grids is that children follow predictable patterns of growth and development. The individual child's measurements are interpreted in relation to the expected norms for other children of the same age, sex, and ethnic group.

There is no clear-cut point on the grid at which "abnormality" starts. Instead, there is a range of normal values within which a healthy child would be expected to fall. The normal range extends from the fifth to the ninety-fifth percentiles (or within approximately two standard deviations of the mean). A measurement that falls outside these limits indicates the need for further evaluation.

A child whose measurements falls outside the normal range is suspected of having disease, although disease should not be regarded as the definite cause of atypical measurements. Atypical status should be taken as an indicator that careful clinical appraisal must be performed. Since growth is a dynamic process, it cannot be accurately evaluated at a single point in time. A single measurement determines the *size* of a child, whereas several measurements must be made in order to evaluate the child's *growth*.

Two important points should be kept in mind in comparing a child's growth to a standardized grid. First, the curve derived from a large group of children naturally obscures the individual patterns of each child in the original sample population and will be a smooth curve with no dramatic peaks; many children will show peaks in the course of normal growth. Second, height is a more stable indicator than weight. Weight is more influenced by environmental factors and may fluctuate more than height. Weight is still a reliable measurement but may be more difficult to interpret, especially in infancy (Falkner, 1962).

Developmental Screening

As was said earlier, humans, unlike other animals, do not inherit a repertoire of instinctive behaviors but must learn behavior; the prolonged period of immaturity enables humans to benefit from their experiences for optimal learning. Behavior is divided into five areas: gross motor, fine motor, language, personal-social, and adaptive behavior.

Gross motor behavior includes the control of the head, trunk, and extremities. *Fine motor behavior* is the control of the movements of the fingers. The acquisition of motor control reflects the integrity of the child's neurological system.

Language behavior includes the production of single or combined words, and the ability to comprehend speech. "Normal speech production will occur if a child hears, understands the language of others, and has no intellectual or neurological defect" (Lowrey, 1973, p. 131).

Wide variations exist in *personal-social behavior,* since it is dependent on the child's interaction with his or her environment and culture. It reflects an integration of biological abilities with the environment. Aspects of behavior that reflect this integration include such daily activities as eating, sleeping, and interacting with others.

Adaptive behavior is significant because it stems from intellectual potential. Adaptive behavior indicates the child's ability to solve problems. It includes the use of motor abilities to execute practical solutions and the use of past experience in the solution of new problems. Learning is dependent primarily on the integrity of the central nervous system but is largely influenced by experience.

The child also undergoes maturational changes in emotional development. This area of behavior is not as easily assessed.

Developmental screening, like screening of physical growth, is an integral part of pediatric health care. A development assessment is a clinical estimate of the developmental progress made in each area of behavior. Accomplishments in the various areas may be achieved at differing rates of progress. The emphasis of the developmental screen is on evaluation of the entire child and all aspects of behavior. It must be kept in mind that developmental screening tests are not intelligence tests. Intelligence tests, which are useful primarily in predicting level of scholastic achievement, are not used as developmental measures in routine health care.

Developmental screening is done at regular intervals in order to identify functional delays in development that can be alleviated by early intervention. Developmental screening is also used to identify mentally retarded children. Although critical periods for treatment of mental retardation have not been established, it is felt that early identification is beneficial to retarded children, as in the case with any deficit. Associated problems can be alleviated or prevented if the abnormality is detected as soon as possible.

Another area of significant clinical use of developmental screening is its use to educate parents about their child's development. Anticipated developmental changes that the child will undergo can be delineated for parents. This information can assist parents in adjusting to their continually changing child. It can also assist parents to adjust the child's environment for optimal support and stimulation of his or her development.

Denver Developmental Screening Test

The Denver Developmental Screening Test (DDST) is widely used in clinical practice and is a typical screening instrument. The axiom underlying the DDST is that development follows predicted patterns common to all, and that the child who does not follow the expected pattern is more likely than the typical child to have disease.

The test consists of 105 tasks selected from existing infant and preschool scales. The test was first administered to a large group of infants and children in Denver, Colorado; the sample population of the children was chosen to match the sociocultural and economic status of Denver's overall population, and is therefore not necessarily representative of populations of other parts of the United States. The ages at which 25, 50, 75, and 90 percent of the subjects passed the items were calculated for the entire sample (Frankenburg and Campbell, 1975).

Children between 2 weeks and 6 years old are tested on only 20 or so simple tasks by means of a few basic testing materials. The items are arranged in order of difficulty and are divided into four major behavioral areas: personal-social, fine motor adaptive, language, and gross motor (see Appendix B). The child's performance is scored "normal," "questionable," or "abnormal" (Frankenburg and Campbell, 1975).

As is true of growth, there is a range of variation in children's development patterns. An abnormal screening at one time is not an indication of disease, but of the need for further evaluation of the child over time. Such a further evaluation must explore whether the abnormal screening was caused by an individual variation or environmental influences affecting the performance. Hunger, fatigue, and many other external factors can influence the child's performance.

The DDST is a useful clinical tool because of its use of an age range in establishing the norms, which allows for a wider variation in individual patterns. Also, the heterogeneity of the original test population, from which the criteria were derived, allows for variations in children of different cultural and socioeconomic backgrounds (Frankenburg and Campbell, 1975).

the predictable stages of infancy, childhood, and adolescence. The physiological processes of growth and development are responsible for the changes in body size and composition and in abilities that the individual undergoes before attaining adulthood.

Growth and development begin at the cellular level and advance under the influence of both the individual's genetic potential and environmental influences. During critical periods of growth and development, the individual is most vulnerable to environmental influences. An environmental insult received during one of these critical periods could produce irreversible, life-long impairments.

Hormones influence growth and development by stimulating protein synthesis. The degree of influence of the individual hormones varies during the different stages of growth and development.

Growth and development follow predictable patterns common to all; they proceed in a cephalocaudal and proximodistal pattern. Growth and development also occur at varying rates in different body parts. The various body parts develop at their individual rates in harmony with the individual's overall growth and development.

The assessment of growth and development is an integral part of pediatric health care. The child's growth and development are compared to established norms in order to determine whether the child's measurements are within the normal range. However, the examiner must keep in mind that individual variations caused by genetic and environmental differences may cause the child to deviate from the norm.

SUMMARY

Humans are born immature. On the path to independent function, they progress through

REFERENCES

Bacon, G., Spencer, M., and Kelch, R. *A practical approach to pediatric endocrinology.* Chicago: Year Book Medical Publishers, 1975.

Bee, H. *The developing child*. New York: Harper and Row, 1978.

Catt, K. *An ABC of endocrinology*. Boston: Little, Brown, 1971.

Falkner, F. The physical development of children. *Pediatrics*, 1962, *29*, 448.

Falkner, F., and Tanner, J. *Human growth, principles and prenatal growth*. New York: Plenum Press, 1978.

Frankenburg, W., and Campbell, B. *Pediatric screening tests*. Springfield, Ill.: Charles C Thomas, 1975.

Gessell, A., and Thompson, H. *The psychology of early growth*. New York: Macmillan, 1938.

Havighurst, R. J. *Developmental tasks and education*. New York: McKay, 1948.

Hoekelman, R. et al. *Principles of pediatrics: health care of the young*. New York: McGraw-Hill, 1978.

Hurlock, E. *Developmental psychology*. New York: McGraw-Hill, 1959.

Hymovich, D., and Chamberlin, R. *Child and family development: implications for primary care*. New York: McGraw-Hill, 1980.

Lowrey, G. *Growth and development of children*. (3d ed.). Chicago: Year Book Medical Publishers, 1973.

Maier, H. *Three theories of child development*. New York: Harper & Row, 1978.

Nellhaus, G. Composite international and interracial graphs. *Pediatrics*, 1968, *41*, 106.

Owen, G. The assessment and recording of measurements of growth of children: report of a small conference, *Pediatrics*, 1973, *51*, 461.

Sinclair, D. *Human growth after birth*. New York: Oxford University Press, 1978.

Smith, D., and Bierman, E. *The biologic ages of man*. Philadelphia: Saunders, 1973.

Timiras, P. *Developmental physiology and aging*. New York: Macmillan, 1972.

Williams, A. *Textbook of pediatric endocrinology*. Philadelphia: F. A. Davis, 1975.

4

Developmental Aspects
of Human Reproduction

GLENDA F. BUTNARESCU

Pregnancy, childbirth, and early parenthood constitute for many young people their first real confrontation with adult life and adult responsibilities. Pregnancy may occur during adolescence, the initial stages of adulthood, or during the mature years; whenever it occurs, it imposes a new and different set of developmental tasks on those already existing. It is for many the transition from dependency to independency or interdependency, a period of profound change and reorganization. "We carry the imprint of this experience for life, even into our dying" (Rich, 1976, p. 1). It has been postulated by some to be a period of crisis, or a period during which crisis may occur, often related to devel-

opmental changes in progress. Motivations for becoming pregnant vary widely, ranging from deliberate conscious efforts to conceive to motives unrecognized by either or both partners. Many are unprepared and least ready to grapple with the demands of pregnancy when it occurs. The pregnant female experiences pregnancy and childbirth as an educative biopsychosocial experience, intensely personal, and one that cannot be totally shared. The male may not associate the sex act with its logical outcome, pregnancy. During the pregnancy, he may be involved only peripherally. For both partners it represents a confrontation with new and often unexpected responsibilities. Many may perceive it as a developmental peak and creative milestone.

The potential for procreation is laid down at the time of conception, when the blueprint for building a new life is first defined, deriving its configuration from the genetic contributions of each of the parents. The blueprint serves the purpose of directing the development of the new life. Just as many factors influence the appearance and functional capability of a building—the availability of materials for construction, the quality of the materials, the influence of the environment, and the personal characteristics of the builders—so is the developing human influenced by a multitude of factors. The absence of quality materials—DNA, nutrients, oxygen—will lead to fetal stunting or abnormality; it may contribute to impairment of vital organ functioning; it will affect size. Some infants persistently lag behind their growth potential because of poor, faulty, or inadequate building materials. Some have abnormalities of such magnitude that they cannot function without technical or human assistance. The pattern of growth and development begun in fetal life is carried well into childhood, often into adolescence and adulthood. If the pattern is flawed, the neonate may not live beyond birth or early childhood.

More and more is being reported about the influence of the intrauterine environment upon the developing organism. So, once more like the building, the construction of which is advanced or delayed because of storms, strikes, costs, etc., the fetus is affected by changes outside of and within the intrauterine environment: infection, maternal cardiovascular problems, trauma, and many other factors. Even as the cornerstone of a building is laid, the structure built, its unveiling or opening scheduled for all to see the beauty of its lines and usefulness, it is vulnerable to stress both from within (as from faulty construction) or without (natural or man-made disaster). Birth, likewise, is traumatic to the neonate. The scars resulting from this

trauma may be transitory, or lifelong. The effects of physical stress associated with childbirth are well known; the extent of psychological stress is not so readily documented since this aspect is still in the early stages of research at this writing, but findings to date confirm that birth maybe of considerable emotional stress to the infant.

The *fact* of human life has been assured through the reproductive capability of the species; the *quality* of life, beginning with conception, is a human responsibility, and this is where the nurse comes into the picture. Armed with knowledge about human development and potential, the nurse can help young people to maximize their potential for reproductive health.

The body of knowledge, derived from a number of disciplines, that is applicable to pregnancy, childbirth, and early parenthood is massive, and growing daily. Only selected concepts have been included in this chapter; it is hoped that the reader will find these concepts to be helpful in preparing to be an effective resource and support for families experiencing these periods of development.

DEVELOPMENTAL READINESS FOR PREGNANCY

Physical Readiness

The most complex and creative biological characteristic of most species is the capability for reproduction. Because of the instinctive nature of sexual drive, conception is often more fortuitous than deliberate. The act of copulation requires only seconds to accomplish; millions of spermatozoa from the male are released, with the potential for union with a female gamete, the ovum, resulting in the creation of a new life. Neither partner is likely to be aware of the precise moment that the new life is begun, and weeks may pass before they comprehend the results of their actions. The act of creating new life is possible only because of a biological master plan into which the partners had no input, and about which they often have a limited understanding.

The foundations of reproduction are human anatomy, physiology, endocrinology, and genetics. Organ growth must be adequate to support reproductive need; functional capability must be present; organ growth and function are largely directed by the neuroendocrine system; and the parents' genetic characteristics largely determine the characteristics of the offspring.

Organ Structure and Function

The study of reproductive anatomy and physiology is circular. The obvious starting point is conception, for that is the beginning of life, but one cannot comprehend conception and its subsequent processes without first understanding how conception becomes possible, necessitating a return to the study of the adolescent and adult periods of life with regard to the processes and structures that underlie the fertilization of an ovum by a spermatozoon.

The structures principally involved in human reproduction are the male and female gonads, the internal ducts and organs, the external genitals, the pituitary gland, and the hypothalamus. Other structures, such as the bony pelvis and the pelvic muscles, are important to the reproductive process but do not contribute directly to reproductive capability. (They may be significant as regards the mode of delivery.) Structures such as the bladder, rectum, and anus are also indirectly related to reproduction, and arise developmentally from similar tissue. Of greatest reproductive significance, however, are the male and female gonads.

There is not a perfect parallel between male and female structure and function; however, some analogies can be made (see Table 4-1). Many of these reproductive structures derive from the same tissues and are differentiated by the influence of the Y-linked genetic sex deter-

minant that is found in the male and is believed to be responsible for stimulating the early testes to produce androgenic hormones.

The female gonads (ovaries) are flat, ovoid or almond-shaped structures approximately 2.5 cm long; 1.5 to 3 cm wide, and 0.6 to 1.5 cm thick. They are located in the upper pelvic cavity and are attached bilaterally to the uterus and the pelvic wall by ligaments (Fig. 4-1). These organs produce and release the female gametes (ova) and secrete the female steroid hormones, estrogen (estradiol) and progesterone. Histologically, the ovaries are made up of two sections, the cortex, or outer layer, and the medulla, or inner layer. They derive from undifferentiated gonadal tissue, as do the male gonads, the testes, and can be identified as an ovary by approximately the seventh to ninth week of gestation. The potential for gamete production exists even at this early stage of embryonic life in the form of oogonia, primitive germ cells. These undeveloped germ cells continue to multiply and develop until about the sixth month of fetal life. Approximately 400,000 primitive germ cells are believed to be present in the ovaries of the female neonate. These germ cells remain in a quasidormant state until puberty, by which time roughly 25%, or approximately 100,000, are believed to have survived. This supply becomes the source from which ova are supplied for human reproduction through a process known as the menstrual cycle.

The male gonads (testes) are ovoid also, but slightly more flattened than the ovaries. The testes are located outside of the pelvic cavity, descending into the scrotal sac after birth. Like the ovaries the testes serve two major functions, the production of the male gametes (spermatozoa), and the secretion of the male steroid hormone testosterone. They arise from undifferentiated gonadal tissue in early embryonic life, and can be distinguished as early as the seventh week after fertilization. The male gonad tends to develop more rapidly than the female gonad in prenatal life. Although the precise

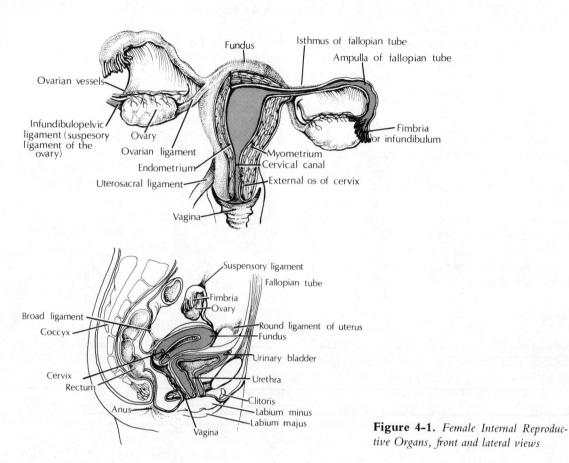

Figure 4-1. *Female Internal Reproductive Organs, front and lateral views*

Table 4-1. Comparison of Male and Female Reproductive Structure and Function

Structure	Function	
	Male	*Female*

Gonads	Testes:
	  Produce male gametes (spermatozoa)
	  Produce male steroid hormone (testosterone) needed for development of male ductile system. Testosterone also promotes and maintains normal sexual drive and promotes development and maintenance of male secondary sex characteristics.
		Ovaries:
		  Produce female gametes (ova)
		  Produce female steroid hormones (estrogen and progesterone).
		  Estrogen stimulates development of female external genitals and ductile system of breasts, and influences distribution of body fat and hair. Progesterone promotes uterine muscle, growth particularly during pregnancy, and stimulates growth of breasts.

Table 4-1. (*Continued*)

Structure	Function	
	Male	*Female*

Ductal System	Epididymides:
	  Storage and maturation site for sperm before ejaculation.
	Vas Deferens:
	  Provides for storage and transport of spermatozoa.
	Seminal Vesicles:
	  Secrete nutritive fluid for nourishment of sperm.
	  Increase quantity of seminal fluid to aid in transport of sperm.
	Ejaculatory Ducts:
	  Transport semen to be ejaculated.
		Fallopian Tubes:
		  Serve as link between ovaries and uterus.
		  Transport extruded ovum to uterus.
		  Outer part is usual site for fertilization.
		Uterus:
		  Muscular organ with inner lining (endometrium) shed during menstruation. Houses and accommodates fetus during pregnancy. Cervix dilates and effaces during labor to permit expulsion of fetus.
		Vagina:
		  Female copulatory organ; receives the penis during intercourse and is the receptacle for ejaculated sperm.
		  Passageway for menstruation.
		  Birth canal for fetus.

Glands	Prostate:
	  Secretes thin fluid to increase volume of seminal fluid to aid in sperm motility and fertility.
	Bulbourethral glands (Cowper's Glands)
	  Secrete a viscous alkaline fluid that may aid in neutralizing acid secretions in the vagina to prevent destruction of spermatozoa.
		Paraurethral glands (Skene's glands)
		  Serves no known function. May play a role in preventing entry of bacteria into the urinary opening.
		Bartholin's Glands:
		  Secrete mucus to lubricate inner surfaces of the labia.
		  Lubricates vaginal orifice during sexual excitement, aiding entry of the penis.

External Genitals	Penis:
	  Male organ of copulation; becomes erect during sexual excitement.
	  Contains tubular canals for transport of sperm and passage of urine. Tip of penis (glans) highly sensitive to touch.
	Scrotum:
	  External pouch containing the male gonads.
		Clitoris:
		  Small cylindrical body in the upper portion of female vulva. Homologue of penis. Contains erectile tissue and is sensitive to touch. Organ of excitement during sexual activity.
		Vestibule:
		  Triangular area covered by two folds of tissue, the labia minora. Contains the urinary opening.
		Vaginal Orifice:
		  External opening to vagina.
		  Point of entry of penis during sexual intercourse.
		  Point of exit of fetus during childbirth.

mechanism for this more rapid development is not fully understood, it is believed to be related to a gene present on the Y chromosome. Production of testosterone begins as early as the ninth week of embryonic life and is believed to be necessary for the development of the male ductal system and external genitals in fetal life. Gamete production in the male begins at puberty instead of during the fetal period, and differs somewhat in that respect from female gametogenesis and development.

The male and female ductal systems, like the gonads, derive from a common tissue in embryonic life, the male system becoming differentiated earlier than that of the female. These systems serve to transport the gametes from their points of origin to a union with each other, and the uterus (womb), developmentally a part of the female ductal system, serves to house and protect the developing fetus during prenatal life. Ova are mature at the time of extrusion from the ovary, and are transported in a mature state, while the spermatozoa leave the testes in an immature state, relying upon aspects of the male ductal system to aid in their maturing.

Morphologically, the female ductal system is much simpler than that of the male. It consists of two tubular structures called oviducts (Fallopian tubes) that are approximately 8 to 14 cm long. The tubes are attached to the uterus at one end and open into the uterine cavity, while the other end reaches toward, but does not connect with, the ovaries and the peritoneal cavity. The Fallopian tube contains two layers and is covered by peritoneum. It consists of four sections: the infundibulum, the ampulla, the isthmus, and the interstitial section. The *infundibulum* is a funnel-shaped part near the ovary containing fimbria (fingerlike projections) that reach out and capture or guide the extruded ovum into the lumen of the oviduct where cilia lining the lumen of the tube act in concert with tubal peristaltic movements to propel the ovum toward the uterus. The *ampulla* is the distal middle part

of the oviduct and is the usual site of fertilization. The end of the tube adjacent to the uterus is called the *isthmus,* and the *interstitial section* is really part of the uterine myometrium, opening into the uterine cavity.

The uterus is a muscular organ often likened to a pear in size and shape in the nonpregnant adult state, but capable of great growth and distention during pregnancy (or in the presence of a pathologic condition such as tumor growth). It contains a cavity that is lined with a mucous membrane similar to that lining the vagina and oviducts. The uterus consists of three parts, the upper rounded top portion, the *fundus,* the body or middle part, the *corpus,* and the lower dilatable portion, called the *cervix.* There are three tissue layers: the endometrium, the myometrium, and the parametrium. The *endometrium,* or inner layer, is highly vascular; parts of it are shed during each menstrual period. It increases in thickness during pregnancy. The *myometrium* is the middle layer, composed essentially of smooth muscle and capable of mild to intense contractility, as during labor. The uterus increases in size during pregnancy through great hypertrophy and some hyperplasia. The outer layer, the *parametrium,* is a fibrous, subserous coating of the uterus.

The lowermost part of the female ductal system is the vagina, a distensible, hollow structure that varies in length from 6 to 8 in. Its walls are corrugated folds of tissue that flatten when the vagina is distended, allowing for extension and expansion of the organ. The vagina is lined with mucous membrane and is capable of enlarging to accommodate the passage of the baby during childbirth; it is often referred to as the birth canal. It serves as the receptacle for the penis during sexual intercourse and receives the spermatozoa following ejaculation, and is the passageway for the menstrual flow.

The male ductal system consists of the epididymis, the ductus deferens (vas deferens), the seminal vesicles, and the ejaculatory ducts

(Fig. 4-2). The epididym is an oblong structure lying along the superior and posterior aspects of each testicle, in the ducts of which the immature spermatozoa are stored and allowed to mature. The *vas deferens* are long (about 18 in.) bilateral ducts connecting the testes and the ejaculatory ducts. Their purpose is to store and transport spermatozoa. The *seminal vesicles* are two pouchlike diverticuli attached to the posterior wall of the bladder that secrete a fluid that nourishes the spermatozoa and increases the volume of the seminal fluid. The *ejaculatory ducts* are located between the ductus deferens and the seminal vesicles, descending between the parts (lobes) of the prostate gland, opening into the male urethra. Their function, as one might infer, is to release semen containing spermatozoa into the male urethra.

Although the external genitals serve a number of functions in both the male and female, the structures most relevant to human reproduction are the male copulatory organ, the penis; the scrotal sac, which contains the male gonads; and the vaginal opening in the female.

The penis is an elongated organ that contains three parallel erectile bodies, the paired corpora cavernosa and the corpus spongiosum. The penis comprises three parts, the root or base of the penis, which is attached to the pubic bone; the body; and the glans penis, which is the distal end. The erectile tissue, containing vascular spaces that become engorged with blood during sexual excitement, provides the penis with its erectile capabilities, allowing it to become turgid, facilitating entry into the vagina during copulation. The penis, which is covered with fascia and skin, contains the urethra, which extends the length of the penis.

The scrotum is a pouch containing two chambers, composed of fascia and some muscle, and covered by skin. It contains the male gonads, the testes; by housing the testes outside the body, the scrotum protects them (and the sperm cells) from the higher temperature of the abdominal and pelvic cavities. Because the testes are protected only by soft tissue, they are particularly vulnerable to physical trauma.

The female external genitals consist of the mons veneris (the upper protion of the female vulva, composed of fatty tissue and covered with hair following puberty), the labia majora and minora (folds of soft tissue that cover and provide some protection to the tissues beneath them), the vestibule (an area shaped somewhat like a triangle and containing the urinary meatus), the clitoris, (the female structure analogous to the penis in embryonic life, and the chief organ of sexual excitement), the vaginal opening (partially covered in the virginal state by the hymen, a thick membrane), and two sets of

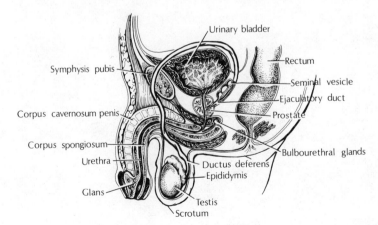

Figure 4-2. *Male Reproductive Organs*

glands, the paraurethral glands, called Skene's glands, and the vulvovaginal glands, known as Bartholin's glands, mentioned earlier (Fig. 4-3).

Neuroendocrine Control

The reproductive processes, foundations for which are laid in fetal life, remain essentially semidormant during infancy and childhood. Puberty, the developmental point at which maturation of primary and secondary sex characteristics begins, brings the reproductive structures and processes to a functional state in most cases. Although all the mechanisms causing the onset of puberty are not presently known, it is believed that this event is principally under the control of the neuroendocrine system, particularly the hypothalamic-pituitary-gonadal axis (Fig. 4-4).

The onset of puberty is predictable, but its timetable is not precise, differing somewhat in each person. Changes associated with puberty in the male and female, and their approximate time of onset, are presented in table 4-2, and are discussed in further detail in Chapter 9.

Before the onset of visible organ growth or body change, the anterior pituitary gland is stimulated by releasing factors produced in the hypothalamus to secrete two hormones, follicle stimulating hormone (FSH), and luteinizing hormone (LH); this occurs in both male and female. FSH in turn acts upon the gonads to stimulate the production of the steroid hormones, testosterone in the male, and estrogen and progesterone in the female. These hormones promote the development of secondary sex characteristics and initiate spermatogenesis in the male and oogenesis and menstruation in the female. Estrogen is not necessary in embryonic life for gonadal differentiation in the female, whereas testosterone is necessary for this to occur in the male. Testosterone decreases to a very low level following birth of the male child, and then begins to rise gradually as puberty approaches.

Neuroendocrine control of the reproductive processes is possible because of a precise feedback system insuring functional "checks and balances." This interrelated system is sensitive to disturbance from a number of sources including biophysical and psychological factors that may disrupt the regularity and preciseness of its functioning.

The *menarche* (onset of menstruation) is recognized as the start of the reproductive period for the female, which spans approximately 30 to 35 years and terminates with *menopause* (cessation of menstruation). There is no precise equivalent of this event in the male.

The menstrual cycle has three phases, the follicular phase, the luteal phase, and the menstrual

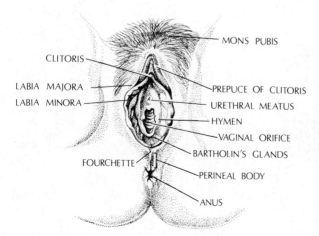

Figure 4-3. *Female External Genitals*

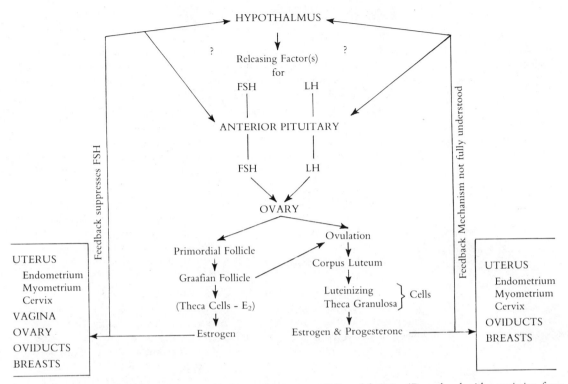

Figure 4-4. *Interaction between the Hypothalamus, Pituitary, and Gonadal Axis. (Reproduced with permission from Butnarescu, G. F.; Perinatal nursing, Vol. 1: reproductive health, John Wiley and Sons, 1978, p. 97.)*

phase, which lasts 3 to 5 days. For the sake of clarity, the menstrual cycle is said to be 28 days long; beginning on the first day of the menstrual period. Reckoned thus, ovulation is said to occur on or about day 14 of a 28-day cycle. However, ovulation cannot be assumed to occur exactly in mid-period since the length of the menstrual cycle varies in different women, and from month to month in some women. It is important for professionals teaching young people and others to be aware that ovulation occurs approximately 14 days before the onset of menses, and not necessarily at mid-cycle. For example, if the cycle is 35 days, ovulation is likely to occur on or around day 21. Ovulation usually precedes menstruation, but some menstrual cycles are *anovulatory*, and menstruation can occur after such a cycle, in which ovulation

is absent. This is particularly true in early puberty, following pregnancy, during menopause, and in the face of emotional or physical illness. Neither the woman nor the health professional can know which cycles are ovulatory or anovulatory without sophisticated study. It should also be kept in mind that pregnancies have been known to occur before the first menstrual period.

The following is a simplified account of the progression of events in the menstrual cycle (see Fig. 4-5).

1. The hypothalamus produces gonadotropin releasing factors that stimulate the production of the anterior pituitary gland hormones FSH and LH.
2. The gonadotropin hormone FSH stimulates

Table 4-2. Male and Female Pubertal Changes

Change	Male	Female
Production of gonadotropin and steroid hormones.	Testosterone produced in prenatal life, as well as gonadotropin hormones, increases in late childhood. Promotes secondary sex characteristics, organ growth, and initiates gametogenesis.	Stimulates reproductive organ growth. Steroid hormones are responsible for development of secondary sex characteristics. Stimulates gametogenesis.
Increased organ growth and function.	Increased size of penis and testes (12 to 17 years); capacity for ejaculation.	Breast enlargement (8 to 13 years); onset of menstruation (10.5 to 15.5 years)
Hair Distribution	Pubic hair appears as straight, fine pigmented hair (12 to 13 years), changing to kinky, coarse pubic hair with adult distribution pattern (14 to 15 years). Axillary hair appears about 14 to 16 years. Facial hair (beard) appears last.	Pubic hair appears as straight, fine pigmented hair (10 to 12 years), changing to kinky, coarse pubic hair with adult distribution pattern at 12.5 to 14 years. Axillary hair appears at about 14 years.
Changes in body size and form.	Increased skeletal growth with increased height and widening of shoulders (12 to 16 years); some breast changes, e.g., increased diameter of areola.	Some increase of skeletal growth. Widening of the hips; redistribution of body fat to breasts, hips, and upper thighs (12 to 18 years); increased breast size, increase in size and prominence of nipples and areolas (12 to 16 years).
Voice changes	Voice deepens late in puberty (15 years).	Voice changes occur, but are less pronounced than in the male.

the maturation of the follicle in the ovary containing the immature oocyte (the Graafian follicle).

3. The cells lining the Graafian follicle (theca interna cells) produce estrogen (estradiol), which promotes the development of the uterine lining (the endometrium).

4. Estradiol feeds back to the hypothalamus or to the anterior pituitary gland, suppressing the production of the releasing factor and FSH, allowing for the stimulation and production of LH by the anterior pituitary gland. LH acts on the ovary to bring about ovulation.

5. The corpus luteum (the shell of the Graafian follicle after ovulation) produces estradiol and progesterone, which further develop the endometrium of the uterus and maintain it.

6. The corpus luteum degenerates, decreasing its levels of estradiol and progesterone. With the withdrawal of estradiol particularly, the endometrium is no longer maintained and menstruation occurs.

7. Menstruation begins, starting the cycle over again.

Biophysical readiness for reproduction is dependent on the integration of the reproductive structures, their functions, and their control mechanisms in such a way that each complements the other. Of course, such readiness is not necessarily accompanied by psychosocial and cognitive readiness for pregnancy and parenthood. Adults (and adolescents) may find themselves biophysically ready yet neither motivated nor prepared to raise a child.

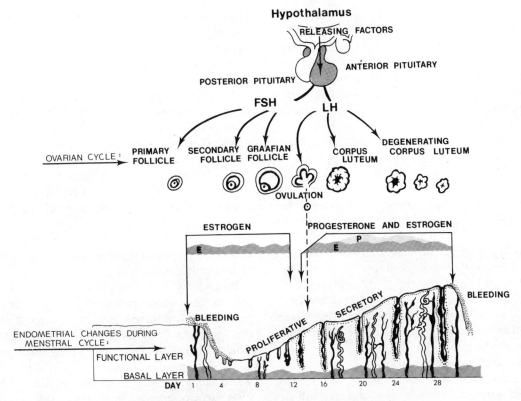

Figure 4-5. *Representation of the Menstrual Cycle*

Psychosocial Readiness for Reproduction

The epigenetic view of human development as an orderly sequence of events in which certain capabilities must be obtained before progression to another level of development is possible, was first described by Wolff in 1759 in regard to embryologic development (Moore, 1977). A true analogy to personality development cannot be made, but, increasingly in the years since the work of Freud, human development has been viewed as an unfolding of the organism in a predictable manner, with clearly identifiable stages (Erikson, 1950), each having distinct tasks that must be completed (Havighurst, 1953, 1972).

A number of factors are likely to influence the child's psychosocial development before the age of procreation is reached. Perhaps the most important influences from a reproductive point of view are those happenings that promote the development of three characteristics that are crucial for pregnancy and parenthood, and that have their roots in early infant-child development: relationship establishment and maintenance; capability for adaptation and coping; and sexual identification and response. These three components, along with motives for becoming pregnant, comprise the nucleus for psychosocial readiness for pregnancy and parenthood.

Establishment and Maintenance of Relationships

Historically, mother-infant relationships have been viewed as having a significant influence upon the adult the infant becomes. Trust, so crucial to the successful marital and parenting relationships, is believed to have its beginnings in early mother/infant relationships, along with the ways in which the female ultimately views herself as a female and a reproductive being. The early parent-child relationship is believed to be reciprocal with each member influencing the other (Brazelton, Koslowski, and Main, 1974; Brazelton et al., 1975). It likely lays the foundation for healthy giving/receiving necessary for satisfactory completion of the pregnancy, and early parenting periods. Caplan has pointed out that the relationship between mother and infant is "influenced by the mother's relationship with her husband and her other children and by their relationship with each other" (1961, p. 133). Deutsch (1945) asserts that the quality of these relationships could be traced back to the mother's relationship with her own mother, reinforcing the idea of the cyclic nature of reproduction.

As the child grows, relationships with others—relatives, other children, and adults—are established, and the advent of puberty and adolescence witnesses the formation of a different type of relationship with a member of the opposite sex. Readiness to assume a mature and responsible caring/giving relationship becomes the foundation upon which the new family will be built. Optimally, a sound relationship will be underway before pregnancy, however, many young people find themselves pregnant before any, other than a sexual, union has been established. Such young people are then faced with coping with the establishment and maintenance of a relationship that may be fragile at best, as well as with adapting to the demands, both physical and psychosocial, that pregnancy places upon them.

Adaptation and Coping

Adaptation is viewed as a person's adjustment to change in his body environment by utilizing personal and other resources to modify that en-

vironment so that his needs are met and his stability is maintained. Coping on the other hand refers to the use of physical and emotional defenses to diminish stresses and overcome their threat.

Adaptation and coping become necessary in early embryonic life when the developing organism must adapt to its intrauterine environment. This is not thought to be a conscious act, but more likely a biological response to forces within the intrauterine environment over which the fetus has no control. Birth thrusts the newborn into a different and more hostile environment to which he must immediately adapt if he is to survive. The necessity, then, for adaptation and coping appears early in life and continues throughout the life span.

Pregnancy, and then parenthood, forces the parents into confrontation with new, yet unexperienced phenomena, and often occurs at a time when both partners are least prepared to cope with it. For the female, pregnancy may come during adolescence, or early adult life, when she is wrestling with the question of her own identity and life goals. "Regardless how planned and how desired the pregnancy, the awareness of the fact that conception has indeed occurred causes the expectant mother to doubt the wisdom of her choice. Another time-any other time—might have been better" (Ziegel, 1978, p. 218).

The ease with which the individual has coped with previous stresses will be reflected in coping capabilities exhibited during pregnancy, childbirth, and early parenthood. Each parent will have developed and matured, to varying degrees, through the resolution of crises that occurred in earlier developmental stages; pregnancy presents another maturational crisis with which they must cope. Unresolved past problems tend to reappear to haunt the woman during pregnancy. If the female is psychologically ready for pregnancy, she will often be able to resolve these and other stresses imposed on her. If readiness is delayed or absent, these unresolved stresses will tend to compound the stress of pregnancy, resulting in a period of growth impairment rather than maturation, and will ultimately affect the woman's relationship with her partner and their relationship with their newborn and other family members.

Sexual Identification and Response

The female sex role has historically been closely associated with reproduction. This traditional point of view is essentially worldwide, and continues in spite of the changing role of women in modern society; the word *mother* connotes to many the pinnacle of femaleness and fulfillment.

In order to reach a level of sexual maturity, the individual must have attained biological sex differentiation and development in prenatal life; sex role identification, including learning the appropriate gender label, acquiring expected sex role standards or behaviors, establishing sex role preference, and identifying with the parent of the same sex; and sex typing of the infant brain, which is the biological imprinting of certain characteristics such as male aggressiveness on the fetal or infant brain. Sex typing is most likely associated with the presence of testosterone and is believed to occur in late prenatal life or early infancy (Money and Ehrhardt, 1971).

Differences between the sexes (other than genital differences) can be seen early in infancy as well as in later life. For example, muscle mass, motor activity, and aggressive behavior are more pronounced in the male infant, while the female is less active but more sensitive to touch.

Puberty brings the individual into biological confrontation with feelings and experiences that are new and often threatening. The need for recognition by and association with the opposite sex often forces relationship realignment. Intimate sexual experience is likely to result in intercourse, perhaps leading to pregnancy, for which neither partner may be ready. On the other hand young people may arrive at the preg-

nancy experience developmentally confident and ready to move through another maturational milestone.

Although pregnancy results from the sexual act, it is not an end point of sexuality; rather, pregnancy is itself a sexual state. Alterations that occur in the female's body because of pregnancy are clearly accentuations of her sexual being: changes in body shape, increased breast size, increased vaginal secretions, and changes in sexual drive.

Caplan (1961) has noted that the expectant mother must be comfortable with herself as a sexual being. Her ability to accept the role of mother is clearly tied in with her ability to accept herself as a woman.

Evidence of discomfort with one's sexuality may be one of the first clues that a woman is not ready to face the maturational crisis of pregnancy. This inability may be evidenced in a number of ways such as refusal to accept the fact of the pregnancy, or embarrassment when discussion turns to sexual activities or sexual anatomy.

The male too, may have ambivalent feelings about his sexual identity. Pregnancy is clear evidence that he is capable of functioning as a sexual being, that he is a man. The adolescent male, who is struggling to find his sexual identity and be comfortable with it, may have mixed feelings of shame, guilt, and anxiety mingled irrationally with the pride of accomplishment. The adult male who perceives his role as husband and father may take pride in his partner's pregnancy. Men are often heard to comment about their wives, "She never looked more beautiful to me than when she was pregnant with our child."

Motives for Pregnancy and Parenthood

Motives for becoming pregnant probably influence whether or not the individual or couple successfully complete the tasks that pregnancy and parenthood impose upon them. Motives for pregnancy are rarely "clearcut" and singular, nor are they always conscious or rational. It is often difficult to separate individual, intrinsic motive from society motive, since society often expects men and women to seek and be rewarded by pregnancy and parenthood. Few will question that perpetuation of the species in reasonable numbers is a worthy goal for humanity, but individual motives may not have (in fact, often do not have) anything to do with that goal.

"Motivations for pregnancy and parenthood seem to fall into three categories: rational reasons, . . . unconscious motivations, . . . and conscious but unsound reasons . . . (Butnarescu, 1977)." One's attitude toward pregnancy and parenthood may very well contain elements in all three categories. Some people desire pregnancy for reasons other than wanting to be a parent; conversely, other individuals want so badly to be parents that they "tolerate" the pregnancy in order to meet the goal of becoming a parent.

Developmentally, the acknowledgment of the goal of parenthood may not come until some of the tasks of pregnancy are completed. There seems to be some evidence that readiness for parenthood evolves, at least partly, from the pregnancy experience itself, as exemplified by maternal behavior at various periods during pregnancy.

The identification of motives for pregnancy and parenthood provides insight into the readiness of the individual to complete in a healthy manner the tasks associated with both. Indications, however slight, of a lack of readiness once the pregnancy is confirmed should not be considered insignificant. On the other hand, apprehension associated with early pregnancy in the absence of other evidence should not be interpreted as pathological.

Cognitive Readiness for Reproduction

Cognitive capability was believed at one time to be almost exclusively under the control of genetic factors, and it was thought that the limits of one's learning ability were established at birth. The study of normal infants and children and of the handicapped has added new dimensions to understanding cognitive functioning and capability, and environment has come to be recognized as having a significant impact on learning ability.

The nature of cognition in early infancy is not yet clearly defined, however, evidence increasingly suggests that learning begins earlier than was once believed, and probably has its early social roots in the interaction between the infant and his caregiver(s).

No standardized test exists for determining cognitive readiness for pregnancy, nor does any law limit the right of parenthood to only those with clearly proven learning capability. The issue of sterilization of persons with learning disabilities or handicaps has long been around because of the incidence of pregnancy among individuals who are unable to parent their offspring. It is not likely that cognitive capability will become a legal prerequisite for pregnancy and parenthood.

Pregnancy requires the individual and often the family to learn quickly about many things. They are often faced with the need to "acquire and assimilate new information . . . about reproduction, family planning, general health maintenance, nutrition, marriage, sexuality, parenting role and task, and growth and development of children" (Butnarescu, 1978, p. 262). At the same time, many other changes are occurring in the woman's mind, some of which support the learning process (e.g., interest in what is happening with the unborn child), others of which impede it (e.g., stress, uncertainty, fear). Learning is related closely to motivation as well as to capability, and the two may or may not be complementary at the point of conception, pregnancy, and parenting periods.

The health worker has the responsibility and opportunity to assess selected aspects of cognitive readiness by eliciting such information as the individual or couple's educational level, their perception of their own needs for information, the ways in which they learn best, and the resources they use for learning. The accomplishment of the developmental tasks of pregnancy and parenthood can be impeded by a lack of motivation or an inability to learn new things. On the other hand, methods designed to be supportive of learning capability and need will enhance the individual and couple's maturation during these periods.

Readiness for pregnancy and parenthood is not an easily diagnosable state, nor is there adequate scientific evidence to define the boundaries of readiness; much is yet inferential. A knowledge of what constitutes readiness for the reproductive experiences is the health care worker's first prerequisite for being successful in helping the pregnant family to anticipate and cope with these periods of life.

Readiness for learning is difficult to assess; however, most authorities agree that some evidence of interest should be present. This may appear in the form of questions asked, or it may be manifested by the woman's returning for prenatal care. Level of education will often (but not always) indicate learning capacity. The person's response to information provided may give further clues to the nurse.

PREGNANCY: A MATURATIONAL CRISIS

Pregnancy has been described as a developmental period in a woman's life, although the preponderance of reported findings derive from psychoanalytic rather than developmental theory (Deutsch, 1945; Caplan, 1961; Bibring, 1961; Rubin, 1961, 1967). Characteristics described by various researchers appear to support the idea that pregnancy may well be developmental in nature. Few will question the clarity of fetal development, and its adherence to developmental principles; what is less clear are the

stages through which the pregnant woman passes in moving from the single, nonpregnant state to the duality of parenthood. Even less is known about how the expectant father experiences pregnancy and early fatherhood. Some findings that support the idea of pregnancy as a developmental period include the following.

1. Development is sequential in nature, each stage proceeding from the previous stage and each affecting the future stages of development. From a biological perspective, pregnancy certainly occurs in an orderly, sequential fashion, and there is increasing evidence that a similar progression occurs from a psychological point of view.
2. Pregnancy creates a state of disequilibrium in which biopsychosocial processes are altered. It has been likened to a crisis state by some. It necessitates increased efforts toward adaptation and coping to reorder old ways and establish new patterns of behavior.
3. Pregnancy is a biological period with descriptive characteristics that can be studied, and during which most women are able to accomplish tasks or make adjustments that enable them to take on an untried role, that of parent, and succeed. There are similarities among women regarding feelings, fantasies, expectations of self and infant, and goals to be reached, supporting the idea that development is consistent and generalizable.
4. Pregnancy affects the total organism and is more than a sequence of biological processes; it is an intermingling of many facets of biopsychosocial functioning. Benedek addressed the psychosomatic aspects of mother and infant and years later wrote:

   > Pregnancy, like puberty, is a biologically motivated step in the maturation of the individual which requires physiologic adjustments and psychological adaptations to lead to a new level of integration that normally represents development (Benedek, 1970, p. 137).

Caplan (1961, p. 73) states that "There is a dynamic unfolding in the psychological area in the same way as the physical area during pregnancy. . . . Delivery is one element of the unfolding process." He further observed hormonal changes during pregnancy that seem to correlate with psychological changes observed in pregnant women. Bibring believed that "Pregnancy, like puberty or menopause, is a period of crisis involving profound psychological as well as somatic changes. These crises represent important developmental steps and have in common a series of characteristic phenomena. . . . The outcome of this crisis is of the greatest significance for the thus initiated phase (maturity in puberty, aging in menopause, and motherhood in pregnancy)" (1959, p. 116). In 1961, Bibring further concluded that the interaction of physiological, emotional, and cognitive processes combined to disrupt the woman's equilibrium, requiring a reorganization of self in order to "accept and incorporate" first the pregnancy, then the developing fetus as a part of herself, and then to view the fetus as an object separate from herself. "This preparedness equals a readiness to establish a relationship to the future offspring and this in turn represents the new developmental achievement" (Bibring, 1961, p. 26).

Some efforts have been made to more closely align the study of the psychological processes of pregnancy with a developmental theoretical framework in order to describe pregnancy as a developmental period of life. Establishing a developmental framework has presented some difficulties for a variety of reasons, one being the dearth of theory adequately describing the developmental aspects of the adult periods of life.

Pregnancy has characteristics that clearly reflect aspects of human development; it also has elements of crisis. For these reasons pregnancy is viewed in this chapter as a *maturational crisis* during which there is evidence of disequilibrium within both the individual and the family, necessitating biopsychosocial adaptation and requiring the use of coping strategies in order to work through the tasks needed to restore the individual and family to a steady state of functioning. Changes occurring during pregnancy fall under two headings, the *biology of pregnancy* and the *psychosociology of pregnancy*. The biology of pregnancy is here discussed first, with clear recognition that the two are almost inseparably interrelated. This order is chosen in accordance with the assumption that the psychosociological aspects of pregnancy derive in part if not in whole from the biological happenings. The content presented must be seen as representing major understandings about pregnancy, but should not be considered an exhaustive report of all that is known about this fascinating period of life.

Biology of Pregnancy

Fertilization

Simply stated, pregnancy results from the fertilization of the ovum, i.e., from the union of the sperm and ovum in the outer part of the Fallopian tube, following sexual intercourse. Fertilization causes no discernible biological response even though some women, particularly in cases of planned, highly desired pregnancies, have reported that they were aware of the moment of conception.

Fertilization is possible when the following conditions are met.

1. The male and female gametes are active and mature.
2. The mature gametes are transported to a site appropriate for fertilization.
3. Timing is appropriate (fertilization is possible only during a small part of the female menstrual cycle because the life of the mature gamete is brief—approximately 24 hours for the extruded ovum and 48 hours for the ejaculated spermatozoa within the female body).

4. The female reproductive tract is receptive to the sperm (alterations in vaginal or cervical secretions that decrease the pH of the vaginal environment, causing it to be more acid, are hostile to the sperm).

5. The sperm is able to penetrate the ovum. Ova are much larger than spermatozoa and are nonmotile; they are covered with a tough membranous coat, the corona radiata and the zona pellucida (Fig. 4-6). Usually only one ovum is extruded from the ovary at the time of ovulation. Spermatozoa, on the other hand, are smaller and highly motile; approximately 120 to 500 million are usually released with each ejaculation of semen. Aside from sperm, semen contains prostatic secretions, mucoproteins, and secretions of the seminal vesicles. The last is believed to counteract the acidity of the vagina. The minimum standards for a probably fertile sperm specimen, according to Page, are:

> in excess of 20 million spermatozoa per ml of ejaculate, and over 50 million spermatozoa in the total ejaculated specimen.
> motility is apparent in more than 50% of the spermatozoa and these motile spermatozoa can move in a foward motion, necessary for them to reach the ovum in the Fallopian tube,

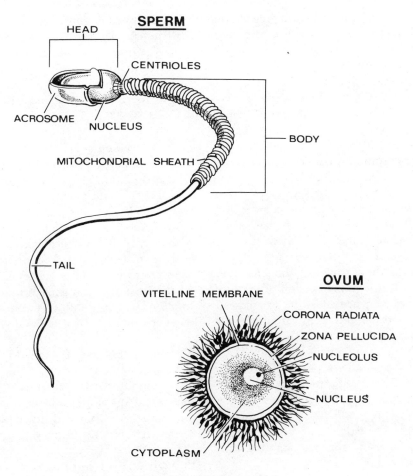

Figure 4-6. *The Ovum and Sperm*

in excess of 50% of the sperm have a normal morphology.

"If any one of these standards is not met, particularly with two or more specimens, the man is suspected of being infertile, but only when motile sperm are completely absent can it truthfully be stated that he is sterile." (Page, Villee, and Villee, 1976, p. 161)

Gametogenesis is the process through which the immature male and female gametes are matured in readiness for the act of fertilization. Meiosis is a form of cellular division that occurs during gametogenesis in which the genetic components of the gametes intermingle and replicate themselves in each gamete in preparation for transmission to another life. A reduction in cell number also takes place during meiosis, in which the usual complement of 46 chromosomes (44 autosomes in 22 pairs and 2 sex chromosomes) is halved, providing each gamete with only one of each kind of chromosome, i.e., 22 autosomes and 1 sex chromosome, X in the female and X or Y in the male gamete. Sometimes a chromosome is lost or damaged during this reduction division, laying the foundation for a genetic abnormality. Fertilization unites the chromosomal pairs once more and usually results in the restoration of the normal number of chromosomes—44 autosomes and 2 sex chromosomes.

Upon entry of the sperm head, the acrosome, into the ovum through penetration of the corona radiata and zona pellucida, the sperm head separates from its tail, moves toward the center of the ovum, and unites with the nucleus of the ovum. At this moment a new life begins. Fertilization initiates another type of cellular division, mitosis, or cell replication, and the new life begins its journey from the Fallopian tube to the uterus where it will attach to the uterine wall and grow. Nidation (attachment to the uterine wall) occurs on approximately the fifth day after fertilization.

Fetal Development

There are three distinct stages of fetal development:

1. The preembryonic period, sometimes called the period of the zygote. This period begins with fertilization and ends with the completion of the third week after fertilization. The major tasks of this period are cell division and germ layer differentiation.
2. The embryonic period, weeks 4 through 8. This is the time of organ growth and development (organogenesis). It is during this time that the rudimentary structures of the organs, face, ears, and limbs become histologically distinguishable.
3. The fetal period, weeks 9 through 40. This is a period of continued organ growth and differentiation. During this period the fetus grows from a weight of approximately 15 gm to that of an average full-size infant, 3,200–3,400 g (7 to 7.5 lb).

Fetal development is possible because of three supportive factors: genetic endowment (parental contribution), growth substrates (essentially glucose, fats, and amino acids), and maternal-fetal vascular support (maternal cardiovascular system, fetal cardiovascular system, including the placenta and umbilical cord). Other factors such as maternal and fetal environment may support or impede fetal development. Maternal nutritional status and general health have been found to be major influences on fetal development, particularly fetal size and weight.

The Preembryonic Period (Fertilization through Week 3)

The period of the zygote initiates the most sophisticated developmental process known, in which the blueprint laid down at fertilization is translated into the beginning of a human life.

Cell division: The single-cell zygote develops through a rapid mitotic process of cellular multiplication called cleavage in which the cells (blastomeres) continue to double their number until they become a multiple-cell solid sphere, a *morula* (the term means "mulberry," to which the morula's appearance is often likened). This early period of cell division occurs within the walls of the fertilized ovum. There is no increase in size of what was once the ovum, for the cells reduce their size as they multiply.

Blastocyst: The appearance of fluid within the center of the morula initiates a cell realignment or rearrangement, sending some cells to the periphery of the sphere to form a single layer of cells lining the sphere (trophoblasts) while others remain clustered in the center (embryoblast).

The central cells form the developing embryo; the trophoblast (*tropho* means "nourishment") attaches to the uterine endometrium, and then divides into three cell layers, the cytotrophoblast (inner layer), which attaches to the base of the uterine lining, the mesoblast (middle layer), which becomes the vascular structures of the placenta, and the syncytium (outer layer), which contains the micro-villi that will form the circulatory link between mother and fetus, i.e., the placenta. The trophoblast also provides nourishment for the developing embryo until the placenta can take over.

Cell Differentiation. Early cell differentiation begins with the reorganization of the cells within the blastocyst, and the separation of the trophoblast and embryoblastic cells. The inner cell mass begins to undergo a process of differentiation, which in turn begins the definition of the developing human being. The inner cell mass separates into two hollow spheres or sacs divided by a barrier of cells, the embryonic disc. The outer sac is the forerunner of the amniotic sac, which will eventually surround the developing fetus, providing a protective water cushion (the "bag of waters"). The inner sac, or hollow sphere, becomes the yolk sac, believed to be important in early blood formation and the transfer of nutrients to the embryo; the yolk sac is the earliest developed part of the embryonic (later fetal) circulatory system. The embryonic disc comprises three layers of tissue, the entoderm, the ectoderm; and the mesoderm. The entoderm (inside layer) gives rise to such structures as the thyroid and thymus glands, the liver, biliary tract, pancreas, intestinal tract endothelium, respiratory tract, bladder, and other small internal structures. The ectoderm (outside layer) gives rise to the skin and its appendages, the central nervous system, the anus, the mammary glands, the external ear, and parts of the eye and nasal passages. From the mesoderm (middle layer) issues the muscles, skeleton, gastrointestinal and genitourinary tracts, heart, and the circulatory and connective systems.

The trophoblastic cells are the cells of the placenta, a flat, round, vascular organ resembling a pancake, approximately 8 in. in diameter at term, and weighing approximately 1.5 to 2 lbs. It has two sides, the basal plate (often referred to as the maternal side), which adheres to the uterine wall following implantation, and the chorionic plate, the fetal side, which is covered with the fetal membranes. The umbilical cord rises from the center of the placental fetal surface. The placenta is divided into sections or lobes (cotyledons). Placental formation begins during the preembryonic period with the event of nidation. Although the process of placental development is not completely understood, it is believed to include the following steps.

1. Implantation by the blastocyst.
2. Invasion and proliferation of the trophoblast at approximately 7 to 8 days of fetal life.
3. Tapping of endometrial venules and capillaires, and establishment of circulation of maternal blood (approximately days 9 to 11).
4. Formation of villi, the forerunner to placental vessels, and establishment of fetal-placental circulation on about day 13 to 21.

Two other components of the fetal-placental unit that develop during the early prenatal period are the fetal membranes and the umbilical cord. The fetal membranes are two opaque, partly fused layers of tissue, the amnion (or inner layer) and the chorion (or outer layer). The chorionic and amniotic membranes form the amniotic sac, which is filled with fluid and contains the developing fetus. These membranes also cover the fetal surface of the placenta, and the amnion covers the umbilical cord. The site of origin of the fetal membranes is not clear; however, it is believed that they develop from the trophoblast or the fetal ectoderm. Their function is not definitively known, but they are believed to be the source of amniotic fluid in early prenatal life.

The umbilical cord is a gelatinous structure normally containing two arteries and a vein that are covered with a mucopolysaccharide substance known as Wharton's jelly and encased in the amniotic membrane. The cord is believed to develop from the hindgut of the embryo. At term, the whitish-gray cord is approximately 55 cm (20 in.) long. It connects the fetus at its lower-mid-abdomen with the center of the fetal aspect of the placental surface, and establishes rudimentary fetoplacental circulation by the twenty-first day of fetal life.

The fetoplacental unit is the life support system of the developing fetus. It provides for the exchange of nutrients and oxygen as well as for the elimination of some fetal waste products. The placenta also produces hormones and provides an anatomical barrier between mother and fetus. Some believe that it functions as an immunological buffer zone (Simmons and Russell, 1963; Billingham, 1964). During pregnancy it provides, through its endocrine function, an index of fetal maturity and well-being.

The Embryonic Period (Weeks 4 through 8)

The embryonic period is particularly critical for the new life; the embryo is especially vulnerable to the influences of both external and internal factors such as maternal infection or disease, drugs, and other assaults, which may cause death or developmental aberrations. Changes occur rapidly during this period, when the amorphous cellular structures assume a distinctly human form. By the end of this period, the embryo has achieved an average weight of 5 g and a crown–rump length of 30 mm. Brain growth is particularly rapid during this period, and there is a concomitant increase in head size. This allows the central nervous system to develop sufficiently to begin to integrate the functions of the other systems. The heart develops from a single tube into a four-chambered organ. Although the pulsation of the heart is not detectable until later, the heart is complete and functioning by the end of the sixth week of fetal life.

By the end of the embryonic period, the embryo has achieved the beginning of organ development and function and taken on a human appearance, with distinct facial features, hands, feet, and a disproportionately large head. The embryo is attached by a thin umbilical cord to the miniature placenta and has survived the basic beginnings of life (Fig. 4-7).

Placental development continues during the embryonic period, although the placenta is not yet completely ready to support the needs of the developing embryo. Its development continues until by the end of the seventh week of prenatal life, its cotyledons and their vascular network are formed and are capable of some sluggish circulatory activity.

The fetal membranes and umbilical cord are developed to a functional level in the early days of the embryonic period. For the remainder of that period, they, like the placenta, increase in size and become structurally refined and more efficient.

The Fetal Period (Weeks 9 through 40)

Entry into the fetal period is a major milestone for the new life that has survived the first two

Week 4
CR = 4 - 5 mm

Week 8
CR = 30 mm (1 inch)
W = 1 gm (1/30 ounce)

Week 12
CR = 75 mm (3 inches)
W = 30 gms. (1 ounce)

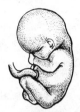

Week 16
CR = 150 mm (6 inches)
W = 120 gms (4 ounces)

Week 20
CR = 25 cm (10 inches)
W = 300 gms (10 ounces)

Week 24
CR = 30 cm (12 inches)
W = 680 gms (1½ lbs)

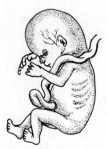

Week 28
CR = 37.5 cm (15 inches)
W = 1135 gms (2 ½ lbs)

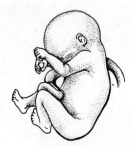

Week 32
CR = 42.5 cm (17 inches)
W = 1816 gms (4 lbs)

Figure 4-7. *Fetal Development*

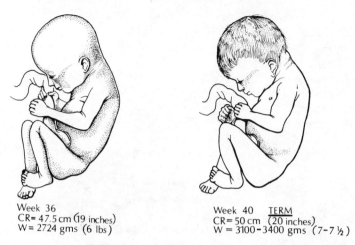

Week 36
CR = 47.5 cm (19 inches)
W = 2724 gms (6 lbs)

Week 40   TERM
CR = 50 cm (20 inches)
W = 3100 – 3400 gms (7 – 7 ½)

Figure 4-7. (Continued)

critical developmental stages. The tasks of the fetal period are to attain an increase in size and weight through continued organ and structural growth as well as through the development and distribution of fatty tissue; to achieve the systemic efficiency necessary for fetal survival and preparatory to extrauterine life; and, of course, to continue to survive in utero.

Fetal movements (quickening) are first appearent to the mother between the sixteenth and twentieth weeks of gestation. Although the heart has been beating since the embryonic period, it can first be auscultated by about the eighteenth to twentieth week. Ultrasonic techniques enable an examiner to determine fetal heart function as early as the tenth week of gestation. (Table 4-3 provides an overview of selected characteristics of fetal development, by trimester.)

Although all fetal systems are important during prenatal life, the cardiovascular system is especially crucial to survival during prenatal life and to the eventual transition to extrauterine life. This system provides for the transport and exchange of nutrients and other essential substances, such as oxygen, without which the fetus cannot survive. The cardiovascular system is the first system established in the new life.

The placenta, along with the maternal and fetal cardiovascular systems, is responsible for ensuring that the fetus receives adequate nutrition, oxygen, and other substances (such as antibodies) from the mother. During the earlier weeks of prenatal life the embryo is nourished essentially through the mechanism of diffusion, the rapid movement of molecules across a membrane. Diffusion is a transfer mechanism with limitations, so there is a need for a more effective system, i.e., the placenta. The transport of substances to and from the fetus is accomplished in the following manner:

1. Nutrients and oxygen from the maternal circulation pass through the placenta and the umbilical vein, and on into the fetal circulatory system.
2. A division of the umbilical vein occurs at the edge of the fetal liver, sending about 50 percent of the blood through the portal vein to the liver and then into the inferior vena cava through the hepatic vein; another 50 percent of the blood is then shunted into the inferior vena cava via the ductus venosus.
3. The blood is then circulated to all parts of the fetus, carrying nutrients and oxygen to vital organs and structures. Only about 10

Table 4-3. An Overview of Fetal Growth and Development

System or Structure	Appearance	Function	Significance
Neuromuscular	*Preembryonic Period* Primitive neuroblast neurola arises from the ectoderm (neuroectoderm), 18–20 days. Brain develops from the endoderm. Two sets of tissue merge.	Potential for integrating functions of all other systems in later fetal development, and to coordinate embryonic, fetal, and neonatal functioning.	The fetus develops in a cephalocaudal direction, with the immature brain and neurological system developing first. The embryo/fetus is vulnerable to assault at any time, but never more so than during this early period of organ-system development.
	Embryonic Period *28–36 Days* Major voluntary muscles develop from 40 pairs of tissue blocks and have nerve associations. By day 33 brain expands from 3 to 5 compartments with all cavities connected.	Potential for coordinated movement. Control of muscular movement apparent by 6 weeks.	
	Fetal Period *12th week* Coordination of neuromuscular junctions present; Reflexes respond as total body unit instead of individual structures. Can oppose thumb and forefinger.	Coordination of movement. G.I. tract and lung tissue movement present.	Some movement may be felt by mother by the 14th week; not definitive. Distinct movement apparent to mother between 16–20 weeks. Probably begins to swallow amniotic fluid about this time. Nomal development in primates.
	Weeks 13–27 Brain and neuromuscular system continue to grow and develop. Muscles and ligaments present but not coordinated.	Controls total organism. Increased reflex responses. Spontaneous stretching and yawning.	Capable of responding to electrical and physical stimuli. Active fetal movement can be felt by the mother or examiner.
	Weeks 28–40: Increase in organ size. Cerebellum less developed than cerebrum.	*Reflexes* rooting grasping sucking stepping	Potential for these reflexes present but not demonstrated until birth.
Cardiovascular	*Preembryonic Period* Day 9–11: Blood-forming elements formed; beginning		No clinical evidence of fetal life.

Table 4–3. (*Continued*)

System or Structure	Appearance	Function	Significance
	circulation established. Day 18–20: Heart is a single tube.		
	Embryonic Period Day 22–24: Heart tubes fuse to form a 4–chambered organ.	Heart begins to beat.	Cannot be determined by clinical methods. Circulatory system vulnerable to adverse influences leading to death or anomaly.
	Day 24: Heart bulge apparent. Early placental development apparent; vascular channels forming. Umbilical cord forms.	Beginning transport of some nutrients and other essential substances.	Cannot survive if born at this time.
	Fetal Period Heart and circulatory structures and vessels increase in size.	Allows increasingly efficient transport of nutrients and other essential substances.	Ten Weeks: heartbeat detectable by ultrasonic technique. Heartbeat can be auscultated 16–20 weeks; Definitive sign of pregnancy. Heartbeat and circulatory function provide some index of fetal well-being.
Respiratory System	*Preembryonic Period* No real indication of respiratory system structures during this period, although the lung buds begin to appear by 4th week. Called the early embryonic period of lung development.		
	Embryonic Period 5 weeks: the pseudoglandular period begins during which the lobar buds branch and bronchial arteries are laid down. 13th week: early beginning of formation of gas-exchange units.	No pulmonary function clearly evident.	
	Fetal Period Continued development of bronchial arteries. Bronchial tree is established.	13th week: Some respiratory activity has been noted. Lung is immature and usually incapable of supporting	Infant born with immature lung function will have great difficulty surviving if born before 26 weeks.

Table 4-3. (*Continued*)

System or Structure	Appearance	Function	Significance
	17th week: major elements of the lung are present except for gas exchange areas. 25th week: Gas exchange units are formed. 24–40 weeks: pulmonary structures are complete; gas exchange is possible. Late Fetal Period: air sacs become thin and more distinct. Lung fluid is present during prenatal life.	life before the 26th week. 24 weeks: surfactant produced in increasing amounts; acts on the surface of the alveoli and is necessary for the stability of the lung after it is aerated. Episodic respiratory movements have been noted during fetal life. Partially inflates the fetal lung during prenatal life.	Not clearly understood. Probably preparatory to extrauterine life. Probably facilitates adaptation to extrauterine life.

Digestive Tract	*Preembryonic Period*
	Beginning of formation of the digestive system derived from the primitive gut late in this period.
	Embryonic Period
	Digestive system arises from the entoderm as a single tube running the anterior-posterior length of embryo; becomes alimentary canal.
	Basis for development of:
	    Gall bladder
	    Thyroid gland
	    Stomach
	    Pancreas
	    Intestines
	    Thymus
	    Liver
	    Spleen
	Fetal Period
	Growth (primarily in size) continues.
	Gut fills with meconium, containing:
	    digestive secretions
		Most structures not functional during this time.
		
		Early lymphocyte development. production of red blood cells.
		Fetus swallows amniotic fluid.
			Thought to play a role in development of fetal immunology. Exact mechanism not certain.
			Decrease in O_2 will cause relaxation of sphincter, releasing

Table 4-3. (*Continued*)

System or Structure	Appearance	Function	Significance
	mucus lanugo desquamated cells vernix caseosa bile	Makes sucking movements by 24th week. Bowels usually not functioning.	meconium, a sign that fetal-placental circulation is impaired.
Other	*Preembryonic Period* Germ layer differentiation lays the foundation for all fetal structures and systems. *Embryonic Period* Sense organs develop—eyes, ears, nose. Paired buds, which become the arms and legs, are apparent. Hands are apparent as paddles. Feet appear as bulges. 31 days: mesonephric duct present; forerunner to the kidneys' collecting ducts and tubules continue to form. Kidneys appear, replacing the earlier metanephros. Bone begins to form. Sexual differentiation is possible. Dentine begins to form. Skin	Movement occurs but not usually apparent. Renal function is present beginning at 8th week. Water reabsorption by 11–12 weeks. Bone formation provides structural form and rigidity. RBC production in bone marrow. Protection and partial containment of internal structures.	All basic organ systems are laid down in the embryonic period. They are all vulnerable to assault. Teratogenic and iatrogenic factors can destroy or severely impair the potential for normal functioning in any or all of these systems. Most pregnancies not confirmed yet; earliest possible diagnosis of pregnancy and start of prenatal care essential.
	Fetal Period Growth and development continues.	The kidneys secrete urine; movements of extremities readily discernible.	Desquamated cells and other substances in amniotic fluid; aids in diagnosis of sex and any fetal distress. Desquamated cells are swallowed and contribute to volume of amniotic fluid. Fetal movement can be felt by mother and examiner; aids in confirmation of pregnancy.

SOURCE: Adapted from Butnarescu, G. F., *Perinatal Nursing, Vol. I: Reproductive Health,* New York: John Wiley & Sons, 1978, pp. 137–139. Used with permission. (Note: Chronological references are given in lunar time.)

percent of the blood goes to the lungs, since the lungs are essentially dormant during fetal life.

4. The umbilical arteries then return the deoxygenated blood through the umbilical cord to the placenta.

The process of fetal circulation is a sustained rather than an interrupted one, except in the event of maternal or fetal disease or trauma such as maternal cardiovascular or renal disease, poor nutrition, general ill health, or placental, umbilical, or fetal anomalies, and numerous other factors that can influence fetal well-being and development. It is important to remember that the fetus can live and grow only when the fetoplacental mechanism is structurally and functionally adequate to provide for transport and exchange of essential substances, when these substances are available in adequate amounts and are accessible for transport, and when the fetus is capable of utilizing and exchanging them.

Maternal Biological Alterations

Pregnancy is a time of considerable biological upheaval for a woman. Every system and many organs and structures are affected by the processes necessary to accommodate the developing fetus, and to prepare the woman both physically and emotionally for childbirth and recovery. The biological and psychosociological aspects of pregnancy are very closely related. The woman's behavioral response to pregnancy may have its roots in one or more of the physiological changes occurring (see Table 4-4).

Psychosocial Aspects of Pregnancy

Pregnancy is a period of new physical symptoms, greater sensitivity to familiar feelings and moods, and an increased awareness of new feelings and new moods. It is characterized by emotional and physical highs and lows. In spite of the joy she may feel, the pregnant woman may also feel trapped; once conception has occurred, she cannot "back out," if only because she is usually not the only one who would be affected by such a decision.

One pregnant woman offered this insight to the author: "Thinking about being pregnant is one thing, but when you become pregnant, you're really pregnant 24 hours a day; it doesn't come and go like your dreams." Another woman, reflecting on the emotional ups and downs of her pregnancy, recalled "I felt so silly when I would burst into tears for no apparent reason. I knew this was normal, but I felt so stupid, and my husband would just stand there looking helpless." Another woman aptly remarked, "No woman without a sense of humor should be permitted to become pregnant, much less be a mother." Pregnancy affords the opportunity to experience the gamut of human emotions—anger, joy, fear, depression, euphoria, anxiety, and many, many others.

Pregnancy usually occurs within some type of family structure, although the characteristics of each family may differ.

> The family is a complete organism, a unity in its own right, as real as an individual. . . . An event in any part of this organized system impinges on every part in the rest of the system; every experience within the organized system belongs to the system. This childbirth and its attendant pregnancy is an experience that belongs to the family as a whole (Howells, 1972, p. 127).

Research into the effect of pregnancy on the family is far from definitive. Along with the influence of ethnic values and cultural practices regarding pregnancy, changing societal expectations about the family and about reproduction in contemporary America make it difficult to see the total picture of the impact of pregnancy on the family.

Crisis and *stress* are relatively new words used to describe events of the sort that have influ-

Table 4-4. Maternal Physiologic Alterations During Pregnancy

System or Structure	Physiological Alteration	Significance

Reproductive	*Uterus*:
	Size: Weight increases from approximately 40 gm in nullipara or 60 to 70 gm in multipara (nonpregnant weight) to approximately 1,000 gm at term, caused by cell growth and appearance of some new cells. Shape of uterine fundus changes from spherical to cylindrical as pregnancy progresses. Uterine muscle increases in thickness for approximately 3 months, than begins to thin until term.
	Increased myometrial contractility during pregnancy.
	Blood Flow: Uterine and ovarian arteries carry arterial blood to the uterus.
	Progressive increase in uterine blood flow during pregnancy.
		Increased size, weight, and shape of uterus help the examiner: diagnose pregnancy; assess length of gestational period, and evaluate fetal growth and development.
		Causes pressure on other organs, causing minor difficulties. (See other systems.)
		Prepares uterus for labor.
		
		Influenced by:
		  maternal position:
		    supine (lying on back) = decrease
		    lateral recumbent (lying on side) = increase
		    uterine contractions = decrease

	Endometrium: Increases in thickness and vascularity; becomes the *decidua* during pregnancy.	The site of implantation of the fertilized ovum; its glandular secretions contribute to nourishment of the ovum prior to development of fetoplacental circulation. Provides the basis for development of new endometrium following termination of pregnancy. Part of this lining is cast off as lochia during the post-delivery period.
	Cervix: Becomes softer and cyanotic because of increased vascularity, increased edema, and hypertrophy and hyperplasia of cervical glands.	Occurs as early as 4 weeks after conception. Aids in diganosis of pregnancy. Mucous plug seals off cervical os until onset of labor. Capable of dilatation and effacement sufficient to allow passage of fetal head.
	Ovaries: Follicle maturation and ovulation cease throughout pregnancy. Corpus luteum is probably maintained in a functional state for about first 4 months of pregnancy. *Fallopian Tube*: Musculature essentialy unchanged.	Produces steriod hormones; thought to contribute little throughout the rest of pregnancy. None.

Table 4-4. (*Continued*)

System or Structure	Physiological Alteration	Significance
	Vagina and Perineum: Increased vascularity and hyperemia of skin and muscles of the perenium. Softening of connective tissue. Increased vascularity of vagina.	Deep pink or violet color of vagina (Chadwick's sign); may aid in diagnosis of pregnancy.
	Increased thickness of vaginal mucosa. Loosening of connective tissue. Enlargement of smooth muscle cells. Increased vaginal secretion; increased acidity of secretion.	Allows for distensibility during labor. Combines with increased cervical secretion to form a thick white discharge; thought to aid in the control of vaginal pathogens.
Breasts	Increased size and vasularity. Increased size and pigmentation of nipples and areola. Appearance of hypertrophic sebaceous glands (Montgomery's tubercles).	Preparatory to lactation. Breast tenderness and tingling in early weeks of pregnancy.
Cardiovascular	*Blood Volume*: Red blood cells increase (approximately 450 ml), increasing maternal iron by about 400 to 500 mg. Plasma increase of 40% or more. Total blood volume increases 30 to 50 percent. Cardiac output (the volume of blood circulated by the heart per minute increases 20 to 50 percent; increases still more during labor.	Hgb below 11.5 gm/100 ml indicates iron deficiency. These changes compensate for the additional cardiovascular work load imposed upon the woman during pregnancy.
	Increased weight and size of gravid uterus compresses vessels and sometimes interferes with return blood flow from the lower half of the body.	Woman may feel faint when standing or remaining in one position for long periods of time.
Renal	Total body water (mostly extracellular) increases.	Some generalized edema of extremities and upper torso in about 25 percent of women. Ankle edema common—edema is not significant unless blood pressure is elevated and/or there is protein in urine.
	Increase in urinary volume. Increase in glomerular filtration rate (GFR). Renal structures, ureters, and renal pelvis are dilated.	Increased voiding particularly in early and late pregnancy because of pressure of uterus on bladder. Increased chance of urinary tract infection.

Table 4-4. (*Continued*)

System or Structure	Physiological Alteration	Significance
Respiratory	Growing uterus displaces diaphragm against lungs, decreasing vertical expansion, necessitating some lateral expansion. Increase in tidal volume, inspiratory capacity, alveolar ventilation, minute volume, and total oxygen consumption. Decrease in residual and total lung volume, expiratory reserve.	During late pregnancy the woman may experience shortness of breath, difficulty with deep inspiration, possibly dyspnea. In an attempt to compensate for this difficulty, she may breathe rapidly, increasing oxygen intake and decreasing pCO_2. As much as 20 percent increase in oxygen in late pregnancy has been measured.
Gastrointestinal	Decreased secretion of gastric acid (essentially in first and second trimesters). Decreased secretory response to histamine. Relaxaton of G.I. tract tissues throughout pregnancy (related most likely to hormones) produces decreased gastric motility, diminished peristaltic action, and poor bowel tone. Gastric emptying time is increased.	Decreased tendency for peptic ulcer.
	Relaxation of the cardiac sphincter along with uterine pressure on stomach may cause heartburn. Saliva becomes more acid. Amount of saliva is increased; cause unknown. Gall bladder emptying time is increased and serum cholinesterase is reduced.	Heartburn (regurgitation of undigested food into lower esophagus). Thought to contribute indirectly to tooth decay. May play a role in formation of gallstones.
Metabolic	*Carbohydrate*: Decreased renal threshold for glucose. Increased insulin production. Increased resistance to insulin by free fatty acids. Increased destruction of insulin by the placenta. *Protein*: Positive nitrogen balance is found in first trimester; increases during third trimester when fetal needs are greatest. Blood loss, lactation, and involution of uterus seem to contribute to a negative nitrogen balance during the puerperium.	Glycosuria present in about 30 percent of pregnant women. Some women develop "gestational diabetes." Maternal protein status can be evaluated through measurement by singla and 24-hour urinary urea nitrogen and total nitrogen ratios (UN/TN ratios).

78

Table 4-4. (*Continued*)

System or Structure	Physiological Alteration	Significance
	Nitrogen reserve is maintained by mother through storage of surplus protein in maternal organs and tissues. *Fat*: Blood lipids are increased (probably related to fetal needs and breast development, and influenced by steroid hormones).	
	Progressive increase in free cholesterol until 30th gestational week. *Mineral*: Calcium and phosphorus requirements double during last half of pregnancy (half used by fetus, half stored by mother). Serum calcium levels usually remain unchanged. *Folic Acid*:	Increased ketonuria in diets low in CHO or high in fat.
	Requirements increase fivefold during pregnancy to 0.5 mg per day.	Necessary for metabolisn of some amino acids annd synthesis of nucleic acids. Deficiency leads to megaloblastic anemia. Dietary supplement of 0.3 mg daily usually adequate.
	Acid/Base Maternal plasma bicarbonate and total base decrease. At term the average CO_2 combining power decreases (cause unknown).	Maternal blood pH remains stable, indicative of a compensatory mechanism at work; may be affected by increased ventilation.
Endocrine	*Thyroid Gland*: Increase in size in 50 percent of pregnant women. (Caused by hyperplasia of glands, new follicle formation, and increased vascularity). Increased circulating thyroid hormone. *Parathyroid Glands*: Hypertrophy occurs, probably because of fetal demands for calcium. Increased production of parathyroid hormone. *Pituitary Gland*: Anterior lobe enlages 20 to 40 percent. This occurs in cells containing prolactin rather than in those containing growth hormone.	Increased basal metabolic rate (BMR) usually apparent by 16th week of gestation. Increases to 10 to 30 percent above nonpregnant rate in third trimester.

Table 4–4. *(Continued)*

System or Structure	Physiological Alteration	Significance
	Posterior Lobe: Shows increased production of oxytocin and possibly vasopressin (antidiuretic hormone).	Stimulation of uterine contractions. Active in lactation (facilitates let-down reflex).
	Adrenal Glands: Increase in size (hyperplasia of the cortex). Production of corticosteroid hormones increases. Two peaks occur, the first in the first trimester and the second around the seventh month.	Concurrent with period of greatest water retention.
Musculoskeletal	*Bones:* Symphysis pubis, sacroiliac synchrondroses, and sacrococcygeal joints are widened slightly and capable of some movement (about tenth to twelfth week of gestation). (Probably caused by secretion of a hormone, relaxin).	
	Alterations in center of gravity contribute to attempts to compensate by moving shoulders, upper spine, back, and abdomen; this causes pronounced lordosis.	Contributes to backache, poor posture, and possible instability of movement; numbness; weakness.
Abdominal Wall and Skin	*Abdominal Wall:* Diastasis recti(verticle separation of the recti abdominal muscles) occurs in some women; usually caused by stress placed on the abdominal wall by the gravid uterus.	Wide separation of these muscles can be palpated. Herniation of the growing uterus can occur. Abdominal muscle tone is weakened and may be less efficient during the second stage of labor.
	Skin: Striae gravidarum (reddish-brown streaks on abdomen, thighs and breasts) can be seen in about half of all pregnancies.	May cause distress in some women because of concern with appearance.
	Brown vertical line (linea nigra) in abdominal midline. Chloasma (dark brown discolorations on face) may occur. Cause not known but may be related to presence of melanocyte stimulating hormone (similar to ACTH), which is present in pregnant women from about 8 weeks' gestation until term. Estrogen and progesterone may play a role since they may have a melanocyte-stimulating effect.	

SOURCE: Adapted from Butnarescu, G. F., *Perinatal Nursing, Vol. I: Reproductive Health.* New York: John Wiley & Sons, 1978. pp. 124–130. Used with permission.

enced the individual and family throughout the course of history. Caplan (1961) sees *crisis* as an occurrence that causes a period of disorganization and upset; usual coping methods are not effective in dealing with it. Crisis has also been described as being of a developmental nature, because it often occurs in conjunction with expected, predictable events associated with the normal maturation of individuals (Erikson, 1953). Pregnancy and parenthood fall into this category, as do adolescence and menopause. Not all crises are maturational in nature; some are situational, i.e., not at all predictable, unexpected events that may be imposed upon an individual or family, but not on every individual and family. Accidents and illnesses fall into this classification. Pregnancy is viewed by most authorities as a period of increased susceptibility to crisis.

Stress and *crisis* are sometimes used to refer to the same event. *Stress* has been variously described as an event itself, the emotional state of the individual experiencing the event, and the stimulus for the event. For purposes of this discussion, however, *stress* is used to mean the factor precipitating a crisis. In that sense, there are characteristics of pregnancy that can be viewed as stressful events, but which may or may not lead to a crisis. Further, there are times during pregnancy when the individual and family are more vulnerable to crisis.

The following factors are among those that may generate stress related to pregnancy:

1. The reason for the pregnancy: to have a child to love; to be loved; to "save a marriage"; to get attention; to prove one's maleness or femaleness.
2. Circumstances surrounding the conception: planned, unplanned, accidental, or forced; marital status; sexual guilt associated with intercourse and pregnancy.
3. Myths and fears about pregnancy and childbirth.
4. Increase in family size, concomitant role changes and increased financial burden.
5. Influence and expectations of other relatives, especially grandparents.
6. Previous experiences concerning pregnancy and childbirth (one's own or another's).
7. Identification of self with pregnancy and parenthood.
8. Parental preferences for the sex of the unborn child.
9. Meaning of the unborn child to the individual and family—extension of self or a separate being?

The timing of the occurrence of these and other stresses is of particular significance, often making the difference in the family's ability to cope. Phases of the pregnancy that may be particularly disturbing to the individual and family are discussed here in chronological order.

Suspicion of pregnancy. A woman's reaction to the earliest suspicion of pregnancy is closely related to the reasons for and attitudes toward pregnancy. For most women, it is a time of anxiety and uncertainty. Many women who are eager for confirmation of the pregnancy find themselves impatient if the physician lacks a definitive answer. Other women respond in the opposite way, ignoring all symptoms in the hope that they will "go away."

Confirmation of pregnancy. A probable diagnosis of pregnancy can be made by about the second month of gestation (or earlier if sophisticated tests are used). Although the better pregnancy tests are approximately 98 percent accurate, a positive diagnosis of pregnancy must be confirmed either through the use of ultrasonic technique, at about 10 weeks' gestation, or by detection of the fetal heartbeat, which can be accomplished at about 14 weeks with the Doppler method, or at about 18 to 20 weeks with a fetoscope. The reaction of the woman and her family to the confirmation of pregnancy is closely related to the reasons for and attitudes toward the pregnancy, as in the case of suspicion of pregnancy. At this time a decision to continue or end the pregnancy must be made.

Appearance of physical changes. If the woman views her pregnancy with pride, she will probably be pleased by her body's evidence of the growing child. Her mate, too, is likely to be proud of these changes in his partner's body. Conversely, if the woman has negative feelings about the pregnancy, or about sex, or if there is a lack of support and acceptance by other family members, particularly the partner, these physical changes may not be acceptable to the woman, creating problems for her and her family.

Late second and early third trimesters. These are the peak fantasy periods, described by Deutsch (1945) and Caplan (1959, 1961). Unresolved problems from the woman's childhood or other early periods of development may emerge. Deutsch found that the woman's relationship with her own mother, and how she viewed her mother, influenced her way of resolving problems during her pregnancy. While this fantasy period may potentiate crisis, it may, on the other hand, be growth-promoting, and may permit the resolution of old problems.

Late third trimester. At this time the woman is confronted with the approaching experience of labor, which some women associate with trauma or death. The woman who is pregnant for the first time is facing an unknown event that may be viewed in the context of myths and phobias. This late point of pregnancy is also a time of physical fatigue because of increased body size, and for some women, it represents weeks of extreme boredom with the pregnancy. Women are sometimes heard to remark facetiously that pregnancy should not exceed seven months in duration.

Labor. The actual confrontation with labor imposes the ultimate reality of the task of pregnancy. Women somtimes express feelings of being trapped: "I can't change my mind now." Feelings of loss of control over one's body are often present. With some of the current health care practices during labor, such as isolation from family, extensive use of analgesia and anesthesia, position during labor, and exclusion of the gravida from decision-making, this feeling of loss of control may have a basis in fact. Some women fear death during labor. Women have been known to be sure that their will is made, and that their affairs are in order before the onset of labor. One woman, facing her fourth labor, told the author, "I know it is silly, but I just have to be sure everything is taken care of just in case I die." Other women may see labor as an exciting climax to a long period of expectation. One woman commented, "I felt like a Broadway star on opening night." Women with this attitude often look forward to their participation in the experience of labor, as well as to its outcome, the child. Increasingly both partners are being prepared to share the labor experience. In some instances, such as in home delivery, other family members are also involved.

Recovery. The recovery period is characterized by rapid physical readjustment, but it may also be a time of discomfort and fatigue. Coping is often diminished if the woman is in pain or lacks energy. The family must now attend to the responsibilities of parenthood, and the woman may feel overwhelmed. Recovery brings with it expectations of fully assuming the parenting role, and the necessity of delineating the responsibilities of each member of the family. Sibling response, if this is not a first child, is often unpredictable, and may pose problems when the mother and new infant come home.

Tasks of Pregnancy

The concept of a developmental task has been used for a number of years. Its precise origin is not clear, although Robert J. Havighurst is usually credited with the first formal use of the concept. (See Havighurst, 1953.) Erikson (1950), in his work on the critical periods in the development of the human being (which he calls "The Eight Stages of Man"), provided a framework into which development tasks could fit. Duvall and Hill (1948) utilized the concept of developmental tasks in relation to family interaction, and Duvall has continued through the years to develop this idea further. The work of some of the psychoanalytic theorists provided data that seemed to reinforce the idea of pregnancy as a period of becoming self-fulfilled. The phrase "developmental tasks of pregnancy" has been suggested by a number of researchers.

In 1969, Tanner studied a group of pregnant women and described three tasks associated with pregnancy: "incorporation and integration of the fetus as an integral part of the woman" (essentially the first trimester), "perceive the fetus as a separate object and not as an integral part of self" (second and third trimesters), and "establish a caretaking relationship with the infant" (third trimester, and early parenting period).

Tanner implied the use of a developmental framework, but she did not utilize any one developmental theory to direct her study. However, she was one of the first to attempt to see pregnancy as a developmental period. Tanner's work has been widely used, particularly in nursing circles.

Clark supports the notion of pregnancy as a maturational crisis. She described (Clark and Affonso, 1979) four developmental tasks for the pregnant woman:

1. *Pregnancy validation,* the facing of the reality of pregnancy and the acceptance of its implications.
2. *Fetal embodiment,* or the incorporation of the fetus into the gravida's own body image.
3. *Fetal distinction,* in which the woman begins to view the fetus as an individual separate from herself, and begins to identify with the mothering role.
4. *Role transition,* when the woman is ready to relinquish the fetus, experience labor and delivery, and take on the tasks related to mothering the newly born child.

Duvall (1977) compared the developmental tasks of the man and woman and looked at them in terms of the ways in which they were complementary. The first two have similarities, while the third is somewhat different for each partner:

1. Starting a family (both man and woman).
2. Becoming a woman (becoming a man).
3. Being responsible as the main support of the family (man); seeing one's chief jobs as mother-to-be as well as wife (woman).

Jensen, Benson, and Bobak (1977) described five phases a man must go through to become a father:

1. Realization or confirmation of pregnancy.
2. Awareness of the mother's increasing body size and fetal movement.
3. Anticipation of approaching labor.
4. Involvement in the birthing process.
5. New parenthood.

As of this writing, the study of the expectant father is still in its early stages; little has been confirmed regarding the father's response to pregnancy. Increasingly, men are insisting upon their rights to participation in the experience of pregnancy, childbirth, and parenthood, while women are in many instances demanding their rights to be recognized and rewarded for accomplishments other than those related to pregnancy and parenthood. Roles are being realigned, and in some instances, reversed from the traditional models. Such role reversal has been the subject of film and television programs in which the care-giver role of the father has been presented.

Much is presumed and little actually known about the developmental nature of pregnancy. What is known supports the idea that pregnancy is truly a period of rapid change, when compared with the total life span, and that it is a period of identifiable growth, particularly in the woman, and perhaps in the man. Pregnancy is a period during which the individual, and to some extent the family, is prepared for the event of labor, and provided the opportunity for gaining the maturity necessary to assume parenting tasks and roles with some degree of success. Failure to achieve the developmental milestones presented by pregnancy is probably the basis for later difficulties in the parenting role, and is a contributing factor to child abuse and neglect. There is also evidence that appropriate parenting behavior facilitates healthy development of the child, whereas parenting aberrations may actually impede it.

BIRTH

Birth is the culmination of the biopsychosocial efforts made during pregnancy to prepare the woman and fetus for labor. Labor is appropriately named, for there is an intense expenditure of emotional and physical energy.

Labor can be defined as a series of processes through which cervical effacement (thinning out) and dilatation (opening) occur and during which the fetus descends through the birth canal and is delivered, usually through the vaginal introitus. In some instances labor may culminate not in a normal delivery but with the use of operative techniques such as cesarian section or vaginal delivery aided by instruments such as forceps.

Readiness for Labor

Labor is possible because of a number of factors that have occurred before and during pregnancy. The ability of the female bony pelvis to accommodate passage of the infant is one such factor. Maternal pelvic size is established before

conception and is dependent upon the individual's body build and development. The joints between the pelvic bones contain a cartilage-like tissue that is softened somewhat by hormones during pregnancy; this provides for minimal give, if not true expansion, during labor, increasing the pelvic dimensions slightly. For the most part pelvic size is determined more by genetic and other factors that are established before the pregnancy.

Preparation of the uterine myometrium for forceful contractions during labor is an essential prerequisite for labor since the uterine muscle is the primary force for dilatation, effacement, and fetal descent during labor. The uterine muscles contract somewhat during pregnancy but these contractions are usually not discernible by the pregnant woman until the latter months of pregnancy. However, the presence of earlier contractions of the myometrium has been confirmed with electronic monitoring techniques. Contractions during pregnancy are believed to help prepare the uterus for labor by causing some cervical changes (such as cervical softening; effacement, and some dilatation) before the clinical onset of labor. The contractions are much more forceful during labor. The uterine myometrium, being a smooth muscle, is capable of great contractility.

All maternal systems become prepared to cope with the events of labor. More demands are placed on certain maternal systems than others during labor, yet all systems are affected in some manner and to some extent. For example, it is known that labor necessitates an increased cardiac work load for the mother, and certainly a great expenditure of physical energy. The muscles become more elastic and ready for work; the soft tissues, particularly of the pelvic and perineum, are also more elastic and ready to give with the birth of the baby; the respiratory system compensates for its displacement by the gravid uterus with alterations that enable the mother to tolerate the period of labor without becoming oxygen-deficient; blood clotting factors are altered in a way that will help prevent hemorrhage during and after labor; and there is apparently an increase in maternal energy level just before the onset of labor as if to prepare the gravida for work.

There is a psychological preparation for labor as well. Women appear to become quieter and more in control of themselves as labor nears. They seem to be more task-oriented, as evidenced by efforts to set their homes in order, to be certain that all is ready for the infant. Although many women are anxious and even frightened to some extent, there is a prevailing attitude of "let's get on with this." There is an element of truth in the comic portrayal, in films and books, of the pregnant woman in early labor getting everything together and then waking her blundering husband.

If the expectant mother (parents) have attended childbirth preparation classes, in which they have learned about labor and how to help themselves during labor, they are often eager to try out the new techniques learned. Their fears are often decreased and they view labor as a challenge with a delightful outcome, rather than as an ordeal to be endured. Other expectant parents may have read extensively, or discussed labor with their nurse or physician rather than attending formal classes. There is no question that preparation for labor is desired for expectant couples, and helps them cope with the demands of the childbirth experience.

The fetus too gives evidence of readiness for labor. The fetus aligns with the maternal pelvis is preparation for descent through this passage. In some instances, particularly with first pregnancies, the beginning of the descent (engagement of the fetal presenting part) is usually apparent at about 4 weeks before the onset of labor. The fetus becomes quieter and less active, almost as if in recognition of what is ahead. Energy will be needed for the abrupt adjustment to extrauterine life. Whatever the reason, decreased fetal activity has been noted before and during normal labor and birth.

Fetal maturity also seems to influence labor and its onset. The precise mechanism of the onset of labor is not known, but it is thought that an interaction between fetal and maternal systems promotes the event of labor. Labor rarely occurs if the fetus is immature, unless there is a pathological cause. Premature labor does occur, but it is an exception to the usual picture.

Clinical Course of Labor

Traditionally, labor has been divided into three stages. The first stage of labor begins with identifiable cervical changes, essentially dilatation, and ends with complete dilatation and effacement of the cervix (Fig. 4-8). The research of Caldyro-Barcia (1960) has shown that cervical changes begin before labor, particularly the effacement (thinning out) of the cervix. Most textbooks continue to define the first stage of labor in terms of the events just described. The second stage, sometimes called the expulsive stage, begins with complete cervical dilatation and ends with the birth of the baby. The third stage starts with the expulsion of the baby and is completed with the delivery of the placenta and membranes (the afterbirth). Friedman (1978) introduced another way of looking at labor, in which the stages are seen as functional divisions rather than sequential events: in the *preparatory division*, contractions become coordinated and the cervix is prepared for intense activity; in the *dilatational division*, cervical dilatation is accelerated; and in the *pelvic division*, essentially the period of fetal descent during which the pelvis is negotiated, the fetus is able to descend and be born.

Since each division is named for its function, there is little question about what happens when the client is in one of these stages of labor. Some analogies between Friedman's classifications and traditional stages can be made. The preparatory division, like the latent or early part of the first stage of labor, is longer than the other

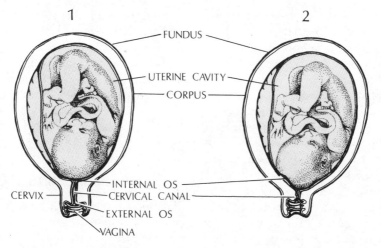

1   FUNDUS   **2**

UTERINE CAVITY

CORPUS

INTERNAL OS
CERVICAL CANAL
EXTERNAL OS
VAGINA

CERVIX

No effacement or dilatation   **Partial effacement, No dilatation**

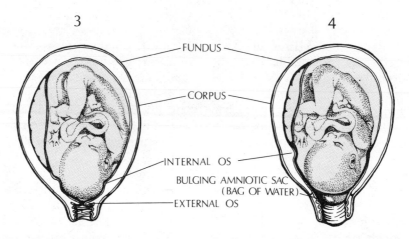

3   FUNDUS   **4**

CORPUS

INTERNAL OS
BULGING AMNIOTIC SAC
(BAG OF WATER)
EXTERNAL OS

Complete effacement, No dilatation   **Complete effacement and dilatation**

Figure 4-8. *Cervical Changes during Labor*

stages. Contractions are felt, but are not usually perceived to be very painful. There is a longer period of time between contractions, during which the mother can relax and recover. The dilatational division corresponds to the last part of the first stage of labor, the active phase, during which contractions are stronger, more frequent, and often more painful. The pelvic division overlaps to some extent the traditional first and second stages of labor since there is

some descent during these stages, and, in a first pregnancy, even some descent before the onset of labor. The pelvic division is most analogous to the third stage.

Forces of Labor

There are two major forces during labor. The primary force, uterine contractions, is active in all three stages of labor, and following labor;

a secondary force is that of the abdominal muscles, which come into use during the second stage of labor to work in concert with the uterine contractions to expel the fetus. The uterine contractions are involuntary, while the abdominal contractions are voluntary. There are factors that tend to counteract the forces of labor and that are considered by some to be negative forces; the bony pelvis and the pelvic floor muscles sometimes act to impede the efforts made by uterine and abdominal forces to bring about descent and expulsion of the baby.

Uterine contractions are intermittent rather than sustained. This enables the mother to rest and relax between contractions, and allows the fetus to receive enough blood through the uterine vessels. Rarely does a contraction completely decrease circulating blood to the fetus, but there is some decrease. As labor progresses, uterine contractions tend to increase in duration and intensity. This accounts for the increase in cervical activity during the dilatational phase of Friedman. A typical uterine contraction in early labor might be described as being 30 to 40 seconds in duration, of mild to moderate intensity, with intervals of 5 minutes between contractions. In active labor, a characteristic uterine contraction pattern is about one contraction every 2 to 3 minutes, of strong intensity, and lasting 50 to 60 seconds. The contraction pattern varies among women, and at different stages in labor. It is one of the best indices of uterine activity and labor efficiency. Electronic uterine and fetal heart rate monitoring have provided much information about the characteristics of contractions during labor, and about fetal response to contractions.

Contractions bring about the uterine cervical changes of effacement and dilatation mentioned earlier, but they also account for changes in the uterus itself. At the onset, the muscular top part of the uterus (the fundus), the body of the uterus (the corpus), and the lower uterine segment are fairly thick. As labor progresses, the fundal portion becomes thicker as the myometrial cells overlap, giving the appearance of thicker, shorter cells, and the lower uterine segment thins out. The tissue of the cervix and lower uterine segment retracts, opening the way for the fetus to descend. This gives rise to two distinct sections of the uterus.

Cervical changes are accompanied by the rupture of small blood vessels in the cervix, causing some bleeding, which mixes with the mucous plug that has sealed off the uterus during pregnancy, and with cervical and vaginal secretions contributing to a pink- to red-tinged vaginal discharge during labor called bloody show. This sometimes frightens the woman and her family since they may associate this discharge with hemorrhage, when in fact it is a normal characteristic of labor. It increases in amount as cervical changes increase during labor.

The efficiency with which the uterus contracts during labor seems to be influenced by:

1. The strength of cervical resistance. Less uterine work is needed to dilate the cervix of a woman who has been through labor and delivered a child previously.
2. The strength of resistance of the pelvic musculature and the lower uterine segment. The upper part of the uterus is the most contractile part, with the lower uterine segment assuming a somewhat passive role. In situations where laboring women are frightened and unable to relax, there is increased muscle tone throughout the body, which is of particular significance when the pelvic muscles and lower uterine segment resist the efforts of the uterine fundus to promote descent of the fetus.
3. The rupture of the bag of waters, which allows the fetal presenting part, usually the head, to press against the cervix, aiding in its dilatation.
4. The position the woman assumes during labor. Lying on her side or walking seems to provide for greater uterine contractile ef-

ficiency during labor than lying in a supine position (on the back).

Delivery

Normal birth is possible if the fetus can pass through the mother's birth canal. This passageway consists of the lower half of the maternal bony pelvis, which roughly resembles a cylinder covered with soft tissue, and the mother's vagina, which has corrugated folds of tissue (rugae) that make it capable of enlarging to accommodate the baby. The ease with which the fetus negotiates this passageway is greatly dependent upon the compatibility of the fetal presenting part (usually the head), the dimensions of the maternal pelvis, and the resistance posed by the pelvic floor musculature, as well as the forces of labor described earlier.

The bones of the fetal head are not fused together as in the adult and are capable of overlapping to a slight degree, decreasing fetal head size. The maternal pelvic joints give slightly and the soft tissues of the pelvis, the uterine muscles, and the vagina have increased vascularity and elasticity, so that they can stretch.

The forces of labor serve as a catalyst for moving the fetus through the birth canal. The pelvic shape requires that the fetus undergo a series of movements called internal rotation to better align the smaller diameters of the presenting part with the pelvic diameters that are large enough to allow passage. Once the fetus reaches the vaginal opening, and the larger part of the body, the head, is delivered, the rest of the fetal body follows easily and the baby is born.

Occasionally, all factors do not combine in such a way that birth is free of problems. At times alternatives to a normal birth are necessary, although most babies can be born in the normal way. In modern obstetric practice, most women are not allowed to labor in the second stage beyond the limits of this period (30 min-

utes to 2 hours) because of increased likelihood of harm to the infant and sometimes the mother.

Shortly after birth of the baby, the placenta and membranes are delivered and birth is complete. This third stage of labor is over in 15 to 30 minutes.

Response to Labor

Responses to labor vary among women. Some responses, however, are characteristic of most women. Uterine contractions may cause some degree of discomfort for the woman, although perception of discomfort and pain are highly personal. The extent of pain during labor arising from exclusively physiological origins is not known. However, discomfort during labor is believed to be of emotional as well as physiological origin, and supportive efforts directed at helping the pregnant woman cope with pain and discomfort during labor tend to decrease her perception of discomfort. Professionals working with pregnant women during labor have observed that knowledge, support, and the use of relaxation techniques tend to help the woman to maintain control of this discomfort, as well as to decrease the discomfort felt.

Fatigue is another maternal response to labor. The first stage, according to Friedman's (1978) study, of primigravid and multiparous labor is approximately 14 and 8 hours respectively, with the second and third stages adding another 30 minutes to 2 hours. Variations of this average length of labor are common. During labor, whatever its length, the woman's body and mind are constantly active. Uterine contractions and maternal efforts to aid in the birth of the child are energy-consuming. Discomfort and pain, if present, and fear, anxiety, and the necessary amount of concentration also consume energy. It is not uncommon for a woman to experience such excitement following the birth of her child that this fatigue may not be immediately apparent to her or those around her. It will tell later in her recovery period, and may influence how she assumes the parenting tasks for her newly born child.

A woman's emotional response to labor is often related to her expectations of herself during labor. Some women are placid in their manner; others respond loudly, releasing tension in this fashion; some are stoic. As labor progresses, a woman tends to exhibit signs of intense concentration, and may be relatively unaware of others around her. Her comments are often brief, and sometimes cryptic, or abrupt, giving the appearance of anger. Expectant fathers who are not prepared for this behavioral change may take the woman's comments (or her silence) personally.

A woman may experience a feeling of loss with the separation of the baby from her body. If the tasks of pregnancy have been accomplished, she will likely be ready to accept this separation; if she has not reached this point of development, she may have problems accepting this loss of the symbiotic tie between herself and her child.

The fetus also responds to labor. The contractions occurring in the uterus may decrease his blood supply, with its essential oxygen, intermittently. Electronic fetal heart monitoring has shown how the fetal heart rate varies with the intensity and duration of contractions. The longer and stronger the contraction, the more effect on the fetal heart pattern, first causing it to increase to compensate for decreased oxygen, and then to decrease if the interruption is sustained. The length and qualities of the labor affect the fetus. The longer, more difficult labors tend to affect the fetus adversely, as do the short, precipitous labors. The normal labor patterns do not seem to affect the fetus detrimentally, provided the status of mother and fetus at birth is a healthy one. The immature fetus, or the fetus whose well-being is in jeopardy during pregnancy, will often have a difficult time during labor, and may not survive the stresses associated with labor and the adjustment to extrauterine life.

Following birth the newborn must make an abrupt adjustment to his extrauterine environment. This adjustment includes physiological adaptation, (cardiovascular and respiratory adjustment being the most crucial) as well as emotional and social adjustment. Much is known about the physiological adaptation of the newborn to life outside the uterus; more is beginning to be known about psychosocial adaptation.

The family, except for the expectant father, may have been on the periphery of activities during labor. Some extended family members may not even be present at labor. Siblings are often not allowed by institutions to participate in the event. The family's role is likely to be one of waiting to find out what the outcome is. Family response to labor and its effect on the family as a developmental event has not been studied extensively enough to allow a knowledge of what effects labor and birth have. Cultural values and mores certainly will be prominent factors in any study of the laboring family.

Recovery

Recovery from childbirth is affected by a number of factors. The health of the mother during pregnancy and labor is of great importance. For the most part the physiological changes that occur during pregnancy reverse themselves (Table 4-5). This occurs over a period of about 4 to 6 weeks. Often this period of recovery, the puerperium or postpartal period, is essentially ignored as being an insignificant part of pregnancy and birth; it is all too often assumed to simply "happen." It is important to recognize that, in most cultures, this recovery period occurs simultaneously with the mother's assumption of both care-giving tasks and the role of mother. Other family members must also face the reality of role change during this period,

Table 4-5. Maternal Biological Readjustment After Delivery

System or Structure	Adjustmment	Significance
Uterus:	*Changes in Size, Weight, and Position:* Decreases in weight from approximately 1,000 gm immediately after delivery to 500 gm one week later, to 300 gm within two weeks; returning to its nonpregnant weight of approximately 40 to 60 gm within six weeks.	Size, position, and degree of descent of the uterus into pelvis can be determined by abdominal palpation and serve as indicators of progress in uterine involution, as well as involutional problems.
	Size after delivery has been compared to that of a large grapefruit. Intrauterine walls after delivery lie close to each other, decreasing markedly the size of the cavity. Walls are about 4 to 5 cm thick. Nonpregnant uterine size is regained in about 4 to 6 weeks.	
	Following delivery, uterus can be felt in the lower abonmen, midway between symphysis pubis and umbilicus. Descends into pelvic cavity until it can no longer be palpated abdominally by the end of two weeks.	
	Myometrium usually remains in a contractile state after delivery. Because of the significant cell hypertrophy that occurs during pregnancy, these larger cells must rid themselves of excess cytoplasm (including protein). The precise mechanism by which this occurs is not known as of this writing.	Uterine atony is abnormal and is a major cause of postpartal hemorrhage. The uterine myometrium responds to massage by contracting, providing a simple clinical approach to the control of excess bleeding. Allows for the uterus to return to its nonpregnant size.
	Uterine Decidua (Endometrium): Spongy layer of the decidua is discarded with the delivery of the placenta and membranes.	Regeneration of endometrial tissue occurs rapidly except in the area of placental attachment.
	Placental site is an open wound that requires about 6 weeks to heal completely.	Improper healing may lead to delayed hemorrhage.
	Size of site decreases to about 3 to 4 cm in diameter by end of second week.	Placental site susceptible to infection.
Cervix	Cervix is flabby after delivery; regains its tone by the end of involutional period (6 weeks).	
	May have small lacerations, especially in the area of the external os.	May be the site of bleeding if cervical lacerations are not detected and repaired following delivery.
	External os does not return to its prepregnant (round) appearance but tends to be wider.	Shape of os identifies cervix as multiparous.

Table 4-5. (*Continued*)

System or Structure	Adjustmment	Significance
Vagina	Rugae are absent after delivery and do not begin to return until about the third postpartum week; they do not return completely to the nulliparous state.	
Abdominal and Pelvic Muscles and Ligaments	Distention of abdominal muscles to accommodate the gravid uterus and rupture of elastic fibers of the skin tends to cause the abdominal wall to remain flabby for a time.	
	There may also be a separation (diastasis) between the two recti muscles.	
	The pelvic musculature may be stretched and remain lax for a peroid of time.	Contributes to weakened support of pelvic organs.
Cardiavascular	*Blood*: Blood volume (essentially plasma) decreases. Hematocrit increases by the seventh day postpartum probably because of decreased blood volume with only a moderate loss of red blood cells at delivery. Nongravid red-blood-cell-volume is thought to be regained within 60 days following delivery.	Most pregnant women with normal hearts can adapt without difficulty to the stresses pregnancy imposes upon them and can adjust to the nonpregnant state, following delivery, with relative ease.
	Blood Volume: Decreases significantly after delivery. Usually reaches normal values by end of third week postpartum.	
	Cardiac Output: Marked changes in cardiac output are believed to occur during and after labor. Most researchers agree that the cardiac output rises significantly immediately after delivery.	Monitoring of cardiac output permits assessment of the cardiovascular status of women after labor; special attention should be given to women completing a difficult labor or twin gestation, or having cardiac disease.
Renal	Kidney function generally returns to normal levels by one month postpartum.	
	Water retention occurring during pregnancy seems to reverse rapidly after delivery, the greater amount of fluid being lost within the first two weeks postpartum and the remainder excreted by the sixth week postpartum.	Postpartal water loss (dieuresis) necessitates close observation of urinary output.
	Dilated ureters and renal pelves usually return to normal within one month.	Increased chance for urinary tract infection because of dilated ureters and

Table 4–5. (Continued)

System or Structure	Adjustmment	Significance
		renal pelves and desensitization of the bladder to pressure, which may result in overdistention of bladder, incomplete emptying of bladder, increased residual urine
Endocrine	Delivery of placenta results in a decrease in placental steroid and nonsteroid hormones. Increased production of anterior pituitary gonadotropins, prolactin, and oxytocin (unless lactation is suppressed). Increased thyroid activity occurs. (Endocrine adjustment to support lactation is discussed under "Breasts.")	Return to normal endocrine functioning in most women within 4 to 6 weeks postpartum unless breastfeeding. The marked decrease in placental hormones with delivery of the placenta may contribute to emotional instability during early postpartal period. Data inconclusive as of this writing. Ovulation may be delayed in some lactating women, however, breastfeeding is not a reliable contraceptive measure.
Respiratory	Emptying of the gravid uterus allows for the descent of the diaphragm to its normal anatomical level.	Difficulty with breathing, often apparent during pregnancy, is no longer a problem. Respiratory problems aggravated by pregnancy usually dissappear or decrease in severity after delivery.
Gastrointestinal	Hormonal influences on the G.I. tract during pregnancy are reversed, although decreased muscle tone of the bowel may continue for several days after delivery. Appetite is usually present and most women can easily tolerate food after delivery.	Postpartum constipation may be a problem for some women because of decreased muscle tone in combination with fasting, enemas during labor, and fear of pain in connection with episiotomy, hemorrhoids, or lacerations.
Skin	The discolorations of the skin (striae gravidarum, linea nigra, and chloasma) present in some women during pregnancy usually fade or dissappear completely.	In some instances in which these discolorations do not disappear completely, women may express concern about appearance or body image.
Breasts	The principal function the breast must serve after delivery is that of lactation. During pregnancy, estrogen, pregesterone, placental lactogen, and pituitary hormones are believed to work together to prepare the breasts for lactation. With the delivery of the placenta, the withdrawal of estrogen and	Breasts provide a natural food supply for the newborn infant. Potential site for infection, especially in breast-feeding women exercising inadequate or poor breast care or improper suckling of baby causing lesions in the nipple and areolar area.

Table 4-5. (*Continued*)

System or Structure	Adjustmment	Significance
	progesterone probably initiates lactation. Colostrum (a yellow fluid containing some nutrients) precedes the production of milk (usually during days 1–3 postpartum). Suckling of the infant seems to stimulate the secretion of milk by the breasts. Sucking of the infant also seems to stimulate the posterior pituitary (neurohypophysis) to secrete oxytocin, which causes the breast to release (let down) milk. In most women who have not received drugs to suppress lactogenic activity, lactation will occur and milk will be present in the breasts by the third to fifth post-delivery day. Venous and lymhatic engorgement may occur by the third or fourth day postpartum.	Usually accompanied by discomfort, feeling of breast fullness and tenderness.

SOURCE: Adapted from Butnarescu, G. F., *Perinatal Nursing, Vol. 1: Reproductive Health.* New York: John Wiley & Sons, 1978, p. 170–173. Used with permission.

when the mother may or may not feel well enough to cope with this change in a positive way. Developmentally, the woman has been prepared through pregnancy for this biological role, and, optimally, for its psychosocial aspects. The father and the other family members have had less intense involvement with the pregnancy and may have some difficulty with role change during this period.

This recovery period is one of constant adaptation for the mother, the father, newborn, and other family members. It is a period of emotional highs and lows, a critical period for parent-infant relationships, and in some instances a vulnerable period for the family generally. It completes the reproductive experience and begins a new cycle of life.

ANTICIPATORY GUIDANCE

Anticipatory guidance, the preparation of an individual or family for an event before it happens, is an integral part of quality health care and an essential ingredient for healthy reproduction and parenting. Many people face their reproductive years with little understanding of reproduction; some harbor gross misconceptions. Human sexuality and pregnancy have long been talked about in whispers; beliefs have been passed on from parent to child and from peer to peer, and only recently, amid much discussion and controversy, has sex education begun to be taught in the public school system. In this era of sexual awareness and liberalism it is surprising that so little is understood about sexual intercourse and pregnancy. Similarly, parenthood is often viewed as merely instinctive, not as involving learned activities.

Where to begin is an unresolved issue, an unanswered question. Perhaps the most logical long-term starting point is with the pregnant couple, preparing them to pass on more accurate information and positive attitudes about sex to their children. There is a more immediate need, however, to better prepare young people approaching their reproductive years to understand and control their own reproductive destiny.

Increasing numbers of young adolescent and even preadolescent boys and girls are having sexual intercourse. With the legal availability of abortion, the precise number of pregnancies occurring in these age groups is not known since some pregnancies are terminated early through elective abortion. This itself raises another issue, that of abortion trauma.

Pregnancy occurring during early adolescence is considered to be a high-risk condition for mother and fetus. A lack of prenatal care among pregnant adolescents is believed to be a major factor. Also, the readiness of an adolescent to assume a parenting role is questionable. Adolescent pregnancies are often unplanned, and while an unplanned pregnancy is not necessarily an unwanted pregnancy, the young parent often faces the developmental tasks of adolescence, pregnancy, marriage, and parenthood all at the same time.

That anticipatory guidance should be given is not the question. Rather, how can it be provided? By whom? How much? When? The physician, the nurse, the social worker, the teacher, the parent, and other responsible adults should all be interested in ways of better preparing persons for healthy reproduction through the promotion not only of healthy physical development, but of healthy psychosocial and cognitive development as well. It is equally important that an effort be made to reach parents who already have children, by means of educational programs to help them with problems in child rearing. Such programs should focus on sexuality and reproduction, along with child discipline and nutrition, and growth and development generally. Some communities have already provided excellent services to meet this need; their funds are often limited as are their material and human resources. Parents frequently do not know about these services. Community groups such as religious organizations and parent groups have done much to publicize and develop such programs. Professional health workers are increasingly being called upon to participate or serve as consultants to them. Public and private funding has made possible some programs for parents, but programs that deal with aspects of parenting beyond pregnancy and childbirth are limited in number and often in scope.

Teachers, nurses, physicians, and others involved with children are key candidates to assume responsibilities for anticipatory guidance. However, teachers are often not prepared through their teacher education program to teach classes in sex and family life education. It would be logical to involve the school nurse in such educational programs, but she is often not available, not interested, or not allowed (because of teacher role definition or certification requirements) to assume this responsibility. Increasingly, collaborative efforts are being made by teaching professionals, health professionals, and parents to fulfill the need for sex education programs, both within and outside of formal educational settings.

Whatever the age of the child, and certainly as the child approaches puberty, a major aspect of anticipatory guidance is the assessment of the child's understanding of human reproduction. What beliefs and misunderstandings does the child hold? When is the child ready to acquire a knowledge of sex and reproduction? Where should data collected through assessment strategies be maintained? How can it be made available for retrieval when needed? Who should have access to this information? There questions are all without clear answers. Information related to sexual understanding and activity is personal, intimate information; the child's privacy must be maintained. Should such a record be continual in the way that developmental records kept by a pediatrician or family physician are? No approach to the assessment of reproductive readiness can be suggested that will suit each person and family in every situation, but some specific aspects to be assessed are suggested in Figure 4-9.

To discuss sexual activity, pregnancy, and childbirth in a nonthreatening, objective way requires that the individual conducting such a discussion have a mature understanding of her own sexuality, and, further, that she have the knowledge and skill to teach others. Jensen, Benson, and Bobak (1977) suggested three factors that influence the ability of a nurse to deal with sexually oriented content:

1. Comfort with one's own sexuality; This is necessary if the nurse is to accept the sexuality of individuals seeking or needing help.
2. Awareness of one's own attitudes and prejudices about sexuality, and an awareness of sexual myths and probias commonly held by others.
3. Ability to discuss sexual matters in language understood by the client.

To these three requirements might be added several others:

1. An understanding of reproductive anatomy and physiology and a comprehensive knowledge of the psychosociology of sexual activity, pregnancy, and childbirth.
2. Knowledge about human growth and development, as well as about stress and crisis; skill in implementing the strategies of crisis intervention, and a knowledge of the circumstances under which it is appropriate to refer the client to other sources.

3. Adequate preparation as a teacher, including not only an understanding of the topics relevant to reproduction, but also a knowledge of what constitutes teacher effectiveness.
4. Knowledge of and skill in interviewing techniques.
5. Continuity of contact with individuals and families. A lack of continuity is often seen in the client's contacts within the health care delivery system. Even after a pregnancy is confirmed, sustained contact with the same health care provider may not occur. Clients using a private physician are not always assured that that physician will deliver them; it may be his partner or yet another physician. Clients using public health care facilities have little control over continuity of health care providers and the information that they impart.

Some of the best efforts in implementing the concept of anticipatory guidance have been seen in the emergence of childbirth preparation programs, which have come into being through the efforts of consumers and professionals alike. Most of the programs emphasize preparation for coping with pregnancy, particularly labor and delivery. Some include selected aspects of preparation for parenthood. A variety of philosophies underlie these programs. The major types of childbirth preparation concepts are presented in Table 4-6.

The post-World-War-II era ushered in heightened consumer interest in health care, particularly with regard to pregnancy and childbirth. This interest came hand in hand with increased open-mindedness about sexuality, pregnancy, and childbirth. This motivation to learn and to be involved in one's own pregnancy constitutes a major support for the success of educational programs preparing families for pregnancy, birth, and parenthood. The time is ripe for the professional health care provider to capitalize on this interest and motivation.

Despite the great need for reproductive health

Individual:

General Impressions:

   Appearance: Description of size, posture, dress, personal hygiene.

   Attitude: Posture, facial expression, presence of absence of eye contact.

   Relationship with others: Did she come to this appointment alone or accompanied? How does she respond to the nurse and/or other health care workers?

Family:

Organization and Composition: Number of family members, sex and age of siblings, identification of significant other, boyfriend, husband, girlfriend, etc.

History of conditions, diseases, genetic disorders, or accidents that may affect reproduction.

Cultural aspects: Ethnic background, religion, cultural practices that may differ from those frequently seen by nurse or different from nurse's background.

Family role relationships and family style: democratic vs. autocratic. Who makes decisions? Activities in which family members participate cooperatively. Sources of affection and support. Family goals.

Living accomodations: Privacy for all family members? Safe environment? Sanitation? Provisions for heat and refrigeration?

Economic status: Source of income and extent to which it supports family life-style and can support a new addition.

Physical condition:

Present diseases or disorders.

Growth and developmental status as compared with established norms.

Nutritional status: weight; eating patterns; elimination patterns; food preferences.

Obstetric history (if applicable): course of previous pregnancy and its outcome.

Birth control method if used, and way in which it was used.

Age at which menarche occurred and characteristics of menstruation (interval, duration, discomforts)

Sexual practices: Intercourse, frequency with one or several partners.

Psychosocial condition:

Response to stress; coping style and use of supportive others.

Perception of family relationships, particularly own role within family; interaction with and expectations of other family members.

Educational level.

Own perception of needs for information.

Ways in which individual learns best.

Resources used for learning: books, newspapers, magazines, teachers, broadcast media, parents, peers, husband or partner.

Ways in which individual communicates with others.

With whom does individual communicate most easily and effectively?

Language level of individual: Is it compatible with growth and developmental expectations for age; does individual speak more than one language; if so, which language, and level of fluency if it is the first language learned.

Reasons for becoming pregnant as perceived by the individual and partner.

Figure 4-9. Assessment of Reproductive Readiness (Adapted from: Butnarescu, G. F. *Perinatal Nursing, Vol. 1: Reproductive Health*, New York: John Wiley & Sons. 1978. p. 263–64.

Table 4-6. Approaches to Preparation for Childbirth

Concept	Focus
Psychophysical (Dick-Read)	Emphasizes education of the pregnant woman to decrease her fear and promote relaxation in order to decrease pain by breaking the fear-tension-pain cycle. Pregnancy and childbirth are seen as normal physiological processes that should not be accompanied by pain. No clearly prescribed series of exercises is identified with this method, although a variety of breathing techniques and exercises are used to promote relaxation. Promotes passive rather than active participation of the pregnant woman generally, although she does actively use mental and physical techniques during uterine contractions.
Psychoprophylaxis in Childbirth (Lamaze)	Emphasizes reeducation of mind and body in order to accommodate the activities of labor; based on Pavlov's theory of conditioned response. The goal of psychoprophylaxis in childbirth is to promote a woman's control over the functions of her mind and body; by controlling these activities, she is able to control pain during childbirth. The techinques advocated require: increased emotional and educative support by a known monitor or coach; increased active participation of pregnant woman in order to decrease or prevent perception of pain; reeducation of body and mind to respond positively rather than negatively to painful stimuli; recognition of the physiological basis for the prescribed relaxation techniques and breathing styles; use of clearly prescribed relaxation techniques and breathing styles during labor, with the presence of a coach for reinforcement; practice of some physical conditioning exercises during pregnancy; practice of relaxation techniques and breathing styles with coach before labor.
Psychosexual (Kitzinger)	Focuses on the use of the sensory memory. Incorporates sexuality as a broad concept that includes family relationships, childbirth, infant care-giving, pregnancy as crisis, and self. Emphasizes tactile and verbal support, usually by husband. Extensive use of touch. Uses contrast of muscle contraction in order to promote relaxation. Advocates conscious use of pelvic floor muscles to facilitate descent of infant.
Communications Theme or Framework (Edwards)	Focuses on the use of a communicational theme or framework. Emphasizes the use of increased education to promote better understanding of own motivations for behavior. Uses the concepts of structural analysis, transactional analysis, games, and scripts in examining the interactions between the pregnant woman, her family, and the health professionals. Promotes control of self and choice of actions. Emphasizes equality of the consumer-helper relationship.

Table 4-6. (*Continued*)

Concept	Focus
	Encourages the use of the husband or partner as a client advocate during labor and delivery.
	Promotes collaboration between the expectant parents.
	Provides guidelines for coping with the institutional delivery situation.
	Encourages the use of "strokes" for reinforcement and reward for behavior.
	Incorporates the use of information giving, comfort skills, body relaxation, breathing techniques, and teacher-learner interaction in the classroom setting.

SOURCE: Adapted from Butnarescu, G. F., *Perinatal Nursing, Vol. 1: Reproductive Health*. New York: John Wiley & Sons, 1978, p. 250. Used with permission.

education, there is a remarkable absence of professional personnel interested in or actively participating in the development of educational services to meet these needs. Although most American women receive quality health care during pregnancy, as is reflected in decreasing maternal and infant mortality rates, not enough attention is given to the questions raised by pregnant families about what is happening to them, or will yet happen. One pregnant woman expressed it as, "I didn't know what I didn't know, so how could I ask?"

Some aspects of pregnancy, birth, and parenthood in which anticipatory guidance is an appropriate approach are:

Before Conception:

Biology of human sexuality, pregnancy, and birth.

Family life education with particular attention to parental responsibility.

Family planning, including contraception and abortion.

Personal hygiene relative to sexuality and reproduction.

Human growth and development, including nutritional demands and the influence of nutritional status on pregnancy outcome.

After Conception:

Biology and psychosociology of pregnancy.

Sexual changes during pregnancy.

Preparation for labor and delivery.

Preparation for assumption of the parenting role, including role changes, fetal/neonatal development and demands, and infant caregiving tasks.

After Birth:

Biological and psychosocial adjustments.

Periods of stress in the early parenting period.

Growth and development of children, particularly the newborn.

Family planning.

Health maintenance for the family, with particular attention to the mother and child.

Preparation for menopause.

Guidance in the areas listed above needs to be made available to all families and individuals. This goal is not likely to be achieved easily, quickly, or without great expenditure of economic as well as human resources. As mentioned earlier, many efforts are already underway. Far too many young people reach their childbearing years with an inadequate understanding of human reproduction, and, all too often, families complete pregnancy and childbirth with little knowledge beyond that with which they entered this period of life, and with even more confusion and fear concerning it. This is then passed on to their children as well as to other family members and friends, perpetuating a cycle of ignorance and fear. How much better to replace this cycle with one of knowledge, understanding, and control that will better enable individuals and families to make knowledgeable decisions about their own reproductive health and the health of their children.

SUMMARY

Conception, pregnancy, childbirth, and early parenthood are developmental milestones in the lives of the individual and the family. The degree of success in mastering tasks of the childbearing period influences the parents' success in nurturing and caring for their child. It may also affect subsequent developmental periods in their lives.

Readiness for pregnancy and parenthood is dependent upon adequate physical, psychological, and social development that has prepared the female and her mate for dealing with the requirements of this developmental period. The nurse has the capability for assuming responsibility for the assessment of reproductive readiness.

The nurse as a professional health care provider is an contact with individuals and their families experiencing pregnancy and early parenthood and is therefore in an excellent position to play a major role in promoting healthy human development during these periods through the use of many educative, therapeutic, and supportive strategies. Foremost among these is the strategy of anticipatory guidance, which prepares the individual and family to meet the tasks of pregnancy, childbearing, and

child rearing. This preparation is essential for promoting and perpetuating a healthy reproductive cycle built on knowledge rather than on fear and ignorance.

REFERENCES

Ainsworth, M. D. S. The development of infant-mother attachment. In B. Caldwell, and H. Riccuiti (Eds.), *Review of child development research: child development and social policy.* Chicago: University of Chicago Press, 1973.

Avery, G. B. *Neonatology.* Philadelphia: J. B. Lippincott, 1975.

Bardwick, J. M. *Psychology of women: a study of bio-cultural conflicts.* New York: Harper & Row, 1971.

Bell, R. O. Contributions of human infant to caregiving and social interaction. In M. Lewis, and L. A. Rosenblum (Eds.), *The effect of the infant on its caregiver.* New York: John Wiley & Sons, 1974.

Benedek, T. The psychosomatic implications of the primary unit: mother-child relatedness. *American Journal of Orthopsychiatry,* 1949, *19,* 642.

Berelson, B., and Steiner, G. A. *Human behavior: an inventory of scientific findings.* New York: Harcourt, Brace & World, 1964.

Bergner, L., and Surser, M. W. Low birth weight and prenatal nutrition: an interpretive review. *Pediatrics,* 1970, *46,* 946.

Bibring, G. Some considerations of the psychological processes in pregnancy. *Psychoanalytic Study of the Child,* 1959, *14,* 113.

——. study of the psychological process in pregnancy and of the earliest mother-child relationship. *Psychoanalytic Study of the Child, 14,* 1961.

Billingham, R. E. Transplantation immunity and the maternal-fetal relation, *New England Journal of Medicine,* 1964, *270,* 720.

Bowlby, J. *Maternal care & mental health,* (2nd ed.) Monograph Series #2. Geneva: World Health Organization, 1952.

——. The nature of the child's tie to his mother. *International Journal of Psychoanalysis.* 1958, *39,* 350.

——. *Attachment and loss. Vol. I, Attachment.* New York: Basic Books, 1969.

Branfenbrenner, U. *The ecology of human development—experiments by nature and design.* Cambridge, Mass: Harvard University Press, 1979.

Brazelton, T. B. Neonatal behavioral assessment scale.

Clinics in developmental medicine, (No. 50). London: Spastic's International Medical Publications, 1973.

Brazelton, T. B., Koslowski, B. and Main, M., The origins of reciprocity in mother-infant interaction. In M. Lewis, and L. A. Rosenblum (Eds.) *The effect of the infant on its caregiver.* New York: John Wiley & Sons, Inc., 1974.

Brazelton, T. B. et al. Early mother-infant reciprocity. In *Parent-infant interaction.* Ciba Foundation Symposium. Amsterdam: Elsevier Publishing Co., 1975.

Brody, S. *Patterns of mothering: maternal influence during infancy.* New York: International University Press, 1956.

Butnarescu, G. F. *Perinatal nursing, Vol. 1—reproductive health.* New York: John Wiley & Sons, 1978.

———. *Perinatal nursing, Vol. 2—reproductive risk.* New York: John Wiley, 1980.

Caldyro-Barcia, R., and Posiero, J. J. Physiology of the uterine contraction. *Clinical Obstetrics and Gynecology* 1960, *3,* 386.

Caplan, G. *An approach to community mental health.* New York: Grune & Stratton, 1961.

———. Psychological aspects of pregnancy. In H. L. Lief et al. (Eds.), *The psychological basis of medical practice.* New York: Harper & Row, 1973.

Cibils, L. A., and Hendricks, C. H. Normal labor in a vertex presentation. *American Journal of Obstetrics Gynecology* 1965, *91,* 385.

Clark, A. L. Overview of crisis intervention with application to childbearing. In *Parent-child relationships role of the nurse.* New Brunswick, N.J.: Rutgers University, November 1968, p. 82.

Clark, A. L., and Affonso, D. D. *Childbearing: a nursing perspective* (2nd ed). Philadelphia: F. A. Davis, 1979.

Cohen, L. B., and Salapatek, P. (Eds.), *Infant perception: from sensation to cognition.* New York: Academic Press, 1975.

Deutsch, H. *Psychology of women, vol. II: motherhood.* New York: Grune & Stratton, 1945.

Diamond, M. A critical evaluation of the ontogeny of human sexual behavior. *Quarterly Review of biology,* 1965, *40,* 147.

Duvall, E. M., and Hill, R. Report of the committee on the dynamics of family interaction. Washington, D.C.: National Conference on Family Life, 1948.

Duvall, E. M. *Marriage and family development* (5th ed.). Philadelphia: J. B. Lippincott, 1977.

Dwyer, J. M. *Human reproduction—the female system & the neonate.* Philadelphia: F. A. Davis, 1976.

Erikson, F. *Childhood and society.* New York: Norton, 1950.

Erikson, E. Growth and crisis of the 'healthy personality.' In C. Kluckhohn, H. A. Murray, and D. M. Schneider (Eds.), *Personality in nature, society and culture* (2nd ed.). New York: Alfred A. Knopf, 1953.

Frantz, R. L., Fagan, J. F., and Miranda, S. B. Early visual selectivity as a function of pattern variables, previous exposure, age from birth and conception and expected cognitive deficit. In L. B. Cohen, and P. Salapatek (Eds.), *Infant perception: from sensation to cognition.* New York: Academic Press, 1975.

Fraser, F. C. Causes of congenital diseases in human beings." *Journal of Chronic Diseases,* 1959, *10,* 97.

Fraser, F. C., and Nora, J. *Genetics of man.* Philadelphia: Lea & Febiger, 1975.

Freud, S. *Complete psychological works of Sigmund Freud* (Vol. 18), standard ed. (J. Strachey, Ed. and trans). London: Hogarth Press, 1955.

Friedman, D. G. *Human infancy. an evolutionary perspective.* New Jersey: Lawrence Erlbaum Association, 1974.

Gesell, A. *The first five years of life: a guide to the study of the preschool child.* New York: Harper & Row, 1940.

Goren, C. C., Sarty, M., and Wu, P. Y. K. Visual following and pattern discriminations of face-like stimuli by newborn infants. *Pediatrics,* 1975, *56,* 544.

Greenberg, M. and Morris, N. Engrossment: the newborn's impact upon the father. *American Journal of Orthopsychiatry,* 1974, *44,* 520.

Hamilton, V., and Vernon, M. D. (Eds.). *The development of cognitive processes.* New York: Academic Press, 1976.

Harding, P. C. R. Fetal growth & nutrition. In J. W. Goodwin et al. (Eds.), *Perinatal medicine.* Baltimore: Williams & Wilkins, 1976.

Harlow, H. F. The nature of love. *American Psychologist,* 1958, *13,* 673.

Havighurst, R. J. *Human development and education.* New York: Longmans, Green and Co., 1953.

———. *Developmental tasks and education.* New York: David McKay, 1972.

Hendricks, C. H., Brenner, W. E., and Kraus, C. Normal cervical dilatation pattern in late pregnancy and labor. *American Journal of Obstetrics Gynecology,* 1970, *106,* 1065.

Holt, K. S. *Developmental pediatrics perspectives and practice.* London: Butterworths, 1977.

Horsley, S. Psychological management of the prenatal period. In J. G. Howells (Ed.), *Modern perspectives in psycho-obstetrics.* New York: Brunner/Mazel Publishing, 1972.

Howells, J. G. Childbirth is a family experience. In J. G. Howells (Ed.), *Modern perspectives in psycho-obstetrics.* Brunner/Mazel Publishing, 1972.

Huff, R. W., and Pauerstein, C. J. *Human reproduction: physiology and pathophysiology.* New York: John Wiley & Sons, 1979.

Hutt, S. J., and Hutt, C. *Early human development.* London: Oxford University Press, 1973.

Illingworth, R. S. *The development of the infant and young child.* New York: Churchill Livingstone, 1975.

Jensen, M. D., Benson, R. C., and Bobak, I. M. *Maternity care: the nurse and the family.* St. Louis: Mosby, 1977.

Jessner, L., Weigert, E., and Foy, J. L. The development of parental attitudes toward pregnancy. In T. Benedek et al. (Eds.), *Parenthood: its psychology and psychopathology.* Boston: Little, Brown, 1970.

Kaluger, G., and Kaluger, M. F. *Profiles in human development.* St. Louis: Mosby, 1976.

———. *Human development—the span of life.* St. Louis: Mosby, 1979.

Klaus, M., Kennell, J. H., and Plumb, N. Human maternal behavior at the first contact with her young. *Pediatrics.* 1970, *46,* 187.

Klaus, M. H., and Kennell, J. H. *Maternal-infant bonding.* St. Louis: Mosby, 1976.

LeMasters, E. E. Parenthood as crisis. *Marriage and Family living,* 1957, *16:*(4). Reprinted in H. J. Parad (Ed.), *Crisis intervention.* New York: Family Service Association of America, 1965.

Levine, L. *Biology of the gene* (2nd ed.). St. Louis: Mosby, 1973.

Lewis, M., and Rosenblum, L. A. (Eds.) *The effects of the infant on its caregiver.* New York: John Wiley & Sons, 1974.

Lidz, T. *The person: his development through the life span.* New York: Basic Books, 1968.

Lipsitz, J. *Growing up forgotten.* Lexington, Mass: D.C. Heath, 1977.

Manaster, G. J. *Adolescent development and the life tasks.* Boston: Allyn & Bacon, 1977.

Maslow, A. *Motivation and personality* (2nd ed.). New York: Harper & Row, 1970.

Money, J., and Ehrhardt, A. A. Fetal hormones and the brain: effect of sexual demorphism on behavior—a review. *Archives of Sexual Behavior.* 1971, *1,* 241.

Moore, K. L. *The developing human: clinically oriented embryology* (2nd ed.). Philadelphia: W. B. Saunders, 1977.

Newman, B., and Newman, R. *Development through life: a psychological approach.* Homewood, Ill: Dorsey, 1975.

Nilsson, A., Uddenberg, N., and Alongren, P. E. Parental relations and identification in women with special regard to para-natal emotional adjustment. *ACTA Psychiatrica Scandinavica,* 1971, *47,* 57.

Niswander, K. R., and Gordon, M. *The women and their pregnancies.* Philadelphia: W.B. Saunders, 1972.

Nyhan, W., and Sakoti, N. *Genetic and malformation syndromes in clinical medicine.* Chicago: Year Book Medical Publishers, 1976.

Page, E. W., Villee, C. A., and Villee, D. B. *Human reproduction: the core content of obstetrics gynecology and perinatal medicine.* Philadelphia: W.B. Saunders, 1976.

Parke, R. D., and O'Leary, S. Father-mother-infant interaction in the newborn period: some findings, some observations and some unresolved issues. In K. F. Riegel, and J. Meacham (Eds.), *The developing infant in a changing world. Vol. 2, social and environmental issues.* The Hague: Mouton, 1975.

Parton, D. A. Learning to imitate in infancy. *Child Development* 1976, *47,* 14.

Piaget, J. *The construction of reality in the child,* (M. Cook trans.). New York: Basic Books, 1954.

Piaget, J. *The origins of intelligence in children,* (M. Cook trans.). New York: Norton, 1963.

Pritchard, J. A., and MacDonald, P. A. *Williams obstetrics* (15th ed.). New York: Appleton-Century-Crofts, 1981.

Rapoport, L. The state of crisis: some theoretical considerations. *Social Service Review,* 1962, *36* (#2).

Rayner, E. *Human development* (2nd ed.). London: George Allen & Unwin, 1978.

Ribble, M. H. *The rights of infants: early psychological needs and their satisfaction.* New York: Columbia University Press, 1943.

Rich, A. *Of woman born.* New York: Norton, 1976.

Rubin, R. Basic maternal behavior. *Nursing Outlook,* 1961, *9,* 683.

———. Attainment of the maternal rate, Part I, processes. *Nursing Research,* 1967, *16,* 237.

———. Maternal tasks in pregnancy. *Maternal-Child Nursing Journal,* 1975, *4,* 143–153.

———. Binding-in in the post-partum period. *Maternal-Child Nursing Journal,* 1977, *6,* 67.

Shaffner, H. R. Objective observations of personality development in early infancy. *British Journal of Medical Psychology,* 1958, *31,* 174.

Simmons, R. L., and Russell, P. S. The immunologic problem of pregnancy. *American Journal of Obstetrics and Gynecology,* 1963, *85,* 583.

Sontag, L. W. The significance of fetal environmental differences. *American Journal of Obstetrics and Gynecology,* 1941, *42,* 996.

Sontag, L. W. Differences in modifiability of fetal behavior and physiology. *Psychosomatic Medicine,* 1944, *6,* 151.

Spelt, D. The conditioning of the human fetus in utero. *Journal of Experimental Psychology,* 1948, *38,* 338.

Spitz, R. Hospitalization: an inquiry into the genesis of psychiatric conditions in early childhood. *Psychoanalytic Study of the Child,* 1945, *1,* 53.

Spitz, R. *The first year of life: a psychoanalytic study of normal and deviant development of object relations.* New York: International Universities Press, 1965.

Stone, L. J., Smith, H. T., and Murphy, L. B. *The competent infant: research and commentary.* New York: Basic Books, 1973.

Sullivan, H. S. *The interpersonal theory of psychiatry.* Published as *The collected works of H. S. Sullivan*, Vol. I. H. S. Perry and M. L. Gowel (Eds.) New York: Norton, 1953.

Tanner, L. M. Developmental tasks of pregnancy. In B. S. Bergerson et al. (Eds.), *Current concepts in clinical nursing* (Vol. 2). St. Louis: Mosby, 1969.

———. *Foetus into man*. London: Open Books. 1978.

Vander, A. J., Sherman, J. H., and Luciano, D. S. *Human physiology—the mechanisms of body function* (2nd ed.). New York: McGraw-Hill, 1975, p. 430.

Waisman, H. A., and Kerr, G. R. *Fetal growth & development*. New York: McGraw-Hill, 1970.

Wallace, H. M., Gold, E., Lis E. F. (Eds.). *Maternal and child health practices*. Springfield, Ill.: Charles C. Thomas, 1973.

Willson, J. R., Beecham C. T., and Carrington, E. R. *Obstetrics & gynecology* (6th ed.) St. Louis: Mosby, 1979.

Wolff, P. H. The causes, controls and organization of behavior in the neonate. *Psychological Issues*. Monograph No. 17. New York: International Universities Press, 1965.

Wolff, P. H. The development of attention in young infants. *Annals of the New York Academy of Sciences*, 1965, *118*, 815.

Woods, N. F. *Human sexuality in health & illness*. St. Louis: Mosby, 1979.

Ziegel, E., and Cranley, M. *Obstetric nursing* (7th ed.), New York: MacMillan, 1978.

5

The Infant

PATTY MAYNARD HILL

For practical purposes, infancy will be defined here as the period from birth through 12 months of age, although it should be pointed out that controversy as to what period comprises infancy exists, and some sources view the period as extending from birth to 2 or even 3 years of age. Infancy encompasses the transition from the intrauterine to the extrauterine environment, with a progression from almost total dependence toward emerging independence.

THE NEWBORN

The *newborn*, or *neonatal*, period encompasses the first 28 days of infancy. The critical event of the neonatal period is the neonate's adaptation to extrauterine existence.

The neonate is in a stressful and somewhat precarious position. The mortality rate for this period is 50 percent greater than that for the remainder of infancy. The most critical period for the neonate is the first 24 hours, the first hour of life being the time of greatest vulnerability. Leading causes of death during the period include respiratory distress related to prematurity, infection, and congenital malformations (see Fig. 5-1). Although the threat presented by each of these has been mitigated through improved prenatal care, the neonatal period continues to be most critical. Each added day of life is directly correlated with an improved survival rate. According to 1977

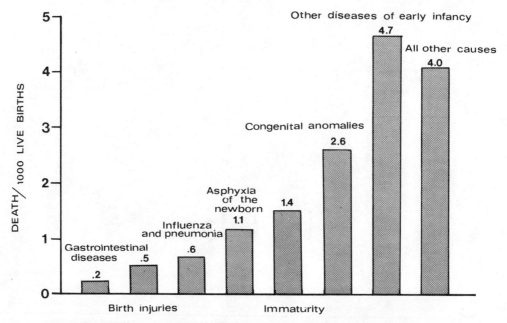

Figure 5-1. *1976 infant mortality in the United States. Based on 10 percent death sample. (Source: National Center for Health Statistics.)*

WHO statistics, the United States ranked 12th in infant survival rates (see Table 5-1).

The neonatal period is characterized by extreme physical changes and adaptations that hinge on the newborn's reaction to certain interrelated stages, as described in observations of newborn behavior by Desmond et al. These steps correlate to the potential hazards of the neonatal period. Step 1 marks the adjustment to uterine contractions. Step 2 consists of additional reactions and adaptations to external stimuli within the extrauterine environment, such as light, cold, and sounds. Step 3 consists of physiological changes associated with the transition from dependence on placental function to independent cardiopulmonary function. Steps 4 and 5 are related to a variety of biophysical changes that occur as organ function begins. Step 6 marks the emergence of homeostasis during the first 4 weeks of life (Desmond et al. 1967).

At birth, the neonate's behavior presents interesting clues to the significance of the birth process, which has been described by some researchers as traumatic, hazardous, brutal, and potentially life-threatening (Janov, 1970; Schwartz, 1973). The newborn is suddenly thrust from a warm, stable, and dependent state into an external environment complicated by sudden temperature changes, bright lights, and harsh, obtrusive sounds. The intensity of distress has frequently been seen in such neonatal behaviors as heightened crying and activity (Arnold, 1965; Desmond et al., 1967; Janov, 1970). Some have speculated that stress reactions during birth, and other early experiences, may influence coping behaviors of later life (Rank, 1973; Janov, 1970). LeBoyer (1975) describes a method of decreasing stressors by controlling stimuli (temperature, sound, light, etc.). The infant's body temperature is stabilized by a warm water bath followed by wrapping in a blanket. The environment is kept as quiet as possible, and dim lighting is used. LeBoyer (1975) proposes that infants born in such environments are less stressed, cry less, and are more alert.

Brazelton (1961) and Klaus, Leger, and Trause (1974) also support more sensitive, controlled treatment of the neonate but feel that some stimulus, particularly temperature change, is necessary for adequate pulmonary function. Ad-

ditional research is necessary to support the speculation.

Appearance

In contrast to the "Gerber baby" appearance, the newborn looks wrinkled, red or even cyanotic (see Fig. 5-2). Typically, facial features include pudgy cheeks, puffy eyes, a flat nose, a high forehead, and recessed mandible. Later

Table 5-1. Infant Mortality Rates and Perinatal Mortality Ratios: Selected Countries, Selected Years 1972–77

Country	Infant mortality rate		Average annual percent change	Perinatal mortality ratio[b]		Average annual percent change
	1972	1977[a]		1972	1976[c]	
	Infant deaths per 1,000 live births			Perinatal deaths per 1,000 live births		
Canada	17.1	14.3	−5.8	19.2	14.9	−8.1
United States	18.5	14.1	−5.3	21.9	17.3	−5.7
Sweden	10.8	8.0	−5.8	14.4	—	—
England and Wales	17.2	13.7	−4.4	22.0	17.9	−5.0
Netherlands	11.7	9.5	−4.1	16.7	14.5	−3.5
German Democratic Republic	17.6	13.1	−5.7	19.4	17.6	−3.2
German Federal Republic	22.7	17.4	−6.4	24.1	19.4	−7.0
France	16.0	[d]11.4	−6.6	[d]8.8	[d,e]17.0	−4.9
Switzerland	13.3	10.7	−5.3	16.3	13.2	−5.1
Italy	27.0	17.6	−8.2	29.6	[e]26.5	−5.4
Israel	21.3	22.9	2.4	20.7	20.9	0.2
Japan	11.7	9.3	−5.6	19.0	14.8	−6.1
Australia	16.7	14.3	−5.0	—	—	—

[a] Data for Canada, Israel, and Australia refer to 1975; data for German Federal Republic, Switzerland, and Japan refer to 1976; all 1977 data are provisional, except for the United States. All data are based on national vital registration systems.

[b] Fetal deaths of 28 weeks or more gestation plus infant deaths within 7 days per 1,000 live births. For all countries, fetal deaths of unknown gestation period are included in the 28 weeks or more gestation. This is not the usual way of calculating the perinatal ratio for the United States, but it was done for the purpose of comparison.

[c] Data for France and Italy refer to 1974; data for Canada, German Democratic and Federal Republics, and Israel refer to 1975.

[d] Excludes infants who have died before registration of birth.

[e] Fetal deaths are of 26 weeks or more gestation.

SOURCE: United Nations *Demographic Yearbook 1973–1974, 1976,* and *1977.* Pub. Nos. ST/STAT/SER.R/2, ST/ESA/STAT/R.3, ST/ESA/STAT/SER.R/4, and ST/ESA/STAT/SER.R/6. New York. United Nations, 1974, 1975, 1977, 1978.

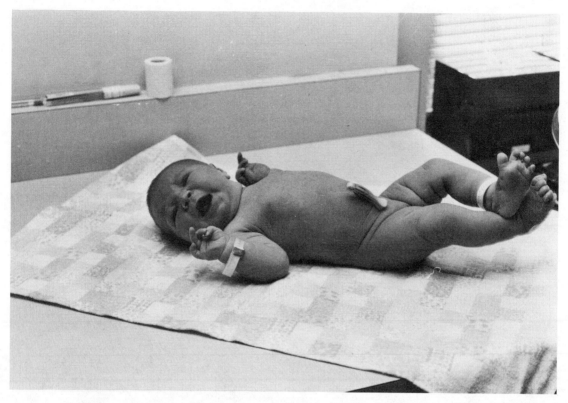

Figure 5-2. *A normal neonate. Note the proportionately large head, and the typical pudgy cheeks and high forehead.*

in development, these features evolve into individualized physical traits that will distinguish the infant more. The body is covered with varying amounts of *lanugo* (fine, downy hair) and *vernix caseosa*, a white, cheesy substance that serves as a protective skin coating, and which tends to be more heavily distributed in the fat folds. Head hair may be scant or abundant. Desquamation (peeling) of the skin is often seen as the newborn loses weight after fluid retention ceases. The skin is soft and blotchy (in reaction to environmental temperature changes). The underlying tissue is firm and elastic. Pinkish epidermis or mottled skin is also seen. Localized cyanosis (*acrocyanosis*) may be present in the extremities as a result of decreased or insufficient peripheral circulation. The musculoskeletal system is immature; the newborn's muscles are soft and small. The total muscular system typically accounts for less than 25 percent of the neonate's body weight. The skeleton is quite soft and flexible because of limited ossification, the bones being composed mainly of cartilage (Fig. 5-3).

Birth Length

Normal neonatal length is 19 to 21.5 in. (47.5 to 53.7 cm); the average is 20 in. (50 cm). The male newborn tends to be slightly longer than the female. Length increases dramatically throughout the infancy period; normal infants increase their length by approximately 50 percent, to approximately 30 in. by 12 months of age. Newborn length appears to be correlated with maternal height and weight. After the infancy period, heredity influences linear growth, and length correlates with the average of the parents' heights. Mortality also correlates with the newborn's size; larger, gestationally mature infants appear to have lower mortality and morbidity rates.

Birth Weight

Term infants on the average weigh between 6.5 lb (2,700 g) and 8 lb (3,850 g). Again, the male tends to be heavier, averaging 7 lb (3,200 gm); this is thought to be related to the higher testosterone level in the male. There also appear to be birth-weight variances related to ethnic differences, geographic location, socioeconomic status, maternal nutrition, and prenatal weight gain. As with newborn length, birth weight appears to correlate directly with the mother's adult weight and her birth weight. No correlation is necessarily seen with the father's weight (Ounsted, 1971).

A study initiated at Johns Hopkins University in the 1950s examined the relationship between birth weight and later intelligence. Measures of intelligence taken in infancy, the preschool period, early and late childhood, and early ado-

---

A deep flush spreads over the entire body if the baby cries hard. Veins on head swell and throb. You will notice no tears, as tear ducts do not function as yet.

The skin is thin and dry. You may see veins through it. Fair skin may be rosy-red temporarily. Downy hair is not unusual. Some *vernix caseosa* (white, prenatal skin covering) remains.

Head usually strikes you as being too big for the body. It may be temporarily out of shape—lopsided or elongated—due to pressure before or after birth.

The feet look more complete than they are. X-ray would show only one real bone at the heel. Other bones are now cartilage. Skin often loose and wrinkly.

The trunk may startle you in some normal detail: short neck, small sloping shoulders, swollen breasts, large rounded abdomen, umbilical stump, and slender, narrow pelvis and hips.

Eyes appear dark blue, have a blank stary gaze. You may catch one or both turning or turned to crossed or wall-eyed position. Lids characteristically puffy.

The legs are most often seen drawn up against the abdomen in prebirth position. Extended legs measure shorter than you would expect compared to the arms. The knees stay slightly bent, and legs are more or less bowed.

Genitals of both sexes will seem large (especially scrotum) in comparison with the scale of, for example, the hands to adult size.

The face will disappoint you unless you expect to see pudgy cheeks, a broad, flat nose with mere hint of a bridge, receding chin, undersized lower jaw.

Weight, unless well above the average of 6 or 7 lb, will not prepare you for how really tiny the newborn is. Top to toe measurement is anywhere between 18 and 21 inches.

The hands, if you open them out flat from their characteristic fist position, have finely lined palms, tissue-paper-thin nails, dry, loose-fitting skin, and deep bracelet creases at wrist.

On the skull you will see or feel the two most obvious *soft spots*, or *fontanels*. One is above the brow, and the other, close to crown of head in back.

Figure 5-3. *Characteristics of the neonate. (Reproduced with permission from American Baby Magazine, July 1973.)*

lescence showed that people of lower birth weight performed significantly less well than those of normal birth weight. Only 45 percent of the lowest-weight (premature) infants included in the Johns Hopkins study had achieved appropriate school placement by age twelve. Similar findings were offered by Caputo and Mandel (1970), who reported that, among those in their study, high school dropouts in contrast to "slow learners" and "normal performers," had the lowest birth weight of the three groups.

Many infants lose approximately 5 to 10 percent of their birth weight in the first few days after birth. This loss is attributed to the passage of meconium (fecal material) and to loss of excess body fluids caused by a decline in estrogenic hormone levels and by the low fluid intake during this period. The loss usually subsides by the third to fourth day, and within 1 to 2 weeks of age the infant will regain the lost weight. Until 6 months of age, the infant gains 5 to 7 oz (150 to 210 gm) per week, or approximately 2 lb (900 gm) per month. During this six-month period the infant will usually double his birth weight.

Head

Head circumference is another useful measurement in assessing growth. The frontal-to-occipital measurement correlates directly with brain size and growth. Accurate base-line data obtained at birth and updated periodically is invaluable in assessing brain development. This measurement is often taken monthly during the first 6 months of infancy, and twice during the next 6 months. Normal newborn head circumference is 13 in. (33 cm) to 14.5 in. (265 cm), averaging 13.5 in. (34 cm). At birth, the head circumference is equal to or slightly larger than the chest circumference. The head is comparatively large, accounting for 25 percent of the total length at birth. Serial measurements of length, weight, and head circumference provide a sensitive indicator of the infant's and older child's general health status and should be taken routinely throughout the first year.

The characteristics of the head include *fontanels*, the soft spots or openings at the sites of the union of the skull bones. The fontanels are protected by thick membranous tissue; they disappear as skull growth and ossification take place. The two fontanels of primary concern are the anterior and posterior fontanels. The anterior fontanel is diamond-shaped and is located at the junction of the parietal and frontal bones. Size varies, the average width being .8 in. (2 cm) to 1.2 in. (3 cm), and average length being 1.2 in. (3 cm) to 1.6 in. (4 cm). The posterior fontanel is triangular and is located at the junction of the parietal and occipital bones. It is smaller but can be detected by following the saggital suture line posteriorly from the anterior fontanel. Patency of the fontanels is important in that it allows for accommodation of the birth process and rapid brain growth. Premature closure (craniostenosis) can result in diminished brain development and mental retardation. The anterior fontanel normally ossifies by 12 to 18 months, while the posterior fontanel has usually closed by 2 to 3 months after birth (see Fig. 3-4).

Deviations of the head often seen immediately after birth include *caput succedaneum* and *cephalhematoma*. Caput succedaneum involves swelling or fluid retention in the soft tissue as a result of trauma during delivery. This edema tends to affect the part of the head that presented during delivery. No treatment is necessary; the condition usually subsides in 3 or 4 days (see Fig. 5-4).

Cephalhematoma is an accumulation of blood between the periosteum and the flat cranial bones. It also is a result of trauma. Being somewhat localized, it does not cross the suture line. This deviation is also self-resolving, by 6 weeks, and requires no treatment. Cephalhematomas should never be aspirated because of the potential for infection (see Fig. 5-5).

Molding of the head occurs very frequently.

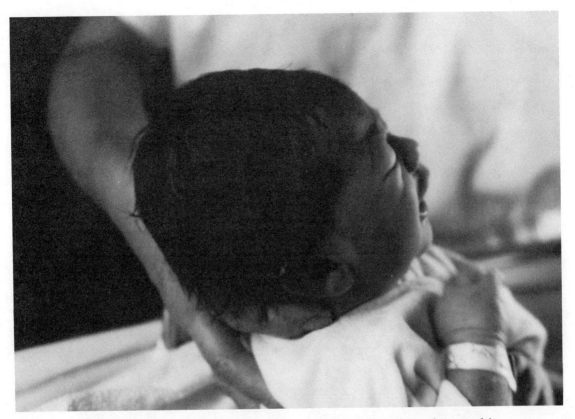

Figure 5-4. *Caput succedaneum is a diffuse, edematous swelling of the soft tissues of the scalp involving the presenting part during labor and delivery. General or localized ecchymotic discoloration may be present. Edema usually disappears within the first few days of life. The swelling is not sharply defined and may extend across the lines of the sutures. (Reproduced with permission from Mead Johnson & Company, Evansville, IN.)*

This is a temporary condition resulting from the overlapping of the lambdoid, sagittal, coronal, and frontal sutures as adaption to the various diameters of the pelvis occurs (see Fig. 5-6 and 5-7).

Gastrointestinal Development

The gastrointestinal tract is somewhat immature at birth although it is capable of secretion and absorption. The stomach's capacity at birth is approximately 30 to 60 cc, increasing to 240 to 300 cc by the end of the first year. Simple carbohydrates and amino acids can be digested and absorbed, in slowly increasing amounts, throughout infancy. Some feel that exposure to certain foods very early in infancy increases an allergic potential because of the antigenic effect of certain proteins in the immature gut. Smith and Nelson (1976) mention anatomical limitations of the stomach and limited acidity of gastric juices. The limited supporting musculature of the stomach, attributable to deficient longitudinal fibers over the fundus of the stomach and deficient elastic fibers in the mucosa of the intestine, gives the newborn's abdomen its dis-

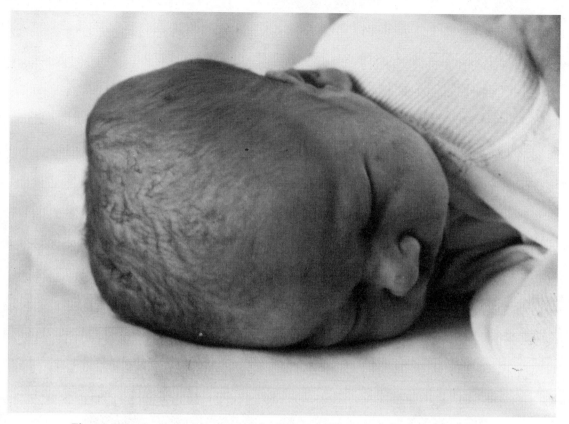

Figure 5-5. *Caput succedaneum may temporarily mask the presence of cephalhematoma, a subperiosteal hemorrhage. Unlike caput succedaneum, swelling never extends across the suture lines, although more than one cranial bone may show a swelling. There is no discoloration of the overlying scalp and visible swelling is not usually present for several hours after birth, since subperiosteal bleeding is a slow process. The swelling usually disappears in an average of six weeks, but roentgen findings persist long after the disappearance of clinical signs. (Reproduced with permission from Mead Johnson & Company, Evansville, IN.)*

tended appearance. Relaxation of the cardiac and pyloric sphincters results in regurgitation or spitting up after feeding. "Air-pocket" formation in the upper curvature of the stomach occurring during feedings and crying necessitates frequent burping. Irregularity in peristaltic motility results in delayed movement of ingested contents through the stomach and large intestines. Therefore, the stomach empties 1.5

to 2 hours after ingestion, with complete digestion occurring 3 or possibly 4 hours later. This necessitates spacing feeding times to allow for adequate stomach emptying. Increased peristalsis in the lower small intestine and large bowel results in frequent bowel movements (up to 8 per day). Neutrality of gastric contents may result in a decrease in the growth of normal bacterial flora of the bowel and stomach, which

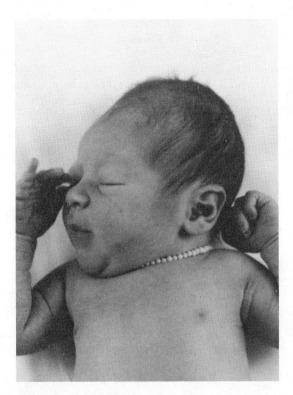

Figure 5-6. *Molding of the skull, an asymmetry resulting from the overlapping of bones of the skull occurring because of compression during delivery. (Reproduced with permission from Mead Johnson & Company, Evansville, IN.)*

in turn can result in a deficit of vitamin K. (Formation of vitamin K is dependent on the bacterial flora.) During the first 24 to 48 hours, the gastrointestinal tract usually functions smoothly, accomplishing the elimination of the pasty, odorless, greenish-black material called meconium. Absence of meconium during this key period should alert one to possible bowel obstruction.

These gastrointestinal characteristics determine the feeding modifications necessary for the neonate. Because of these physiological characteristics, the newborn requires small, frequent feedings and frequent burping.

Genitourinary Development

The neonate's renal system is functionally immature; adult functioning potential is achieved by 1 to 2 years of age. This functional immaturity is a result of decreased circulation to the kidneys (a result of low arterial blood pressure at birth and increased renal vascular resistance). The consequences of this are poor glomerular

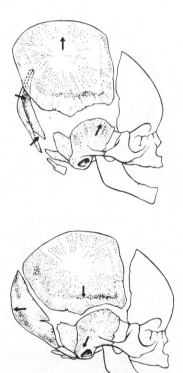

Figure 5-7. *Above, the neonatal skull on the first day of life, demonstrating molding of the bones of the calvarium with overlapped edges and narrowed sutures caused by the normal process of compression during labor and delivery. Below, the bones of the skull on the third day of life. The cranium has reexpanded and the sutures and fontanels have widened; the parietal, occipital, and temporal bones have returned to normal position. (Reproduced with permission from Mead Johnson & Company, Evansville, IN.)*

filtration and immature tubular function. In the first 24 hours, 15 to 60 cc of fluid is filtered, increasing to 400 cc by the third or fourth day of life. These limitations result in the kidneys' being relatively unable to concentrate urine; increased susceptibility to water imbalance (because of limited ability to retain or excrete water as needed); and instability in maintaining acid-base balance. (Limited excretion of electrolytes, particularly sodium, increases neonatal vulnerability to acidosis, dehydration, cardiopulmonary overload, and hyperkalemia.)

Urination occurs in the first 24 hours in almost all neonates. Occasionally, the neonate's urine will leave an orange-rust stain on the diaper; this is a result of uric acid crystals and is of no consequence. It is important for the care provider to note the character, quantity, and frequency of voiding, particularly during the first 24 to 48 hours.

The female genitals may show some enlargement related to maternal estrogenic effects. The labia minora appear larger than the labia majora. A vaginal mucoid discharge (which may be blood-tinged) may be present and is also related to the influence of maternal hormones. The term male neonate has a pendulous scrotum with rugae (ridges). Both testicles will have descended. In the uncircumcised male, the prepuce (foreskin) is partially retractable.

Immunological Characteristics

Passive immunity characterizes the immune system of the newborn's first 6 to 12 months of life. During this period, the infant is temporarily protected against many of the common viral communicable diseases of childhood (e.g., rubella, rubeola, varicella) by maternal antibodies. The newborn produces very limited amounts of immune globulins initially. This production, along with lymphoid tissue enlargement, increases with exposure to a variety of antibody-stimulating antigens.

The Central Nervous System and Sensory Development

Changes within the central nervous system proceed in an orderly, sequential pattern. Initially the newborn responds to certain stimuli with almost the entire body (e.g., the Moro reflex). With increased maturation, more localized reflexes emerge. (see Table 5-2 for a description of the most common reflexes.) The presence or absence of these reflexes is an important indicator of neurological maturation as well as of any dysfunction.

A distinct characteristic of the neonatal central nervous system is the *blood-brain barrier.* Because of limited myelination throughout CNS, the possibility exists that substances will penetrate the blood-brain barrier and be absorbed in the blood. As a result, excessive drug absorption and limited renal excretion may produce toxic effects.

The newborn's vision is somewhat developed. Although the eye is structurally immature, the newborn is able to distinguish light, dark, and color, and has increasing visual acuity. The eye undergoes maximum growth during the first year, finally attaining anatomical maturity by age three. Tear glands (lacrimal ducts) begin to function by two weeks of age. The iris of the white infant is a bluish-gray color, while in blacks and orientals it is dark brown. Permanent eye color is established between 6 and 12 months as pigmentation of the iris takes place. Subconjunctival hemorrhage, chemical conjunctivitis, and a bluish tint of the sclera sometimes occur in the newborn's eyes.

The newborn demonstrates increasing sensitivity to brightness and requires less and less intensity of light to elicit a response. Additionally, at birth the infant is able to fixate momentarily on objects and within a week can fixate for longer periods on objects within close range (10 to 20 in.). Bower (1974) believes that at less than 2 weeks of age the infant is able to react to an object moving toward him. Other

Table 5-2. Normal Neonatal Reflexes

Reflex	Description	Significance
Corneal (blinking)	Tearing or blinking in response to corneal stimulation (touching).	Presence indicates intactness of central nervous system (particularly 5th cranial nerve).
Pupillary	Pupillary constriction in response to bright light.	Presence indicates intactness of central nervous system.
Doll's-eye	Eyes do not move when head is moved laterally.	Presence indicating intactness of central nervous system.
Sucking	Sucking elicited by stroking the lips.	Persistence indicates possible CNS damage.
Rooting	Touching one cheek will cause the baby to turn his mouth toward that side.	Presence indicates possible central nervous immaturity or dysfunction.
Extrusion	Outward movement of tongue after it has been touched or depressed.	Presistence indicates possible central nervous system damage; may also be seen in Down's syndrome.
Cough	Coughing in response to irritation of lining of tracheobronchial tree.	Protective reflex.
Grasp	Strong, grasping motions elicited by light pressure on the palms or on the soles of the feet.	Continued presence may indicate CNS dysfunction. This reflex must disappear for normal grasping to occur.
Babinski	Dorsiflexion of the great toe with fanning of the other toes.	Persistence is seen in central nervous system lesions.
Moro (startle)	Extension of the trunk and extension and abduction; followed by flexion and abduction, of the arms, with index finger and thumb forming a "C"; may be accompanied by weak flexion of the legs. Elicited by striking the surface on which the infant lies.	Persistence may indicate CNS damage.
Tonic neck	Extension of arm and leg on the side to which the face is forcibly turned; limbs on opposite side are flexed.	Persistence may indicate CNS insult.
Dance (stepping)	Stepping movements elicited when the infant is held upright and inclined forward with soles of feet touching a flat surface.	Persistence indicates CNS damage.

research shows that the newborn can follow slowly moving objects with his eyes (McGurk, Turnure, and Creighton, 1977). Peoples and Teller (1975) have demonstrated that the infant exhibits color preferences, responding best to brightly colored and contrasting objects. Bower (1971, 1974) speculates that the neonate actually has a "built-in" or innate perceptual response mechanism. Visual acuity may be better than 20/150 (Dayton et al., 1964).

Research (Fantz, 1963; Fantz, Fagan, Miranda, 1975) has shown that newborns between

5 and 10 days of age are able to distinguish among various visual patterns. Fantz (1961) found that infants from 4 days to 6 months old would single out the human face. Neonates are more attracted to patterns than to color differences, demonstrating a preference for facelike patterns. They tend to prefer complex figures over simple ones, and oval and curved contours rather than straight contours (Fantz and Miranda, 1975).

This has significant implications for facilitating attachment even in the period immediately after birth. The infant appears well-equipped to perceive his new environment, including immediate parental contact. Also, these factors should weigh heavily in planning for meeting the sensory needs of the neonate and older infant.

Research indicates that neonates are capable of hearing even in utero. Fetal heart rates have been shown to increase in response to various musical tones (MacFarland, 1977). It should be noted, however, that the reception of auditory stimuli at birth is reduced because the eustachian tube and middle ear are partially filled with amniotic fluid and mucus. The neonate's hearing rapidly improves within hours or days after birth, as is evidenced by physiological responses to auditory stimuli (Ousbs, 1934; Weiss, 1934). Brazelton (1973) has observed that, almost immediately after birth, neonates react to auditory stimulation by moving toward the stimulus. Also, the neonate demonstrates different responses to different sound intensities (Bortashuk, 1964) as well to changes in pitch. For example, the neonate will demonstrate more alert activity (such as the Moro response) in response to high frequencies, such as the sound of a whistle, while lower-pitched sounds appear to have a soothing effect (Eisenberg et al., 1964; Eisenberg 1978; Brackbill, 1970). Intermittent low-pitched sounds arouse the neonate to motor activity, while sharp, high-pitched sounds produce a "freeze" response. Thus, the neonate is more responsive to a higher-pitched voice, like

that of his mother. Mendelson and Haith (1976) and McGurk et al. (1977) attempted to show that neonates possess interdependent auditory and visual perception. However, while the newborns responded to sound by visual scanning (Mendelson and Haith, 1976), they appeared to be uninfluenced by the addition of sound to a moving visual target.

Although more work is needed in the study of the neonate's perceptual powers, it appears that the newborn's ability to receive and process environmental stimuli increases gradually. Gustation, or taste, is quite well developed at birth also, although it may not be as important as the other senses. Jensen's classic study (1932) demonstrated that the infant appears to have taste preferences, preferring milk and glucose water to more salty, acetic solutions. He also correlated the sensitivity of taste to the degree of hunger in the infant; moderately full infants appeared to have greater sensitivity than hungrier infants.

Olfactory capacity is present in the newborn. Although much research remains to be done in this area, as of this writing it seems that the neonate is capable of odor discrimination as early as 1 to 2 days of age, and that sensitivity to odors increases directly with age (Engen, Lipsitt, and Kaye, 1963; Engen and Lipsitt, 1965). Also, some work supports the hypothesis that young infants can detect the odor of breast milk (McFarland, 1975).

The sense of touch is an important response mechanism for the neonate, so important that Frank (1966) described it as the language or communication system of the infant. The newborn is apparently soothed and quieted by a variety of forms of touching (such as stroking, holding, and swaddling) (Brazelton, 1969). The newborn is sensitive to pressure, temperature, and pain.

Some research on pain perception has supported the hypothesis that little pain sensitivity is present in the neonate at birth, but that sensitivity increases significantly within the first week. Potentially painful procedures such as circumcision should therefore be performed as soon after birth as possible. Certain body parts appear more sensitive than others (e.g., eyelashes, lips, soles, and nasal membranes). In addition, the infant is sensitive to temperature changes, reacting most dramatically to colder temperatures (Lipsitt and Levy, 1959; Pratt, Nelson, and Sun, 1930).

Evaluation of the Neonate

In the early 1950s anesthesiologist Virginia Apgar developed a simple tool for immediate gross assessment of the neonate. The Apgar score, as it is called, is a quantitative expression of the infant's condition and is derived from an evaluation of the following five criteria: heart rate, respiratory effort, muscle tone, reflex irritability, and color. A maximum score of 2.0 is allotted to each criterion; with a maximum total score of 10. Evaluations are made at one and five minutes after birth. Total scores of 0 to 3 represent severe distress, 4 to 6 indicates moderate distress, and 7 to 10, absence of distress. Greater than 90 percent of term neonates score 9 or 10 at the five-minute scoring. Half of all infants who score 0 or 1 at the five-minute evaluation do not survive (see Table 5-3).

Cognitive Development

What Piaget calls the *sensorimotor period* begins with birth and extends to approximately 18 to 24 months of age; this period is characterized by profound *egocentrism* (i.e., inability to differentiate between self and other). The young neonate does not differentiate himself from his environment, and his actions are body-centered.

Piaget divided the sensorimotor period into six distinctive substages, the first being the *reflexive* substage, which extends from birth to approximately 1 month of age. (The other substages are discussed later in this chapter.) It should be remembered that Piaget neither sees the stages as discrete nor correlates them with specific ages, rather, he perceives them as gradually emerging, with overlaps. *Sequence* is the predictable factor in that the attainment of each stage, according to Piaget, is contingent upon successful completion of the preceding stage.

The processes of assimilation (absorption of new information) and accommodation (alteration in behavior to adjust to the requirements of the object or event being assimilated) are the means by which the infant is able to respond to his environment. According to Piaget, in the first substage of the sensorimotor period, the *reflexive* stage, adaptation and organization take place primarily by means of reflexive responses, the most important of which are the rooting, sucking, swallowing, head-rotation, crying, Moro, grasping, and smiling reflexes. Initially, the neonate will suck reflexively on anything; shortly after birth, he learns to adapt and accommodate to various stimuli, and will suck

Table 5-3. Assessment of the Newborn: APGAR Scoring System

Sign	0	1	2
1. Heart rate	Absent	Below 100 per minute	Above 100 per minute
2. Respirations	Absent	Slow, irregular	Cry; regular rate
3. Muscle tone	Flaccid	Some flexion of extremities	Active movements
4. Reflex irritability	None	Grimace	Cry
5. Color	Body cyanotic or pale	Body pink, extremities cyanotic	Body completely pink

SOURCE: Murray, R. and Zenter, J. *Nursing Assessment and Health Promotion Through the Life Span*. Englewood Cliffs, N.J.: Prentice-Hall, Inc., 1979. Reproduced with permission.

on a nipple or pacifier mainly when hungry. Piaget (1952) sees this stage as the foundation of the intelligent behavior later to come.

Wolff (1966) demonstrated that the neonate is more responsive to a stimulus when he is alert but inactive, and that, if he is already active, he becomes less active in response to certain stimuli. Along with other researchers, Wolff helped to establish that the neonate's responsiveness depends on his degree of activity. Of equal interest is the fact that infants as young as 1 day old are capable of learning. Three- and four-day-old neonates were given a paired presentation of a nipple and the sound of a tone, while others were presented only the tone. Classical conditioning was observed in those who were given both the nipple and the tone; they eventually began to suck when given only the tone (Wolff, 1966; Papoušek, 1961, 1967). An interesting facet of this learning was that after a point the infants appeared to lose interest. That is, the infants ceased responding to certain stimuli after repeated exposures, demonstrating habituation.

Individuality of the Neonate

Although there are many characteristics common to all neonates, there are equally important behavioral differences. Temperament may be purely genetically determined, as hypothesized by Chess (1967). In attempting to classify newborn temperament types, she identified three general temperament types: "easy babies" (babies who are biologically regular and rhythmical); "difficult babies" (babies who adapt slowly to change, withdraw from new stimuli, and frequently seem to be in a negative mood); and the "slow-to-warm-up babies" (those who show interest in new situations only when allowed to adapt slowly). Chess found no relationship between temperament types and parental behavioral influences, and suggested that temperament appears to be as genetically determined as eye color.

Health care providers and parents can gain some insight into a newborn's personality type by observing how much time the baby spends in various behavioral states. Table 5-4 describes several common states as well as some characteristic behavior responses seen during this period, and Fig. 5-8 shows percentages of time spent by the newborn in five states.

Research on behavioral reponses and changes supports the notion that newborns exhibit individuality, i.e., detectable differences at birth in the manner in which behavior is organized (Wolff, 1969). There appears to be a wide range in individual behavioral patterns. Neonates also appear to differ in the distinctiveness of their behavior style, some giving very clear cues and others more subtle, ambiguous ones (Thomas, 1968).

In assessing differences in the response patterns of male and female newborns, Berg et al. (1973) found that the females had longer periods of alert inactivity. Other researchers have demonstrated ethnic differences in neonatal changes from one behavioral state to another; for ex-

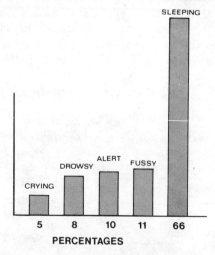

Figure 5-8. *Time spent by neonates in five behavioral states. Adapted with permission from: Berg, W., Adkinson, C. and Strock, B. Duration and Frequency of Periods of Alertness in Neonates. Developmental Psychology. 9, 3, 1973.*

Table 5-4. Neonatal Behavioral States

	Description of Behavior
Regular Sleep (RS)	limited motor activity, muscle tonus low, eyelids firmly closed with no observable spontaneous eye movement; no facial muscle movement; even, regular respiratory rate; infant demonstrates no response to milk stimulus
Irregular Sleep (IS)	increased muscle tonus, motor activity greater, varying from slight limb movement to general stirring involving limbs and trunk; facial grimacing, including smiling, mouthing motions, and crying; occasional rapid eye movement; respirations irregular, infant more responsive to external stimuli
Periodic Sleep (PS)	decreased body movement in contrast to (IS); facial responsiveness decreased; respirations are periodically rapid, shallow followed by slow, deep respirations; infant less responsive to external stimuli than in (IS)
Drowsiness (D)	less active than in (IS) and (PS), brief writhing seen as infant wakes; eyes dull, glazed, intermittently open; respirations vary, usually regular; infant is responsive to external stimuli
Alert Inactivity (AI)	relative inactivity, relaxed face with no grimacing or smiling, eyes are open, appear "bright, shining"; respirations variable and faster; infant responsive to specific stimuli
Waking Activity (WA)	periodic spurt of diffuse motor activity, involving trunk, limbs; face is relaxed or in a "cry grimace"; moaning, grunting, whimpering; eyes open, no "shine" noted; respirations very irregular
Crying (C)	diffuse motor activity accompanying crying; contracted in cry grimace, eyes closed or partially.

SOURCE: Adapted with permission from Vander Zander, J. W. *Human Development*. New York: Alfred A. Knopf, Inc., 1978.

ample, Chinese neonates changed less frequently than white neonates from states of contentment and upset. They were also more easily consoled when picked up (Friedman and Friedman, 1961). Newborns in Zinacanteco, Mexico, remained in quiet alert states for longer periods of time than American newborns and moved slowly from one state to another (Brazelton, Robey, and Collier, 1969).

THE INFANT

Appearance

As the neonate enters infancy, he offers constant reminders that he is not a miniature adult in appearance or response. The head continues to comprise approximately one-fourth of the body length. The frontal cranial area is proportionately longer than the rest of the head, while the mandible continues to remain small, with some recession. The eyes appear proportionately large and have almost reached maturity in size. The nose continues to look flat and small. The mouth is also small and may show a sucking tubercle on the inner aspect of the lips resulting from vigorous sucking.

The upper and lower extremities are noticeably different from those of adults. The neck is short, seeming almost absent, while the arms are also short and the shoulders are narrow. The trunk appears proportionately large, with a large, bulging abdomen. The lower extremities appear much too small for the rest of the body. Infancy is characterized by rapid physical growth and motor development (Fig. 5-9). Gesell, et al. (1974) proposed that in all normal infants these rapidly occurring physical changes progress in an orderly, predictable sequence, even though the chronological age at which the changes occur varies greatly among infants.

Gesell and his colleagues also concluded that growth is cephalocaudal (i.e., it proceeds from head to foot) and proximal-distal (progressing from the center outward), and that it proceeds from the general to the specific. Thus the infant develops head control before he can sit; he is able to wave his arms before being able to de-

velop finger control; and he can employ a gross palmar grasp before before being able to grasp an object with the fingers.

Weight

The infant continues to gain 5 to 7 oz per week, or a little less than 2 lb per month, through 6 months of age. Many infants double their birth weight by 4 to 6 months, usually tripling it by 12 months. The one-year-old's median weight is 22.5 lb (10.15 kg) for boys and 21 lb (9.53 kg) for girls. As the infant ages his weight may

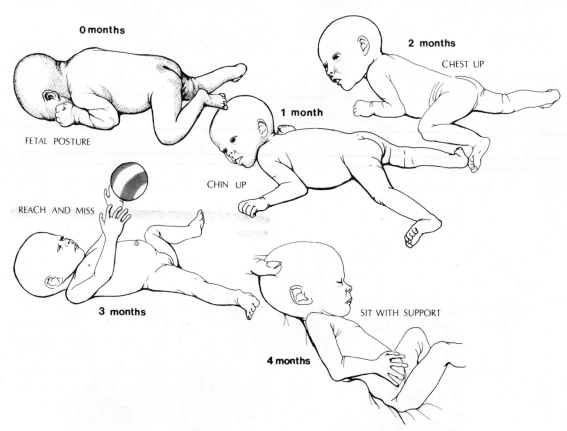

Figure 5-9. *Milestones of motor development and the average ages at which they occur. (Adapted with permission from The First Two Years of Life, by Mary M. Shirley, v. 7, Child Welfare Monograph Series. Minneapolis: University of Minnesota Press. © Copyright 1933 by the University of Minnesota.)*

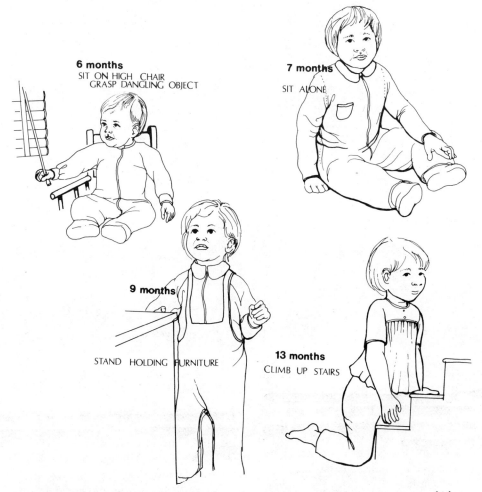

6 months
SIT ON HIGH CHAIR
GRASP DANGLING OBJECT

7 months
SIT ALONE

9 months
STAND HOLDING FURNITURE

13 months
CLIMB UP STAIRS

Figure 5-9. (Continued) *Milestones of motor development and the average ages at which they occur.*

vary more while remaining within the normal range.

Height

During the first 6 months the infant's height increases by approximately 1 inch per month so that by six months of age a median length is 23.75 in. (67.8 cm) for males and 26 in. (67.9 cm) for females is attained. As with weight, the increase tapers off so that in the remaining 6 months the increase is approximately .5 in. per month. By 12 months of age the normal infant has increased in length by approximately 50 percent, the male being 30 in. (77.1 cm) and the female 29.25 in. (74.3 cm) in length.

The average stature of children (including infants) in the United States, Japan, and Western Europe increased steadily from the 1880's through the 1970's, resulting in an approximate 4-in. gain in height for both males and females, or approximately .5 in. per decade. Schmeck (1976) found that 10-year-olds in the 1950s were

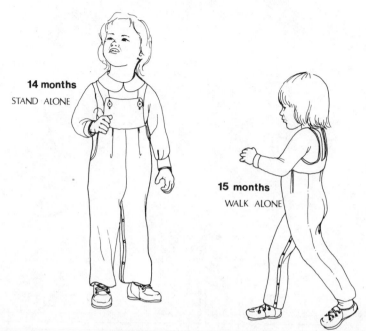

14 months
STAND ALONE

15 months
WALK ALONE

Figure 5-9. (Continued) *Milestones of motor development and the average ages at which they occur.*

approximately .5 in. taller than those in the 1940s. Factors such as improved dietary intake, improved prenatal and general health care, and immunizations are thought to be partly responsible for the increase. Hybrid vigor resulting from increased interbreeding across local, regional, and national populations has also been postulated to be a factor.

Head Circumference

Rapid increases in brain weight and, consequently, head size take place during infancy as cells hypertrophy, acquire elongated, branched processes, and gain myelin sheathing. The increase in head circumference is an estimated 4–5 in. per year. The difference between males and females in median head circumference throughout infancy is insignificant; the norm for both at 12 months is 18 to 18.5 in. (45.6 to 47.0 cm). The major portion of brain growth is attained during this first year, the brain reaching 70 percent of its adult size by 12 months.

Dentition

Most infants experience tooth eruption between 5 and 7 months, the average time being 6 months. The lower central incisors erupt first (at 6 months), followed by the lower lateral incisors (7 months), the upper central incisors (7.5 months), and the upper lateral incisors (9 months). Tooth eruption in black and oriental infants occurs sooner than in white infants. Occasionally, a supernumerary (extra) tooth will erupt as a result of aberrations in the initiation of tooth germ development (see Fig. 5–10).

The physical changes of the infancy period occur with some variations resulting from heredity, environmental conditions, general health status, and nutrition. Infants and children who enjoy good health and adequate nutrition and nurturing appear to attain height maximums, dentition, and sexual maturity earlier. Racial differences can be seen not only in dentition

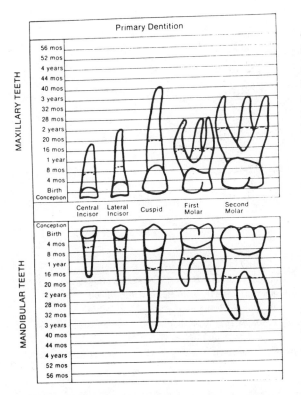

Figure 5-10. *Chronology of the development of primary dentition. (From Children Are Different: Developmental Physiology, Johnson, T. R., Moore, W. M., and Jeffries, J. E., eds. Copyright © 1978 by Ross Laboratories, Columbus, Ohio. Reproduced with permission.)*

patterns, as was just mentioned, but in bone growth; ossification centers develop earlier in black infants than in white infants, and black infants mature faster and attain larger sizes.

The Central Nervous System and Sensory Development

The order of growth and development seen in the infancy, toddler, and preschool years is related to an identifiable and sequential order in which the general functional areas of the brain develop (see Fig. 5-11). This order appears to derive from gradual cellular changes within the brain.

The brain is composed of two different types of cells, *neurons* and *neuroglia*. The *neurons*, or nerve cells, are the cells that transmit impulses. Each neuron is composed of a cell body with a nucleus and cytoplasm drawn out into many fine, wirelike processes called *dendrites*. In most neurons, there is one process, the *axon*, that is longer and composed of many smaller branches. Interestingly, the neurons are distributed less densely in the brain than in other tissue. The gaps between the neurons are filled with tissue fluid. Connections with other cells are made by axon branches that extend close to, but do not touch, dendrites of other cells. These connections occur as chemicals are released at the nerve endings or myoneural junctions. Chemoelectrical messages are transmitted through the connections of the neurons made with other cells.

The *neuroglia* (glial cells), comprise almost half the brain's volume. Instead of carrying chemoelectrical messages, they appear to function as support links. They act as a connector between neurons and the blood supply; another possible function of glial cells is the transmission of glucose, amino acids, and other substances to neurons for energy production and manufacture of protein and the chemoelectrical messages. Additionally, they play a part in the manufacture of the myelin sheaths that surround the axons and serve as an insulator allowing for the transmission of messages. Without myelination, messages would be dissipated in the surrounding fluids (see Fig. 5-12).

As previously noted, brain growth and skull development are reflected in head circumference. This measurement, however, offers no information on the differential maturation of selected areas of the brain. It has been established that different parts of the brain grow at different rates and reach their maturational peaks at different times. For example, the spinal cord, pons, medulla, and midbrain are most advanced at birth, followed by the cerebrum. The structures of the cerebrum mature in this order: (1) the primary motor area (precentral

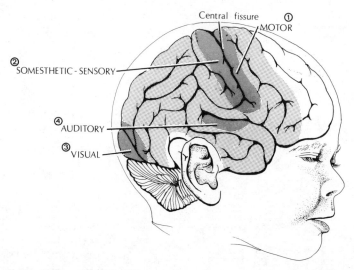

Figure 5-11. *Different areas of the cortex develop at different rates. Darker shadings indicate those areas that mature earlier, and the numbers indicate the sequence of development. (Adapted by permission of Alfred A. Knopf, Inc. from Elements of psychology, Third Edition, by David Krech, Richard Crutchfield, Norman Livson with the collaboration of William Wilson, Jr. Copyright © 1958 by David Krech and Richard S. Crutchfield. Copyright © 1969, 1974 by Alfred A. Knopf, Inc. Reprinted by permission of Alfred A. Knopf, Inc.)*

gyrus), (2) primary sensory area (postcentral gyrus), (3) primary visual area (occipital lobe), and (4) the primary auditory area (temporal lobe). More precisely, the sequence of cerebral maturation progresses thus:

One month of age: Cortical development is limited; beginning myelination of cells in the upper arm and trunk has begun.

Three months of age: All primary areas of the cortex are relatively mature; the most advanced areas are those that control the movements of the hand, arm, and upper trunk.

Six months of age: Cortical maturation continues; myelination of exogenous fibers running to the cortex occurs.

Six through 15 months: Maturation is most advanced in the temporal lobe, cingulate gyrus, and insula; maturation is slower, the brain having almost peaked in growth; the auditory area is slowly myelinating; motor area is still more

highly developed, although lower-extremity function is still delayed.

Two years of age: Primary sensory areas are as mature as primary motor areas; the integration of sensory and motor function has developed further; visual acuity remains greater than auditory acuity.

Evidence of this sequence of maturation can be seen in the primary motor area; development of cells controlling the head, upper trunk, and arm movements precedes the development of cells controlling leg movement. Similarly, hearing and vision develop more rapidly than taste.

A timetable for the acquisition of selected motor skills is shown in Fig. 5-13. It should be noted that the timings shown are approximations, and that individual maturation timings vary. Cortical maturation (arising from myelination, cellular changes, and increased neuron density) is prerequisite to the developmental

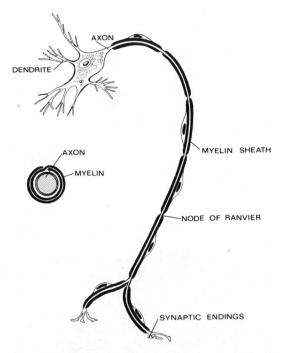

Figure 5-12. *Diagram of a motor neuron; the myelin sheath appears darkened.*

readiness that makes possible these various motor (as well as cognitive) behaviors. For example, the infant can sit only when the cortical area responsible for lower-trunk control is sufficiently mature.

Musculoskeletal Changes

The accelerated growth of infancy brings about dramatic changes in the body's musculature. Skeletal growth, controlled by the pituitary somatotropic hormone, is characterized by an increase in trunk size and sitting height; 60 percent of the head-to-heel height increase is accounted for by this trunk increase.

Facial and cranial ossification, which begins during fetal development, progresses significantly throughout early infancy, during which time the sutures of the skull close. Other areas ossify gradually, progressing from connective tissue to a cartilaginous state to bone. Primary ossification in the wrist involving the capitate and hamate bones take place by approximately 6 months of age.

Muscular development after birth consists mostly of the enlargement of prenatally formed muscle cells. The growth of preexisting muscle fibers and changes in intercellular substances, are controlled by somatotrophic hormone, thyroxin, and insulin. During late fetal development and infancy, growth consists chiefly of building up the cytoplasm of muscle cells. Salts are incorporated and contractile proteins are acquired. These muscle cells continue to increase in size up to about 3 years of age as most of the intercellular substances disappear and the water in the cells diminishes. Fat deposits are evident, particularly during early infancy, as a result of fat accumulation during the thirtieth through fortieth week of fetal development; the fat serves as an energy resource during periods of critical need.

Convergence (coordinated movement of both eyes) and *binocular vision* (the fusing of 2 visual images into 1 cerebral picture) are absent at birth but develop as myelination of the neural pathways of the brain occurs. Coordinated movements of the eyes (binocular fixation) occur by 2 to 3 months of age. Binocularity is well developed by 11 to 12 months.

Accommodation (adjustment of ciliary muscles to enable visual perception of objects at various distances) is relatively limited during the neonatal period; the neonate appears to have a constant focal point approximately 8 in. from his face. By 2 months of age the infant demonstrates the accommodating function, and by 4 months his accommodation capacity equals that of an adult (Hayes et al., 1965).

Visual constancy, or the tendency to perceive an object as a whole rather than to concentrate on its aspects, appears to be an innate human trait. Work by Bower (1971) suggests that per-

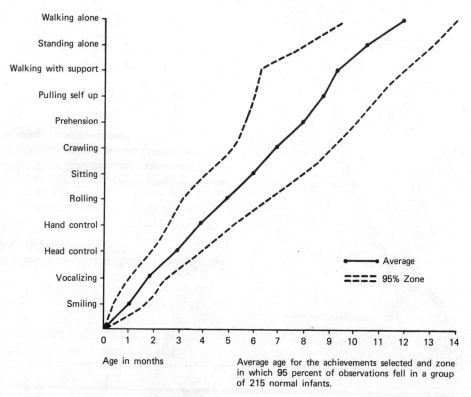

Average age for the achievements selected and zone in which 95 percent of observations fell in a group of 215 normal infants.

Figure 5-13. *Average age of selected achievements of the first year of life. (Courtesy of Ross Laboratories, Columbus, Ohio.)*

ception is oriented toward things, not such aspects as color or texture, although one is aware of these.

Gibson and Walk (1960) demonstrated that the 6- to 14-month-old infant possesses depth perception. In order to do this, they created an experimental "visual cliff." The cliff consists of a heavy glass runway extending from a table at one end, across an open space, to a pair of supporting legs at the other end. The table and the open space are covered by checkered material in such a way as to create the illusion of depth when viewed by an infant crawling across the glass. Gibson and Walk found that the 6-month-old infant avoided the apparent drop. Others (Campos, Langer, Krowitz, 1970) have speculated that this depth perception is innate, not acquired. Infants between 1.5 and 4 months of age demonstrated a significant cardiac response when placed facing the "cliff" side (see Fig. 5-14).

As the infant matures, myelination of the cortical auditory pathways beyond the midbrain and reabsorption of the middle ear ossicles take place. By 2 to 3 months the infant will turn his head in the direction of a sound. As myelination increases, infant and young child demonstrate sound discrimination, imitation, and integration of the meaning of sound. Infants from 6 to 12 months of age appear to be more attentive to new sounds or those that differ from the sounds they know. Kagan (1971), in a study of

8-month-old males, found that infants reacted most to typical, normal speech, in contrast to nonsense words or phrases. The adult level of auditory acuity is not fully developed until 6.5 to 7 years of age.

Motor Capabilities

The purposeless, reflexive responses that dominate the neonate's behavior begin to diminish with the emergence of more voluntary behav-

Figure 5-14. *An infant on the "visual cliff." (From E. J. Gibson and R. D. Walk, "The 'visual cliff,'" Scientific American, April 1960, p. 65. Reproduced with permission. Photo by William Vandivert, by permission.)*

iors (see Table 5-5). Some reflexive responses, including the knee jerk, blinking, and even portions of the Moro reflex, are retained throughout adulthood. Generally, however, most reflexive responses have completely disappeared by 12 months of age (see Table 5-6).

One should note also that some new reflexes emerge as the infant ages; they are important in that their presence or absence serves as an indicator of neuromuscular maturation, or of possible deficiencies. The Landau reflex (flexion of legs against the lower body as the head is flexed against the trunk) appears at 3 months of age and is present up to 2 years. The parachute reflex (extension of the arms in a protective manner as the prone infant is suddenly thrust forward) appears at 7 to 9 months and persists indefinitely. The labryrinth-righting reflex (raising of the head when the infant is in a prone or supine position) emerges at 2 months and possibly serves to facilitate rolling over.

The important milestones for developmental

Table 5-5. Normal Development in Infancy and Childhood

Age in Months[a]	Motor	Social	Hearing and Speech	Eye and Hand
1	Head erect for few seconds	Quieted when picked up	Startled by sounds	Follows light with eyes
2	Head up when prone (chin clear)	Smiles	Listens to bell or rattle	Follows ring up, down, sideways
3	Kicks well	Follows person with eyes	Searches for sound with eyes	Glances from one object to another
4	Lifts head and chest prone	Returns examiner's smile	Laughs	Clasps and retains cube
5	Holds head erect with no lag	Frolics when played with	Turns head to sound	Pulls paper away from face
6	Rises on to wrists	Turns head to person talking	Babbles or coos to voice or music	Takes cube from table
7	Rolls from front to back	Drinks from a cup	Makes four different sounds	Looks for fallen objects
8	Sits without support	Looks at mirror image	Understands "No" and "Bye-bye"	Passes toy from hand to hand
9	Turns around on floor	Helps to hold cup for drinking	Says "Mama" or "Dada"	Manipulates two objects together
10	Stands when held up	Smiles at mirror image	Imitates playful sounds	Clicks two objects together in imitation
11	Pulls up to stand	Finger feeds	Two words with meaning	Pincer grip
12	Walks or sidesteps around pen	Plays pat-a-cake on request	Three words with meaning	Finds toy hidden under cup

[a] Conceptual rather than chronological age.

SOURCE: Wood, B., (Ed.). *A Pediatric Vade-Mecum*. London: Lloyd-Luke (Medical Books) LTD. 9th edition, 1977. Reproduced with permission.

Table 5-6. The Disappearance of Infantile Reflexes

Reflex	Time of disappearance
Sucking	Persists throughout infancy; particularly during sleep
Rooting	3 to 4 months while awake; may persist during sleep for 9 to 12 months
Palmar grasp	5 to 6 months
Plantar grasp	9 to 12 months
Moro reflex	4 to 7 months
Stepping	3 to 4 months
Placing	10 to 12 months
Incurvation of trunk	2 to 3 months

SOURCE: Chinn, P. L. *Child Health Maintenance.* ed. 2, St. Louis: The C. V. Mosby Co., 1979. Reproduced with permission.

assessment include head control, locomotion, sitting, standing, walking. These are accomplished by 15 to 24 months of age as major myelinization and cellular maturity are reached.

Integumetary Changes

The skin is structurally mature at birth. It is sensitive to irritation despite the toughness of the epidermis (the outer skin layer). The infant is prone to inflammation or skin breakdown (e.g., friction blisters) because the epidermis is only loosely bound to the dermis.

Inflammation resulting from bacterial invasion of the epidermis can also take place when the skin of the inguinal and axillary folds becomes moist. The epidermal pH normally serves as a deterrent to bacterial invasion until it is altered by this moisture.

During early infancy, seepage of fluid into the outer epidermal layer results in vulnerability to minimal water loss as evaporation takes place. The young infant does not sweat much because of low endocrine gland activity; the amount of sweat increases very gradually through infancy and middle childhood. This limited perspiration helps to protect skin fluid loss.

The skin is a poor thermoregulator in early infancy because it is unable to contract and shiver in response to cold. Its regulatory capability increases throughout infancy as peripheral capillary responses and eccrine gland productivity increase. Subcutaneous fat tissue assists in temperature maintenance by providing insulation from heat loss.

Melanin production, which is responsible for skin pigmentation, occurs in the epidermal layer and begins at birth. It is somewhat limited throughout infancy and results in sensitivity of the skin to sunlight. As melanin production increases, the skin of the infant comes closer to attaining its adult level of pigmentation; this is true of black, white, and oriental infants.

The sebaceous glands, which are active during the late fetal and neonatal period, begin to produce sebum in such quantities that *cradle cap*, a condition characterized by the formation of flat, adherent, greasy scales frequently develops, particularly around the fontanel areas. *Milia* (pinhead-size papules caused by sebaceous gland occlusion) form on the nose, chin, and other facial areas during the first 2 months of age. These papules are of no consequence and will disappear by 6 to 8 weeks of age as sebaceous gland function diminishes. The progressive decrease in sebum production throughout infancy and childhood accounts for dryness of the skin in low temperatures and humidity.

The hair and nails also demonstrate growth changes throughout infancy. The lanugo present in the neonate disappears early. The hair present at birth is often lost in the first 6 months, and is replaced by a coarser and perhaps differently colored permanent hair. The nails of the fingers and toes of the term neonate are well-developed and continue to increase in size and length rapidly during infanncy. Even at such an early age, many individual variations occur in the development of the integumentary system.

The infant's skin should be kept warm and dry to help maintain body temperature and prevent excoriations that could lead to infection. Clothing should be appropriate for the environmental temperature; caution should be taken to not overclothe the infant. Adequate hydration is important since there is greater insensible fluid loss in early infancy. Frequent bathing of the head will prevent or decrease cradle cap.

Respiratory Changes

Respiratory changes are less dramatic and more gradual in infancy than in the neonatal period. The infant's respiratory rate is more stable and has slightly decreased from the neonatal period because of increasing maturation of the medulla. *Abdominal breathing* (characterized by the predominant use of the abdominal muscles) is characteric throughout infancy. The respiratory tract, composed of fragile tissue, remains small, and is susceptible to respiratory infections. The small size of the tracheobronchial tree makes it vulnerable to easy bacterial infiltration, since bacteria can move more easily between points in the body. Similarly, in the infant and young child, the eustachian tube is both shorter and straighter than that of an adult; it communicates directly with the middle ear and pharynx, allowing for easy bacteria movement from pharynx to middle ear. The tracheobronchial tree remains proportionately small during infancy in contrast to the increase in lung size; this, coupled with the large volume of dead air space (the amount of air needed to fill the respiratory passages), necessitates an increased respiratory rate. The rate is approximately twice that of adults (see Table 5-7 for norms). Nose breathing is characteristic of the young infant; the transition from nose to mouth breathing occurs as infancy progresses. Patency of the nasal passages is important since even a simple mucus plugging can throw the young infant into respiratory distress. At about 12 months of age, chest circumference is approximately equal to

Table 5-7. Vital Signs: Norms for Infants at Various Ages

Vital Sign	Birth	6 Months	12 Months
Heart Rate per minute	120–160	110–120	100
Respiratory Rate per minute	40–60	30	28
Blood Pressure (mm Hg)	60–80/30–50	90/60	90/60

* Diastolic blood pressure is approximately 30–40 mm Hg less than systolic pressure.

head circumference, and the epithelial lining of the tracheobronchial tree has matured to an almost adultlike state.

Circulatory Changes

It should be noted that complete closure of the foramen ovale and ductus arteriosis may not occur for several months; this may result in murmurs. The heart rate progressively slows throughout infancy to approximately 120 beats per minute by 12 months. Heart rhythm is normally irregular throughout infancy. There is a negative correlation between the irregularity (arrhythmia) and the respiratory rate; the arrhythmia diminishes with an increase in the respiratory rate and increases with a decrease in the respiratory rate. By 12 months the heart has doubled in weight and takes up 50 percent of the chest cavity width.

Significant blood pressure changes also occur in infancy. The greatest change occurs in the systolic pressure; it rises to 80 mm Hg by 1 to 2 months of age. The predictable evening blood pressure increase seen in older children and adults, and the subsequent drop later at night, is not present in the infant.

The rate of hemopoiesis (blood cell formation) changes throughout infancy, and anemia is prevalent at critical points. Physiological anemia (below 30 mg percent) is frequent at 2 to 3 months of age as the number of red blood cells decreases. By mid-infancy maternal iron stores are depleted, possibly resulting in a lower hemoglobin (see Table 5-8 for normal infant blood values).

Gastrointestinal Changes

The gastrointestinal tract is relative immature at birth. During the first 3 months digestive capacity remains limited by the small amount of saliva secreted. Through this period swallowing continues to be a reflexive, rather than voluntary, response. Lack of voluntary control results in excessive saliva accumulation and drooling. Reverse tongue action (uncontrolled protrusion) is an indication of limited myelination.

Digestion is accomplished mainly in the stomach by hydrochloric acid and rennin, an enzyme responsible for milk curd formation. Hydrochloric acid production reaches adult levels by 4 months of age.

Pancreatic juices and bile in the large intestine assist in fat protein breakdown. Pepsin, a gastric juice enzyme responsible for protein catabolism, is present at 4 months of age. Amylase, a pancreatic enzyme necessary for the breakdown of complex carbohydrates, is not produced until age 3 months, and then in limited amounts. Fat digestion, especially that of highly saturated fats, is altered because of limited production of lipase in the early months. By the end of the first year enzymatic production, essential for the digestion of proteins, fats, and carbohydrates, approaches adult levels. The exaggerated peristaltic movement that is responsible for multiple loose stools in the neonate has begun to slow, although defecation remains involuntary and reflexive. Changes in stool consistency and frequency are noted as solid foods are introduced and as the peristalsis slows. The stools generally become more solid, although diarrhea and vomiting can occur readily because of the delicacy and immature immunological response of the intestinal mucosa.

Stomach volume, highly variable among infants, increases as overall size does, reaching a capacity of as much as 360 cc by the end of the first year. This volume increase, along with slower peristaltic movement, is responsible for an extended emptying time. Thus, the older infant can go for longer periods between feedings.

Endocrine Changes

The endocrine system is one of the most important systems of the body. This system is primarily responsible for growth, maturation,

Table 5-8. Normal Peripheral Blood Values for the Infant

Age	Hemoglobin[a] (gm/100 ml)	Hematocrit %	WBC (Total mm^3)	Polys %	Lymphs %	Platelets (Total mm^3) (thousands)
1–4 days	15–22	53–65	9,000–30,000	61	31	140–300
5–9 days	11–20	43–54	5,000–21,000	45	41	150–390
10–30 days	11–20	43–54	5,000–19,500	35	56	150–390
1 mo–2 yr	10–15	30–40	6,000–17,500	32	61	200–473

[a] During the first 30 days of life; hemoglobin measurements from capillary blood are 2–3 gm/ml greater than those obtained by venipuncture.

SOURCE: Vaughan, V., McKay, R. S., Jr., and Behrman, R. *Nelson's Textbook of Pediatrics.* Philadelphia: W. B. Saunders Co., 1979.

metabolic processes, reproduction, and the integration of the body's response to stress. The system itself is functionally interrelated, each gland directly affecting the functioning of the others. Functionally, the endocrine system is extremely immature at birth. Because of this immaturity, the infant is vulnerable to imbalances of fluids and electrolytes, amino acids, and glucose up to 12 months of age or beyond.

The hypophysis, or pituitary gland, secretes six hormones that control the metabolic functions of the body. Somatotropin, the growth hormone, is produced by the anterior pituitary and is responsible for tissue growth. It plays a key role in the accelerated growth of infancy, controlling the growth of the skeleton and other structures. It also increases the synthesis of protein, conserves carbohydrates, and affects the metabolism of fats. This hormone is produced in greater amounts during infancy and childhood than in adulthood.

Adrenocorticotropic hormone (ACTH) is produced in limited amounts during infancy. It acts on the adrenal cortex to produce hormones effective in controlling fluid and electrolyte balance during stressful situations. Also, it plays a role in the metabolism of fats, proteins, and carbohydrates.

The pituitary also produces antidiuretic hormone (ADH). Throughout infancy its production is very limited, accounting for limited regulatory function during fluid imbalance.

Thyrotropin, the thyroid-stimulating hormone, controls the function of the thyroid gland in secreting thyroxine, which aids in tissue respiration and cellular metabolism, and has a direct effect on growth. Epinephrine and norepinephrine production is limited during infancy and the early schoolage period. These hormones stimulate the sympathetic nervous system and facilitate homeostasis in response to stress.

Behavioral and Cognitive Development

As the neurological and musculoskeletal systems mature, the infant acquires motor abilities that allow him to explore more of the world. As the infant experiences and explores the environment, cognitive processes emerge that enable him to begin to understand his experiences.

Many theories exist to explain the process whereby the child reacts to the things he experiences. Behaviorists, who feel that one's behavior and developments are determined almost solely by the environment, attach little or no significance to hereditary and biological factors. Theorists such as Guthrie, Hull, Skinner, and Watson perceive very complex behavioral responses as purely conditioned responses developed from environmental exposures. Robert White examined infant motivation and developed his "drive reduction theories," based on the hypothesis that human behavior is motivated by certain unsatisfied basic physiological needs (White, 1959). Abraham Maslow, with a similar philosophy, postulated that behavioral responses are determined by a set hierarchy of basic needs (Maslow, 1970).

Habituation is thought by some to be stimulated in the infant by the diversity and novelty of the stimuli infants are continuously orienting themselves to. As the infant responds to new stimuli, he tunes out old and unchanging stimuli. This can be observed in the infant who no longer kicks or babbles with excitement when a mobile with which he has become familiar is presented again and again.

Kagan (1972) described certain mental representations of events that infants acquire with age as a result of interactions with the environment. His theory centers around both the novelty principle—the idea that the infant's attention to events is aroused because of the events newness, and the discrepancy principle—the idea that more attention to events is elicited if they differ, but differ moderately. In contrast to newborns, who demonstrate a distinct preference for moving, sharply delineated, or contrasting contours, 2-month-old infants, Kagan found, were more responsive to stimuli that differed only moderately. For example, they were responsive to phrases slightly similar to those they had already heard than to those very similar or extremely different.

According to Kagan, at 2 months, the infant shows less interest in masks of the human face. Kagan postulates that since most faces the infant is exposed to deviate little from the mental representation of the general form of the face, the infant loses interest.

The infant at 9 months of age, Kagan said, enters yet another stage in attention development. Responsiveness to the human face appears to increase and continues to do so until approximately 3 years of age. There appear to be no cultural differences in the stages of attention; Latin American and African children respond with the same attention curve as North American children. Kagan feels that at this third stage attention is not so much the result of novelty or discrepancy as of a higher cognitive function, that is, the formation of hypotheses in interpreting discrepant events. Kagan suggests that at 1 year the infant begins active thought; his view is based on the monitoring of cardiac rates of infants, which increase when the infants are presented with masks of human faces (Kagan, 1972).

Although not enough empirical data exist to confirm all specific aspects of theories of learning and cognition, there is sufficient evidence to support some general principles: (1) Early learning appears to be related to both biophysical maturation and environmental influences. (2) Cognitive abilities are developed as the infant processes situations and events he experiences. (3) Early learning is a critical process in the development of higher cognitive abilities.

Jean Piaget is renowned as a leading authority on cognitive development. His four-stage theory of cognition is designed to explain the origin of cognition as well as its subsequent development. As noted earlier, part of Piaget's theory is based on the idea that the newborn's reflexes, such as the sucking and grasping reflexes, serve as the foundation for more complex forms of behavior; they are, according to Piaget, the roots of intelligent behavior. These

reflexive responses gradually evolve into the higher forms of cognition that make use of logical, conceptual, and deductive thinking (Piaget, 1952). Although Piaget's infant studies have come under much criticism because of the limited sample (three babies, all his own) and the mode of data collection (mostly direct observation), replicated studies have essentially validated his findings (Decarie, 1965).

Because of the importance of Piaget's ideas and their continuing influence on cognitive theory, an overview of his basic concepts pertaining to the infant is presented here.

"Sensorimotor Stage" (Birth to 18 or 24 months)

In this period the infant's functioning is limited mostly to sensory and motor experiences. As stated earlier, the neonate, in the earliest part of this period, is characterized by profound egocentrism, i.e., the inability to differentiate the self from the outside world. The sensorimotor stage is divided into the following six substages:

Substage I (birth to 1 month): Reflexive Stage (See the section on Cognitive Development in the neonate.)

Substage II (1 to 4 months): Primary Circular Reaction. In this substage, the infant's reflexes undergo many changes brought about by increased maturation and interaction with the environment. "Primary" refers to the fact that the infant's consciousness is centered on his own body; egocentrism is at its height. "Circular" refers to the infant's preoccupation with behaviors, discovered purposely or accidentally, that motivate the infant to repeat the behaviors. For example, the infant develops a finger-sucking behavior after accidentally experiencing finger-to-mouth contact.

Recognition of objects is achieved through repeated stimulations, which lead gradually to an awareness of functional relationships between things and events. The infant will reach for an object within his field of vision and will look for the source of a sound he hears. Gradually, random movements give way to purposeful ones.

Substage III (4 to 8 months): Secondary Circular Reaction. In this stage the infant attempts to duplicate events that he originally caused by accident; this is called *intentionality*. For example, he will purposely kick a mobile in order to see it move. He is gradually becoming less egocentric, viewing objects as separate from himself. He is establishing a fundamental grasp of what Piaget calls *object permanence* (i.e., the awareness that an object's existence is independent of its visibility), although this awareness will be more developed near the end of the sensorimotor stage. (Some speculate that separation anxiety, which is at its height at 6 to 8 months of age, indicates an awareness of the parent even when he or she is out of sight.) The infant begins to use a deliberate and systematic approach to finding objects that are placed out of his sight. The infant vocalizes when he hears his mother's voice, indicating his awareness of and pleasure in her proximity. This new depth of awareness gives rise to games of repetition, such as pushing a toy off a table to have the toy picked up and replaced on the table so that it can be pushed off again.

Substage IV (10–12 months): Coordination of Secondary Schemata. In this period the infant is concerned more with the external, or secondary, environment than with his own body (Fig. 5-15). Intentionality is demonstrated at a higher level. No longer is the infant diverted easily from his original intended task; rather, he pursues the task with intensity, almost attacking it. The infant will knock a pillow away in pursuit of a red ball behind it. He can conceptualize the means as separate from the end, and this conceptualization serves as a basis for emerging play activities. When he purposefully rolls a musical ball, he hears an appealing sound. Rudiments of problem-solving begin here. The infant begins to arrange many newly acquired *schemata* (Piaget's term for units of thought) into sequences, and rituals emerge to become com-

Figure 5-15. *At 10 to 12 months, the infant is in what Piaget termed the substage of "coordination of secondary schemata," during which the infant is more concerned with the external, or secondary, environment than with his or her own body. (Photo by Sam Gray, M. Photog., Cr.)*

pelling aspects of his life. The infant more clearly than ever perceives that there are causes of events that occur outside himself; his mother and father, and others, can cause things to happen, and the infant will wait for them to do so. The sense of object permanence continues to solidify; however, when the infant is presented with a complex series of movements of an object involving numerous displacements, the infant is inclined to look for the object where it was last found. Imitating abilities are perfected in this stage, and the infant will engage eagerly in "patty-cake," etc.

Substage V (12 to 18 months): Tertiary Circular Reactions. The beginning of trial-and-error behavior is seen in this stage, along with rational judgment. Piaget cites the example of the infant who, in an attempt to reach an object placed beyond his reach on a rug, pulls the rug toward him. At this period the infant is also very concerned with how things feel and are put together, and he will continually touch objects. With the development of object permanence (also called object constancy), the infant no longer will search for an object in the place where he first found it if he sees it moved. Still lacking, however, is the understanding that objects can be moved while he is not looking. The infant now seeks to create novel events rather than to simply reproduce those that accidentally occurred. In dropping a spoon from a highchair, the infant will drop it at various positions and heights.

Imitation develops further in this substage also. The child will now perform behaviors that have not been part of his response repertoire before, imitating the action immediately after watching the model, or a day or two later.

Substage VI (18 to 24 months): Invention of New Schemata. This substage is characterized by the beginning of simple thought processes and problem solving. The toddler begins to learn that problems must be thought out before they are attacked. New solutions begin to emerge as part of goal attainment. For example, when confronted with a rug obstructing the path of a pedal car, the toddler will stop, look around the car, and then raise the wheel. Infants at this stage are trying to think about problems and develop solutions on a mental rather than a physical level (Piaget and Inhelder, 1969; Piaget, 1952; Rosen, 1977). (See Table 5-9.)

Piaget's work has done much to establish that the development of intelligence occurs concurrently with biophysical and psychosocial maturation. Cognitive growth occurs as a result of both biolgical and environmental factors. The gradual but distinct emergence of new mental structures, or schemata, and the modi-

Table 5-9. Piaget's Sensorimotor Period of Cognitive Development

Stage	Behavior	Examples
1. Reflexive schemata (0 to 1 month)	Use of reflexes	Sucking is most salient reflex
2. Primary circular reactions (1 to 4 months)	Extension of reflexes	Sucks fingers, puts out tongue
3. Secondary circular reactions (4 to 8 months)	Earliest stage at which "intention" is distinguished	Moves in crib to make toys on crib shake
4. Coordination of secondary schemata (8 to 12 months)	Application of familiar means to new situations; "means" differentiated from "ends"	Holds a block in each hand and drops one, picking up a third one just presented to him
5. Tertiary circular reactions (12 to 18 months)	Discovery through active experimentation	Devises different ways of making something fall or slide "to see what happens"
6. Invention of new means through mental combinations (18 to 24 months)	Emergence of capacity to respond to or think about things not immediately observable	Uses a stick to reach out beyond arm length to pull something closer

SOURCE: Kaluger, G. and Kaluger, M. F. *Human Development: The Span of Life.* ed 2, St. Louis: The C. V. Mosby Co., 1979. Reproduced with permission.

fication of preexisting ones, is brought about in attempts to accommodate the existing structures to the contours of external reality. Piaget's theory has been empirically validated, although some discrepancies have been found in the chronological ages at which specific schemata are acquired. Previously cited studies suggest that infants are capable of mental processes earlier than Piaget speculated. It should be noted that the chronological ages at which various stages defined by Piaget are attained will vary. Several internal and external factors will determine the individual's progress through the stages of cognitive development; what remains constant is the sequence in which this progression occurs. While each stage is characterized by the most recently acquired mental capabilities, behaviors and capabilities that preceded it continue to be a part of the response repertoire.

Another behaviorist, Hebb (1958), maintains that physiological maturation of the brain is directly affected by sensory experiences. In Hebb's view, behavior is influenced not only by the sensory experience itself, but by the cellular system of the brain that receives the perceptual information. Rich sensory experience results in development of the neural cellular system. It appears that critical periods of learning occur early in the course of neural maturation and that these periods serve as a foundation for subsequent cognitive function.

Communication and Language Characteristics

Through such nonverbal forms of communication as eye contact, touching, patting, kissing, cooing, and laughing, the young infant receives both sensory and emotional nurturing, and in turn facilitates parental attachment. The communication provides the parents with an opportunity to convey to the infant their behavioral expectations and, of course, their emotions. Even more important, these experiences serve as primary self-expressive functions. Reciprocal communication at this nonverbal level is a key to parent-infant bonding.

Various theories have attempted to define the process of language development, although some of these have created more questions than they have answered. Discrepancies between them exist, but each has some validity. Although controversy exists regarding the process of language formation, it is generally accepted that this complex acquisition is closely correlated with central nervous system maturation and cognitive development.

Common Theories of Language

As of this writing, three theoretical approaches are widely used as a framework for explaining the phenomenon of language acquisition: learning theory, linguistic theory, and cognitive theory.

Learning theory as set forth by Skinner involves operant conditioning. Skinnerian language theory suggests that the initial occurrence and progression of language use is directly related to the level of reinforcement received. In other words, Skinner speculated that, as the infant vocalizes, positive reinforcement is usually received from parents through smiles, touches, pleasant sounds, etc. In turn, the infant continues to repeat this vocalization in order to elicit the desired reinforcement. This language repertoire increases as the infant and young child matures and begins to perceive and mimic words, phrases, and sentences received from others. The utterance of these new phrases is also perpetuated by positive reinforcement (Skinner, 1953, 1957).

Linguistic theory states that the infant is born possessing innate models of language that prepare him to determine the rules of syntax of any language to which he is exposed. Inherent in this theory is the concept of "linguistic universals," the elements common to all language (for example, noun-verb sentence structure) (Chomsky, 1957).

Cognitive theory postulates that language skills

develop in response to the individual's interactions with all aspects of his environment. The interactions begin in the sensorimotor stage when the infant deals with objects principally by manipulating them. Action experiences are thus said to be the basis of language formation.

The infant begins, in the sensorimotor stage, to relate objects to actions or functions. In subsequent stages of development, the child learns to relate objects to symbols. These symbolic representations are incorporated into the child's response structure, or schemata. This cognitive theory suggests that at the point of internalization of symbolic representations, speech patterns begin to emerge.

Repetition of syllables is prevalent at 8 or 9 months, when the infant may begin to say "mama" or "gaga." By 9 to 12 months, word comprehension (*receptive language*) emerges. The infant now begins to attach meanings to the sound he produces and may use them appropriately by 12 months. Single words at this point may represent whole ideals or sentences; this is termed *holophrastic speech* (for example, "bye-bye" may represent "I want to go bye-bye"). Some studies have documented that 8- to 9-month-old infants comprehended their mothers' use of "no" (Church, 1966). By the end of the first year the infant has a vocabulary of approximately 6 words (see Table 5-10).

Psychosocial Development

The infant's social response repertoire gradually emerges during infancy. Sensory capabilities continue to serve as the focal point for interactions. Eye contact, touching, smiling, and laughing are ways in which the infant participates in social exchanges (Fig. 5-16). In response to these behaviors the parent or caretaker may reciprocate with a smile, cuddle, or hug. A social exchange between two individuals in which each directly influences the behavior of the other is described as being *bidirectional*.

One's response repertoire is directly related

Table 5-10. Vocal Patterns of Infancy

Age	Characteristics
Birth to 8 wks	crying of various kinds
2 mos	cooing
3½–6 mos	one-syllable babbling
6–9 mos	repetition and reduplication of self-produced sounds
9–10 mos	imitation of sounds produced by others; variable intonation patterns
10–12 mos	understands some words; use of own "jargon"; associates words with objects; says first words; babbling still present

to cultural, family, and class values that begin to be communicated verbally and nonverbally from birth. The infant learns to realize the impact of the "social smile" as early as two months, taking his cue from the delighted response it elicits from the caretaker. North American children at very early stages are often directed toward individuality and uniqueness. Society's value system for them centers around rights and privacy. The core initiators and enforcers of such qualities are the family. This typical American belief system can and does vary from family to family, however. Some speculate that this adds to conflict within the child, as consistency appears to be lacking in the values communicated. As a result of this variable value system, the young child experiences incongruence in learning socially responsive behaviors such as how to respond and what to respond to. In cultures such as China, adults agree on values and consistently communicate these to their offspring, emphasizing empathy and commitment toward others (Comer and Poussaint, 1975; Chan, 1975; Mead, 1928).

Some behaviors emerge in early infancy that assist the individual in meeting future social demands; these include smiling, babbling, laughing, and crying. Smiling is perhaps the

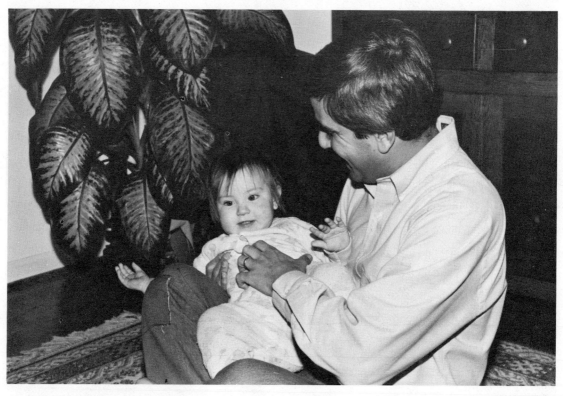

Figure 5-16. *Eye contact, touch, smiling, and laughing are among the ways in which the infant begins to participate in social exchange. (Photo by Sam Gray, M. Photog., Cr.)*

most potent social stimulus the infant develops. Some have speculated that it adds to the appeal of the infant and may, in fact, increase survival potential (Bowlby, 1957). The emergence of different types of smiling apparently correlates with nervous system maturation. First, in spontaneous smiling—composed primarily of grimacing—there is no apparent correlation with any external stimulus; the response occurs often during sleep and inactive states. Stroking the infant's cheek or the sound of the human voice can elicit these limited facial contractions. These responses may extend to 5 to 6 weeks of age. In selective social smiling, the infant begins to smile in response to particular stimuli; the human voice and face are the most successful stimuli. The maturing smile involves more facial muscles, including the muscles surrounding

the mouth and the periorbital tissue of the eye. This selected smiling is seen by two months of age. The infant still is unselective, however, with regard to the recipient, smiling at strangers as well as at the parent or caretaker. As maturation proceeds the infant demonstrates response patterns that are distinctly his.

It appears that very early in his development the infant acquires a beginning awareness of the potential social influence of a smile. Smiling discriminately and indiscriminately, the infant learns an acceptable social behavior that will increase control over the social environment.

Laughter

The infant's laughter may exhibit a predictable maturational course. A longitudinal study by Sroufe and Wunsche (1972) demonstrated that infants of 4 to 6 months of age were more likely to laugh in response to social and tactile stimulation than to visual or auditory stimulation. In the 7 to 9 months age group, social stimulation continued to elicit the greatest laughing response, with visual and then tactile stimulation following. By 12 months of age the infant laughed most in response to visual stimulation. From these behaviors Sroufe and Wensch concluded that early laughing appeared to be related to tension experienced by the infant when an incongruous event occurred appropriate to the cognitive level of perception but inappropriate to the infant's emerging cognitive schema. Those authors felt that the laughter served as a venting mechanism for the felt tension. In fact, they concluded that it might well allow the infant additional opportunities to process the inconsistency, and move toward developing a congruent schema.

Crying

Crying is a most effective method by which the infant communicates with the caretaker. It is certainly received negatively by parents and caretakers, and is very effective because of this displeasure. Immediate steps are taken by the parent or caretaker to terminate the crying. Crying in early infancy is somewhat generalized although it is most often associated with distress. Its effectiveness in eliciting a parent or caretaker response is excellent; picking up the infant, talking to him, patting, repositioning, or offering a pacifier are often attempted. Infants appear to be most readily comforted when they are picked up and held to the shoulder. Talking to them in attempting to soothe them apparently has least effect (Bell and Ainsworth, 1972; Korner, 1974).

Although crying obviously persists throughout infancy, its use decreases as the infant acquires additional means of communication. There appears to be an association between the responsiveness of the caretaker to the crying and the adeptness of the infant at developing other means of communicating. Immediate

responsiveness to the infant's crying results in diminished crying later on. Thus, infants with more responsive parents appear faster to develop other means of expressing themselves (Bell and Ainsworth, 1972).

Attachment

The social behaviors discussed thus far represent a variety of signals or responses that facilitate the emergence of a unique, close, specific, and enduring relationship between the infant and parent or primary caretaker. This formation of an affectional bond or attachment begins soon after birth and is thought to be completed by the end of the first year.

The attachment process is thought to unfold in three stages: asocial, protest separation, and selective separation. In the asocial stage the infant is almost solely aware of the environment itself, being unaware of the significance of individuals as part of this environment. During the protest separation stage, the infant has acquired an awareness and appreciation for individuals and has become very anxious about separation from them. This response is nonselective however, and is seen when the infant loses close contact with any individual.

As the infant identifies his primary caretakers, a more selective response is seen when the threat of separation is experienced. This selective separation period usually begins at the point when the infant has a beginning awareness of himself as separate from his environment. With this awareness comes an increasing vulnerability and dependence on those individuals who have best met essential needs. These individuals are most likely to be parents and/or other primary caretakers. During this third stage, beginning at 6 to 9 months of age, "separation anxiety" or "stranger anxiety" is first observed. The infant may well resort to withdrawal behaviors or crying either when separated from the parents or caretaker, or when confronted by strangers.

Brazelton, Schod, and Robey (1966) established the first minutes and hours after birth as an especially sensitive period in the attachment process. Additionally, the 6- to 12-month-period of infancy has been implicated as the critical period for attachment in humans (Ainsworth, 1973; Lamb, 1977). In examining the attachment process in rhesus monkeys, Harlowe established the probability that tactile stimulation was an integral part of tension and anxiety reduction. It therefore appears that attachment is enhanced by a need for comfort, warmth, and support provided through physical contact, and that the nature of the contact is important throughout infancy.

In examining the effects of separation from the parent, Spitz and Wolff (1946) found that infants in institutionalized settings who were given impersonal care by several caretakers, and who received inadequate tactile stimulation, experienced symptoms of depression. This *anaclitic depression*, as it is called, is characterized by delayed physical, motor, psychosocial development.

Psychosocial Theories

The founder of psychoanalytic theory was Sigmund Freud, and his work, although controversial, remains the most influential contribution to the field. Freud postulated that human behavior could be categorized in terms of certain fixed patterns emanating from the stages of psychosexual developed. Determinants of this behavior were thought to be unconscious. Freud felt that biological drives such as sex and aggression were primary behavior forces. According to him, psychosexual behavior is the root of personality development.

The infancy period was cited by Freud as one in which major phases of psychosexual development occur. The oral stage, from birth to 18 months, is the period in which the mouth is the primary area of gratification. Sucking, biting, or eating are behaviors through which this gratification occurs. Theoretically, insufficient oral gratification results in oral fixations.

Erik Erikson, a psychoanalytically oriented psychologist, has received much attention for his development of Freudian theory. Erikson placed more emphasis on the individual's environment, particularly his social environment. It is Erikson's belief that the basic task of infancy in the development of the personality is that of acquiring *trust*. The quality of interactions between the infant and his parent determines whether this sense of trust is acquired. If the infant experiences a nurturing, close, secure, and consistent relationship with his parents or caretaker, a sense of trust and security will develop. If, on the other hand, that relationship is inconsistent, insecure, and uncaring, the infant is likely to develop a sense of *mistrust*, which will negatively influence development.

COMMON HEALTH AND DEVELOPMENTAL PROBLEMS OF THE NEONATE AND INFANT

Oral Aggressiveness

A focal point of interaction with the environment during the first year is the mouth. Oral and facial responses characterize the period.

Because of limited neuromuscular control, the infant is primarily the recipient of sensory stimulation, being somewhat immobile and limited to upper extremity control. Consequently, "reaching" is performed with the eyes, and touching is done mostly with the mouth. As increased motor control is acquired, the infant exercises curiosity through more mature behaviors. During the last 6 months of infancy, the infant can competently and purposefully reach and even retrieve. At this point oral behaviors are heightened. Capable now of more sophisticated motor and social behaviors, the infant no longer passively receives the environment, but actively and aggressively meets it. "Oral aggressiveness" is a common behavior (Fig. 5-17). The infant takes anything into his

Figure 5-17. *During much of infancy, "reaching" is done with the eyes, and touching is done with the mouth; "oral aggressiveness" is a hallmark of later infancy. (Photo by Sam Gray, M. Photog., Cr.)*

mouth. It is at this point that finger sucking may also begin. Additional modes of sucking should be provided—through the provision of a pacifier, bottle or breast feedings, and soft, pliable toys. This behavior will subside as the infant acquires additional modes of sensory experimentation.

Separation Anxiety

As previously noted, separation anxiety appears as the infant develops a beginning awareness of object permancence. Simply stated, the infant, having once distinguished himself from his environment, apparently has some sense of the omnipotence of those important persons who meet his needs. The infant is fearful of any separation from them. There appears to be no relationship between culture and this phenomenon, since 50 to 60 percent of the children from all populations in which separation anxiety has been studied cry upon separation; this separation response may begin at 6 months of age, peaking between 9 and 24 months, and it is experienced upon separation from either the mother or father regardless of the quantity of interactions with each (Maccoby and Field, 1972). A tempermental disposition toward fearfulness may influence this behavior. It has been suggested that unless the child has specific fears, this anxiety will disappear when the parent or caretaker's absence is no longer a puzzle or when the child feels he has control over some aspects of his environment (Scarr, 1970).

In assisting the infant to overcome this anxiety, the supportive approaches listed below may be helpful.

1. Provide warm, nurturing behaviors, e.g., frequent cuddling, verbalizing, touching, and other tactile stimulation.
2. Be consistent and prompt in meeting care needs. Remember that rituals become a part of the infant's response repertoire as cognitive schemata emerge. Efforts to maintain these bring order and calmness to the infant.
3. Provide a consistent substitute caretaker and environment during absences.
4. Ensure that the caretaker uses the supportive nurturing behaviors described in No. 1 above and is sensitive to the infant's feelings.
5. Provide personal articles that the infant can associate with parents during their absence, e.g., an item of clothing, pocketbook, etc.

Feeding Difficulty

Feeding is considered a very important element in the attachment process. Optimally, it fosters reciprocal attachment between caretaker and infant. It is a time at which a variety of sensory experiences, particularly visual and tactile experiences, offer the infant data about his world and caretakers. The infant's feelings about himself and the world are influenced by the manner in which he is handled, and feeding is a part of that.

Visceral sensations related to feeding and gastrointestinal function are apparently fairly acute during infancy. Hunger and abdominal fullness or distention account for much infant distress.

Colic is characterized by paroxysmal abdominal pain and cramping, loud crying, reddened face, and a rigid abdomen. The infant's legs are rigidly drawn up against the abdomen. Typically, colicky responses are seen after feedings. Occurring most commonly during the first 3 months of infancy, colic has a significant emotional impact on attachment relationships, creating feelings of inadequacy, frustration, and anger in the parent or caretaker.

The cause of colic is not yet known, although some theories have proposed inadequate feeding techiques in which the infant ingests the feeding too rapidly, overeats, or swallows excessive air. Inadequate burping, malposition, or anxious or inadequate feelings of the caretaker communicated to the infant have also been cited as suspected causes. Other possible causes are immaturity of the central nervous system, particularly of the parts that control the gastrointestinal tract, and allergic responses.

While the infant usually thrives well, the great amount of frustration and concern directly affects the development of a relationship between the infant and parents or caretakers. Although the exact cause of colic is unknown, the following interventions may be useful:

1. Observation of the actual feeding process by a health care provider.
2. Formula substitution if allergy is suspected.
3. Feeding modifications:
   a. smaller, more frequent feedings, including burping properly;
   b. positioning to diminish excessive air intake during feeding;
   c. positioning in upright position after feeding.
4. Decreased sensory stimulation of the infant before and after feeding.
5. Emotional support of caretaker during the colicky period; encouragement of parent or caretaker to talk about frustrations related to the colicky behavior.
6. Assurance that the infant's behavior is not reflective of inadequate parenting.

Failure-to-Thrive

Failure-to-thrive (FTT) is a condition of children who fall below the third percentile on growth gradients with no apparent pathological cause. Also referred to as "maternal deprivation," it is thought to be related to several factors, including nutritional inadequacies, such as Hirschsprung's disease, congenital anomalies, including cardiac and renal abnormalities, and psychosocial and emotional deprivation.

No organic factor can be found in approximately 10 percent of FTT cases. In these instances it is thought that inadequate social and emotional nurturing, particularly involving close one-to-one relationships, is a causative factor. Parental deprivation is likely to occur during infancy in a variety of situations, one of which may be feeding.

The personality of the infant has definite effects on the attachment process. In situations in which the infant's responses do not meet the expectations of the parent or caretaker, conflict can emerge. It may be that the infant's individuality makes him more difficult to relate to, and the parent or caretaker is (for whatever reason) unable to give the additional nurturing that may be needed. The parenting (and therefore the role-model) of the parents themselves has been inadequate in many instances.

Another possible factor may be the infant's innate response patterns. Just as the infant can "turn on" mothering or parenting behaviors, these can also be "turned off." This is particularly true in instances in which the parent feels inadequate in her role to begin with and perhaps has inappropriate expectations of the infant.

Interventions should center around a careful family assessment. Observations of infant-parent interactions are useful and can be carried out in the home, hospital outpatient clinic, or during a hospitalization. A careful diagnostic workup should be completeted on the infant to identify or rule out any organic problems. The same staff memebers should be assigned to the family for assessments and for meeting the care needs of the infant and parent. Parents need to be taught to read their infant's cues, and to set appropriate behavioral goals for the infant.

Accidents

The last 6 months of infancy are characterized by crawling, cruising, creeping, and, in a small number of infants, walking. These motor skills create an increased potential for accidents; accidents are the leading cause of death in children above 1 year of age, but parents must begin to think about prevention earlier. The accidents most frequently involving infants are falls, poisonings, burns, and automobile accidents. Frequently, accidents of infancy and later periods occur because of an underestimation of motor potential; for example, 3- to 4-month-old infants roll off a bed because of inadequate restraining measures. A frequent comment from the parent is that the infant has never rolled over; "I didn't think she could."

Measures should be taken by the nurse and other health team professionals to orient the parent and caretaker to specific developmental milestones of infancy, including those that will occur predictably. An effort should be made to make the infant's environment as safe as possible (see Table 5-11).

Infant morbidity and mortality related to automobile accidents occur chiefly because of inadequate or improper restraining and protection. Seats specifically designed for infants through 6 months of age, in most instances, are designed to face toward the rear of the car so that, with any impact, the infant will be thrown backward, not into the dashboard. Only seats that have been safety-approved by the Federal Motor Vehicle Safety Standards should be used.

Infections

Infections are the leading cause of morbidity and mortality throughout infancy. In the neon-

Table 5-11. Safety Measures and Rationale

Dangers	Normal Behavior	Precautions
Falls	Moves self by pushing with feet. Later able to roll over and crawl. Crawls, creeps, cruises, and walks.	Carry baby firmly. Never leave unprotected on bed or table. Keep one hand on baby if reaching for equipment. Keep crib rails up and secure when infant in crib. Obtain sturdy high chair or car seat. Keep fasteners in place when baby in seat. Place rails or gates at top and bottom of stairs.
Burns	Crawls, hitches, pulls self erect. Investigates environment by touching. Has prehensile ability.	Refrain from using tablecloths that hang over table. Cover all electrical outlets with safety devices. Turn pot handles inward on stove. Use vaporizers with caution when child is ill. Do not place crib next to a radiator or fireplace; screen both if possible. Avoid excess exposure to sunlight. Do not smoke around baby.
Inhalation of foreign objects or poisoning. Suffocation	Oral investigation; puts anything and everything into mouth. Rolls from side to side, later back to stomach. Reaches for objects with hands.	Close and remove safety pins from crib when changing diapers. Do not give small beads, coins, or toys to infant or leave them within reach. Check toys and furniture for presence of lead-containing paint. Keep medicines, household cleaners, and cosmetics out of reach. Keep venetian blind or other cords out of reach. Fit mattress snugly in crib. Remove plastic bags from reach. Avoid excessively tight bedcovers, pillows, and restrictive clothing around the neck. Check slats of playpen or crib to determine that head cannot get through. Avoid easily aspirated foods or toys—nuts, raisins, hard candy, popcorn, balloons. If vomiting, hold with head down, hips lifted slightly; do not pick up.
Drowning	Helpless in water,	Never leave alone in tub or sink. Do not leave young sibling in charge with infant in bath or swimming pool.

SOURCE: Murray/Zenter et al. *Nursing Assessment and Health Promotion through the Life Span.* Englewood Cliffs, N.J.: Prentice Hall, Inc., 1975. Reproduced with permission.

ate, infections are likely to involve the umbilical stump and conjunctiva of the eyes. Causative organisms include E. coli, staphylococci, and pseudomonas. Stringent handwashing and other sterile techniques are essential for nurses and others working with infants in order to avoid cross-contamination among infants in the nursery and at home.

Acute respiratory infection is the most common infectious disease in infancy. By completion of the first two years the infant and toddler may have had as many as 8 or 9 infections. Causative organisms include pneumococci and H. influenzae and other viruses. These infections of the upper tracheobronchial tree are significant because of the anatomical and physiological characteristics mentioned earlier in this chapter. Respiratory distress related to simple mucus-plugging in the nasal passages can result because of the nose breathing that is normal in the newborn and young infant. Otitis media is an equally frequent infection because of the shorter and straighter eustachian tube in the infant and its direct connection between the middle ear and pharynx.

Croup, or acute laryngotracheobronchitis, can occur during infancy, usually following an acute upper respiratory infection, although it is seen most in children between 1 and 3 years of age. It is usually of viral origin and is characterized by inflammation and spasms of the larynx, the spasms being more severe than the degree of inflammation. The infant has difficulty breathing and can develop severe signs of respiratory distress. This condition should be considered potentially serious. Parents should be taught to use a vaporizer and positional techniques, which may reduce the spasms and make possible more effective breathing. Mild sedation may be prescribed if the infant or young child is apprehensive and anxious. Usually the child can be treated at home.

Acute bronchiolitis occurs commonly in the infant under 6 months of age. Causative agents are usually syncytial viruses, adenoviruses, and parainfluenza viruses. Respiratory distress may occur because of bronchial occlusion by tenacious exudate and mucus.

Diarrhea, although a less prevalent contributor to infant mortality, is a frequent gastrointestinal infection. In the infant the most likely causative agents are E. coli and the adenoviruses and interoviruses, including echovirus and coxackievirus. Diarrhea may result in fluid and electrolyte imbalance.

Sudden Infant Death Syndrome

Sudden infant death syndrome (SIDS), or "crib death," is one of the leading causes of infant mortality. Between 2 weeks and 4 months of age, its incidence is very high, striking as many as 10,000 babies per year nationally. SIDS is defined as "the sudden and unexpected death of an infant who was either well or almost well prior to death and whose death remains unexplained after an autopsy" (Beckwith, 1970). Several theories (none of them validated as of this writing) seek to explain the cause of SIDS; speculation has centered on suffocation, pneumonia, laryngeal inflammation, allergy (particularly to cow's milk), immunoglobin abnormalities, parathyroid inadequacy, electrolyte imbalance, viral infection, and sleep pathophysiology involving prolonged apnea.

Some authorities give credence to the "prolonged periods of apnea" theory. It has been postulated that immaturity and abnormality of the central nervous system lead to the prolonged periods of apnea. Viral infections have also been associated with the syndrome; half of all SIDS victims have histories of upper respiratory infection. Histological findings often indicate signs of inflammation in the tracheobronchial tree. Significant other factors include a higher incidence in premature infants, particularly males. The incidence is also greater in infants from lower-class families and in infants whose mothers smoked while pregnant (Bergman, 1976). SIDS is not genetically determined although there is a slightly greater chance that it will recur in a family that has already lost one infant to it.

Death from SIDS usually occurs during the infant's routine sleep. There may or may not be signs of a struggle. Obviously, the parent or caretaker who finds the child can be overwhelmed by the appearance that the infant was in distress. Grief over the event is then even more heightened.

Support of the family immediately after the death and during the subsequent grieving period is the primary goal of the nurse. The key goal is to alleviate the feelings of guilt. It should be stressed that this disorder cannot be predicted or prevented. A confirmation of SIDS is made by autopsy; the nurse can play a vital role in assisting the family in deciding to request an autopsy so as to rule out other causes of death. Follow-up by public health nurses and counseling of the family are needed in order to restore family integrity and facilitate the grieving process.

Child Abuse and Neglect

Physical abuse has been suspected by some to be the leading cause of death in children. Although it seems that most victims are toddlers, abuse is second to sudden infant death syndrome as the leading cause of mortality during the first 6 months of infancy and is the number one killer during the 6-to-12-month period.

Child abuse is defined by the 1974 Child Abuse Prevention and Treatment Act as "the physical or mental injury, sexual abuse, malignant treatment or maltreatment of a child, under 18 years of age, by a person who is responsibile for the child's welfare, under circumstances which indicate that the child's health or welfare is harmed or threatened." *Abuse* takes many forms but in general it refers to acts of commission such as beating or excessive chastisement, while *neglect* refers to acts of omission

such as failure to provide adequate food or emotional care.

The abused child is likely to be in a physically active developmental period. Remember that the infant from 6 to 12 months of age no longer passively receives the environment but has the motor skills to actively meet it. Abused children come from all socioeconomic groups; slightly more are males. The child in most instances has been in a home setting in which parenting is characterized by hostility, aggression, rigidity, compulsion, passivity, and dependence.

Kempe, Silverman, and Steele, have devised an assessment schema for predicting the possibility of abusive patterns in parents and caretakers. According to this schema, three factors must be examined: the parents, the child, and a precipitating crisis. Regarding the parents, Kempe (1962) found the following to be common:

1. Thirty to sixty percent of abusive parents were abused children themselves.
2. Inadequate trust; abusive parents (particularly the mother) tend to have been unable, as children, to trust their own parents.
3. The marital relationships of abusive parents are often characterized by instability and passivity.
4. Parental perceptions of the child are often unrealistic from a developmental standpoint.

The child is viewed as "different" by the parent; that is, the child may be retarded, exceptionally bright, excessively active, etc. In any event, the child is incapable of responding in the manner expected by parents.

A crisis or series of changes is usually necessary to set the abusive action into motion. This crisis may seem minor but because of other contributing factors generally proves overwhelming.

Detection of abuse is a key nursing responsibility. Major indicators of physical abuse include bruises, welts, scars, burns, central nervous system injuries manifested in cerebral hemorrhage and subdural hematoma, bone injuries, and failure to thrive. One is likely to find these injuries inconsistent with the history and developmental potential of the child. For example, a six-month-old infant with a fractured humerus that is supposedly the result of a fall from a high chair merits a close history assessment. An infant at this stage is more likely to acquire a head injury because he lacks the parachute reflex. The history provided by the parent may be vague and filled with gaps, and evidence of old injuries may be seen.

The nurse and allied health care professionals are responsible for detecting and preventing the abuse. Knowledge of prevalent lesions and their characteristics is needed, along with an understanding of the dynamics of the abusive cycle. Utilization of this knowledge is critical in assessing the family and specific abuse.

Equally important is prevention. Screening for high-risk families before the abuse occurs is optimal. The nurse should be aware of the parent's degree of competence as a parent. Answers to questions such as "When your child cries, what do you do?" and "What do you like best about your child?" can provide excellent clues. Parents who abuse are anxious and reluctant to give information. They may also be very deficient in and anxious about child-rearing and their parenting role. The mother who says upon discharge from the maternity floor, "I don't know how to put on a diaper" is signaling great potential inadequacies. What else does she not know about her parenting role? Anticipatory teaching of child care, developmental norms, and child-rearing practices is essential.

When abuse is suspected, the nurse and other specified professionals (including teachers, physicians, dentists, and day-care workers) are legally directed to report either to the appropriate social services, the police, or the district attorney's office. The lay person is also legally bound to make a report when abuse or neglect is known to exist.

Support groups like Parents Anonymous, Crisis Line, and Child Teams can be very helpful in functioning to detect crises or support the family during crises. The ultimate goal of all these groups is ensuring the child's safety and maintaining the integrity of the family unit.

ANTICIPATORY GUIDANCE

A knowledge of biophysical and psychosocial maturation is essential to the understanding of behavior as the infant moves from an almost totally dependent state toward one of increasing independence and "self-proclamation." Central to this progression are specific motor, social, and emotional behaviors that emerge as the central nervous system matures.

The nurse must be cognizant of the principles of biophysical and psychosocial change in order to provide anticipatory guidance to parents and other caretakers before and during the emergence of these developmental stages. Although anticipatory guidance related to certain developmental areas will be presented here in separate sections, the readers should remember that development is an interrelated whole the aspects of which are interdependent. Therefore the nurse should take a holistic approach; health maintenance and illness prevention in the infant necessitate consideration of all developmental aspects of the emerging being.

Neuromuscular Function

The Denver Developmental Screening Test (Appendix B) is a useful gross screening tool in assessing neuromuscular maturation. Neuromuscular intactness and maturation can most reliably be established, however, through observation, interviewing, and careful history-taking. Often, particularly with the infant and young child, motor skills can be assessed during the nurse-parent interview and history-taking. The nurse assess during the interview whether the 8-month-old, given a tongue blade to play with, transfers it from one hand to the other while sitting on his mother's lap. Does the 5-month-old sit when placed on the examining table for undressing?

Research studies reveal that stimulation can speed up delayed development and even encourage early acquisition of skills once neuromuscular readiness is attained. For example, walking skills was found to occur 2 to 3 weeks earlier in infants having access to "walkers."

Psychosocial Function

With the development of socialization patterns come other predictable behaviors. The infancy period is marked by infant anxiety, specifically, stranger anxiety, appearing at 5 to 7 months of age, and separation anxiety, beginning between 6 and 10 months. To alleviate the stress, a consistent, sensitive attitude on the part of caretakers is important. Any new interactions with strangers should occur in the presence of the parent or caretaker. The infant should be allowed to "warm up" to the new acquaintance. Physical contact should not be forced; slowly offering an outstretched hand is less threatening. Verbal reassurance and cuddling or holding by the parent during the introduction can do much to reassure the infant.

Support for separation anxiety involves substitute caretaker roles. The nurse should be informed about the infant's idiosyncracies and should be sensitive to his needs. Maintenance of routines offers much security, as does consistent concern in caretakers.

Parents should be made aware of the importance of encouraging social responses. This can be accomplished with frequent talking to the infant; tactile stimulation, such as holding, cuddling, stroking; eye contact with the infant; social exposures involving all of the family; and the creation of a nonthreatening environment for interactions—a limited number of people, with the caretaker present, in familiar surroundings (see Fig. 5-18).

Prevention of Infections

Having lost much of the natural immunity of placental protection, the 6- to 12-month-old infant is susceptible to numerous infections (discussed earlier in this chapter). Several precautions are necessary during infancy to prevent infections. Protective measures include clothing the infant properly, being careful not to over- or underclothe, and limiting his exposure to others. Extreme changes in temperature and exposure to infected individuals should especially be avoided. Both parents and health care providers should give special attention to hand-washing before handling the infant.

Immunizations are one of the most efficacious means of disease prevention. The American Academy of Pediatrics recommends that all normal infants receive the complete immunization series against diphtheria, pertussis (whooping cough), tetanus, rubella, mumps, and polio (see Appendix E).

Accident Prevention

Accidents persist as a leading cause of death in infancy, particularly in the sixth through twelfth months. Teaching parents about accident-proofing the infant's environment should start before or at birth.

The most common accidents in the first months of life are falls. The nurse should educate parents and caretakers about anticipated motor capabilities as well as those already existing. Proper restraint in federally approved cribs, playpens, etc., is essential. Side-rails should always be secured in the up position

Birth to 6 Months:
Language and Personal/Social Development:

1. Talk and sing to the infant (even though he cannot understand).
2. Repeat the noises the baby makes.
3. While the infant is awake, put him in a place where it is possible for him to see and hear what is going on.

Visual, Auditory, and Tactile Stimulation:

1. Hang pictures on the wall or crib where the baby can see them.
2. Hang dangling toys above the baby.
   A. Attach ribbon, bright cloth, colored paper, balls, a shiny spoon, painted spools, measuring spoons, etc., to a string or coat hanger and hang across crib.
   B. Cut a circular piece of cardboard or plastic; using thread, tie on colorful cutouts to make a mobile, hang from light fixture or ceiling above crib.
3. Provide a rattle.
   Fill small cardboard boxes, plastic salt shaker, beer cans, etc. with large spoons, bottle caps, small wooden blocks, spoons, etc. and tape the end shut. (Use objects too large to swallow in case rattle comes apart.)
4. Provide soft, cuddly toys.
   Sew two pieces of cloth together and stuff with rags.
5. Put the infant on a blanket on the floor; this enables him to explore and exercise.

Six to 12 Months:
Language and Personal/Social Development:

1. Talk to the baby; explain what you are doing (even though the baby cannot understand).
2. Point to objects and people and name them repeatedly.
3. Play with the baby (peek-a-boo, patty-cake, etc.).
4. While the infant is awake, keep him in the room with you; allow him to crawl on the floor or move about in a walker.
5. Let the infant look in a mirror. Talk to him while he is looking; "Look at baby's nose," etc.

Visual, Auditory, and Tactile Stimulation:

1. Provide soft, cuddly toys.
2. Provide noise makers.
   Make a drum from an empty oatmeal box, give the infant a spoon to hit it with.
3. Provide objects (unbreakale and too large to be swallowed) to handle and explore. When the infant is in his walker, tie these to the walker to make it easy to play with them:
   A. Spoons, rolling pin, boxes, bowls, pots, pans, cans, cups, etc.
   B. Provide a variety of textures (hard, soft, fuzzy, smooth, etc.) and materials (sponges, velvet, fur, cotton, wool).
4. Provide fill-and-dump toys.
   Provide a container with a large opening and small objects to place in container and dump out (spools, measuring spoons, clothespins, corks).
5. Put a favorite toy in a paper bag and have the infant find it.

Figure 5-18. *Stimulating activities for infants.*

147

when the infant is in the crib. Caution should be used when placing the infant on an open bed; when necessary, adequate pillow restraining and protection should be provided.

The 6- to 12-month-old infant is vulnerable to falls, spill burns, car accidents, poisoning and even drowning. The environment should be freed of hazards such as potentially poisonous cleaning agents, plug-in appliances and other items that can be pulled off of high surfaces and onto the infant, and small items that can be ingested. Restraining gates are recommended for stairways and other potentially dangerous areas. When bathing the infant, minimal levels of water should be used, and the infant should *never* be left unattended while in the water. Toys should be of such sizes and shapes that neither the whole toy nor any part of it can be swallowed.

Feeding Needs

Nutrient needs for the infant are expressed in terms of the amount of calories per kilogram of body weight needed to maintain growth. Meeting nutritional needs during the first year is critical for the maintenance of growth, development, and health. Deficiencies acquired during this period can have lasting effects on later growth. At birth the nutritional and general health status of the infant is directly related to maternal intake. Assuming that the intake has been adequate, and that the mother has not been adversely affected by other factors, the infant's birth weight will be within the normal weight range. (Recommended dietary allowances for the infant from birth to 12 months are listed in Table 5-12.) Since the infant's body has a far greater proportion of water than the adult and the potential for water loss is greater, 100 to 165 ml of water per kg of body weight is needed to meet daily fluid requirements.

Whether parents decide to breast-feed or bottle-feed their child, a key nursing role is to provide information on which to base that decision.

Whole cow's milk contains more than twice the protein (21 percent) of human milk (8 percent). Casein is a constituent of the protein in both, but casein content is much higher in cow's milk. When casein reacts with hydrochloric acid in the stomach, a tough, cheesy, hard-to-digest curd is formed. Brand-name formulas are more

Table 5-12. Recommended Dietary Allowances for Infants Through 12 Months of Age

	0 to 6 mo	6 to 12 mo
Weight	6 kg (13 lb)	9 kg (20 lb)
Height	60 cm (24 in)	71 cm (28 in)
Kilocalories	kg × 115	kg × 105
Protein, g	kg × 2.2	kg × 2.0
Fat-soluble vitamins		
Vitamin A, mcg r.e.	420	400
Vitamin D, mcg (IU)	10 (400 IU)	10 (400 IU)
Vitamin E, IU	4	5
Water-soluble vitamins		
Ascorbic acid, mg	35	35
Folacin, mcg	30	45
Niacin, mg	6	8
Riboflavin, mg	0.4	0.6
Thiamine, mg	0.3	0.5
Vitamin B_6, mg	0.3	0.6
Vitamin B_{12}, mcg	0.5	1.5
Minerals		
Calcium, mg	360	540
Phosphorus, mg	240	360
Iodine, mcg	40	50
Iron, mg	10	15
Magnesium, mg	50	70
Zinc, mg	3	5

NOTE: Allowances for vitamins A and D are currently being expressed in mcg although many food conposition tables state values in international units (IU); 1 retinol equivalent (r.e.) = 3.33 IU of retinol or 10 of B-carotene.

SOURCE: Adapted from *Recommended Dietary Allowances*, 8th rev. ed., National Research Council, National Academy of Science, Washington, D.C. 1980.

comparable to breast milk and are a recommended choice.

It is generally recommended that the infant be fed breast milk or formula for the first 6 months. The rationale for this is to avoid the potential for allergic reaction and overfeeding that exists when solids and formula are given. In the second 6 months solids should be gradually introduced. Cereals should be introduced first, followed by the pureed fruits, yellow vegetables, green vegetables, and meats. The foods should be introduced one at a time, a minimum of 3 or 4 days apart; this allows for easy detection of food allergens. By 1 year, most infants tolerate finger foods as well as some table foods.

The need for supplementation of iron and fluoride depends on the type of feeding and should be assessed by the nurse. As of this writing, the American Academy of Pediatrics recommends that all infants on commercial formulas receive formulas that are fortified with iron. Iron-deficiency anemia is the leading nutritional problem in infancy and adolescence. It is recommended that breast-fed infants receive an oral iron supplement (1 mg per kg of weight) (National Research Council, 1980).

A fluoride supplement is essential for sound dentition and the prevention of caries. Breast-fed infants especially need a fluoride supplement since fluoride is not absorbed in breast milk. Supplementary fluoridated water is usually not necessary and can result in mottling of the teeth. When water contains greater than 0.7 parts fluoride per million, no supplementation is necessary.

SUMMARY

By most definitions, infancy comprises the first 12 months of life, and the first 28 days are regarded as the neonatal period. Infancy rivals adolescence as one of the most dramatic periods of growth in the life span. The infant's birth weight doubles by 4 to 6 months and triples in a year, by which time the birth length will have increased by about 50 percent. The neonatal period of infancy is an especially vulnerable time, characterized by a mortality rate that is 50 percent greater than that for the rest of infancy; leading causes of neonatal death include respiratory distress as a complication of prematurity, infections, and congenital malformations.

Binocular vision and convergence, which are absent at birth, develop gradually as myelination of the neural pathways occurs. Binocular fixation is seen by 2 to 3 months, at which time the infant's hearing has also developed to the extent that he will turn his head toward a sound. By 4 months, the infant's capacity for visual accommodation is mature. Throughout the first 12 months of life and beyond, the infant's endocrine, gastrointestinal, musculoskeletal, respiratory, renal, and central nervous systems are immature, making him more vulnerable to infection, fluid-electrolyte imbalance, and digestive limitations, among other problems.

In Piaget's terms, the first 18 or 24 months of life comprise the sensorimotor stage, which is characterized by the infant's profound egocentrism, which arises from his being limited largely to sensory and motor experiences. The infant engages in social exchanges via eye contact, touching, laughing, and crying. The "social smile," a milestone of psychosocial development, emerges at about 2 months. All of these behaviors facilitate the parent-infant bonding process, which begins at birth and continues throughout the first year. Closely related to the success of the bonding process, i.e., the quality of interaction between infant and parent, is the resolution of the major developmental conflict of the period as Erikson sees it, that is, the development of trust versus mistrust.

Common health and developmental problems of the period include oral aggressiveness, separation anxiety, colic, failure to thrive, accidents (the most frequent being falls, poisonings, burns, and automobile accidents), sudden infant death syndrome, and child abuse. Anticipatory guidance should be directed toward the prevention of infections, play needs (i.e., appropriateness and safety of toys), feeding needs, and acquainting parents with the physiological and psychosocial norms of the period.

REFERENCES

Ainsworth, M. D. S. The development of infant-mother attachment. In B. M. Caldwell, and N. C. Ricciuti (Eds.), *Review of child development research* (Vol. 3). Chicago: University of Chicago Press, 1973.

American Academy of Pediatrics Committee Statement. The ten-stage nutritional survey: a pediatric perspective. *Pediatrics*, 1973, *51*(6), 1095–1099.

Arganian, M. Sex differences in early development. *Individual differences in children*. New York: John Wiley & Sons, 1973, pp. 46–63.

Ariès, Philippe. *Centuries of childhood*. New York: Vintage Books, 1962.

Arnold, H. W., et al. Transition to extrauterine life. *American Journal of Nursing*, 65, October, 1965, 77–80.

Babson, S., et al. Growth and development of twins of dissimilar size at birth. *Pediatrics*, 1964, *33*, 327–333.

Beckwith, P. Discussion of terminology and definition of sudden infant death syndrome. In A. Bergman and associates (Eds.), *Proceedings of the second international conference on causes of sudden infant deaths*. Seattle: University of Washington Press, 1970.

Bell, S. M., and Ainsworth, M. D. S. Infant crying and maternal responsiveness: a rejoinder to Gerwitz and Boyd. *Child Development*, 1977, *48*, 1208–1216.

Bell, R. Q., and Weller, G. M. Basal skin conductance and neonatal state. *Child Development*, 36, 3, 1965, 647–657.

Bellanti, J. A. Allergy and clinical immunology: an interdisciplinary concept. *Annals of Allergy*, 1978, *41*(3), 129–135.

Berg, W. K., Adkinson, C. D., and Strock, B. D. Duration and periods of alertness in neonates. *Developmental Psychology*, 9, 3, July 1973, 9434.

Bergman, A. B. Studies of the sudden infant death syndrome in King County, Washington. *Pediatrics 49*, 3. June, 1972, 860–870.

Bergman, A. Relationship of passive cigarette smoking to sudden infant death syndrome. *Pediatrics*, 1976, *58*(5), 665–668.

———. Sudden infant death syndrome. *Nursing Outlook*, December, 1977, 777.

Bortoshuk, A. K. Human neonatal cardiac responses to sound: a power function. *Psychonomic Science*, 1964, *151*, 52.

Bower, T. J. The object in the world of the infant. *Scientific American*, 225, 4, October, 1971, 30–38.

Bower, T. G. *Development in infancy*. San Francisco: W. H. Freeman, 1974.

Bowlby, J. A symposium on the contribution of current theories to an understanding of child development. I. An etiological approach to research in child development. *British Journal of Medical Psychology*, 30, 4, 1957 320–240.

Bowlby, J. The nature of the child's tie to his mother. *International Journal of Psychoanalysis*, 1958, *39*, 350–373.

———. Beginnings of attachment behavior. In J. Bowlby (Ed.), *Attachment and loss* Vol. 1. New York: Basic Books, 1969.

Brackbill, Y. Continuous stimulation and arousal levels in infants: additive effects. Child Development, 42, March, 1971, 17–26.

Brazelton, T. B. Psychophysiological reaction in the neonate. II. Effects of maternal medication on the neonate and his behavior. *Journal of Pediatrics*, 1961, *58*, 513–518.

———. The early mother-infant adjustment. *Pediatrics*, 1963, *32*, 931–938.

———. *Neonatal Assessment scale*. Philadelphia: J. B. Lippincott, 1973.

Brazelton, T. B., School, M. L., and Robey, J. S. Visual responses in the newborn. *Pediatrics*, 37, Feburary. 1966, 284–290.

Brazelton, T. B., Robey, J. S., and Collier, G. A. Infant development in the Zinacanteco Indians of South Mexico. *Pediatrics*, 1969, *44*, 274–293.

Brazelton, T. B., Kowlowski, B., and Main, M. *The effect of the infant on the caregiver*. New York: John Wiley & Sons, 1974.

Campos, J. L., Langer, H., and Krowitz, A. Cardiac responses on the visual difference in prelocomotor human infants. *Science*, 1970, 196–197.

Caputo, D. V. and Mandel, W. Consequences of low-birth weight. *Developmental Psychology*, 1970, *3*, 363–383.

Chan, I. Letter to the editor. Newsletter of the Society for Research in *Child Development*, Winter, 1975.

Chess, S. Temperament in the normal infant. *The exceptional infant*, (Vol. 1). Seattle: Special Child Publications, 1967.

Chess, S. A. and Birch, H. G. Behavior problems revisited: findings of an aterospective study. In S. Chess, and A. Thomas (Eds.). *Annual progress in child psychiatry and child development*. New York: Brunner/Mazel, 1968.

Chinn, P. *Child health maintenance*. St. Louis: C. V. Mosby Co., 1978.

Chomsky, N. *Syntactic structures*. The Hague: Moulton, 1957.

Church, J. (Ed.), *Three babies: biographies of cognitive development*. New York: Random House, 1966.

Churchill, J. H. The relationship between intelligence and birth weight in twins. *Neurology*, 1965, *15*, 341–347.

Clark, A. L., and Affonso, D. P. *Childbearing: a nursing perspective*. Philadelphia: F. A. Davis, 1979.

Comer, J. P., and Poussaint, A. F. *Black child care*. New York: Simon and Schuster, 1975.

Conel, J. L. *The postnatal development of the human cerebral cortex* (Vol. 1–8). Cambridge, Mass.: Harvard University Press, 1939.

Corbin, C. *A textbook of motor development*, Dubuque, Iowa: W. C. Brown, 1973.

Craig, G. J. *Human development*. Englewood Cliffs, N.J.: Prentice-Hall, 1976.

Dayton, G. O. et al. Developmental study of coordinated eye movements in the human infant. *Archives of Ophthalmology*, 71, June, 1964, 865–870.

Decarie, T. G. *Intelligence and affectivity in early childhood*. New York: International University Press, 1966.

Desmond, M. M. et al. The Clinical Behavior of the Newly Born. *International Journal of Pediatrics*, 1963, 62, 307–325.

Dickason, E. J., and Schult, M. D. *Maternal and infant care: a text for nurses*. New York: McGraw-Hill, 1975.

Eimas, P. Auditory and linguistic processing of cues for place of articulation by infants. *Perception & Psychophysics*, 1974, *16*, 513–521.

Eisenberg, R. B. Stimulus significance as a determinant of infant response to sound. In E. B. Thomas, and S. T. Trotter (Eds.). *Social Responsiveness of Infants*. New Brunswick, New Jersey: Johnson & Johnson Baby Products, 1978.

Eisenberg, R. B. et al. Auditory behavior in the neonate. *Journal of Speech & Hearing Research*, 1964, 7, 245–269.

Engen, T., Lipsitt, L P., and Kaye, H. Olfactory response and adaptation in the human neonate. *Journal of Comprehensive Physiological Psychology*, 34, June, 1963, 73–77.

Engen, T, and Lipsitt, L. P. Decrement and recovery of responses to olfactory stimuli in the human neonate. *Journal of Comprehensive Physiological Psychology*, 59, 1965, 312–316.

Fantz, R. L. The origin of form perception. *Scientific American*, 1961, 204, 5, May, 66–71.

———. Pattern vision in newborn infants. *Science*, 1963, *140*, 296–297.

———. Vision perception from birth as shown by pattern selectivity. In L. J. Stone, *The research and commentary*. New York: Basic Books, 1973.

———. Complexity and facial resemblance as determinant of response to facelike stimuli by 5 and 10 week old infants. *Journal of Experimental Child Psychology*, 1974, *18*, 480–487.

Fantz, R. L., and Miranda, S. B. Newborn infant attention to form of contour. *Child Development*, 1975, *46*, 224–228.

Fantz, R. L., Fagan, J. F., and Miranda, J. B. Early visual selectivity. In L. B. Cohen, and P. Salapatek (Eds.), *Infant perception: from sensation to cognition*. New York: Academic Press, 1975.

Farrar, C. A. Assessing individuality in the newborn. *Nursing Digest*, July–August, 1975, 16–19.

Fein, G. G. *Child development*. Englewood Cliffs, N.J.: Prentice-Hall, 1978.

Flanell, J. H. *The developmental psychology of Jean Piaget*. Princeton, N.J.: Van Nostrand, 1963.

Frank, L. K. *On the importance of infancy*. New York: Random House, 1966.

Friedman, D. S., and Friedman, V. C. Behavioral differences between Chinese-American and European-American newborns, *Nature*, 1961, *224*, 1227.

Frodi, A. M., and Lamb, M. E. Child abusers responses to infant smiles and cries. *Child Development*, 1980, *51*(1), 238–241.

Gerwitz, J. L. The cause of infant smiling in four child-rearing environments in Israel. In R. M. Foss (Ed.), *Determinants of infant behavior* (Vol. III). London: Methuen & Co., 1965.

Gerwitz, J. L., and Boyd, E. F. Does maternal responding imply reduced infant crying? A critique of the 1972 Bell and Ainsworth report. *Child Development*, 1977. 48, 1200–1207.

Gesell, A. F. et al. *Infant and child in the culture of today* (Rev. Ed.). New York: Harper & Row, 1974.

Gibson, J. *Growing up: a study of children*. Reading, Mass.: Addison-Wesley, 1978.

Gibson, E. J. and Walk, R. D. The "visual cliff." *Scientific American*, 1960, *202*, 64–71.

Harris, J. When babies cry. *Canadian Nurse*, 1979, *75*(2) 32–34.

Hayes, H. Visual accommodation in human infant. *Science*, 1965, *148*, 528–530.

Hebb, D. O. The motivation effects of exteroceptive stimulation. *American Psychology*, 1958, *13*, 109.

Helfer, R. N. The etiology of child abuse. *Pediatrics*, 1973, *51*(4), 777–779.

Helms, D. B. and Turner, J. S. *Exploring child behavior*. Philadelphia: W. B. Saunders, 1978.

Hetherington, E. M., and Parke, R. D. *Child psychology: a contemporary viewpoint*. New York: McGraw-Hill, 1975.

Heirlock, E. B. *Developmental psychology: a life span approach*. New York: McGraw Hill, 1980.

Hletko, P. J., and Hletko, J. O. Auto safety: preventive medicine for a pediatric epidemic. *Pediatric Basics*, 1978, *21*, 10–14.

Hughes, J. G. *Synopsis of pediatrics*. St. Louis: Mosby, 1971.

Janov, A. *The primal scream*. New York: Dell, 1970.

Jensen, K. Differential reactions to taste and temperature stimuli in newborn infants. *Genetic Psychology Monographs,* 1932, *12,* 363–479.

Johnson, T. R., More, W., and Jeffries, J. (Eds.) *Children are different.* Columbus, Ohio: Ross Laboratories, 1978.

Kagan, J. *Change and continuity in infancy.* New York: John Wiley and Sons, 1971.

Kagan, J. Do infants think? *Scientific American,* March 1972, 74–82.

Kalmis, I. B., and Brunce, J. S. Infant sucking used to change the clarity of a visual display. In L. J. Stones, *The competent infant: research and commentary.* New York: Basic Books, 1973.

Kaluger, G., and Kaluger, M. F. *Human development: the span of life.* St. Louis: Mosby, 1979.

Kastenbaum, R. *Humans developing: a lifespan perspective.* Boston: Allyn and Bacon, 1979.

Kempe, C. H., Silverman, F. N., and Steele, B. F. Battered child syndrome. *Journal of American Medical Association,* 1962, *181,* 17.

Kennedy, W. *Child psychology.* Englewood Cliffs, N.J.: Prentice Hall, 1975.

Klaus, M. and Kennell, J. H. *Maternal–infant bonding.* St. Louis: Mosby, 1976.

Klaus, M., Leger, T., and Trause, M. A. *Maternal attachment and mothering disorders.* Johnson & Johnson Baby Products Co., New Brunswick, N. J., 1974.

Korner, A. F. The effect of the infants state, level of arousal, sex and autogenetic stage on the caregiver. In M. Lewis, and L. A. Rosenblum (Eds.). *The effects of the infant on its caregiver.* New York: John Wiley & Sons, 1974.

Korones, J. B. *High-risk newborn infants.* St. Louis: Mosby, 1972.

Labov, W. et al. *A study of the nonstandard English of Negro and Puerto Rican speakers in N.Y. City.* New York: Columbia University Press, 1968.

Lamb, M. E. A re-examination of the infant's social world. *Human Development,* 1977, *20,* 65–85.

Lathan, H., and Hecbil, R. *Pediatric Nursing* (3rd Ed.). St. Louis: C. V. Mosby, 1977.

LeBoyer, F. *Birth without violence.* New York: Alfred A. Knopf, 1975.

Lenneberg, E. H. *Biological foundations of language,* New York: John Wiley & Sons, 1967.

Liebert, R. M., Poulos, R. W., and Strauss, G. D. *Developmental psychology.* Englewood Cliffs, N.J.: Prentice-Hall, 1974.

Lipsitt, L. P., and Levy, N. Pain threshold in the human neonate. *Child Development,* 1959, *30,* 547–551.

Lowery, G. H. *Growth and development of children* (7th Ed.). Chicago: Yearbook Medical Publishers, 1978.

Maccoby, E. E., and Field, S. S. Mother-attachment and stranger reactions to the third year of life. *Monography of the society for research in child development,* (Serial 146), *37,* 1972.

MacFarland, A. Olfaction in the development of social preferences in the human neonate. In *Parent-Infant Interaction Ciba Foundation Symposium.* New York: Elsevier, 1975, *33,* 103–117.

MacFarland, J. A. What a baby knows. *Human Nature,* 1978, *1*(2), 74–81.

MacFarland, J. A. *The psychology of childbirth.* Cambridge, Mass.: Harvard University Press, 1977.

Marlow, D. *Textbook of pediatric nursing.* Philadelphia: Saunders, 1977.

Maslow, A. H. *Motivation and personality.* New York: Harper & Row, 1970.

Maurer, D. M., and Maurer, C. E. Newborn babies see better than you think. *Psychology Today,* 1976, 10, November, 85–88.

McGurk, H., Turnure, C., and Creighton, S. S. Auditory-visual coordination in neonates. *Child Development,* 1977, *48,* 138–143.

Mead, M. *Coming of Age in Samoa.* New York: Morrow, 1928.

Mendelson, M. J., and Haith, M. M. The relation between audition and vision in the human newborn. *Monographs for the society for research in child development,* 1976, Serial No. 167.

Moore, M. L. *The newborn and the nurse.* Philadelphia: Saunders, 1972.

Murray, A. D. Infant crying as elicitor of parental behavior: an examination of two models. *Psychology Bulletin,* 1979, *86*(1), 141–215.

Murray, R., and Zentner, J. *Nursing assessment and health promotion through the life span.* Englewood Cliffs, N.J.: Prentice-Hall, 1975.

Murray, R., and Zentner, J. *Nursing assessment and health promotion through the life span.* Englewood Cliffs, N.J.: Prentice-Hall, 1979.

Mussen, P. H., Conger, J. J., and Kagan, J. *Child development and personality.* New York: Harper & Row, 1974.

Neumann, C. G. and Alpaugh, M. Birth-weight doubling time—a fresh look. *Pediatrics,* 1976, *57*(4), 469–473.

Ounsted, M. Fetal growth. In D. Gordner, and D. Hill (Eds.) *Recent advances in pediatrics.* London: Chureheld, 1971.

Papalia, D. E., and Olds, S. W. *A child's world: infancy through adolescence.* New York: McGraw-Hill, 1979.

Papoušek, H. Conditioned head rotation reflexes in infants in the first months of life. *Acta Paediatrica Scandinavica,* 1961, *50,* 565–576.

———. Experimental studies of appetitional behaviors in human newborns and infants. In H. W. Stevenson, E.

H. Hess, and H. Rheingold (Eds.), *Early behavior: comparative and developmental approaches*. Huntington, N.Y.: Krieger, 1967.

Pederson, F. A., Rubenstein, J., and Yarrow, L. J. Father absence in infancy. (Paper presented at the Meeting of the Society for Research in Child Development, Philadelphia: March 1973.)

Peeples, D., and Teller, D. Color vision and brightness discrimination in two-month-old infants. *Science*, 1975, *189*, 1102–1103.

Piaget, J. *The origins of intelligence in children*. New York: International University Press, 1952, Morton, 1974.

Piaget, J., and Inhelder, B. *The psychology of the child*. (H. Weaver trans.), New York: Basic Books, 1969.

Post, P. et al. Skin reflectant of newborn infants from 25 to 44 weeks gestational age. *Human Biology*, 1976, *48*(3), 541–557.

Pratt, K. C., Nelson, A. K., and Sun, S. C. The behavior of the newborn infant. Ohio State University Study Center, Psych. 1930, *10*.

Pratt, K. C. *The effects of repeated visual stimulation on the activity of newborn infants* (2nd ed.). New York: John Wiley & Sons, 1954.

Rank, D. *The trauma of birth*. New York: Harper & Row, 1973.

Robson, K., and Moss, A. Patterns and determinants of maternal attachment. *Journal of Pediatrics*, 1977.

Rosen, H. *Pathway to Piaget*. Cherry Hill, N.J.: Postgraduate International, 1977.

Rheingold, H. L. The modification of social responsiveness in institutional babies. *Monographs of the Society for Research in Child Development*, Serial No. 63, 1956.

Scarr, S. Patterns of fear development during infancy. *Merrill-Palmer Quarterly*, 1970, *16*, 53–90.

Schmeck, H. M., Jr. Trend in growth of children logs. *New York Times* June 10, 1976, 13C.

Schuster, C. S., and Ashburn, S. S. *The process of human development: a holistic approach*. Boston: Little, Brown, 1980.

Schwartz, P. Birth injuries of the newborn. *Archives of Pediatrics* 1956, *73*, 12, December, 429–450.

Scipien, G. M., et al. *Comprehensive pediatric nursing*, New York: McGraw-Hill, 1975.

Sidel, R. *Women and child care in China*. Baltimore: Penguin, 1973.

Sigueland, E. R., and Lipsitt, L. P. Conditioned headturning in human newborns. *Journal of Experimental Child Psychology*, 1966, *3*, 356–376.

Sherman, M., and Sherman, I. C. Sensorimotor responses in infants. *Journal of Comparative Psychology*, 1925, *5*, 53–68.

Skinner, B. F. *Science and human behavior*. New York: Macmillan, 1953.

Skinner, B. F. *Verbal behavior*. New York: Appleton-Century-Crofts, 1957.

Slobin, D. I. Cognitive prerequisites for the development of grammar. In C. A. Fugerson, and D. I. Slokin (Eds.), *Studies of child language development*. New York: Hold, Reinhart, and Winston, 1973.

Smart, M. S., and Smart, R. C. *Children development and relationships*. New York: Macmillan, 1977.

Smart, R. C., and Smart, M. S. Readings in child development and relationships. New York: Macmillan, 1977.

Smith, C. A., and Nelson, N. W. *The physiology of the newborn infant*. Springfield, Ill.: Charles C Thomas, 1976.

Smith, D., Bierman, E., and Robinson, N. *The biologic ages of man: from conception through old age*. Philadelphia: Saunders, 1978.

Spelke, E. et al. Father interaction and separation protest. *Developmental Psychology*, 1973, *9*, 83–90.

Spitz, R. A., and Wolff, K. M. Anaclitic depression: an inquiry into the genesis of psychiatric conditions in early childhood. In A. Freud (ed.), *The psychoanalytic study of the child* (Vol. II). New York: International University Press, 1946.

Starr, B. D., and Goldsten, H. S. *Human development and behavior: psychology in nursing*. New York: Springer, 1975.

Stone, V. (Ed.). *Physiology of the perinatal period* (Vol. I, II). New York: Appleton-Century-Crofts, 1970.

Sroufe, L. A., and Waters, E. The ontogenesis of smiling and laughter: a perspective on the organization of development in infancy. *Psychological Review*, 1976, *83*, 173–189.

Sroufe, L. A., and Wunsch, J. P. The development of laughter in the first year of life. *Child Development*, 1972, *43*, 1326–1344.

Tanner, J. M. *Foetus into man: physical growth from conception to maturity*. London: Open Books, 1978.

The National Research Council. *Recommended dietary allowances* (9th ed.). Washington, D.C.: National Academy of Sciences, 1980.

Thomas, A. *Temperament and behavior disorders in children*. New York: New York University Press, 1968.

Turner, J. S., and Helms, D. B. *Life span development*. Philadelphia: Saunders, 1979.

United Nations Demographic Yearbook. 1973–74, 76 and 77. New York: United Nations, 1974, 1975, 1977, 1978.

Vander Zanden, J. *Human Development*. New York: Knopf, 1978.

Waechter, E. H., and Blake, F. G. *Nursing care of children*. Philadelphia: Lippincott, 1976.

Watson, J. S. Smiling, cooing and "the game." *Merrill-Palmer Quarterly*, 1972, *18*, 323–339.

Watson, E. H., and Lowery, G. H. *Growth and development of children*. Chicago: Yearbook Medical Publisher, 1962.

Watson, R. I. *Psychology of the child*. New York: John Wiley & Sons, 1962.

Whaley, L. F., and Wong, D. L. *Nursing care of infants and children*. St. Louis: Mosby, 1979.

White, R. W. Motivation reconsidered: the concept competence. *Psychology Review*, 1959, *66*, 297–333.

Wilkerman, L., and Churchill, J. A. Intelligence and birthweight in identical twins. *Child Development*, 1967, *38*, 623–629.

Volff, P. H. Observations and experiments. *Psychological Issues*, 1966, *5*(1).

———. Observations on the early development of smiling. In B. M. Foss (Ed.), *Determinants of infant behavior* (Vol. II). New York: John Wiley & Sons, 1963.

———. The causes, controls and organization of behavior in the neonate. *Psychological Issues*, 1966, *5*(1) New York: International University Press.

———. The development of attention in young infants. *Annals of the New york Academy of Sciences*, 1965, *118*, 815–830.

———. The Natural history of crying and other vocalizations in early infancy. In B. Foss (Ed.), *Determinants of infant behavior* (Vol. II). London: Lloyd-Luke, 1974.

Yakonlender, P. I., and Lecours, A. R. The myelogenetic cycles of regional maturation of the brain. In A. Minkowski (Ed.) *Regional development of the brain in early life*. Oxford: Blackwell, 1967.

Yarrow, L. J. Attachment and dependency: a developmental perspective. In J. L. Gerwitz (Ed.), *Attachment and dependency*. Washington, D.C.: Winston, 1972.

Yarrow, L. J. *Research in dimensions of early maternal care*. Merrill-Palmer Quarterly, 1963, *9*, 101.

6

The Toddler

PATTY MAYNARD HILL

The toddler (the child from age 1 to age 3) typically has a large head, a long trunk, and short, stubby legs (Fig. 6-1). The head still accounts for much of the child's length, although less than the 25 percent of infancy. Head circumference is equal to or smaller than the chest circumference. Weak abdominal muscles give the toddler a "potbellied" look, and the child retains adipose tissue deposits that result in "baby fat". The toddler is noticeably swaybacked. The young toddler, with his short legs, appears "close to the ground" and may appear pudgy. The toddler is top-heavy and does indeed "toddle" about, using short tottering steps in a wide-based gait. Because of increased trunk size, the toddler has a greater sitting than standing height.

As the toddler moves into the second year, the growth of the legs and arms accelerates. The child acquires a more erect posture and appears less pudgy because of increased muscle development and increased prominence of the chest; he begins physically to look somewhat like a "miniature adult."

PHYSICAL CHARACTERISTICS

In contrast to the rapid growth of the preceding 12 months, growth in toddlerhood is less accelerated.

Figure 6-1. *The toddler typically has a large head, long trunk, short, stubby legs, and a somewhat "potbellied" look. Note the rudimentary grasp with which this child holds the marker. (Photo by Sam Gray, M. Photog., Cr.)*

Weight

By the second year, many infants weigh more than 20 lb. The average weight gain during toddlerhood is 7 oz per month, or 5 to 6 lb. per year. Typically, by the end of the second year, the child weighs a bit less than 30 lb. The birthweight is quadrupled by 2.5 years. Weight gain during toddlerhood tends to be somewhat erratic, with periods of little or no weight gain followed by periods of significant increase.

Height

The rate of linear growth also decelerates. Length increases by approximately 25 percent of the birth length, or approximately 3 to 5 in. (7.5 cm to 12.5 cm) per year. As a rule, predictable adult height is roughly twice that at age 2, although there is some deviation in adult males. Linear growth is steadier than weight changes.

Head Circumference

Because of the rapidity of brain development in the fetal and infant periods by 2 years of age the brain has acquired 75 percent of its adult weight and size. In toddlerhood, then, the growth of the head also decelerates; head circumference increases by approximately 1 inch (2.5 cm) per year. The anterior fontanel closes by 15 to 18 months.

The sphenoid and maxillary sinuses enlarge during the second and third years. Pneumatization of the mastoid antrum generally occurs by three years. Middle-ear infections often occur during this period because the eustachian tube is straighter and shorter than in the older child and adult.

Chest

Chest size and shape change after infancy. Changes in shape begin to occur as the transverse diameter of the toddler's chest increases beyond the anteroposterior diameter. At 1 to 2 years of age, chest circumference is less than abdominal circumference, but beyond two years, this is reversed.

Dentition

Primary (deciduous) tooth eruption is completed during the toddler period. At 10 to 12 months of age, eruption of the upper lateral incisors takes place; the lower lateral incisors erupt at 12 to 15 months. The first molars erupt at 12 to 16 months, and the canines, at 16 to 20 months. The second molars complete the eruption pattern at 20 to 30 months of age. Normally, primary dentition is completed by three years of age.

Teething is accompanied by increased salivation, irritability, sleep disruption, and painful gums. Calcification of the permanent teeth also occurs during toddlerhood; adequate dental care is essential.

Variations among toddlers in height and weight are slight. Most toddlers will weigh within 3 lb (1.3 kg) of each other and will vary by only 1 to 2 in. (2.5 cm to 5.0 cm) in height (Fig. 6-2). With regard to the assessment of physical data, it should be stressed that individual differences begin to emerge during this period; genetic influences, particularly, begin to be apparent. Attention should be directed more toward individual growth *patterns* than toward height, weight, or head circumference measurements of a given time. The nurse or health care provider should watch for a consistent progression of growth.

Central Nervous System Development

Development and myelination of the primary motor areas of the cortex (located in the precentral gyrus) occurs during the first 2 years. By the end of the second year the primary sensory areas have caught up with the primary motor areas. Myelination of the spinal cord and the primary motor areas (particularly those responsible for gross motor skills) is basically complete by this time, although leg control is not complete until somewhat later. Myelination of the areas responsible for fine motor skills continues during this period and is not complete until 4 years of age.

Fibers of the auditory system myelinate very gradually and the process is not complete until approximately 4 years of age. Areas of the forebrain (insula, cingulate gyri, hippocampus) myelinate slowly during the toddler period and continue myelinating until puberty. These areas are primarily responsible for consciousness and the maintenance of attention.

Myelination of the center responsible for light reception occurs rapidly. The eye itself also matures rapidly, and by 3 years it has completed much of its development. Visual maturity does not occur as rapidly; the 2-year-old has 20/70 vision, and the 3-year-old, about 20/40 vision. Hand-eye coordination is sufficiently refined to enable the 15-month-old to scribble spontaneously on paper, and by 3 years of age, cortical maturation is sufficiently advanced to allow the child to appraise the size and location of an object. Increasing central nervous system maturation results also in an improvement in the senses of taste, smell, and touch. Refinement of these senses continues throughout the toddler period.

Handling and visual inspection replace the oral-aggressive behaviors of infancy as tactile receptiveness increases. Taste matures in in-

Figure 6-2. *Usually, toddlers' heights vary by only 1 or 2 inches. (Photo by Sam Gray, M. Photog., Cr.)*

fancy, and by toddlerhood the child has definite preferences. The sense of smell, present in infancy, continues to develop throughout the toddler period. The toddler uses these sensory capacities in an integrated fashion; the combination of touching, looking, tasting, and smelling enables the toddler to explore and assimilate his environment (Fig. 6-3).

Musculoskeletal Development

Muscular changes are characterized primarily by hypertrophy (i.e., the muscle fibers increase in size). Changes in the skeleton and musculature of the toddler are reflected in the outward appearance. The adipose tissue so prevalent in infancy is gradually replaced by muscle mass. Heredity and muscle utilization appear to be key determinants of the rate of muscular development.

Skeletally, significant changes occur which

Figure 6-3. *The increased use of the senses to explore the environment is a milestone of toddlerhood, and playing with sand or water is especially gratifying. (Photo by Sam Gray, M. Photog., Cr.)*

contribute to proportionate changes in the toddler's height. The toddler's trunk becomes less prominent as leg growth accelerates. Leg height of the toddler approximates 35 to 40 percent of the total height in contrast to 30 percent of infancy.

Also near the end of the toddler period there is a beginning downward sloping of the ribs and drooping of the shoulders, with a beginning increase in neck length.

Numerous ossification centers develop during toddlerhood. By 12 months, new centers include the capital epiphyses of the femur and humerus as well as the distal tibial epiphyses. By 2 years the distal epiphyses of the fibulas has ossified and by 3 years the phalangeal and metacarpal epiphyses have also ossified.

The typical lordosis, or lumbar convexity, often seen in the young toddler is noticeable when the child walks. Additionally, the legs may appear bowed. Spontaneous overcorrection of this bowing results in knock-knee, occurring between 3 and 6 years of age. The lordosis generally disappears by 3 or 4 years of age.

Weight-bearing, which results from walking, causes characteristic anatomical changes of the foot. Throughout much of infancy the foot appears fat, thick, and archless because of the presence of the plantar fat pad. As the child begins to walk, this pad diminishes; the result is the formation of a longitudinal arch and the disappearance of the flat feet characteristic of infancy.

Generally, then, by 2 years of age the child has acquired sufficient maturation and myelination of the central nervous system and musculoskeletal system to be able to perform gross motor skills, such as walking, climbing stairs, and rudimentary running. Fine motor skills remain to be developed. The timing of the acquisition of motor skills, as with so many other aspects of human growth and development, is highly variable.

One of the milestones of the toddler period

is the achievement of walking. This skill opens avenues of exploration the toddler has never experienced before. Never again will he have to sit where placed and passively experience the environment presented by others. With this newly acquired skill, the child can actively and aggressively explore his surroundings. By roughly 12 to 15 months the toddler will walk, and by 18 months he will fall less frequently.

Early walking is characterized by a wide-based gait resulting from the imbalance caused by the toddler's "top-heaviness" and immature muscular development. Walking may not begin as early in infants and toddlers who are avid crawlers, since their mode of movement often serves them very effectively. Should this skill not emerge by the end of the second year, however, a thorough physical (including neurological) examination should be done.

As the skill of walking is being acquired, the child should wear soft-soled shoes such as tennis shoes. Going barefoot also facilitates walking, aiding in balance and control. Orthopedic shoes and expensive children's shoes should be discouraged, since the average period of use is 3 to 4 months. High-top shoes are felt by many authorities to be unnecessary, since low-tops are felt to facilitate muscle development and strength.

The child will have some running skill by 18 months. By 2 years the child walks up and down stairs alone, and at 3 years he can proficiently alternate feet in walking up and down stairs (Fig. 6-4). Table 6-1 summarizes the motor skills of the toddler period.

Figure 6-4. *By 2 years of age, most children can negotiate a flight of stairs without assistance. (Photo by Sam Gray, M. Photog., Cr.)*

Cardiopulmonary Development

Cardiopulmonary function increases significantly during the second and third years. The heart has doubled its weight by the beginning of the second year, and at age 2 or 3 years the cardiac shadows on x-rays resemble those on an adult's x-ray. Functional (innocent and ac-cidental) murmurs are common during toddlerhood and throughout middle childhood (until 7 years of age).

The width of the toddler's heart comprises half of the chest width. The apex of the heart remains in the 4th intercostal space, lateral to the midclavicular line throughout this period, until approximately 4 years of age.

Vascular changes include increased growth of the lumen of pulmonary vessels, an increase in the thickness of the walls of the great vessels (pulmonary veins, arteries, and aorta), and growth of the capillaries throughout the extremities. These changes coupled with central nervous system maturation (specifically, the

Table 6-1. Motor Acquisitions of Toddlerhood

Age	Gross Motor	Fine Motor	Adaptive
13 months	Walks with hand held Stands momentarily	Has refined pincer grasp	Attempts to build tower of 2 cubes Releases cube in cup after demonstration Plays serially with objects
15 months	Walks or toddles alone Creeps upstairs	Places round objects in a hole	Builds tower of 2 blocks Puts 6 cubes in and out of cup Scribbles spontaneously
18 months	Walks proficiently Attempts to run (falling often) Throws ball overhanded Seats self in small chair, climbs into adult chair	Turns pages of book together	Builds tower of 3 to 4 blocks Imitates a stroke with a crayon Dumps objects from bottle
2 years	Walks up and down stairs, using nonalterating steps (placing both feet on step before next step) Runs with little falling Kicks large ball	Turns single pages in book Turns doorknob Unscrews lid	Builds tower of 6 blocks Builds train with blocks Imitates circular and vertical strokes
2.5 years	Walks up and down stairs, alternating feet Stands on 1 foot for short period Takes few steps on tiptoe Jumps using both feet		Builds tower of 8 or more blocks Adds chimney to "train"
3 years	Stands on foot Walks on tiptoe Rides tricycle Jumps from bottom step	Grasps crayon with fingers	Builds tower of 9–10 cubes Imitates 3 cube bridge Copies circle, cross

maturation of the hypothalamus) result in more stable thermoregulation.

By the beginning of the second year blood volume has doubled, averaging 40 ml per lb (90 ml per kg). This increase in blood volume, with the increase in body size, results in changes in normal vital signs during this period. (Table 6-2 lists the normal blood pressure, pulse, and respiratory rates for the toddler.) Accuracy in obtaining vital signs is often a challenge since

Table 6-2. Normal Vital Signs of the Toddler

Age	Pulse Rate Per Minute	Respiration Rate Per Minute	Systolic Blood Pressure (mmHg)
1 yr.	100	28	90
2 to 3 yrs.	95	25	100

SOURCE: Adapted from Hughes, J. G. *Synopsis of Pediatrics.* St. Louis: C. V. Mosby, 1971.

the toddler is so active. These signs should be obtained while the child is in a resting state to ensure that the findings are valid. The pulse should be checked for a complete minute. Sinus arrhythmia is considered a physiological (i.e., normal) phenomenon of this period as well as of infancy.

The tracheobronchial tree grows rapidly during the second and third years, after which growth slows until puberty. The lung itself has tripled in weight by the beginning of the second year. During toddlerhood the short and funnel-shaped pharynx of infancy becomes elongated and wider. By 4 years of age tracheal bifurcation occurs at the third or fourth vertebral level. The anteroposterior and lateral diameters increase during this period but continue to remain proportionately small, contributing to infections of the tracheobronchial tree.

Breathing in the toddler continues to be mainly diaphragmatic, with less intercostal muscle involvement. Respiratory rates should be checked for a full minute in case of physiological irregularity in rate. Cheyne-Stokes respirations (in which the respirations increase in rapidity and then subside or even cease for a few seconds) may be found early in toddlerhood. Respiratory patterns, depth, and rate should all be assessed.

Blood pressure norms vary significantly in the toddler from day to day. Increases are not constant from year to year, and consequently there is a wide normal range. Accuracy in obtaining a pressure requires sound technique. The cuff width should equal two-thirds the length of the upper arm or leg. A cuff that is too small will result in an erroneously elevated reading.

Peripheral blood values of toddlerhood approach adult norms (see Table 6-3). The hemoglobin at 2 years of age is approximately 11.6 to 13 g and the red blood cells number approximately 4.8 million.

Lymphatic Changes

Lymphatic tissue continues to increase greatly throughout toddlerhood and into later childhood. This increase is characterized by hyperplasia and hypertrophy as a response to acute respiratory and gastrointestinal infection.

Tonsils and adenoids increase through normal hypertrophy throughout this period until about age 5, at which point atrophy begins to occur. The tonsils and adenoids apparently serve as a reservoir for phaygocytes and probably play a role in the child's natural immune mechanism.

Gastrointestinal Development

Growth of the gut occurs generally throughout the toddler period. Stomach capacity increases to approximately 500 cc by two years and the stomach has acquired the characteristic "cow's horn" shape. Intestinal length has increased, the small intestinal reaching roughly 450 to 475 cm by 12 months of age.

Gastrointestinal function nearly attains adult levels by the toddler period. The salivary glands have increased their size fivefold by age 2, and they are histologically and functionally mature.

Table 6-3. Normal Peripheral Blood Values for the Toddler

Age	Hemoglobin (gm/per 100 ml)	Hematocrit %	WBC (Total mm³)	Polys %	Lymphs %	Platelets (Total mm³)
1 yr	11.6–12.5	35.2	5000–17,000	30	60	200–473,000
3 yr	11.8–13	36	4800–15,000	30	60	150–450,000

Gastric secretions are near adult levels throughout the toddler period although the adult acidity level is not attained until puberty.

Pepsin and rennin are present in the gastric juices of the toddler is sufficient amounts for digestion of proteins. Amylase activity increases during this period, enhancing starch digestion. Trypsin, an enzyme formed in the intestine responsible for protein digestion, is present in sufficient amounts while lipase activity (responsible for fat breakdown) remains low during this period. The presence of these enzymes permits a more varied dietary intake.

Peristalsis of the tract decreases, bringing a delay in the emptying time of the stomach. This, coupled with the increased volume of the stomach, allows for the traditional adult eating pattern of three meals a day. By 2 to 3 years saliva secretion and gastric juice production respond to emotional stimuli.

The child of 2 years is likely to have stools similar to the more formed, darker stools of the adult. This is a result of the decreased peristalsis, which allows for more water absorption, as well as of a progressively varied diet featuring diminished amounts of milk.

The changes in the gastrointestinal tract are related to the acquisition of control of the bowel and bladder. Complete central nervous system myelination, including that of the spinal cord, allows sufficient stimulation of the autonomic nervous system for sphincter control of the bowel and bladder. The age of control is usually 2 to 3 years. (Bowel and bladder training are further discussed later in this chapter.)

Genitourinary Changes

The kidneys have reached anatomical maturity by the end of infancy. The glomerular filtrate is sufficiently developed to allow for adequate concentration of urine, and thus, water conservation. During stress this water conservation is controlled entirely by the endocrine system, which remains immature at this time. Consequently, although the vulnerability to fluid and electrolyte imbalance seen in infancy has decreased, it still remains a threat in stressful situations such as illness and hospitalization.

Kidney size has tripled by the end of the first year, and average daily urine secretion increases from 400 to 500 cc during the first year to 500 to 600 cc from 1 to 3 years of age.

Endocrine Changes

Except for reproductive hormones, the endocrine system is functionally mature by the toddler period. Production of somatotropin, the growth hormone, along with insulin, thyrotropin, and corticotropin, is elevated during toddlerhood. (Somatotropin production is 1.5 times greater than in adulthood.) Somatotropin secretion results in tissue development through increases in the size and number of cells.

Insulin functions in carbohydrate metabolism, while thyrotropin plays a role in controlling the basal metabolic rate. Corticotropin indirectly affects the metabolism of fat, glucose, and protein.

Cellular Changes

During the second and third years a shift toward more adult norms occurs in extracellular and intracellular fluid consumption. The intracellular fluid increases as growth of new cells occurs, while extracellular fluid decreases as development proceeds. The toddler's percentage of body fluid is 64 percent, as compared to 59 percent in the adult male. Consequently, the toddler is less vulnerable to electrolyte imbalance or dehydration than the infant.

Integumentary Changes

The skin achieves both the appearance and qualities of maturity during the toddler and preschool period. The skin appears somewhat dry because of limited sebum production. Eccrine sweat is produced, but in limited qualities.

Immunological Changes

Both IgG (immunoglobulin G) and IgM (immunoglobulin M) have reached adult levels by this period, so that the toddler is better equipped to combat infections. Naturally acquired passive immunity, however, has disappeared, and immunizations are a necessity if they have not been previously administered.

COGNITIVE DEVELOPMENT

Although toddlerhood spans only the first to third years, cognitive gains during the period are immense. While in what Piaget calls the sensorimotor stage, the child makes tremendous progress in developing simple forms of reasoning through the use of object symbolism. The sense of object constancy increases, and more sophisticated levels of imitation occur. The child, even at this young age, can follow complex rearrangements of objects within his sight. The plateau in the sensorimotor stage is the emergence of simple problem-solving.

At this point, the young child has already made great strides in developing more complex mental functions in response to increasing environmental demands, and in developing meaningful, coordinated behavioral organization (denoting more complex and meaningful mental function). In Piaget's terms, the toddler's cognitive grasp is enhanced by the appearance of the faculties of *assimilation* (incorporation of new events into *schemata*, i.e., mental categories, already present) and *accommodation* (modification of an existing schema or the development of a new one in response to new stimuli). Assimilation is seen (for example) when the young infant incorporates a pacifier as an object to be sucked. In contrast, accommodation is demonstrated by the infant who, after attempting to suck a wrapped piece of candy, will remove it, attempt to unwrap it, and place it again in his mouth.

As the child becomes less egocentric, he evidences "decentering" behavior. In other words, the older infant and toddler's actions become less bodily centered as the ability to differentiate oneself from others develops. Much more object manipulation and environmental exploration are seen.

The toddler period as defined here includes substage V of Piaget's *sensorimotor period* (12 to 18 months), and substage VI (18 to 24 months). In substage V, the substage of "tertiary circular reactions," the toddler characteristically seeks out new experiences rather than merely duplicating those that have occurred accidentally. For example, instead of repeatedly dropping a spoon from his highchair, the infant will try dropping it at different heights and expand this testing to other objects as well.

The ability to discovery a new means to an end is developed in this substage also. Initially solutions are arrived at accidentally, but once proven effective they can be transferred to attain other, similar goals. Piaget and Inhelder (1969) describe a situation in which an object is placed on a rug out of reach of a child. Several times the child tries unsuccessfully to reach the object. Then accidentally the child grasps an end of the rug, pulls, and sees the object move. At this point the child processes a relationship between the movement of the object and the rug. Through trial-and-error behavior the child arrives at a solution. Confronted with a similar situation, the child will use learned behavioral patterns in problem-solving.

Imitation progresses to a more refined level in this period. Piaget describes how at this substage his daughter was able to imitate a series of touching motions (touching tip of tongue with index finger) demonstrated by himself.

A developing sense of object constancy is seen although it is not fully developed until substage VI. The young toddler will follow and look for objects that have been displaced several times. This repeated displacement, however, must be seen by the child. The 12- to 18-month-old child will look for an object that has, for instance, been hidden in a closed hand and then displaced in several places, looking for it in the place where the object is most frequently seen, not in the place hidden. The toddler is likely to seek the toy from the closed hand or a shelf where it has been put before because he does not comprehend that objects can be moved around while he is not looking.

A keen interest in the human face is present at substage V; the child will continually touch the faces of parents and others. Piaget postulates that the young child is exploring facial construction (i.e., the relationship of the parts of the face to one another) (Piaget, 1967; Piaget and Inhelder, 1969).

Substage VI (18 to 24 months) is characterized by the invention of new schemata, or mental units. In this substage the young child is considered capable of true thought, for he has the potential for problem-solving.

The emergence of thought using symbolic representation or mental images and the attainment of true object permanence are the two outstanding features of this substage. By using symbolic representation, the toddler is able to conceive of an unseen object or sequence of events, and is able to visualize and figure out a solution to a problem without physically testing the solution. Evidence of memory and imagination is also seen.

At this substage, the toddler is able to imitate an event that occurred days before without the presence of the model. Piaget described such "deferred imitation" in his daughter, who duplicated a temper tantrum she had observed two days before in a visiting child (Piaget, 1967; Piaget and Inhelder; 1969). Many positive behaviors can be imitated at this period, but numerous negative ones may be as well.

During substage VI the child acquires a fully developed sense of object constancy (object permanence). The toddler is able to make inferences about the location of hidden objects. When presented with an object to be placed in a closed hand, which is then sequentially placed

under three different objects, the child has the capability of following the succession and finding the object where it was apparently last hidden, not last seen. True thought is demonstrated in that the toddler infers where the object is likely to be.

A sense of *causality* is another cognitive ability achieved in this period. For example, the child can comprehend that a tricycle is immobilized because a wheel is caught on a table leg or hindered by a rug. This cause-and-effect comprehension is one of several abilities that will equip the toddler for more complex thought processes that will emerge later. The sensorimotor stage is completed with the attainment of goal-directed behaviors.

The *preoperational stage* as described by Piaget includes the period from 2 to 7 years and is subdivided into the *preconceptual period* (2 to 4 years) and the *intuitive period* (4 to 7 years). It is during the preoperational stage that the child uses objects symbolically to represent parts of his environment. A 2- or 3-year-old may use a stick in play as a candle and blow it out, or a block as a truck, pushing it about or sitting on it, pretending to drive it. This emerging representational thought allows the child to deal increasingly with the past and future, not just the present. The child is freed from the present or immediate environment, and he can create a greater psychological distance between himself and his social and physical environment.

In the preconceptual period, at around 2 years, the toddler is capable of creating mental images that represent objects or actions that are not present or are not being experienced. In Piaget's terminology, the toddler has acquired the ability to distinguish *signifiers* (symbols, such as words and images) from *significates*, the actual objects, actions, or events represented by the signifiers.

Piaget felt that the young toddler develops mental symbols by using motor behaviors as signifiers. Describing his young daughter observing him ride a bicycle, Piaget says that she began to move forward and backward; he felt the motion was a primitive motor symbol, or signifier, for the bicycle. Eventually the mental image and the word *bicycle* alone, without the motor symbol, represent the same thing (Piaget, 1967; Rosen, 1977).

The toddler's symbol assimilation is very personal, egocentric, and subjective; the symbol may evoke a negative memory or a positive one. For example, the toddler who has an accident involving a particular toy may develop an aversive view of the toy.

Centration—the inability to consider more than one aspect of a thing at a time—is a characteristic cognitive limitation of the preoperational stage. For example, after seeing a horse, the 3-year-old is likely to identify all four-legged animals as horses, the child cannot consider other qualities (size, shape, color, type of coat, etc.) at the same time while bearing in mind the one aspect (number of legs) that he is "centered" on. Obviously, this kind of thinking leaves the toddler vulnerable to false conclusions.

Another cognitive limitation of the period is *irreversibility*: the child can follow a process from beginning to end but not in reverse order. For example, if balls of three different colors are rolled into a closed cardboard tube, the toddler can give the order going in, but when asked about the order in which the balls would come out, the child would be unable to say.

Transductive reasoning, or "particular-to-particular thinking," is another hallmark of the preoperational period. Transductive reasoning, according to Piaget, dictates that the child will be incapable of truly logical thinking because of an inability to identify similarities, and therefore patterns, in objects and events. The child sees a similarity in one aspect of two situations and concludes that all aspects are similar; for example, if the child sees his mother starting the car in preparation for a trip to a particular destination, he reasons, the next time he sees her start the car, that her destination is the same.

Egocentrism is another limitation on the toddler's cognitive ability. Unable to differentiate his viewpoint from that of others, the toddler assumes that his view is shared by everyone; this greatly influences language and behavior. The use of pronouns in "telegraphic speech" (two to four words representing whole sentences) is common. For example, "me go" may represent the toddler's request to go outside or to go to the potty; there is no awareness of what information is needed by others, and, unfortunately, this kind of "communication" may lead to frustration for all.

Animism is an outgrowth of egocentrism characterized by the attribution of life and purposefulness to inanimate objects such as the sun, trees, or a chair. The toddler might be heard to say, upon bumping into a table, "Table, get out of my way."

Artificialism and *realism* are often expressed by the child in the preconceptual period. Artificialism refers to the belief that all objects and events have the sole purpose of satisfying human needs, including personal ones. "There is snow to play in and there are beds to sleep in." Realism is the confusion of physical realities with psychological events. For example, the child insists that his dreams are real; he may describe a "bunny rabbit that was in the house" after dreaming of one.

LANGUAGE DEVELOPMENT

As noted in Chapter 5, numerous theorists have attempted to explain the language acquisition process. Key aspects of these theories are briefly summarized below.

Chomsky (Linguistic Theory)

1. The human brain is innately programmed to comprehend and create language.
2. The programmed system is dependent upon cerebral cortex maturation, which occurs throughout infancy.

3. Certain aspects of the central nervous system contain universal features that are innately determined and applicable to all human language (Chomsky, 1957).

Skinner (Behavioral Theory)

1. Speech occurs as a result of selective reinforcement of spontaneous babblings and vocalizations.
2. These reinforcements are most probably delivered when babbling sounds more clearly approximate adult speech.
3. The infant's verbal behavior is shaped through "successive approximations" eventually attaining more adultlike patterns (Skinner, 1957).

Lenniberg (Biological Theory)

1. Central nervous system maturation serves as the basis for language development.
2. Language acquisition milestones occur in a regular, fixed order.
3. Milestones generally occur at a similar rate in all normal children.
4. Culture and environment appear to have a limited effect on the order and rate of the attainment of verbal milestones.
5. Minimal exposure to a variety of language examples appears to be the only prerequisite for the order of emergence (Lenniberg, 1967).

The nature of the process of language acquisition remains unknown as of this writing. It is assumed to be extremely complex, yet occurs in an extremely short period of time (birth to 4 years of age).

Many children speak by 15 months, and almost all normal children speak by 2 years of age. Individual factors, environmental stimulation, and certainly central nervous system maturation appear to be determinants. Word comprehension is controlled by Wernicke's area of the temporal lobe of the brain, which myelinates more quickly than the area controlling the action of speech (Broca's area of the motor cortex). The toddler therefore understands many more words than he can say.

The child commands a vocabulary of approximately 19 words by age 15 months, and possibly 50 words by 18 months (Smith, 1926; Nelson, 1973). By 2 years of age the child has acquired a vocabulary of 300 words; by 3 years, it is 900 words. The 2-year old commonly uses two-word patterns; three-word sentences emerge at around 2.5 years. The sentences of the 2-year-old are usually composed of nouns and verbs only. Consequently, key words that would add clarity to the sentence are often missing.

At 2.5 to 3 years, the word *why* begins to be heard. It is thought that the child is attempting to elicit adult speech or attention more than he is seeking answers. First-person pronouns are also used frequently.

Speech usually develops in the female earlier than in the male. Although studies have implied that female infants receive more stimulation by being talked to, there is also evidence that myelination of the dominant cerebral hemisphere (usually the left) occurs earlier in females (Geschwind, 1975).

THEORIES OF DEVELOPMENT

Francis Ilg and Louise Bates Ames (1955) point out that the child's behavior develops in predictable stages. There are several periods in which the child is in a state of disequilibrium—confused, frustrated, egocentric, and unhappy. These periods appear to alternate with more balanced, happier states of equilibrium. At the ages of 2 years and 5 years, the child is in an equilibrium state; the child of 2.5, 6, and 11 tends to be in a disequilibrium state (Ilg and Ames, 1955).

Robert Havighurst (1963) postulated that specific developmental tasks occur throughout the life span. He assigned four tasks to toddlerhood: learning to take solid foods, learning to walk, learning to talk, and learning to control elimination.

Increased central nervous system maturation facilitates anal sphincter control. From approximately 18 months to 4 years, the young child focuses on both the pleasurable and unpleasant or frustrating experiences associated with bowel evacuation, and is said, in Freudian terms, to be at the anal stage of psychosexual development. Freud felt that difficulties in bowel training at this stage correlated with fixations on orderliness, possessiveness, or parismony in the adult (Freud, 1949).

Robert Sears, a learning theorist, combined psychoanalytic and behavioral theory and postulated that personality development is greatly determined by social influences to which the infant and young child is exposed. Behavioral patterns are primarily influenced by parent-child relationships, established initially because of biological and emotional dependency. For the infant, behavior is controlled by unmet physical needs. Meeting these needs, usually by the parent or primary caretaker, results in emotional dependency, which serves as a motivating force for later infant behavior.

According to Sears, behaviors are learned by following a sequence of observation, imitation, and consistent reinforcement. During early childhood these behaviors are externally (usually parentally) motivated, but eventually they become self-motivated. This theoretical observation is particularly true of eating and toileting skills that emerge during toddlerhood (Sears, Maccoby, and Levin, 1957).

Mussen, Conger, and Kagan (1974) speculate that elements other than behavior reinforcement and imitation are involved in personality development. It is their feeling that some aspects of personality—such as motives and mores— are acquired spontaneously without being taught. Such aspects are relatively stable and enduring. The degree to which they are acquired is variable and it appears to be related to the desira-

bility of the traits, the attractiveness of the model, rewards offered by the model, and similarities felt between the child and model (Mussen, Conger, and Kagan, 1974).

Erik Erikson, a psychoanalytical psychologist, sees psychosocial development as being influenced by environmental as well as biological factors. It is Erikson's view that the quality of interactions in one's social environment determines the probability of successful completion of developmental tasks.

The developmental task designated by Erikson for the toddler period is that of acquiring *autonomy*, as opposed to a sense of *shame and doubt*. The infant, having successfully acquired a sense of trust through a variety of interactions with parents or primary caretakers, enters toddlerhood with a sense of confidence.

"Holding on" versus "letting go" is a key conflict, played out in behavioral patterns such as eating jags, ambivalence in going to the potty, resistance in changing clothes, and unpredictability or instability in social interactions. Positive interactions lay the groundwork for feelings of self-confidence essential for subsequent psychosocial gains. Erikson (1963) says: "This stage, therefore, becomes decisive for the ratio of love and hate, cooperation and willfulness, freedom of self-expression and its suppression. From a sense of self-control without loss of self-esteem comes a lasting sense of good will and pride; from a sense of loss of self-control and of foreign overcontrol comes a lasting propensity for doubt and shame, p. 254." It is thus essential for parents and caretakers to deal with the child firmly but flexibly, and to nurture the child while at the same time encouraging independence.

PSYCHOSOCIAL DEVELOPMENT

Characteristically much ambivalence is experienced by the toddler, and often by the parents too, as the increasingly complex environment

opens new domains but also exerts new demands. The toddler is often torn between choices—whether to come when Mommy calls or to continue playing, whether to go peacefully to bed or to protest. Striving for independence through the use of newly acquired skills often creates stress and conflict, but the exercise of these skills will ultimately lead to confidence, self-control, and a positive self-image (Fig. 6-5).

As was noted earlier, the child of 1 to 3 is extremely egocentric in both cognitive and language development. This egocentrism is reflected as intensely in social behaviors. The inability to compromise, self-centeredness, and possessiveness are common features of the toddler's social behavior. In this period, the child continues to struggle with "when to hold on" and "when to let go." "Holding on" or "taking" behaviors are a typical part of the toddler's repertoire; from the toddler's egocentric perspective, it seems appropriate that others do the "giving." In his own view, the toddler is the most important person in the world.

In more intense or complex social situations the toddler can and often does resort to negative behaviors. These are most likely to emerge in threatening or challenging situations—those in which the toddler may feel little or no control and those that involve an adult authority figure. Negativism, characterized by the frequent use of *no*, is used to challenge authority and test limits. Ambivalence often takes the form of the child saying "no" and then proceeding to comply with the request.

In instances in which the toddler feels extremely threatened or helpless, he may respond with temper tantrums. Such behavior stems from the toddler's cognitive and social inadequacy as well as from emotional instability. Temper tantrums should be ignored if possible,

since acknowledging them tends to reinforce the behavior; however, parents should be sure the child is physically safe.

Dawdling is another manifestation of ambivalence in the toddler. The toddler may be torn between coming to dinner and continuing to play, desiring to do both. In other words, he wants to please his mother but also wants to assert himself, and therefore responds lackadaisically.

Ritualism begins to emerge at around 12 months of age and is prevalent throughout toddlerhood, peaking at around 2.5 years. These rituals serve to provide security, predictability, and a sense of control, and should not be interfered with unless the ritual poses a threat to the child's health or safety.

During particularly stressful or threatening situations (such as separation from parents or hospitalization) the child may resort to thumb- or finger-sucking. Habitual thumb-sucking may be present up to 18 to 24 months, but it usually subsides, except in stressful situations, by age 2 or 2.5, by which time it is seen more when the child is anxious although it can serve both as a means of sensory gratification and as a means of channeling stress arising from hunger, fatigue, or frustration (Ilg and Ames, 1967).

At approximately 2.5 years the toddler commonly sucks his thumb or fingers while holding a security object (a favorite blanket, teddy bear, or other special toy). Repetitive manipulation of certain body parts, particularly those of the face, may be seen. The toddler may suck his thumb while pulling at an ear, twisting his hair, or rubbing his nose. Withdrawal of such objects or verbal discouragement of such behaviors is met by intense frustration in the toddler. Crying, screaming, and other tantrum behaviors may be observed.

A brief description of social patterns in the

Figure 6-5. *The toddler's cognitive and motor development allow him to explore the world around him. The struggle for autonomy inevitably leads to conflicts with parents and others in authority; optimally, a firm but flexible approach to discipline will ultimately encourage the emergence of the child's confidence and self-control.*

toddler follows; the intent of this description is to depict the typical "personality" of each age.

Characteristic Social Behaviors of the Toddler

Twelve months (tranquil period). The infant is friendly, self confident, enjoys an audience. He seeks approval by repeating positively sanctioned behaviors; imitates simple social actions (waving "bye-bye", etc.). The child enjoys such games as "patty-cake" and "peek-a-boo", and asserts himself through motor skills, wanting to feed himself with a spoon, etc.

Fifteen months (unsettled period). The toddler is testing new motor skills; egocentrism is greater than at 12 months, and more "taking" behavior is seen. The child resists commands such as "no," is very uninhibited, explores anything. He experiences unpredictable but short-lived mood swings, and is easily diverted.

One and a half years (testing period). The toddler responds negatively to many requests or commands. Use of the word *no* emerges; the child is impatient ("now"-centered); "taking" dominates social interactions.

Two years (equilibrium period). The toddler is more self-assured, more cooperative, and makes fewer demands. He can tolerate longer periods of waiting, handles frustration better, wants to please, and is pleased by others. Although he does not yet share; he will provide a playmate with another toy. He is loving, affectionate, and good-natured.

Two and one half years (peek disequilibrium period). The middle toddler resists many commands and requests; he is rigid, inflexible, "taking"-oriented, impulsive, very ritualistic, domineering, and demanding. He shows violent emotions, indecisiveness, and wants things to go on and on; he is vigorous, enthusiastic, and energetic.

Three years (quieted period). The older toddler is cooperative, sharing, easy-going, less ritu-

alistic, and gives positive verbal responses. He seeks more verbal interactions and is more responsive to language.

Play

The manipulative form of exploratory play observed in infancy is replaced with a more diverse form of exploration—investigation. Having increased motor capabilities, the toddler is challenged to investigate all that he can reach. The toddler's play focuses on sensory and more complex manipulative experiences; the toddler enjoys water, sand, playdough, blowing bubbles, and finger painting in particular. Push-pull toys and trucks are favorites. The toddler also delights in simple puzzles, paints, large crayons, and blocks.

Record players and musical and talking toys are also of particular interest. Although the attention span of the toddler is still somewhat short, television programs such as *Sesame Street* and *The New Zoo Review* provide stimulation through exposure to new words and visual images. Toddlers like being read to and looking at books with pictures, and often enjoy imitating sounds such as a train whistle or a firetruck's siren (Fig. 6-6).

Imitative behavior is also seen. The toddler will duplicate behaviors of parents and other important role-models. Parents should be urged not to sex-type toys such as dolls, playhouse equipment, cars, trucks, and balls, so that boys and girls can experiment with a variety of roles.

Another characteristic of the toddler's play is the shift from the typical solitary play of infancy to *parallel play*. Because of his limited communication skills and intense egocentrism, the toddler plays alongside another child but not with him.

Many play items can be cheaply improvised at home. Large boxes, crates, etc., can be converted into play stores and sinks. A modeling compound similar to the commercial product

Figure 6-6. *Although the toddler's attention span is still rather short, children of this age enjoy being read to and looking at pictures. (Photo by Sam Gray, M. Photog., Cr.)*

playdough can be made from a flour, oil, and food-coloring mixture.

Safety should be considered in choosing age-appropriate toys for the toddler. Sharp or pointed objects are to be avoided. Stuffed animals and dolls should have features (eyes, nose, mouth) that cannot be removed and swallowed. Close supervision is essential for toddlers.

COMMON HEALTH AND DEVELOPMENTAL PROBLEMS OF THE TODDLER

Stranger Anxiety

Stranger anxiety, which is at its height from 6 to 12 months of age, continues in toddlerhood, although the toddler is more likely to react more subtly, showing signs of frustration, anxiety, or discomfort. Proximity of the stranger to the child also appears significant. More intense negative reactions are elicited as the stranger approaches, and physical contact is typically met with an even more negative response (Sroufe and Wunsch, 1976). Using toys and imitating the mother's behaviors in the approach to the child result in less intense negative reactions (Eckerman, Whatley, and Kutz, 1975; Rheingold and Eckerman, 1970). Proximity of the mother to the infant has been found to decrease or delay negative child responses. If the mother is nearby, the child may even tolerate being picked up by the stranger.

In approaching the toddler, the health care provider should impinge very gradually on the child's territory, smile, and speak softly. *Person permanence,* that is, an awareness of a person as a separate entity, appears to heighten the child's reactiveness; this awareness emerges between 12 and 24 months. The presence of a parent is comforting to the toddler and should be encouraged. These approaches are particularly relevant to strange places such as a hospital.

As the toddler grows, stranger anxiety remains but diminishes. There is some willingness to meet strangers, although the support of the parent's presence is usually needed.

It is probable that the child is reacting not so much to separation from the mother or father as to fear elicited by unfamiliar situations, or situations in which the child doesn't understand the parents' absence and can do little about it anyway. Eventually the toddler becomes better able to understand and tolerate these separations. As the child matures, anxiety arises only in more complex situations.

Reaction to Hospitalization

Hospitalization is particularly stressful because the toddler has little control over and understanding of it. Bowlby (1966), Robertson (1958), and Spitz (1945) have studied the child's response to hospitalization as it relates to separation from parents or caretakers. The findings of all three indicate that toddlers are likely to demonstrate certain defensive behaviors in protecting themselves against the loss, and in attempting to "settle in." These have been designated as behaviors of protest, despair, and denial. It is felt that these behaviors emerge in phases that often overlap with one another, each lasting as little as a few hours to as long as several days (Bowlby, 1966; Robertson, 1958).

In the first phase, *protest*, the child is restless, cries frequently, and calls desperately for Mommy or Daddy. Being cognitively incapable of understanding the separation and the hospitalization, the toddler may perceive the experience as punishment or even abandonment. Irregular sleep patterns (and more crying) may be observed as the toddler anxiously and hopefully anticipates any sign of the parents' return.

After this protest phase the toddler will evince signs of *despair* or depression. The demonstrative behavior that occurred initially is replaced by passivity and apathy. The child displays a sense of hopelessness through diminished appetite and activity. Often the child will not engage in play or social interaction. Often, such a child is categorized in pediatric unit reports as being "a good kid; she'll give you no trouble," when in fact the child is in deep depression and intense mourning. Regression to an earlier state may be seen in the form of thumb-sucking, refusal to drink from a glass, and lack of bowel and bladder control. The parents' return may cause distressed crying, creating guilt or even avoidance on the part of the parent.

The child in the *denial* phase appears to have "adjusted well." In fact, however, the child has become resigned and detached from the situation as a means of self-protection. Repression of feelings about the parents and of mental images of them is thought to occur. Only superficial relationships are developed with the nurse and other care providers in order to avoid a recurrence of such an intense hurt. In this phase the child may acquire some aspects of a normal routine and may even begin to play, eat, and sleep. The parents, upon returning, may be greeted coolly and with declarations of attachment to and affection for substitute caretakers.

During the protest phase, the nurse and other care providers must bear in mind the therapeutic aspects of crying and other aggressive behaviors, especially in light of the fact that the toddler lacks more subtle or sophisticated means of expressing his pain and displeasure. Crying should therefore be encouraged through comments such as "It's all right to cry; I know you don't like being here." If possible, the child should have the same nurse at all times. Parents should be allowed and indeed encouraged to visit as often as possible.

Sensitivity to the needs of the child is essential in the despair phase. The symptoms may demand alterations in pediatric unit regulations. Parents should be encouraged to visit the toddler even more often, and to participate in as much of the care as possible. When parents are unable to visit, grandparents or substitute caretakers should be encouraged. Play can help the child vent feelings of hostility, fear, or aggression; both staff and visitors should encourage the child to play, and should play with the child whenever possible.

During the denial phase, the child needs consistent support from the nurse or other care provider. Frequent contact may restore the child's sense of trust. Despite the child's apparent rejection of his parents, it is essential that they visit as frequently as possible.

Parental support throughout the hospitalization is vital. Explanations of the child's reactions should be given to parents and reinforced so that parents will gain a better understanding of their child's reactions and their role in providing assistance and support.

Pediatric units can do much to diminish or even eliminate such negative reactions to hospitalizations. Flexible regulations should provide for rooming-in or parent participation in care whenever possible. Visiting hours should be liberal, encouraging visits by parents. The parent should be recognized as an asset, as the key support person for the child. Supportive, understanding nurses can do much to encourage and assist parents in participating in care.

Fears

Although separation anxiety gradually decreases after 24 months, other fears remain common throughout the toddler and preschool periods. These early childhood fears are somewhat unpredictable, reflecting marked individual differences. Some children are more responsive to certain stimuli than others. For example, the toddler who has had constant contact with a dog may be completely at ease when confronted with other dogs, while another toddler may be extremely frightened of any dog. The child has a tendency to develop fears seen in the parents; these may include fears of storms, doctors, dogs, or insects (Hagman, 1932). Fears of 2- to 6-year olds identified in a study by Jersild and Holmes included fear of loud or sudden noises, strangers, shadows, and unexpected movement. Some of these fears eventually decrease while others apparently intensify. Fear of loud noises comprises 25 percent of all fears of the 2-year-old. This fear decreases dramatically by middle schoolage (Jersild and Holmes, 1935).

The toddler is also likely to fear darkness and strange or unfamiliar objects or situations. Flushing of the toilet results in much anxiety in the toddler, who possibly fears being flushed down.

With cognitive and social maturation, as many as half the fears disappear spontaneously within one to two years. The child, however, should be provided with parental support, reassurance, and understanding during this period. Practical aspects such as night lights, security objects (blankets or toys), and consistency in the physical arrangement of the home can provide added security for the child. Introduc-

ing the child gradually to new situations will help to alleviate fears arising from sudden changes. If necessary, reconditioning can be used. For example, a fear of the doctor's office or hospital might be overcome by taking the child on a pleasant, nonthreatening orientation visit, acquainting him with the physical layout, equipment, and personnel.

Discipline

The toddler, because of the cognitive and social traits described above, is unusually difficult to guide. Until a child establishes an identity and achieves self-control, the parent must act as ego and conscience. The parents must teach the child standards of acceptable behavior, safety measures, and concern for other people and the environment.

Ilg and Ames (1955) note that the most effective way for parents to guide the child is through adapting individual techniques of discipline to the child's abilities, interests, weaknesses, and developmental stage. The approach of using developmental techniques is difficult however because of the knowledge it requires. It necessitates understanding the characteristics of the child's age level so thoroughly that the parent acquires a sense of how the child can be motivated toward the desire behavior. Simplifying the environment and making sure it is safe and not overly stimulating is helpful. "Hands-off" items are best placed out of reach of the child. Use simple, positively phrased language; do not confuse the child by providing too many options.

The 2-year-old is somewhat more cooperative than the younger toddler. Dawdling emerges at this age; giving notice of the next activity as well as talking about it may allow time for the child to prepare for it. Encouragements like "say good-bye to Grandmom" are effective, as are diversionary or enticing means such as "Let's go inside and find our new blocks."

Two-and-a-half years is the peak age of aggressive, asserting behaviors. The child's behavior shifts from one extreme to another. Allowing choices in important situations is usually not effective, although simple ones ("Do you want to wear your yellow or green dress today?") may be successful. The child of this age is likely to be selfish and egocentric, so sharing is not an appropriate expectation. However, the child is often willing to find a substitute toy for a visiting friend (Fig. 6-7).

Using positively stated commands or suggestions is helpful: "Let's hang up our coats" or "We drink our juice in the kitchen." Using simple, understandable directions for the child also encourages success, as does a communi-cation of the expectation that the child will obey, e.g., "We will be going home after you finish your snack." Avoid questions that can be answered by "No." "Do you want to take your medicine now?" is much less effective than "It's time to take your medicine; do you want to take it with juice or water?"

Another disciplinary approach is reasoning, which involves explaining about an unacceptable behavior, explaining why certain outcomes are the result, encouraging the child to behave in a certain fashion, communicating how the child's action makes one feel and removing the child from the situation if necessary. Finally, the parent might resort to intense reactions such as shouting, threatening, or actually hitting the

Figure 6-7. *Toddlers are typically unwilling to share toys, and parallel play, in which children play alongside one another but not together, is characteristic of the stage. (Photo by Sam Gray, M. Photog., Cr.)*

child. Although this is perhaps the easiest approach, it does little to encourage consistently desirable behavior.

Ultimately, the parent or caretaker's philosophy of disciplining is an individual decision, based on personal experience, education, and sociocultural factors. The decision is influenced markedly by how one was disciplined in one's own childhood.

Once a disciplinary approach is chosen, consistency in the approach is essential. The parent, nurse, or caretaker creates frustration and insecurity when inconsistent approaches are used. The child gets conflicting messages about acceptable and expected behavioral patterns. Setting of reasonable limits is also essential. This involves one's personal value system. Knowing how to "give a little" and when to "stand firm" presents the disciplinarian with an immense challenge. Through this "limit setting," however, the child learns the boudaries of safety, acceptable behaviors, and sociocultural standards. A sense of control and security is acquired as the child responds to the parent or caretaker's guidance and experiences as sense of love even though at times his behavior has been disapproved of. Effective discipline is a major contributor to the toddler's sense of autonomy. By pleasing his parents and others, the child comes to feel competent and good about his emerging self.

Sibling Rivalry

The birth of new sibling often creates great anxiety in the toddler, who may fear the loss of his parents' love. Anger and frustration are frequently acted out toward the new sibling or even the parent. *Sibling rivalry*, as it is referred to, is particularly evident in the toddler who is oldest in the family and therefore not initially accustomed to sharing parental affection and attention. Sibling rivalry is evident between 18 months and 4 years of age, peaking at around age 2 (Church and Stone, 1973; Koch, 1956).

The child is likely to react both overtly and in more subtle ways. The parent may hear "Don't like baby" or "Want baby go away." One may also hear comments that reflect subconscious desires, such as "Let's cover her" (while putting a pillow over the baby's face) or "Don't drop baby."

The parent can also expect overtly hostile behavior, such as pinching, hitting, and pulling at the infant. The infant should never be left alone with the toddler nor be accessible without adult supervision. Regressive behavior such as disturbances in sleeping, eating, and toilet-training are often seen.

Preparing the child before the birth of the sibling can help to facilitate adjustment. However, the toddler has little understanding of time; that a baby will be born in five months has little meaning for the toddler, nor is the idea likely to be retained. Usually a month's notice, or in some instances two, is adequate for preparing the toddler. The timing, however, should be determined by the toddler's individual response pattern. If the toddler anticipates well and enjoys looking forward to things, then perhaps more notice can be given.

The specifics of the preparation should be gauged to fit the child's level of understanding. Often it is useful to involve the toddler in preparations for the baby's coming. Frequent reference by the parents to "our" baby can reinforce a feeling of mutual involvement.

Providing contacts with young infants or showing the toddler pictures of newborns can help him anticipate some typical characteristics, particularly size. The toddler should be informed that newborns spend most of their time sleeping, eating, and crying; it is useful for the toddler to know this so that he does not expect a new playmate or companion. Allowing the toddler to see the physical changes that the mother is experiencing can help the child understand how babies grow and where they come from.

Even with excellent preparations, parents can

expect some regression and anxiety in the toddler. If toilet-training has not been started or completed during this period, it is useful to delay any initiation or postpone continuation until the toddler is less stressed and more secure. Similarly, if the child is making the transition from crib to bed, it should be completed well in advance of the arrival of the baby. Additional changes, especially if they involve the belongings of the toddler, should be minimized but, if necessary, they should be carried out before the birth so that the toddler does not view the baby as "taking my things." The toddler's surroundings and routine should remain the same, as far as possible.

Sibling visits to the hospital should be encouraged if at all possible. Such visits provide a less threatening opportunity to see what the new baby looks like. Presents to the toddler "from baby brother" (or sister) can aid in a smoother adjustment; a comment like "You are so special to her that she bought you this surprise" can do wonders for the transition.

Once the infant is home, parents and visitors should be particularly attentive to the toddler. Bringing simple presents to the toddler along with the baby's present is helpful. Comments to the toddler such as "Your new sister is so lucky to have a big brother who's such a good helper" are supportive.

Involving the toddler in simple care procedures can be helpful also. Gathering equipment or observing the care are excellent methods. During these care periods, feeding in particular, the parent may find that it is less stressful for the child to be near. In some instances, however, the very young toddler may become too easily upset and should be removed or diverted during this period.

Demands on the toddler when the new infant enters the home may need to be altered. Parents or caretakers may find that even with all their preparation, the toddler is more explosive, volatile, and quick to cry. Gradually, the toddler will regain a secure, comfortable feeling within

the family unit and will become more accepting of the new baby.

Toilet Training

Toilet training is one of several critical training situations in which the social learning that takes place affects the personality development of the individual as a child and adult (Dollard and Miller, 1950). Some degree of toilet training or control is imposed in all societies. The varying cultural impositions reflect different value systems regarding cleanliness and control of bodily functions. The ease with which these values are enforced also appears to differ from culture to culture. In warm-climate cultures, fewer, more flexible demands appear to be made on the child. Often a sibling or peer is the primary model for training the child. In a peasant family in India, for example, by the time the toddler is walking, he learns toileting by squatting, with assistance if necessary, and by observing and imitating older adults who may be seen spontaneously squatting in a variety of places. With little clothing and few restrictions, the toddler in such a society has little difficulty learning bowel and bladder control (Smart and Smart, 1977). More industrialized countries like Russia and the United States promote toilet training at an earlier age, resulting at times in parental frustration and shame and hostility in the child.

The toddler's level of neuromuscular and psychosocial maturity is a basic factor in successful bowel and bladder training. Neuromuscular maturation occurs when myelination of the spinal cord is complete. At this point the toddler will have voluntary sphincter control. Walking generally indicates that the myelination process is complete. Additional indicators are the ability to retain urine for a minimum of two hours, and awareness of the wet sensation after urination. Awareness of the urge to urinate usually develops later.

Readiness is indicated when the toddler demonstrates interest in the process. The young child may grunt and point to the diaper, or to the potty. Cooperation and participation once the training is begun are additional readiness signs. It should be noted that the necessary maturation occurs at different times in different children. As a rule, training is most successful at around 2 years of age. In fact, mothers who have made little effort to train their children report that the children "trained themselves" around 2 years (Ilg and Ames, 1955).

Bowel control usually precedes bladder control, occurring as early as 18 months. At this point, the toddler usually has associated a word such as "pooh-pooh," "stinky," or "push" with the function. The child requires much as-sistance with undressing but by 2 years many toddlers are capable of using the potty by themselves if their pants are removed.

Some authorities advocate the use of "potty-chairs" because of the control and security they afford the child. The feet rest securely on the floor, and the lower height is much less intimidating (Fig. 6-8). The toddler's concern about being flushed away with the feces is reduced as well. Others feel that an adaptable seat that fits on the regular toilet aids in the transition to the regular facility. Flushing the toilet after the toddler is off it (or out of the bathroom) may decrease anxieties related to flushing.

Regardless of the type of facility used the young child should not be forced to sit or stay

Figure 6-8. A "potty-chair," rather than a seat that adapts to a standard toilet, permits the toddler's feet to touch the floor and thereby may provide more control and security. (Photo by Sam Gray, M. Photog., Cr.)

on it. The experience should be as pleasant as possible; even reading a book may be helpful. Much positive reinforcement should be given, e.g., "Good girl, Susie, Mommy really likes it when you push in the potty." If after these approaches are tried the toddler resists, the training should be stopped and started later when the toddler is more ready. This may be a few weeks or even months. Attempts at forcing the child to comply before he is ready to do so invariably fail.

Bladder control is acquired similarly. However, it is usually more difficult for the child to attain, since the sensation of needing to urinate is more subtle, and the toddler may not become fully aware of the need in time to prevent accidents. By 2 years of age the child may remain dry throughout the day although nocturnal enuresis (bedwetting) may continue until 4 or 5 years of age. During periods of increased activity, daytime accidents can occur. The parent or caretaker should be instructed to provide regular periods for bathroom breaks.

Training pants can be useful in helping the child become more aware of the training process and in making him feel more grown-up. Along with other loose-fitting, easily manageable clothes, these allow the child to go to the bathroom independently.

Sleep Patterns

Sleep needs vary considerably in individual toddlers. During sleep the child's basal metabolic rate is slowed, making conserved energy available for cellular growth and repair. The year-old usually sleeps through the night and requires one to two hours less sleep than the 6- to 11-month-old. The year-old child needs about 14 to 16 hours of sleep per day. By 2 years of age, sleep needs may have decreased to 12 to 14 hours per day.

The 15- to 18-month-old may take only one daytime nap which may follow lunch and last up to two hours. The 21-month-old toddler may sleep two to two and a half hours and usually requires a longer time to fall asleep. By 2 years the toddler may nap two hours per day, although naps may not be taken everyday. When they cannot sleep, most 2 year olds will engage in solitary play during the nap period.

The two-and-a-half-year-old often has great difficulty with naps. The child is slow in falling to sleep and may be irritable upon waking. In contrast, the 3-year-old takes a shorter nap, is more amenable to falling asleep, and wakes pleasantly. The 3-year-old may sleep for a one to two hour period. Even as the child decreases or eliminates naps in this stage, a quiet resting time can benefit the child by allowing for a period of decreased stimulation.

Night and day sleeping can most easily be established through some specific interventions. The most important factor is establishing a routine. Optimally, this routine should be introduced early in infancy and consistently reinforced thereafter. Of course, some children will vary in their responsiveness to these approaches.

The presleep period is also important maximizing the likelihood that the child will fall asleep. Overstimulation stemming from activity, television, or other sources should be avoided. Books, puzzles, and other quiet activities can help the child "gear down" and can lead to less resistance. Allowing the child to take a favorite toy or object to bed may serve to increase security.

Behaviors such as head-rolling, rocking, talking, singing, and thumb- or finger-sucking are common sleep-time behaviors in the toddler. Ignoring these behaviors is usually most effective; by 3 to 4 years of age they have often disappeared.

Nightmares and dreams may begin to emerge by 2.5 or 3 years of age. Confident, calming, and reassuring support by the parent or caretaker is most useful. Verbal reassurance, along with holding the cuddling, can be soothing. The child will usually return to sleep after such reassurance.

The child should be dressed properly for sleep. In colder months, the child needs clothing heavy enough to allow for the possibility of becoming uncovered during the night. In warmer months, the clothing should be light and loose-fitting. The toddler should ideally have his own bed and room, and the room should be well ventilated but free of drafts.

Accidents

Automobile Accidents

Forty-five percent of deaths in the 1 to 14 age group in the United States are attributed to accidents. While the American childhood mortality rate has declined significantly since 1950, the United States is still slightly higher than other industrialized countries in the rate for this age group (see Figure 6-9) (USDHEW, 1979).

Major causes of death in the 1 to 14 year age

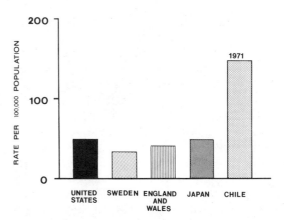

Figure 6-9. *Death Rates for Ages 1–4 Years: Selected Countries, 1975. (Source: United States data, National Center for Health Statistics, Division of Vital Statistics; data for other countries, United Nations.)*

group differ from those of infancy. As can be seen in Figure 6-10, accidents, cancer, birth defects, infections, and homicide are the primary causes.

Automobile accidents involving toddlers as

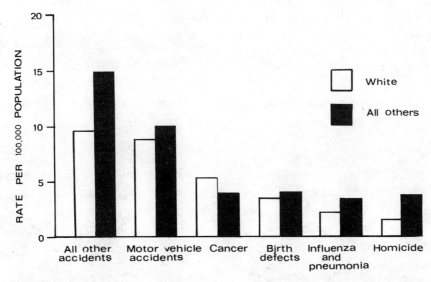

Figure 6-10. *Major Causes of Death Among Children Aged 1–4 Years, United States, 1976. (Source: Based on data from the National Center for Health Statistics, Division of Vital Statistics.)*

passengers result from the absence or improper application of restraints (see Figure 6-11). A 1974 survey by the Insurance Institute for Highway Safety found that proper restraints were used for only 7 percent of approximately 9,000 passengers under age ten (USDHEW, 1979). A child who weighs less than 40 pounds (as is usually true of the child under 4 years of age) lacks sufficient ossification of the pelvic structures to allow safe use of the lap belt only. Internal organ injuries often occur from impact. For the protection of the toddler the use of specially designed seats is essential; in contrast to infant seats, the toddler seat faces forward and has both lap and shoulder restraints, sized appropriately for the child. In the absence of such a seat, regular seat belts, including the shoulder restraint, are better than nothing. Consistent use of these systems will not only protect the child, but will establish a habit of use that can extend into adulthood.

Because the use of restraining systems is not widespread, other approaches have been proposed to reduce automobile-related deaths in children. Some European countries have had significant success by changing school hours so that they do not coincide with rush hours.

Federal law requires that by 1983 all new cars be equipped with protective devices (airbags and belts) that are automatically positioned when the car door is closed, and which provide protection for front seat passengers during a frontal collision.

The nurse should mention the prevention of automobile injuries as part of all well-child assessments and follow-ups. The consistent use of appropriate safety belts should be encouraged. Many civic clubs and other groups are now lending safety seats to families of infants and young children. If car belts are necessary, both lap and shoulder restraints must be used. The child is much safer in the back seat, away from the windows and doors. Locks should always be used when traveling with a child.

Of course, the child must learn pedestrian safety also. Parents should be encouraged to supervise the toddler closely while crossing or even playing near a street. The child's hand should be held, and instructions on proper crossing (looking both ways, not crossing behind parked cars) should be given.

Burns

Thermal burns are the most common type of burn in the young child; among these, scald burns are most prevalent. Because of their in-

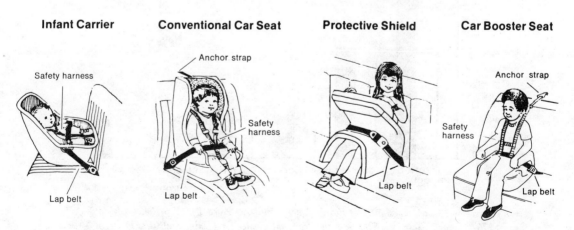

Figure 6-11. *Acceptable Auto Restraint Systems for the Young Child. (Courtesy of Physicians for Automotive Safety.)*

nate inquisitiveness, toddlers are very prone to pulling things like filled coffee-pots and hot pots and pans off the stove and onto themselves, or they may touch hot burners or other hot objects. The resulting injuries can be minor or extreme. Small localized surface burns can result if the child is lucky, or full-thickness injuries involving a large percentage of the body can occur. Burns resulting from spills commonly involve the face, trunk, and upper extremities.

Electrical burns are also prevalent. Biting an electrical cord can cause a deep, caverous burn on the lip or face. The child may insert objects (including his own fingers) into outlets. It should be noted that with electrical burns there may be two areas of involvement, with tissue injury occurring where the electrical current entered and exited.

Again, nursing interventions should focus on prevention. Parental teaching regarding developmental aspects of this age group and appropriate prevention should be part of the nurse's anticipatory guidance for this age period. (See Figure 6-12 for measures to prevent burns.)

Drownings

Fifty percent of all drownings occur in the under-5 age group, usually in inadequately supervised situations. The toddler must be constantly supervised when around water.

The child should be taught water safety as early as possible. This will not replace the need for constant supervision, but will prepare the child to enjoy the water safely and sensibly when he is older. Swimming instruction can often be started during this period.

Constant observation in the bathtub is also essential. The water should be kept at a safe level. A frightened child may be unable to lift his head out of as little as three inches of water.

Teaching parents resuscitation techniques is particularly useful. Since time is so critical in preventing cerebral anoxia, resuscitation by the parent or others may be essential.

Use flame-retardant clothes (launder according to directions).

Do not position crib or child's bed near radiator, fireplace, furnace.

Use safeguards around fireplace, radiator, stove.

Have available for use fire alarms, sprinklers, extinguishers.

Reduce thermostatic control on tub or shower water.

Keep all matches out of reach of children.

Maintain bath water temperature below 105°F.

Do not pour or have hot liquids around child (coffee, tea).

Use cool mist vaporizers (avoid those that are heated).

Avoid excess exposure to sunlight.

Avoid smoking when in proximity to toddler.

Cover electrical outlets with safety plugs.

Place all electrical appliances out of reach of child.

Avoid any hanging cords from electrical appliances.

Utilize nonhanging tablecloths or decorations.

Position all handles of pots and pans on stove so that they point inward.

Avoid pouring hot liquids directly into glass containers.

Figure 6-12. *Safety Measures for the Prevention of Burns.*

Falls

Head injuries and fractures resulting from falls are common in the toddler period. The child will climb up dangerous stairs, into cabinets, and even out of windows. Many times, this kind of accident occurs because the parent or caretaker underestimated the child's motor potential.

Making the environment as safe as possible is essential; windows and doors should be equipped with gates and locks. Although the child can be introduced to certain safety rules at this stage, he should never be trusted to follow the rules, for the lure of the unknown is too great. Adequate supervision is a must.

Poisonings

The incidence of fatal poisonings among the 1- to 5-year age group has dropped dramatically since 1970. Once the leading cause of accidental death in that age group, poisonings now account for 6 percent. In 1977, 4,000 fatal poisonings occurred with less than 1 percent involving children under five. Children most frequently involved were 2-year-olds (US-DHEW, 1979).

Morbidity, however, remains quite significant; approximately 2 million ingestions of poison by American children occur yearly. Perhaps this is not surprising, since it is estimated that some 850 toxic substances are commonly used in American homes as of this writing. This, coupled with improper storage and placement of these substances, is a direct cause of accidental ingestions (Arena, 1970).

The nurse or other health care provider can do much to aid in prevention through parental teaching. Figure 6-13 outlines preventive measures. Additionally, poison control centers have been set up in many states to provide immediate information on drug or chemical antidotes as well as other emergency directions.

Federal regulations have mandated the use of "child-proof" or "lock-top" caps on all prescription drugs. These tops in many instances require the capability of palmar rotation, which the typical child under 4 or 5 does not have. Although all medications should be kept out of the reach of children, in instances in which the child has access, these caps may often prevent ingestion.

Congenital Anomalies

Congenital defects contribute to the mortality rate of toddlerhood although to a much smaller extent than accidents. Common defects involve the central nervous system and closure of the neural tube. These defects vary in severity and are characterized by an inadequate or incom-

Keep all medications, cleaning agents, and other toxic substances out of reach of the child, in a locked cabinet.

Keep all medications, cleaners, and toxic agents in safety-locked containers.

Replace safety-lock caps immediately after use.

Avoid taking medicine in presence of the child.

Never refer to medicine as candy.

Use medicines only as prescribed.

Store all medicines and toxic agents in properly labeled containers.

Avoid storage of surplus of medications, cleaning agents, and toxic substances.

Always have syrup of ipecac available and be familiar with administration directions.

Keep hydrocarbon agents (gasoline, charcoal lighter fluid) in proper containers.

Store hydrocarbons under lock at all times.

Anticipate motor capabilities of the child.

Provide routine snacks to prevent exploration for food or drink.

Always provide adequate supervision of the child.

Have information on nearest poison control center always available.

Figure 6-13. *Prevention of Poisonings.*

plete closure of the vertebral column, possibly involving meningeal protrusion.

The most common type of neural tube defect is *spina bifida cystica*, which is characterized by a defective closure of the vertebral column and a cystic protrusion through the defect of the meninges, spinal cord, or both. Myelomeningocele (involving both the cord and meninges) accounts for about 75 percent of the cases of this anomaly. The lesion tends to occur in the lumbar or lumbosacral area, since this is the last segment of the neural tube to close. Many times lower extremity involvement results in hemiplegia.

Hydrocephalus, characterized by an abnormal collection of cerebrospinal fluid in the cranial vault, accompanies myelomeningocele in 90 percent of cases. It is heralded in the toddler and older child by an increase in occipital-frontal measurement.

A potentially less severe defect is that of *meningocele*, which is a deformity of the vertebral column with a protruding sac encasing the meninges and cerebrospinal fluid. Without spinal cord involvement there is limited neurological dysfunction. These defects occur around the fourth week of embryological development and are suspected of being caused by viral, radiation, pharmacological, and other factors.

Musculoskeletal defects are common anomalies as well. Congenital hip dysplasia, including sublaxation and dislocation of the hip, is one of the most common defects. These conditions are more common in females, and they are characterized by impaired development of the ossification centers of the acetabulum and femoral head.

Hip dysplasia presents with limited abduction, either bilateral or unilateral. There may be shortening of the affected limb with accompanying assymetrical fat-fold deposits. In the walking toddler, one is likely to see lordosis, a waddling gait with bilateral dislocation, or, when the child is walking downstairs, unilateral involvement and Trendelenburg's sign (dropping of the gluteal fold on the normal side when the affected limb is bearing weight).

Although the actual cause is unknown, genetic factors and fetal positioning are thought to contribute to abnormal joint development. Abnormal relaxation of the acetabular capsule and joint ligaments, caused by increased estrogenic production, has also been identified as a factor. It is noteworthy that in cultures in which infants are carried on their mothers' backs in such positions that a consistent abducted position is maintained, little hip dislocation is seen. The disorder is common among Eskimos and Navajo Indians.

As with all congenital defects, the nurse's role focuses on prevention and detection. Early detection of hip dislocation is important for weight bearing in the toddler period can result in additional hip, acetabular or femoral deformity.

Infections

Upper respiratory infections such as the common cold, acute coryza, and otitis media generally occur in toddlers without complications, although pneumonia and meningitis may appear as complications. The pathogens mostly likely to be involved are Hemophilus influenzae and penumococci. Meningococci are often the causative agents of meningitis.

Common minor respiratory infections must be treated promptly. If antibiotic therapy is needed, parents should be informed of the purpose of the drug and the importance of administering it for the entire prescribed period.

Otitis media (middle ear infection) is the most common bacterial infection of toddlerhood. It results from a dysfunctional eustachian tube in which inadequate opening and closing permits the passage of bacteria from the pharynx to the middle ear. Secretions from the middle ear mucosa can also be trapped, providing an excellent medium for the growth of bacteria. Otitis media can occur as a primary infection or secondary to viral nasopharyngitis, tonsillitis, or other infections. Otitis media merits prompt treatment because of possible secondary infections. Only the mastoid bone separates the middle ear from the brain. An untreated infection can develop into mastoiditis, or, even worse, a brain abscess or meningitis.

Signs of otitis media in the toddler may include nasal congestion, irritability, vomiting, or diarrhea. Fever may or may not be present. If the tympanic membrane has ruptured, drainage from the ear may develop. An ear examination will often reveal a reddened, bulging tympanic membrane; neither the light reflex nor the umbo will be clearly visible.

Because the young child may not be able to communicate pain or other symptoms verbally, nonverbal behaviors should be closely assessed. The toddler may pull at an ear or frequently roll his head from side to side. Anorexia or sleep interruptions may also be seen.

Croup is also prevalent in the toddler up to three years of age. Croup is a viral infection in which relatively severe laryngeal spasms accompany a less serious inflammation. The young child typically has a sudden onset of inspiratory difficulty, often at night. An inspiratory stridor or high-pitched crowing sound usually develops. Symptoms of respiratory obstruction develop as laryngeal spasms become more frequent. Cyanosis, anxiety, and a dry, barking cough may also develop. Total respiratory obstruction can occur.

Prompt relief of acute respiratory distress accompanying croup is important. Parents can be instructed to take the child to the bathroom and turn on the hot water to create a mist for temporary relief. Hospitalization may be required. Warm or cool mist is effective in relieving spasms, and usually a vaporizer placed by the bed results in effective relief.

Nursing Bottle Caries*

By 2 years of age, a majority of children in the United States have dental caries. Some children may even experience caries during the nursing years if use of the bottle is prolonged. This condition has been labeled "nursing bottle syndrome," "Nursing bottle caries," or nursing bottle mouth." Use of the nursing bottle should be discontinued by no later than 12 months of age; if a child is permitted to use the nursing bottle beyond the first birthday, and especially

* The section on nursing bottle caries was written by Gary P. Hill, DDS, MS, Durham, N.C.

if the child takes the bottle in a lying position, dental caries may result. Milk or any liquid containing sugar can cause the decalcification process to begin.

Many times the parents are not aware of a problem until there is much decay present. Parents believe the situation occurred "almost overnight." Normally, caries begin on the tongue side of the upper front teeth; the tongue permits the pooling of liquid around the back of these teeth. With time, decay gradually encircles the tooth.

It is not uncommon for one or more of these teeth to be broken off at the gum line because of trauma resulting from a fall or bump; frequently, this is the first occasion for dental care, and it is under these circumstances that the parent first realizes that another problem exists. Sometimes the parent will blame the fall itself for causing the decay.

Although the tongue promotes the decay process in the upper anterior teeth, it actually protects the lower anterior teeth, which are never involved in nursing bottle caries. If the child is still on the bottle by the time the upper and lower primary molars erupt, these teeth will likely be affected also. Of course, treatment of these teeth is essential. The severity of the dental infection determines the extent of treatment necessary. Hospitalization and the use of general anesthesia is often required in order to properly treat the child. Other treatment ranges from simple restorations to root canal therapy and crowns or extractions.

Although it is not usually recommended that a child visit a dentist until he is 2.5 to 3.5 years of age, it is essential that health care professionals who see young children help educate the parents concerning the use of the nursing bottle. Nursing bottle caries may result in the need for early extraction of the infected primary teeth. Speech impediments and psychological problems might possibly ensure. With proper parental education, the condition can easily be prevented; it is a trauma that a child need not suffer and a financial expenditure that a parent can avoid.

NUTRITIONAL NEEDS

The toddler requires 100 calories per kilogram of body weights (45 calories/lb) in contrast to the 110 calories per kilogram of infancy (50 calories/lb). Protein needs drop from the 2.0 to 3.5 g per kg needed in infancy to 2.0 to 2.5 g per kg. The toddler needs 125 ml of water per kilogram, or two ounces per pound (see Table 6-4). Because the toddler's weight continues to increase (although more slowly than in infancy), the needs for the nutrients just mentioned increase as well, even though caloric, protein, and water requirements decrease.

Eating Habits and Manipulation of Utensils

By 12 months the child can tolerate almost all table foods. Also by this age, the child can grasp a spoon and attempts to use it for self-feeding; this is usually unsuccessful because of immature ulnar deviation of the wrist at this age. However, the toddler can grasp a cup if it is sized appropriately. If a bottle is still being used, it can be grasped. In any event, nonbreakable utensils are a must.

By 15 months of age some refinement in the use of a spoon has occurred, although the toddler is not always successful in scooping food with a spoon or in getting it to his mouth. Much spilling occurs because of poor wrist rotation. Although a cup can be managed, lifting and lowering problems exist.

In contrast, the 16- to 17-month-old has acquired a proficient ulnar deviation (rotation), allowing for excellent wrist control. Less spill-

Table 6-4. Recommended Daily Allowances for the Toddler

	Toddler (1–3 Years)
Weight	13 kg (28.6 lb)
Height	90 cm (35 in.)
Protein (g)	23
Fat-soluble vitamins:	
Vitamin A, mcg r.e.	400
Vitamin D, mcg (I.U.)	10 (400)
Vitamin E, I.U.	7
Water-soluble vitamins:	
Vitamin C, mg	45
Thiamine, mg	0.7
Riboflavin, mg	0.8
Niacin, mg	9
Vitamin B_6, mg	0.9
Folacin, μg	100
Vitamin B_{12} mcg	2.0
Minerals	
Calcium, mg	800
Phosphorus, mg	800
Magnesium, mg	150
Iron, mg	15
Zinc, mg	10
Iodine, mcg	70

Note: Allowances for vitamins A and D are currently being expressed in mcg although many food composition tables state values in international units (IU); 1 retinal equivalent (r.e.) = 3.33 IU of retinal or 10 of B-carotene.
SOURCE: Adapted from *Recommended Dietary Allowances*, (8th rev. ed.), Washington, D.C.: National Research Council, National Academy of Sciences, 1980.

ing occurs in attempts at bringing spoon to mouth. Spilling has greatly diminished by 18 months to 2 years, so that the child of this age is more successful in feeding himself. By this age the cup can be manipulated with much more ease since wrist and finger control has improved.

Finger-foods remain a favorite for the toddler. Attempts may be made to finger-feed all foods. Some finger foods such as cheese, bits of meat, or green beans should be offered at each meal.

The toddler's preferences include meat, cereal, grains, baked products, fruit, and sweets. Toddlers also like carbohydrate-rich foods (potatoes, cereals, bread, or rice) because they are easily chewed.

Protein-rich foods such as cheese, yogurt, and meat become increasingly popular during this period. Meats that are eaten most willingly appear to be the less fibrous types (chicken, ground meats, frankfurters) that can be chewed with greater ease.

Food dislikes also emerge in the toddler period. Toddlers tend to dislike liver, mixed dishes such as casseroles, and many cooked vegetables. Toddlers prefer raw vegetables. Milk, which has been the major food source through much of infancy, may also be rejected in the second half of infancy (Beal, 1957). Throughout toddlerhood and the early preschool period, milk intake is often as low as 1 to $2\frac{1}{2}$ cups per day. Periodically, milk may be rejected altogether. If this rejection persists, allergy should be considered.

The toddler is also susceptible to food jags or fads, which surface between 2 and 4 years. The toddler may demonstrate a strong preference for only a few foods or just one food and want to eat it exclusively for days. The child might wake up one morning and declare that he will eat only foods that rabbits eat, and will proceed to ingest lettuce and carrots. These jags are usually temporary and can best be handled by calling little attention to them. It has been noted that parents and siblings serve as key role models regarding food preferences. The toddler is particularly responsive to the likes and dislikes of older siblings (Eppright, 1969; Pipes, 1977). When the toddler is outside the home in a day-care center or preschool, peers play an influential role. Television likewise plays a key role; most young children are constantly exposed to food commercials that emphasize sweet foods.

Despite the strong preferences and fads, the child's overall nutritional status is usually good since, over a period of days, he tends to select what amounts to a balanced diet. The parent often needs to be made aware that the child can be allowed a wide variety of foods without concern for nutritional deficiencies as long as eating is satisfactory over a long period. Reports of parental dissatisfaction with the eating patterns of 2- to 4-year-olds evidence the need for both parental education and support during this period (Eppright, 1969; Beal, 1957) (see Table 6-5).

Table 6-5. A Daily Meal Plan for Toddlers

Breakfast	Sample Menu
1. One high-protein food	1. Egg
2. Bread with butter or margarine	2. One slice white enriched toast with butter (1 tsp)
3. Milk	3. 4–8 oz milk
4. Fruit	4. 4 oz orange juice
Lunch	*Sample Menu*
1. One high-protein food	1. (1 oz) cheese slice
2. Vegetables	2. few pieces of raw carrots
3. Bread	3. One slice of whole-grain bread (may be prepared as cheese toast)
4. Milk	4. 4–8 oz milk
5. Dessert	5. One apple, sliced
Dinner	*Sample Menu*
1. One high-protein food	1. 2 oz baked or fried chicken
2. Vegetables	2. 4 oz string beans
3. Bread	3. 4 oz mashed potatoes or rice
4. Milk	4. 4–8 oz milk
5. Dessert	5. 4 oz ice cream

SOURCE: Adapted from Pipes, P. *Nutrition in Infancy and Childhood.* St. Louis: C. V. Mosby, 1972.

The toddler's ritualism is seen in eating patterns as well as in the behaviors described earlier. Foods must often be arranged on the toddler's plate in a certain way. Many toddlers do not like having foods mixed, or even touching. Sandwiches or slices of bread may need to be cut in a specified way, perhaps diagonially instead of halved longitudinally. This ritualism may show up in very strong perferences for a certain plate or cup, or even for seating arrangement. To facilitate the child's eating, these ritualistic preferences should be honored as much as possible.

Serving simple foods, creating a tension-free atmosphere, and setting an appropriate time for meals help ensure an adequate nutritional intake for the toddler. He should not be subjected to the stress of having to conform to a rigid mealtime schedule, sufficient time should be allowed for eating, but additionally, the toddler should not be expected to sit for long periods at the table. A comfortable chair of proper height should be provided; ideally, it should have a foot rest. The food should be colorful and portions small. Conversations should be pleasant, and should include the toddler. Controversial discussions should be avoided and mealtime should not be used as a punishment period.

ANTICIPATORY GUIDANCE

Accidents, including burns, poisonings, and falls occur in almost all instances because parents or caretakers have underestimated the child's motor potential. It must also be stressed that the toddler requires almost constant supervision. With his ability to move about and his keen desire to explore and manipulate, the toddler is frequently victimized by his curiosity.

Prevention of common communicable viral diseases can be ensured through immunization programs (see Appendix E). The MMR (measles, mumps, rubella) vaccine is administered at 15 months, while at 18 months a DPT (diphtheria, pertussia, tetanus) inoculation is given. Parents should be informed that immunizations are essential.

The nurse should direct parents toward fostering the child's independence. Encouraging self-help skills, social interactions, and cognitive growth will better prepare the toddler for the challenges of later life (Fig. 6-14). More assertive, decisive, and competent behavior will emerge from this independence.

Parental control is a necessary aspect of parenting. Democratic, not autocratic, control is most conducive to the attainment of autonomy. The wise parent is one who dominates the child enough to socialize him, but at the same time allows the child some self-control, providing stability through predictable authority (Hughes, 1979).

With half of the mothers of toddlers and preschoolers entering work situations, parental roles within the family have been altered (National Council of Organizations for Children and Youth, 1976). The impact of the father on the child's development is also quite significant, particularly in these early years. Absence of or lmited exposure to the father during these years has been associated with more dependent, less aggressive children (Biller, 1979, Hetherington, 1973). The father's impact appears to be greater on the male child at this period. Parents should be made aware of the child's need for role models.

Parenting during the toddler period is demanding even under optimal conditions. Crises such as separation, illness, or divorce create additional stress for both the child and parent. Insensitivity on the part of the nurse to such needs can do little to facilitate the parenting role or the continued growth of the toddler. Assessment of the family's level of functioning, readiness to listen, and ability to implement suggestions appropriate to the family's life-style are essential for successful anticipatory guidance.

COGNITIVE DEVELOPMENT

1. Talk to the toddler; listen with interest to what is being said. Use complete thoughts—not "Pick it up," but "Pick up the ball from under the table."
2. Read to the toddler or tell stories.
3. Have the toddler tell you stories about pictures in books or magazines. Have the child name objects. Make a picture book: Cut out large pieces of paper bag for the cover and pages; fold them in half, tie them together with string or yarn, and paste in pictures from magazines, cereal boxes, newspapers, etc.
4. Name parts of the body and pictures of people.
5. Play singing games: "Ring Around the Rosy," "Row, Row, Row Your Boat" "Three Blind Mice," or nursery rhymes.
6. Play telephone with the toddler.
7. Play with puppets; have a conversation using puppets.
   A. Paper bag puppets: Fill the end of a small bag with cotton or crumpled newspaper; insert a stick or pencil; tie a string around the stuffed area and stick; paint or draw a face on the bag
   B. Potato puppets: Insert a stick in a small potato; facial features can be created by painting the surface or pasting on bits of paper.
   C. Potato-finger puppets can be made by cutting a small hole in the bottom of the potato.
   D. Old gloves make good puppets; cut off the fingers and thumb and stuff them with cotton, nylon hose, old rags, etc. The thumb becomes the head and body of the puppet. Two fingers become the arms when sewn to the thumb section. Bind the head and waist section with string or yarn. Decorate with pieces of cloth, yarn, or ribbon.
8. Let the child look in a mirror and point out facial features and body parts.

FINE MOTOR DEVELOPMENT

1. Fill-and-dump toys:
   As the child's coordination improves, decrease the size of the container and the opening (e.g., plastic milk bottle, small jars, and cans) and give him small objects to put into the container (e.g., buttons, bottle cpas, peas, beans, macaroni).
2. Sorting Activity:
   Give the child two or three containers and bottle caps, buttons, and beans and have him put all the caps in one container, buttons in another, and beans in a third.
3. Stacking toys:
   Build a pyramid with different size boxes or cans.
4. Nesting toys:
   Use boxes, bowls, pots, pans, cups, etc., that fit inside one another. Start with three sizes.
5. Clothespins and a coffee can or loaf pan:
   Have the child put the clothespins on the edge of the can. (This can also be used as a sorting activity.) Paint the pins different colors and have the child sort them by color.
6. Blocks:
   Make from milk cartons (thoroughly washed), boxes, wood scraps; cut off tops of two milk or cream cartons, push them together; make different sizes. Show the child how to build; have him copy what you build.

Figure 6-14. *Stimulating Activities for Toddlers.*

7. Large Pencil or Crayon:
   Have the child copy a line or simple shape; then allow him to draw whatever he wishes. Start teaching the child about colors; use paper bags or cut-up cardboard boxes to draw on.

8. Stringing Objects:
   Use old shoe laces or heavy string and various size spools. As the child's coordination improves, give him macaroni and small beads.

9. Puzzles:
   Make puzzles by pasting pictures from magazines to a piece of cardboard and cutting it into pieces. Start with simple pictures of one object and cut into 3 to 5 large pieces. With an older child, use a more complex picture and 5 to 10 small pieces.

PERSONAL AND SOCIAL DEVELOPMENT

1. Have the child play with dolls and stuffed animals.
2. Take the child to a store, neighbor's house, riding in a car or on a bus, to a park, zoo, etc. Point to and name people, objects, animals, etc.
3. Play games such as hide-and-seek.
4. Dress up, role-playing:
   A. Provide old hats, dresses, shoes, and purses.
   B. Make hats and masks out of paper bags or paper plates. Make dresses out of blankets or cloth.
   C. Allow the use of cooking utensils and house-cleaning equipment.
   D. Make a house by putting a blanket over a high table.

GROSS MOTOR DEVELOPMENT

1. Push-pull toys.
   A. Attach a string to a large box and fill it with lightweight objects.
   B. Tie a cord to each end of an oatmeal box or coffee can and fill with bells, spools, bottle caps, stones, hair rollers, old jewelry, etc.
   C. Make a train; tie boxes (shoes, milk cartons, salt boxes, etc.) together with heavy string and make a pull string.
2. Cardboard tunnel to crawl through—cut ends out of large cardboard boxes and attach several boxes to make a tunnel.
3. Climb stairs.
4. Walking board—rest a board one foot wide and three feet long on bricks. Encourage the child to walk forward, backward, and sideways, and to jump down. As the child's coordination improves, decrease the width of the board.
5. Play catch with a ball or bean-bag. Make a large ball out of two pieces of cloth sewn together and stuffed with rags, cotton, nylons, etc.
6. Sand play (use unbreakable things).
   A. Give the child spoons, cans, bowls, boxes, cups, glasses, sieves, and funnels to play with.
   B. Cut a bleach bottle in half; use the bottom for a pail (make a handle out of heavy string), and use the top for a funnel.
   C. Use an inner tube as the frame for a sand box and fill the inside with sand.
7. Water play (in a large tub, or in a sink or bathtub): give the child unbreakable containers, sponges, cork, bar of soap.
8. Encourage tricycle riding.

Figure 6-14. (Continued)

SUMMARY

Toddlerhood, the period from 12 to 36 months, is physically and psychosocially as distinctive as infancy. The rate of growth decreases from that of infancy. The toddler gains approximately 5 to 6 pounds per year, and grows about 3 to 5 inches per year. By 2 years of age the brain has acquired 75 percent of its adult weight, and the head increases in size by approximately 1 inch per year.

Motor development opens new avenues of exploration to the toddler. Walking typically occurs by 12 to 15 months, running by 18 months. At three years the toddler can walk up and down stairs, alternating feet.

Cognitively, the toddler is in Piaget's sensorimotor period, which is characterized by simple forms of reasoning involving symbolism. The toddler is incapable of truly logical thinking, however, being hampered by the characteristic limitations of the stage, i.e., include centering, irreversibility, transductive reasoning, and egocentrism.

Language development progresses significantly during the toddler period, although the toddler comprehends much more than he can say. Two-word sentences dominate speech up to approximately 2.5 years, when three-word patterns emerge. Pronouns such as *I, me,* and *mine* dominate speech at this period.

Dramatic psychosocial growth occurs in toddlerhood. The conflict between the desire to please and the desire to assert onself (the latter being stronger than even before) leads to ambivalence, which in turn manifests itself in negativism and dawdling. Temper tantrums, ritualism, and thumb- or finger-sucking are other characteristic behaviors of the period.

Common health problems include accidents, infections, and other illnesses. Other areas of concern include appropriate play activities, toilet training techniques, nutritional guidance, and stranger and separation anxiety.

REFERENCES

American Academy of Pediatrics. Report of the Committee on Infectious Diseases, 1977. Evanston, Ill.: American Academy of Pediatrics. 18th Edition.

Arena, J. M. *Poisoning: toxicology, symptoms, treatments* (2nd ed.). Springfield, Ill.: Charles C Thomas, 1970.

Ariès, P. *Centuries of childhood.* New York: Vintage Books, 1962.

Batter, B., and Davidson, C. Wariness of strangers: reality or artifact? *Journal of Child Psychology and Psychiatry,* 1979, *20,* 93–109.

Beal, V. A. On the acceptance of solid foods and other food patterns of infants and children. *Pediatrics,* 1957, *20,* 448.

Biller, H. B. *Father, child and sex role.* Lexington, Mass.: Lexington Books, 1979.

Bloom, L. *Language development, form and function in emerging grammars.* Cambridge, Mass.: MIT Press, 1970.

Bloom, L. Language development review. In F. Horowitz (Ed.), *Review of child development research* (Vol. 4) New York: Russell Sage, 1975.

Bowlby, J. *Maternal care and mental health.* New York: Schocken, 1966.

Brown, R. *A first language.* Cambridge, Mass.: Harvard University Press, 1973.

Chinn, P. *Child health maintenance.* St. Louis: Mosby, 1979.

Chomsky, N. *Syntactic structures.* The Hague: Mouton, 1957.

Church, J., and Stone, L. J. *Childhood and adolescence.* New York: Random House, 1973.

Dale, P. S. *Language development: structure and function* (2nd ed.). Hensdale, Ill.: Dryden Press, 1976.

Dollard, J., and Miller, N.E. *Personality and psychotherapy: an analysis in terms of learning, thinking and culture.* New York: McGraw-Hill, 1950.

Drillien, C. M., and Drummond, M. B. *Neurodevelopmental problems in early childhood.* London: Blackwell Scientific Publications, 1977.

Easterbrooks, M. A., and Lamb, M. E. The relationship between quality of infant mother attachment and infant competence in initial encounters with peers. *Child Development,* 1979, *50,* 380–387.

Eckerman, C. O., Whatley, J. L., and Kutz, S. L. Growth of social play with peers during the second year of life. *Developmental Psychology,* 1975, *11,* 42–49.

Eppright, E. S.; Fox, H. M.; Fryer, B. A.; Lamkin, G. H.; and Vivian, V. M. Eating behaviors of preschool children. *Journal of Nutrition Education,* 1969, *1,* 16.

Erikson, E. *Childhood and society.* New York: Norton, 1963.

Fein, G. *Child development.* Englewood Cliffs, N.J.: Prentice-Hall, 1978.

Freud, S. *An outline of pschoanalysis.* (J. Strachey, trans.). London: Hogarth, 1949.

Geschwind, N. Language and cerebral dominance, the central nervous system (Vol. II). In Tower, D., and T. Chase (Eds.), *The clinical neurosciences.* New York: Raven Press, 1975.

Gibson, J. *Growing up: a study of children.* Reading, Mass.: Addison-Weley, 1978.

Godfrey, S., and Baum, J. D. *Clinical pediatric physiology.* London: Blackwell Scientific Publications, 1979.

Hagman, R. R. A study of fears of children of preschool age. *Journal of Experimental Education,* 1932, 110–130.

Harlow, H., McGaugh, J., and Thompson, R. *Psychology.* San Francisco: Albion Publishing Co., 1971.

Havighurst, R. Successive aging. *Gerontologist,* 1961, *1,* 8–13.

———. *Human development and education.* New York: Longmans, 1963.

Hayes, R. L. Deprivation of body pleasure: origin of violent behavior: a survey of the literature. *Child Welfare,* May 1980, 287–297.

Helms, D., and Turner, J. *Exploring child behavior.* Philadelphia: Saunders, 1978.

Hetherington, E. M., and Porke, R. D. *Child psychology.* New York: McGraw-Hill, 1975.

Hetherington, E. M. Girls without fathers. *Psychology Today,* 1973, *6* (9), 46–52.

Hughes, J. *Synopsis of pediatrics.* St. Louis: Mosby, 197.

Hurlock, E. *Developmental psychology.* New York: McGraw-Hill, 1980.

Ilg, F., and Ames, L. B. *Child behavior.* New York: Harper and Row, 1967.

Jersild, A. T., and Holmes, F. B. *Children's fears.* New York: Bureau of Publications, Teachers' College, Columbia University, 1935.

Kagan, J. The concept of identification. *Psychology Review,* 1958, *65,* 296–305.

Kaluger, G., and Kaluger, M. *Human development—the life span.* St. Louis: Mosby, 1979.

Kastenbaum, R. *Humans developing a lifespan perspective.* Boston: Allyn and Beacon, 1979.

Kennedy, W. *Child psychology.* Englewood Cliffs, N.J.: Prentice-Hall, 1975.

Klein, R., and Durfee, J. Prediction of preschool social behavior from social-emotional development at one year. *Child Psychiatry and Human Development,* 1979, *9* (3), 745–751.

Liebert, R. M.; Poulos, R. W., and Strauss, G. D. *Developmental psychology.* Englewood Cliffs, N.J.: Prentice-Hall, 1974.

Koch, H. L. Some emotional attitudes of the young child in relation to characteristics of his siblings. *Child Development,* 1956, *27,* 383–426.

Largo, R., Stutzle, W. Longitudinal study of bowel and bladder control by day and at night in the first six years of life. I. *Developmental Medicine and Child Neurology,* 1977, *10,* 598–606.

Lenniberg, E. *Biological foundation of language.* New York: John Wiley & Sons, 1967.

Lidz, T. *The person.* New York: Basic Books, 1968.

Lowery, G. H. *Growth and development of children* (7th Ed.). Chicago: Yearbook Medical Publishers, 1978.

Maacoby, E. E., and Field, S. S. Mother-attachment and stranger reactions in the third year of life. *Monographs of the Society for Research in Child Development,* 1972, *37,* No. 146.

Maier, M. W. *Three theories of child development: the contributions of Erik Erikson, Jean Piaget and Robert R. Sears and their application.* New York: Harper and Row, 1965.

Marlowe, D. *Textbook of pediatric nursing.* Philadelphia: Saunders, 1977.

Morris, D. The hazards of toilet training. *Nursing Mirror.* September 1978, 26–27.

Murray, R., and Zentner, J. *Nursing assessment and health promotion through the life span.* Englewood Cliffs, N.J.: Prentice-Hall, 1975.

Murdock, L. L. Spoiling: an ambivalent response to growing up. *Maternal-Child Nursing Journal,* Spring, 1978, 47–49.

Mussen, P. H. *Child development and personality.* New York: Harper & Row, 1974.

National Council of Organizations for Children and Youth. *America's children 1976.* Washington D.C.: National Council, 1976.

National Research Council. *Recommended dietary allowances.* Washington D.C.: National Academy of Sciences, 1980.

Nelson, K. Structure and strategy in learning to talk. *Monographs of the Society for the Research in Child Development,* 1973, *38,* No. 149.

Petrillo, M., and Sanger, S. *Emotional care of hospitalized child.* Philadelphia: Lippincott, 1980.

Piaget, J. and Inhelder, B. *The psychology of the child.* (H. Weaver, trans.). New York: Basic Books, 1969.

———. *Six psychological studies.* New York: Random House, 1967.

———. *Play, dreams, and imitation in childhood.* (C. Gattegno and F. M. Hodgson, trans.). New York: Norton, 1962.

Pipes, P. *Nutrition in infancy and childhood.* St. Louis: Mosby, 1977.

Rheingold, H. L. The modification of social responsiveness in institutional babies. *Monographs of the Society for Research in Child Development,* 1956, *2,* No. 63.

Rheingold, H. L., and Eckerman, C. O. The infant separates himself from his mother. *Science,* 1970, *168,* 78–90.

Robertson, J. *Young children in hospitals*. New York: Basic Books, 1958.

Robinson, C., and Lawler, M. *Normal and therapeutic nutrition*. New York: Macmillan, 1977.

Rosen, H. *Pathway to Piaget*. Cherry Hill: New Jersey: Postgraduate International, 1977.

Ross Laboratories. *Children are different: development physiology*. Columbus, Ohio: Ross Laboratories, 1978.

Sapala, S., and Mead, J. Pediatric management problems (toilet training). *Pediatric Nursing*, July–August, 1980, 31.

Scarr, S., and Salapatek, P. Patterns of fear development during infancy. *Merrill-Palmer Quarterly*, 1970, *16*, 53–90.

Schuster, C. S., and Ashburn, S. S. *The process of human development: a holistic approach*. Boston: Little, Brown, 1980.

Sears, R. R.; Maccoby, E. E., and Levin, H. *Patterns of child rearing*. Evanston, Ill: Row, Peterson and Co., 1957.

Skinner, B. F. *Verbal behavior*. New York: Appleton-Century-Crofts, 1957.

Slobin, D. I. Cognitive prerequisites for the development of grammar. In C. A. Ferguson, and D. I. Slobin (Eds.) *Studies of child development*. New York: Holt, Rinehart and Winston, 1973.

Scipien, G. *Comprehensive pediatric nursing*. New York: McGraw-Hill, 1975.

Sears, R. R. Identification as a form of behavioral development. In D. B. Harris (Ed.), *The concept of development*. Minn.: Minn. University of Minnesota, 1957.

Slobin, D. I. *Psycholinguistics*. Glenview, Ill: Scott, Foresman, 1970.

Smart, M. S., and Smart, R. C. *Children*. New York: Macmillan, 1977.

Smith, D., Bierman, E., and Robinson, N. *The biological ages of man*. Philadelphia: Saunders, 1978.

Smith, M. E. An investigation of the development of the sentence and the extent of vocabulary in young children. Iowa City University of Iowa Student Child Welfare, 1926, *3*(5).

Spelke, E., Kagan, J., Zelazo, P., and Katelchuck, M. Father interaction and separation protest. *Development Psychology*, 1973, *9*, 83–90.

Spitz, R. A. Hospitalism. In O. Fenichel et al., (Eds.), *Psychoanalytic study of the child* (Vol. 1). New York: International University Press, 1945.

Sroufe, L. A., and Wunsch, J. P. The ontogenesis of smiling and laughter: a perspective on the organization of development in infancy. *Psychological Review*, 1976, *83*, 173–189.

Starr, B., and Goldstein, H. *Human development and behavior*. New York: Springer, 1975.

Stangler, S., Huber, C. J., and Routh, D. K. *Screening growth and development of preschool children*. New York: McGraw-Hill, 1980.

Tanner, J. M. *Foetus into man: physical growth from conception to maturity*. London: Open Books, 1973.

Trause, M. A. Stranger responses: effects on familiarity, stranger's approach and sex of infant. *Child Development*, 1977, *48*, 1657–1661.

Turner, J. S. and Helms, D. B. *Life span development*. Philadelphia: Saunders, 1979.

U.S. Department of Health, Education, and Welfare. *Healthy people*. Washington D.C.: U.S. Government Printing Office, 1979.

U.S. Department of Health, Education, and Welfare. *Health United States 1979*. Washington D.C.: U.S. Government Printing Office, 1979.

Vander Zanden, J. W. *Human development*. New York: Knoff, 1978.

Vaughan, V., and McKay, R. J., (Eds.). *Nelson textbook of pediatrics*. Philadelphia: Saunders, 1979.

Waechter, E., and Blake, F. *Nursing care of children*. Philadelphia: Lippincott, 1976.

Whaley, L. and Wong, D. *Nursing care of infants and children*. St. Louis: Mosby, 1979.

7

The Preschooler

PATRICIA HUMPHREY

As compared to toddlerhood, the preschool years—ages 3 through 5 years—comprise a much calmer period in the child's development. Along with infancy and toddlerhood, the preschool period forms the crucial part of life during which the child begins to establish himself as a unique person.

PHYSIOLOGICAL DEVELOPMENT

The growth rate remains relatively steady during the preschool years. During each year from 3 through 5 years of age, the average child gains about 2 kg (4.5 lb) and grows 7 cm (2.75 in.) (Vaughan, 1975). The 3-year-old may still have the prominent potbelly of toddlerhood but will slim down over the next few years. The trunk elongates along with the arms and the legs. The brain and skull continue to grow rapidly and will reach almost adult size by 5 years of age. Although the head is still relatively large, the rest of the body is catching up, and the child gradually assumes a more adultlike appearance (Fig. 7-1). By 3 years of age, most body systems are functionally mature.

Skeletal Growth

The child's posture improves as the earlier forward tilt of the pelvis straightens out and the abdominal muscles become better developed and stronger. Toeing-in (pigeon-toeing) may

Figure 7-1. *During the preschool years, the proportions of the child's body gradually become more adultlike. (Photo by John Elkins.)*

be present in the preschooler if tibial torsion (inward rotation of the lower legs), which is normal in infancy, is still present; this condition usually disappears without treatment. Corrective shoes that influence the direction of the bones can be prescribed if necessary (Smith and Bierman, 1978). A mild degree of knock-knee may also remain in late preschool years but is usually temporary and does not require treatment.

While ossification of bones continues until full stature is reached, the rate of ossification gradually declines. During the preschool years fusion of some of the ossification centers occurs, resulting in the formation of bone. The age at which bones appear and fuse differs for boys and girls; for example, the bony center of the patella appears at 3 years in the girl and between 4 and 5 years in the boy (Baer, 1973).

Neuromuscular Development

Neuromuscular maturation continues throughout childhood and allows the growing child to perform increasingly complex tasks requiring gross and fine coordination. Voluntary movement is often accompanied by involuntary movements on the other side of the body. The use of the muscles stimulates an increase in the size and strength of muscle fiber.

Motor Development

Three-year-olds have much more motor control than toddlers. Three-year-olds are more sure on their feet, walk more erectly, and can balance on one foot; many will dance to music. By the latter part of the third year most children are able to ride a tricycle and build towers of nine or ten blocks. Most children can pour from a pitcher, string large beads, work puzzles with large pieces, and use paste. Three-year-olds can usually unbutton clothes, put on shoes, and wash their hands. Eye-hand coordination increases steadily, and by three years most children are fairly accurate in hitting large pegs.

By age 4, most children can run smoothly and quickly. They can usually negotiate stairs without using the handrail, walk backwards, hop on one foot, and steer a tricyle at full speed. The 4-year-old is also able to use scissors to cut out pictures, copy a square, and make a drawing of a person that shows more than one body part (Fig. 7-2). Four-year-olds will try to tie their shoes although they still have difficulty with this task.

By 5 years of age, most children learn to skip, play running games, jump rope, and jump from heights of several steps. They begin to use the fingers (rather than the palms) in catching a ball although they still have some difficulty catching

Figure 7-2. *By 4 years of age, most children can copy simple figures, and can use paste and scissors. (Photo by John Elkins.)*

it. Five-year-olds can march rhythmically to music, and most can dress and undress themselves. The 5-year-old's drawing of a man will include the main body parts. Some children of this age can also copy a triangle and letters. The preference for one hand or the other (handedness) is fully established by around 4 years of age.

Sensory Development

The tendency toward hyperopia (farsightedness) gradually continues; in spite of the fact that the eyeball grows rapidly during the preschool years, there is a relatively greater increase in the curvature of the cornea and lens. The lacrimal glands are fully developed by 4 years.

Central acuity may reach 20/30 by 3 years but, during the preschool years, vision of 20/40 to 20/50 is not considered abnormal; by age 5, most children have 20/20 vision. Depth perception and color vision are established during the preschool years, although the child will still be unable to recognize subtle differences in shades of color.

By 3 years, frontal lobe control is sufficiently developed to afford both accommodation and convergence. Late-developing strabismus (deviation of the eye) may present between 18 months and 4 years; strabimus should be identified and treated as early as possible so as to prevent further visual difficulties.

Amblyopia (dimness of vision) can result from disuse of an eye. It can develop secondarily to strabimus, or it may take the form of *ambylopia anisometropia*, in which the refractive power of each eye is different. This condition is reversible if it is detected early and treated by patching the good eye. Treatment should be instituted before school age, or the child may sustain permanent damage.

The internal hearing organs are essentially complete at birth, and few changes occur afterward. The normal preschooler should be able to hear all frequencies audible to an adult. Preschool children should have screening audiograms, since mild hearing defects may not be evident until the preschool years.

Immunological Development

One of the first systems to reach functional maturity is the immunological system. The levels of most immune factors are age-related, and many reach adult levels in the early years (see Fig. 7-3). IgG, IgA, and IgM all reach adult levels by 3 to 5 years of age. Children develop antibodies not only to agents that they are exposed to but also to normal flora in the body even though these antigens do not cause disease. Development of IgG antibodies provides relatively permanent immunity to specific organisms such as those that cause polio or mumps. IgM antibodies are developed in initial response

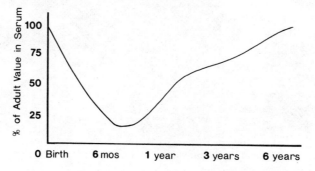

Figure 7-3. *Changes in the body's ability to produce antibodies, birth to 6 years.*

to infection and are important in controlling gram-negative infections, which often cause diarrhea and certain respiratory infections. IgA antibody is important in controlling and localizing infections in the body; it allows the child to localize respiratory and gastrointestinal disease. IgE antibody is responsible for producing localized allergic reactions. Children build up immunity to common pathogens as they are exposed to them; thus, the preschooler may easily succumb to colds, intestinal infections, flu, and other common contagious diseases upon entering a day-care center or nursery school. Thanks to immunizations, most children are spared from the more serious infectious diseases. The basic immunizations should be completed before 2 years. Booster immunizations of diphtheria-pertussis-tetanus (DPT) and oral polio vaccine (OPV) are needed before the child enters kindergarten (Fig. 7-4).

Circulatory and Respiratory Development

Although the circulatory and respiratory systems are operating at birth, their functional capacity is influenced by size. Decreases in the heart rate, increases in the blood pressure, and changes in the vascular resistance of various parts of the body (in response to growth of the vessel lumen) continue gradually.

Soft physiological (innocent) murmurs, caused by the thinness of the chest wall and the changing relationship between the chest and the inner organs, occur in 50 percent of all children at some time (Smith and Bierman, 1978). These murmurs are of no medical consequence and usually disappear as the child grows older. As the capillary beds mature, their capacity to respond to heat and cold in the environment gradually increases, thus improving thermoregulation.

The pulse rate is normally 80 to 100 beats per minute, and the blood pressure, about 90/60 mm Hg. Any preschooler with a blood pressure

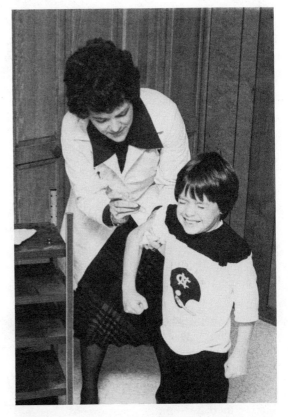

Figure 7-4. *Grimacing bravely, this preschooler receives a diphtheria-pertussis-tetanus (DPT) booster. Because of the decreasing incidence of these infectious diseases, the level of immunity in the general population has decreased, and inoculation is therefore more important. (Photo by John Elkins.)*

above 110/70 mm Hg should be referred for medical evaluation (Londe and Goldring, 1976).

The respiratory rate gradually increases and should be about 25 to 30 respirations per minute for the 3- to 5-year-old. The vital capacity of the lungs increases gradually as the lungs grow; by the preschool years, vital capacity is already slightly greater in males. During this period, the proximity of the respiratory structures to one another begins to be less of a factor in respiratory infections. While the anatomy of the ear and throat continue to resemble that of the infant more than that of the adult, the gradual increases in the size of the structures lessen the probability of communicating infection from one area to another. The tonsils and adenoids are relatively large, contributing to the high incidence of upper respiratory infections during the preschool years.

Endocrine Development

Most of the endocrine system (except the reproductive components) becomes functionally mature during early childhood; there is little change in endocrine function during the preschool years. The most vital hormones for normal growth and development during this period are probably the growth hormone, thyroid, insulin, and corticoids.

Genitourinary Development

The kidneys are well-differentiated, and specific gravity and other urine findings should be similar to adult norms. From 3 to 5 years of age the length, weight, and filtration capacity of the kidney increase slightly (Johnson, Moore, and Jeffries, 1978). The renal system should be able to conserve water and concentrate urine at a level similar to that of an adult. Under conditions of stress this ability is decreased, and the child's renal system takes longer to reestablish homeostasis than does the adult's.

Maturation of the neurological pathways that are necessary for bowel and bladder control allows for good control during the third year of life. There may still be an occasional accident in toileting when the child is busy at play and fails to heed the body's signal to eliminate, and nighttime control may not be as good, especially if the child is a sound sleeper or is ill. Failure to achieve control is often embarrassing and upsetting to the child and very upsetting to parents. A careful medical evaluation should be made in order to rule out the possibility of organic dysfunction. Evaluation for psychological causes may be advisable if there is no organic cause and the wetting continues into the fifth year.

Gastrointestinal Development

The preschooler's gastrointestinal tract is functionally mature and can tolerate most adult foods. A mature swallowing pattern (without a forward thrust of the tongue) is established during the second and third year. Food is digested at a slower rate in the preschooler. The glands that secrete digestive juices approach maturity during the preschool years.

Integumentary Development

By 3 years of age the child's skin is tougher, more resistant, and more effective in protecting the body from invasion and loss of fluids. Only small amounts of sebum are secreted during the preschool years; the preschooler is therefore prone to dry skin. Gradually the eccrine sweat glands increase in function, but the amount of eccrine sweat produced remains minimal. The apocrine sweat glands are nonsecretory at this age. Hair changes (which may have begun in toddlerhood) may continue, the hair becoming more coarse, darker, and straighter than in earlier years.

Hematological Development

Cellular constituents of the blood approach adult concentrations by 3 years. The normal hemoglobin is 13 g in children of preschool age.

Nutritional Requirements and Eating Habits

Caloric requirements remain relatively high during the preschool years (see Figure 7-5 for recommended daily dietary allowances for preschoolers). Like the toddler, the preschooler is

kcal: 1,300 to 2,300; mean, 1,700

calcium: 800 mg

protein: 30 gm

iron: 10 mg

ascorbic acid: 45 mg

vitamin D: 400 IU

Figure 7-5. *Recommended daily dietary allowances for preschoolers. (Source: Food and Nutrition Board, National Academy of Sciences. National Research Council, Recommended daily dietary allowances. Revised 1979.)*

prone to food ritualism and enjoys eating food in a certain sequence. Preschoolers do not like foods mixed together and will often demand a separate plate for each food. The early preschooler may still have food jags, eating only a certain food for a few days.

By this age children have definite food preferences. Cooked vegetables are often disliked but the child will readily eat the same vegetable raw. Preschoolers prefer mild flavored foods and will frequently refuse strong flavored foods. They like brightly colored foods. At this age the child responds better to small servings and can be allowed to ask for seconds if still hungry.

PSYCHOSOCIAL DEVELOPMENT

Preschoolers are more socially adept and much less "mother-bound" than during the toddler years. The preschool child begins to emerge as a social being and is capable of forming relationships with teachers, neighbors, or babysitters, and can feel secure without a parent or other family member being present.

Increases in physical growth, maturation, advances in cognitive development, and greater social abilities make it possible for the preschool child to begin to master the developmental tasks of this period (Figure 7-6).

Emotional Development

The 3-year-old is still struggling to gain and maintain what Erikson calls a sense of *autonomy*. Three-year-olds are no longer totally attached to their parents but still identify with them. The major developmental crisis at the end of the third year is that which Erikson sees in terms of *initiative versus guilt*. Achieving initiative means the child can plan and control his activity and has the ability to reassure himself. Failure to achieve a sense of initiative results in the development of a rigid superego (Erikson, 1963). The child may develop a sense of guilt, characterized by his feeling that he is responsible for things he is not really responsible for, and by feeling he deserves punishment. In this stage, children are curious about how much they can do, when, and where. By pretending to be other people, they seek answers to these questions. Creativity, which links the child's fantasies to reality, should be encouraged, but perfect results should not be expected or demanded. The child's attempts at building, drawing, and other creative activities should be acknowledged and praised. Recognition of children's efforts makes them feel good about themselves and encourages further creativity. Failure to recognize and encourage the child's attempts at creativity at this stage may leave the child afraid to engage in creative activities. The child's behavior during this period should not be belittled since this

Mastering physical skills requiring large and small muscle coordination

Learning to communicate effectively with people outside the family

Becoming a participating family member

Learning to express emotions acceptably

Developing a conscience and foundations for moral development

Developing control of actions

Figure 7-6. *Developmental tasks of the preschool years.*

may result in guilt feelings about foolish behavior.

The social view of emotional development stresses the influence of membership in groups as the key element. According to the social view, human beings are goal-directed and all behavior is purposeful; interactions between the child and the family are influential in determining the goals the child sets for himself in the development of a life plan. The birth order also influences the child's role in the family. Dreikurs (1964) observed that the child may feel less worthy than other family members and may therefore adopt one of four compensating roles: (1) forcing those around him to pay more attention to him by being cute, lazy, or annoying; (2) seeking more power by being rebellious or stubborn; (3) hurting others by seeking revenge; (4) giving up altogether and assuming a passive role. Dreikurs's theory suggests that the child needs positive interactions within the family in order to have successful interactions with peer and societal groups later in life.

In Sullivan's interpersonal theory, development of the person's self-system and security operations (behaviors that the child employs to feel secure) commences during childhood (Chapman, 1976). By the middle of the fourth year of life the child has acquired some of the features that will characterize him in adulthood. During what Sullivan calls the juvenile period (4 to 12 years) the child develops a strong need for interpersonal relationships with children of his own age other than siblings. Sullivan views this as an important period; in social environments outside the home the child can develop new ways of viewing himself and of interacting with people.

The behaviorist view of emotional development stems mostly from the work of B. F. Skinner (1948) who introduced the idea of the "outside-in" shaping of the individual; according to Skinner the individual's behavior is shaped by the social environment, which is constructed by those who have gone before us. Behavior is seen as a function of antecedents, or cues, that precede the child's responses, and the positive or negative consequences following the behavior. The implications of behaviorist theory are useful in child-rearing practices. For example, according to the behaviorists, a child who misbehaves should be removed from the room and allowed no interactions with people or the environment. Cooperative behavior should be reinforced, i.e., rewarded.

Language Development

The prerequisites of language development include an intact central nervous system, adequate psychological nurturing, specific linguistic stimulation, and a certain level of cognitive development (Lewis, 1971). The preschool years are characterized by an almost explosive development of language skills. Between 3 and 4 years the child becomes very conscious of the importance of speech and the power it provides. Between 2 and 5 years the child becomes increasingly proficient at rapidly forming new concepts (mental images) that are associated with labels (words and symbols). Preschoolers are constantly "labeling" (applying names to objects) and frequently use 20 to 30 words in labeling each day (Nelson, 1977).

By 3 years of age the child has mastered all the vowels and can use the consonants *h, m, n, p, b, k, g, t,* and *d.* Three-year-olds use three-word ("telegraphic") sentences that are intelligible 70 percent of the time. They ask many questions and can give and follow simple commands. Three-year-olds use the pronouns *I, you,* and *me* correctly, as well as plurals and the past tense. Four-year-olds use four-word sentences that are 100 percent intelligible. They use the prepositions *over, under, in, on,* and *behind;* verbs are used more frequently than nouns. Five-year-olds use five to six words per sentence. They are able to use blended sounds such as *tr, bl, pr,* and *gr* as well as *f, v, r,* and *l.* Speech is now more gramatically correct and children of this age are using syntactically complex sentences. By 5 to 5.5 years children are able to use such constructions as "You are going, aren't you." Five-year-olds use more conjunctions, prepositions, and articles in speech. Language becomes less egocentric and more socialized. Most 5-year-olds have a vocabulary of 2,000 to 2,500 words.

At the beginning of the preschool period speech and language are unstable and the process of communication can be interfered with; speech problems may have their origin at this point in development. *Nonfluency*, a verbal gap occurring while the child searches for words, is common; for example, the child repeats "I-I-I." Stuttering may originate during the nonfluency period although 99 percent of children pass through this period with stable speech. If sound repetition continues into the fifth year the child should be referred for a speech evaluation.

Play

Play serves many purposes for the young child; through play children learn and grow. Play is an important factor in social and emotional development and enhances physical and cognitive development. Many of the tasks of play require repetitive acts that facilitate musculoskeletal development and increase coordination (Fig. 7-7). Mastery of many motor skills is achieved during play; language skills are also enhanced, since children talk to one another during play, often repeating new words they have learned. Children often act out adult roles in play and thereby increase their understanding of social roles (Fig. 7-8).

According to Catherine Garvey (1977), four developmental trends are evident in all aspects of play: (1) Play entails biological maturation and increasing skill; (2) play becomes more elaborate and tends to evolve into more complex events in which two or more resources are used simultaneously; (3) as play develops its

Figure 7-7. *The preschooler's play often involves repetitive acts, such as swinging, that enhance coordination and musculoskeletal development. (Photo by John Elkins.)*

forms become less determined by properties of the materials at hand and come increasingly under the control of plans or ideas; and (4) as the child's world expands to include more people, situations, and activities, these new experiences tend to be incorporated into play.

Preschool children relate to their peers differently than toddlers do, and make the transition from parallel play to *cooperative play*, which involves simple games, sharing of toys, or play involving pretending. Imaginative and symbolic play are common. At 3 to 4 years the child is imaginative, creative, and struggling with fact and fantasy. Children from middle-class families are much more likely than poor

Figure 7-8. *Dress up play stimulates the preschooler's imagination and provides the opportunity to act out adult roles. (Photo by John Elkins.)*

children to engage in sociodramatic or pretend play (Rosen, 1975). Many children have an imaginary playmate during the preschool years. The imaginary playmate, a normal manifestation of childhood, is often blamed for the child's mistakes and is able to accomplish tasks the child is fearful of, such as climbing a tree. By 5 years the child is less self-centered and is ready to take part in group activities and obey rules and regulations (see Table 7-1).

Sex-Role Identification

During the period identified by Freud as the phallic stage (3 to 7 years), children become

interested in their genitals, they talk about their bodies, and references to bathroom activities are incorporated into play. The child becomes aware of anatomical differences between the sexes and becomes very definite about his or her own sex. The development of sexuality comes to the forefront with what Freud called the *Oedipus complex* (3 to 6 years). Up to this phase of development the child accepts the love and attention of both parents and does not care what else the parent does or whom else the parent loves. Now the child wants to be the exclusive recipient of the love of one parent (usually of the opposite sex), and becomes ex-

Table 7–1. Characteristics of Play Among Preschool Children

Age	Type of Play	Activities
3 yr	Parallel play	Uses combination of toys; dramatic play involving family activities, etc.; imaginative; fond of fairy tales; may play in shifting groups of two or three.
4 yr	Associative play	Increased dramatic and imaginative play; dresses up; accepts taking turns but may be bossy; puts toys away when reminded to do so; silly during play; no day-to-day continuity in themes of play; plays in groups of two or three.
5 yr	Cooperative play	Varied, more realistic activities; enjoys cutting and pasting; wants to finish projects; interested in rules; enjoys active play involving running, jumping, large wheel toys; plays in groups of five or six; friendships become important.

tremely jealous of any demonstration of affection toward anyone else. While this phase of development occurs in almost all children, no clear explanation for it exists. One theory is that the child has to play at being Mommy or Daddy long enough to get a sensitive perception of what that role may be like (Erikson, 1963). Children imagine themselves "big," and play at being Mommy or Daddy. They try to act and talk like adults and at some point in this process the child realizes that Mommy and Daddy sleep in the same bed, while the child has to go to his or her own room. Thus, the child begins to see that Daddy is Mommy's man or Mommy is Daddy's woman, and, in the desire to be like Mommy or Daddy, the child wants an exclusive love object. While the child feels rivalry with the parent of the opposite sex, the child also has ambivalent feelings that arise from wanting the love and approval of that parent (Whipple, 1966). When handled with understanding and sensitivity, the phase does not last too long. It is extremely important for parents to let children know they are loved in spite of the ambivalence they are experiencing.

Sex-role identification is influenced by the family and the dominant cultural values. In a family or culture with very traditional values, it is very important that the child identify strongly with the parent of the same sex. In contemporary American society, in which males and females are slowly being treated more equally, there is much speculation about the effect this phenomenon will have on the identification process.

COGNITIVE DEVELOPMENT

According to Piaget (1932), children remain in the *preoperational* stage from the age of 2 to 7 years. Intellectual functioning is seen as preoperational in the sense that the preschool child still lacks many of the cognitive faculties that adults have. While this period of cognitive de-

velopment covers a span of 5 years there is considerable variation in the intellectual ability of children at the different ages within the stage.

Between 2 to 4 years of age the child acquires *symbolic function*, i.e., the ability to conceive of a mental symbol, a word, or even an object as representing something else that is not present. The child can now differentiate between the *signifiers*, the symbol of an object or event, and the *significates*, the actual object or event. Symbols are very important to young children since they think first in symbols and continue to think in them after they become proficient in language. The preschooler uses a mental image of a tricycle, the word *tricycle*, or a toy tricycle to stand for a tricycle when a real one is not in view. Thus the child is no longer restricted to reacting to things in the immediate environment. Symbolic function allows the child to recall the past, and can be seen in the child's play. In symbolic play, an object represents something else. For example, a doll may be assigned the role of "Mommy" or "baby," or a sand bucket may become a garbage can.

Symbolic function is also evident in *deferred imitation*; for example, children can frequently be observed at play talking on the telephone. Closer observation reveals that the children use the telephone in play at ordering groceries, making appointments, and so on. They associate the telephone with its function and copy the behavior they have observed.

Preoperational thought is characterized by *egocentrism, transductive reasoning, centration,* and *irreversibility*. (See the discussion of these qualities in the Cognitive Development section of Chap. 6.)

The child's grasp of quantity, one of the more difficult concepts to learn, improves with age and learning. Terms such as *more, less, all, some, few,* and *many* are confusing to the young mind (Elkind, 1974).

Preschoolers also have difficulty with *time concepts* and are unable to separate morning and afternoon before 4 years of age; 5-year-olds can correctly identify the day of the week (Ames, 1946). Development of time concepts progresses from a concrete, action-oriented level toward a more abstract level. Time is first learned in relation to activities in the child's schedule, e.g., meals. Later the child learns to tell time by the clock, beginning with hours, then half and quarter hours (Springer, 1952).

Number concepts are also difficult for the preschooler; the child is better at rote counting than at object counting. There is a significant gap between the child's ability to say numbers aloud and the ability to understand conceptually what is being counted.

The urge to know and comprehend influences the preschooler's thought. The persistent "why" of preschoolers reflects their curiosity and desire to learn.

Piaget's followers have applied his theories to education, and as a result, changes have taken place in preschool education since about 1970. Piaget views children as active learners building an understanding of themselves and their world. Increasing acceptance of this view has resulted in improvements in preschool educational programs that help children to grow and develop (Furth and Wachs, 1975).

MORAL DEVELOPMENT

The development of the child's views of right and wrong and good and bad begins in the preschool years. As in all other matters, the parents exert the strongest influence on the child during these years. The parents' attitudes, moral values, spiritual and religious beliefs, and life-style convey to the child what is good and bad. Many of these values will be held throughout life.

According to Piaget (1932), children from 4 to 7 do not know or follow rules but insist that they do. Piaget observed a group of boys play-

ing marbles and found that each boy had a separate set of rules for the game. Each child had in effect an individual game. Piaget concluded that having a good time was most important to the children. Children from 4 to 6 years of age believe in the sanctity of rules but are willing to accept changes in the rules.

During the school years the child is in the stage Piaget refers to as *moral realism*. In this stage the child considers only facts, not the motives. For example, the child believes that the boy who broke a dozen eggs accidentally is more guilty than the boy who purposely broke a single egg.

Kohlberg (1969), inspired by Piaget's work, has provided further information on how children develop morally. Preschool children remain at what Kohlberg calls the *preconventional level* and have little concept of what constitutes morally acceptable behavior. As they progress through the two substages of the preconventional level they demonstrate initial signs of moral development. The first stage, *punishment and obedience orientation*, is characterized by avoidance of punishment and unquestioning deference to power. In the second stage, *egoistic orientation*, the child acts to satisfy his own needs; later in this stage *instrumental reciprocity* develops, i.e., "you do something for me and I'll do something for you."

Ilg and Ames (1955) studied the child's concept of a diety and found that not until around the age of 4 is a religious sense evidenced. Piaget believed that preschoolers exhibit the "religion of the parents" (Piaget, 1932). Small children view their parents as all-knowing and all-powerful and may not need the concept of a supreme being until around the age of 4.

Most 4- and 5-year-olds can memorize short prayers, but it is doubtful that they fully understand the meaning of them. Throughout the preschool years the child's religious and moral development is influenced by the parent's religious teachings and the practices in the home.

DAY-CARE, NURSERY SCHOOLS, AND OTHER GROUP EXPERIENCES FOR PRESCHOOLERS

A great many preschool children will spend at least part of their day in some type of group (Fig. 7-9). Approximately half of the preschool children in the United States spend up to nine hours a day away from home. Group care may range from several children being cared for in a private home to a formal preschool experience such as a Montessori preschool. Early group experiences should offer more than just custodial care for children while parents are working.

Figure 7-9. *About half of American preschoolers spend up to nine hours a day away from home in such settings as daycare centers, nursery schools, and Montessori preschools. (Photo by John Elkins.)*

Day-care centers originated with the work programs established by Roosevelt in the 1930s. They were expanded in the 1940s to provide care for children of mothers who went to work during World War II. During the 1950s and early 1960s day-care was less popular, since the majority of mothers of preschoolers were staying at home with their children. By the mid 1960s more and more mothers were entering the work force, and the need for day-care centers again arose. Day-care centers are licensed by state licensing agencies. While the requirements vary from state to state, certain minimum standards regarding the ratio of children to caretakers, educational background of the caretakers, space per child, health supervision, and some plan of activities for the children are required. In some areas of the country, day-care centers offer care for infants as young as 6 weeks of age and will continue to offer services until the child is of school age.

Nursery schools became popular after World War II. The early nursery schools were often affiliated with colleges or universities and provided a setting for training student teachers and research in child development.

In the 1960s America became concerned about its underprivileged children. Project Head Start, introduced in 1965, was designed to teach academic and language skills to compensate for the deprived home conditions of poor children. These programs also provided medical and dental care, and part of the child's daily nutritional needs. The development of creativity, self-discipline, and self-confidence was stressed, and parents were encouraged to participate in the program through special activities.

Following World War II free half-day kindergartens for 5-year-olds were introduced in some elementary schools. Since children who attended kindergarten before entering school were often able to perform better, kindergarten has become increasingly popular, and in some parts of the United States, attendance at kindergarten before entering the first grade is mandatory.

THE FAMILY AND THE PRESCHOOLER

Even though the preschooler is forming new relationships outside the family, the relationship with the family remains of prime importance. Family relationships provide opportunities for children to become aware of the emotions of others as well as their own. It is important to foster a sense of competence and independence while at the same time providing guidance and protection; maintaining this balance is at times difficult. At certain times children may seek freedom they are not yet ready for, such as in crossing a street by oneself. At the other extreme, parents may become so accustomed to setting limits to protect the child that they do not recognize when it is appropriate to lift certain restrictions.

The presence of siblings offers opportunities for interpersonal relationships that contribute to the child's developing self-concept. Older siblings may serve as role models for the preschooler. The older sibling may or may not assume a protective role toward the preschooler. Sometimes a conflict arises if the preschooler wants to participate in the same activities as the older sibling, while the sibling wants to relegate the child to a lesser role. For some preschoolers this is not a problem, and they become willing "servants" of the older sibling.

The arrival of a new baby changes life for the preschooler as he or she is no longer the center of attention. The preschooler may have difficulty accepting the family's affection toward the new baby and may in fact resent it. Sometimes when this occurs the preschooler may regress and display a need to be babied. Regression sometimes takes the form of enuresis or encopresis (discussed later in the chapter). The preschooler may also attempt to harm the baby.

Some preschoolers accept the arrival of a new baby with no difficulty and even feel very protective toward the new family member. It is helpful to prepare the preschool child for the arrival of a new baby carefully. Involving the child in the preparations for the baby is very important. Providing a doll or pet for the preschooler to "care for" while Mother is caring for the new baby is also helpful. Recognizing the child's feelings is extremely important. Often this can be done by reading the child stories about other children with similar feelings and allowing the child to express his or her feelings. Puppets or dolls can also be used to act out similar situations.

SIGNIFICANT OTHERS

Preschoolers often form significant relationships with other adults. Grandparents, aunts, uncles, and cousins may become very important to the preschooler. The influences of these extended family members vary with the type of relationship that exists. Some grandparents serve as surrogate parents and may be as influential in the child's development as the parents. Neighbors, babysitters, or teachers at the nursery school or the day-care center may also become significant others for the preschooler (Fig. 7–10).

COMMON HEALTH AND DEVELOPMENTAL PROBLEMS OF THE PRESCHOOLER

As is true throughout childhood, accidents are the leading cause of death of preschoolers in the United States. Table 7–2 (page 208) presents the mortality from the leading types of accidents for children ages 1 through 4 years. While the death rate from all other causes combined has decreased since 1970, the accident mortality for

Figure 7-10. *During the preschool years, children often form relationships with adults other than their parents. Neighbors, babysitters, nursery school teachers, and day-care workers may become significant others to the preschooler. (Photo by John Elkins.)*

preschool children has remained virtually unchanged.

Many factors combine at this stage to contribute to accidents. The preschooler is ambulatory and strives for independence, therefore, he or she explores and comes into contact with a variety of environmental hazards. At about this time parental anxiety and supervision often diminish. It is important to maintain the correct balance of parental protection.

Death among preschoolers caused by motor vehicles has increased since 1970. Pedestrian deaths in traffic accidents are especially common at ages 3 and 4. Often the child rides a tricycle into traffic or darts into the street after a ball.

Approximately 20 percent of all accidental deaths during the preschool period are caused by drowning; boys are more likely to be involved in drownings, while girls are more likely to die in fires. Fires and burns are the third-ranking cause of death in American children between the ages of 1 and 13, although the increased use of flame-retardant clothing has reduced the severity and number of injuries from burns (Wheatley, 1977). Fire and smoke detectors installed in dwellings can give early warning and allow escape in case of fire. Since the child's cognitive ability improves during the preschool years, it is both appropriate and important for parents to explain *why* certain activities are dangerous.

During the preschool period, accident-proneness may emerge. Accident-prone children may suffer an excessive number of cuts, fractures, burns, or accidental poisonings. These children should be referred for psychological evaluation.

Infectious Diseases

Respiratory infection is the most common cause of illness during the preschool years and accounts for approximately two-thirds of the visits to health care providers by preschoolers. Since the respiratory tract is continous, the child who has trouble in one part is likely to have trouble in the rest of the respiratory tract. As the body's defenses become more proficient a great majority of the causative organisms become trapped in the adenoids and tonsils, and fewer get through to the lower respiratory tract. Coughing, sneezing, and blowing one's nose are part of the body's local defenses against respiratory pathogens. As the child grows older these symptoms increase as the body is better able to localize infections.

Respiratory infections range from relatively mild ones that cause little change in the preschooler's behavior and activities to serious and

Table 7–2. Mortality from Leading Types of Accidents Among Children Aged One to Four in the United States, 1972–73

Type of Accident	Average Annual Death Rate per 100,000									
	Boys at ages					Girls at ages				
	1-4	1	2	3	4	1-4	1	2	3	4
Accidents—all types	38.0	44.5	40.7	36.5	30.1	25.4	32.9	27.9	21.9	18.9
Motor vehicle	13.6	12.7	13.3	14.7	13.9	10.2	10.8	11.2	9.3	9.4
Pedestrian (in traffic accidents)	6.5	3.6	6.3	7.9	8.3	3.3	2.3	3.5	3.6	4.0
Drowning[a]	8.3	8.8	10.2	8.4	5.6	3.6	5.1	4.6	2.8	1.8
Fires and flames	5.6	5.8	7.1	5.5	4.1	4.6	4.9	5.1	4.2	3.9
Inhalation and ingestion of food or other objects	1.8	4.1	1.8	0.9	0.5	1.4	3.2	1.2	0.7	0.6
Poisoning	1.6	3.8	1.4	0.9	0.5	1.1	2.4	1.2	0.6	0.3
Falls	1.5	2.8	1.4	1.2	0.8	1.2	2.1	1.0	1.1	0.5
Firearm missile	0.7	0.4	0.8	0.7	1.1	0.4	0.2	0.4	0.7	0.4
Accidental deaths as a percentage of all deaths:	42%	34%	44%	49%	49%	36%	30%	38%	40%	41%

[a] Exclusive of deaths in water transportation.

Source of basic data: Reports of the Division of Vital Statistics, National Center for Health Statistics.

Adapted from *Statistical Bulletin*, Metropolitan Life, Volume 56, May, 1975.

SOURCE: *Pediatric Annals 6*, (11), Nov. 1977, 691/15. Insight Publishing Co., Inc., New York, N.Y. Reproduced with permission.

life-threatening cases. The preschooler who is attending a day-care center or nursery school for the first time may experience a great many infections, since the body must gradually build up immunity to new pathogens. The preschooler may experience 7 to 10 respiratory infections per year. Children attending day-care centers or school are likely to introduce the illness into the home. The agents most likely to cause respiratory illness in preschoolers include Streptococcus pneumoniae, Hemophilus influenzae, respiratory syncytial (RS) virus, parainfluenzae type I and II, and the rhinoviruses.

For many children, recurrent or occasional ear infections will continue to be a problem during the preschool years. Most acute purulent otitis before 5 years of age is caused by Hemophilus influenzae. Acute and recurrent otitis media tends to be familial, since the facial configuration and the length of the eustachian tube is congenitally determined. Children who have repeated ear infections most often have a shorter, straighter, wider, and more pliable eustachian tube, resembling the eustachian tube of the toddler which allow microorganisms easier access from the nasopharynx to the middle ear cleft. Any child with two or more episodes of otitis media in one year should be referred for further evaluation.

The major concern regarding repeated episodes of otitis media is the risk of developing serous otitis media (SOM), or "glue-ear syndrome." The most serious consequence of SOM is chronic and fluctuating hearing loss. SOM may be brought to the health care provider's attention by the parents' complaint that the child "just doesn't seem to listen anymore" or the child may complain of a crackling or popping sensation in the ear. Preschool children occasionally present with mild speech retardation caused by an undetected hearing loss from SOM. As many as 60 percent of children with SOM have a hearing loss detectable by audiometry (Lewin, 1978).

A much less serious ear infection that often occurs during the preschool years is otitis externa, characterized by scaly erythema of the concha and external acoustic meatus. Children often scratch and pick at the ear, creating fissures and superficial ulcers that promote colonization of bacteria. Symptoms include swelling, pain, itching, and seropurulent discharge. This condition is easily treated with antibiotic drops. Otitis externa is sometimes referred to as "swimmer's ear" because it most frequently presents during the summer months as a result of water remaining in the ear after swimming.

Impetigo, superficial skin lesions created by invasion of staphylococcal and streptococcal organisms, is highly contagious. This disease is characterized by subcorneal vesicles and bullae that rupture and form a distinctive yellow crust. It is easily transferred to other areas of the child's body, or to other children, by the child's own contaminated hands. If begun early, local treatment is often effective and can prevent serious complications.

Behavioral Problems

Some behaviors that normally emerge during the preschool years (including many that are also seen in toddlerhood) may be disturbing to parents and others. Fortunately, most of these behavior problems are transient and do not require medical intervention.

Three- and four-year-olds are often fond of telling long, elaborate, and mostly imaginary stories. Adults often become very upset because the child is not telling the truth, and they need reassurance that this is a normal part of development and will pass quickly.

Preschoolers tend to vent their aggressions on toys, and adults may become upset when they observe the child beating a doll or banging toy trucks together. Such behavior is normal for the preschooler and is an acceptable way for the child to deal with aggressive feelings.

Preschool children, like toddlers, sometimes appear contrary, and balk at parental directions. Preschoolers may resent limitations or directions imposed on them because they represent a threat to the recently acquired sense of autonomy. Preschoolers are trying to decide things for themselves and need to be given time to linger, to be indecisive, or to slowly carry out a task.

Preschool children occasionally awake at night with the feeling that something bad is happening or is about to happen. When nightmares occur frequently it may mean that the child has many negative feelings. Usually children can be reassured by the parents' turning on the lights and letting them see that they are in a familiar place; most children will calm down and go back to sleep. Children should not be overfed or overstimulated at play or story-time before bed. If nightmares persist the family may need help in determining the cause. The nurse can help the family review the situation, identify the cause, and find ways to correct the problem. Often a little extra attention and affection will provide the reassurance the child needs.

The child of this age may have many fears since his imagination is vivid and his understanding is limited. Preschoolers tend to distort and magnify things and often see danger where there is none. Shadows resembling the outline of a frightening stranger, loud noises such as thunder, or the appearance of a strange animal sometimes arouses strong fear. When children are fearful they need comforting and assurance; they need the opportunity to discuss their fears without fear of ridicule.

Although temper tantrums are more common in the toddler, some preschool children may still throw temper tantrums if they do not get their way. Children of this age are still in the process of learning to control themselves;

since they are still immature, their goals are very short-sighted and they are mostly concerned with their own immediate needs and desires. Parents and other caretakers need to set firm limits. Avoiding conflict by giving into demands will only intensify the problem. Children need to be taught to respect the rights of others. For the preschool child, an inappropriate demand or an infringement on another person's rights can be met with an attempt to distract the child with some more appropriate interest.

Nail biting and thumb sucking may persist in some preschoolers. Such behaviors are usually the child's way of dealing with tension. Parents and caretakers need to attempt to decrease the tension. It is not helpful to nag or punish children for these behaviors. Jealousy, another common problem, may take the form of bickering, rivalry, or hostile acts. These behaviors may be directed at siblings or at other children in a group setting. Giving the child some special attention or assigning the child some special tasks often decreases the incidence of the undesired behavior.

Sex play is common among preschoolers, since they are fascinated by anatomical differences between the sexes. Preschoolers often want to touch their parents' bodies and may want to bathe with their parents. They may want to watch others use the toilet and may simply burst into the bathroom in order to do so. Sex play with other children, involving assuming the roles of mother and father or doctor and patient, is common. Preschoolers may talk about making babies and may remove their clothes and examine each other's bodies. Another common form of sex play during the preschool years is masturbation. Parents and caretakers are often very distrubed by these behaviors. The child's actions and curiosity about sex differences are quite normal. Questions about sex should be answered simply and truthfully but without elaborate explanations. When the child's behavior infringes on the parents' privacy, they must decide what limits to set. If a child is dis-

covered in sex play with other children the situation can be handled by asking the children to get dressed and encouraging some other play activity.

Enuresis may occur in some children. *Primary enuresis* is that which exists in children who have never had control of urination; the child who has had at least 3 to 6 months of control followed by involuntary wetting is said to have *secondary enuresis*.

Enuresis is more common among males, and in families of lower socioeconomic status and lower educational level. Often there is a history of enuresis in one of the parents or in a sibling. Children with enuresis frequently have a smaller functional bladder capacity and experience increased frequency and urgency more often than nonenuretic children. The exact cause of enuresis is not known, although it is thought that some children do not have the neurological maturation necessary for adequate sphincter control. The high incidence of spontaneous resolution occurring a few years after the problem appears supports this developmental view (Cohen, 1975).

Encopresis is involuntary defecation. The term is used only with reference to cases occurring after age 3 and in the absence of organic causes. As with enuresis, it is more common in males. Affected children are often shy and may have developmental delays. There is often a history of premature toilet training or a situational stress at the time of training. Behavior modification often obtains positive results in a short time (Erikson, 1976).

PREVENTIVE DENTAL CARE*

Preventive dentistry should be a part of the child's total health care. Parents should be re-

---

* This section on preventive dental care was written by Gary Hill, D.D.S., M.S., Durham, North Carolina.

sponsible for brushing and flossing a child's teeth until the child develops the dexterity to clean the teeth properly. It is essential that a parent become involved daily in assisting young children in oral hygiene care.

A good preventive program should include diet evaluation and counseling if dental caries is a problem with the child. The consumption of refined carbohydrates is the cause of dental caries. There are certain foods that obviously contain sugars. However, there are many processed foods that contain "hidden sugars." The detrimental effects of refined sugars on dentition should be clearly explained to parents and to children, if they are old enough to comprehend.

The proper use of fluoride is essential in the prevention of dental caries. Fluoride has greatly decreased the incidence of caries in the general population. It is important that children receive systemic fluoride either in the water supply or in a supplemental form as prescribed by a dentist or physician.

It is hoped that the child's first visit to the dentist will not result because the parents have noticed decay or because the child has suffered a traumatic injury. Instead, it should be a pleasant experience. A child's dental experience beginning on the first visit should be based on prevention of problems and not treatment after a problem exists. Ideally a child's first dental visit should be between 2 and 3 years of age. A child should be prepared for dental visits. However, a parent's apprehension about dentistry should not be transmitted to the child. A child should not be overprepared; instead, in a matter-of-fact way, what will be accomplished at the first visit should be explained to the child. The child should be told that on this first visit the dentist will count the teeth with a special "tooth counter." Many dentists use the first visit as a time to begin preventive dental treatment; brushing and flossing instruction is introduced. At the conclusion of the visit, fluoride is normally applied to the teeth. The fluoride application may be described as a flavored "tooth toughener." The proper application of the fluoride requires a certain amount of discipline from the child. If the child is in need of further dental treatment, the first visit is quite helpful in preparing the child for future visits, since a certain amount of discipline is required for dental treatment.

ANTICIPATORY GUIDANCE

As with other periods in the lifespan, anticipatory guidance is of great importance during the preschool years. Since there are a variety of changes in both physical and psychosocial development during the preschool years it is important to prepare parents and caretakers for these changes. This should be done before the change happens, by providing verbal and written information regarding normal growth and development during the preschool years. Discussions provide an opportunity for the health care worker to assess what parents already know and to build on existing knowledge.

Most children proceed through the preschool years with no difficulty. In cases of children who do not successfully resolve the developmental crises of the period, the nurse must be prepared to help the parents find the resources necessary for further evaluation. A knowledge of community resources is a must. Allow the parents to discuss their concerns, and provide emotional support. If the child has a problem in only one area of development, it is helpful to point out to parents the child's accomplishments in other areas.

Another important aspect of anticipatory guidance is to provide parents with information about immunization schedules and continued health supervision. Many parents can also benefit from information on when and how to use primary health care facilities.

Sudden and unexplained high fevers should

also be discussed with parents. Children may get up feeling well and continue with play until midday, when they suddenly lie down and say they do not feel well. The temperature may be 104° F. Since the mother will not be able to identify a cause for the fever, she may become extremely (and understandably) alarmed. It is most appropriate to contact the health care provider in cases of such sudden high fever; many childhood illnesses such as chickenpox, ear infections, and strep throat begin in this manner. Most sudden fevers in the healthy child do not signal serious illness but warrant attention from the health care provider.

The child who has symptoms of an uncomplicated respiratory infection (without earache or sore throat) and a low-grade fever can usually be managed at home. It is helpful to review a plan of treatment with the mother stressing the need for increasing the child's fluid intake and providing additional rest and quiet play until the child is better. Although there is controversy regarding the effectiveness of over-the-counter medications for cold relief, many parents choose to use such preparations. The health care provider can suggest preparations that are safe for children. Parents should be cautioned not to give adult cold preparations to children.

Providing information regarding safety and accident prevention is imperative for parents of preschoolers. One helpful way to do this might be to use a home safety checklist.

As the child moves toward the end of the preschool years, many parents have mixed feelings about the child's starting school. Some parents are saddened by the child's entrance into school since they see themselves as losing the child in some way. While some sadness about the child being away from home for most of the day is normal, a great deal of concern over this event indicates a need for further assessment since it may signal problems within the family.

Some parents may be concerned with the child's readiness for school. A simple screening test such as the School Readiness Checklist may be used to assess school readiness. If the child does not appear ready for school you may offer suggestions to parents regarding activities which may help to facilitate readiness. Occasionally, children who are otherwise normal are socially immature and are therefore just not ready for school; parents may elect to delay the child's entrance into school by one year. These parents may need support in making such a decision, and may need continued support after the decision has been made.

HINTS ON WORKING WITH THE PRESCHOOLER

When working with a client of preschool age, consistency is the best approach. Whether the setting is an inpatient or an outpatient facility it is desirable to have the same members of the nursing staff care for the child if possible. The preschooler can relate better to a few familiar people and may become upset and uncooperative if his caretakers are constantly changing. If the child can have the same nurse for several days he or she can develop a relationship with the nurse. As with clients of any age it is important to be truthful. A nurse should never tell a child that a procedure such as an injection will not hurt; it becomes obvious to the child that the nurse cannot be trusted. It is far better to be honest with the child and say the injection may hurt a little. Relate the procedure to something the child is already familiar with; for example, an injection may be related to a pinch or pinprick. Distraction and relaxation techniques can be used with children. Even at a young age, many children can be coached to relax with breathing techniques. If the child is good during a procedure, positive feedback should be given frequently. On the other hand, many children cry during or after a procedure and should be assured that it is all right to do so.

If a child must be hospitalized, the child's parents should be encouraged to participate in the child's care whenever possible. This may help to alleviate the parents' feeling that they are losing control of their child, and it often helps children to see that the parents approve of the hospital routine. Parents of extremely ill children are sometimes afraid to do things for their children for fear of hurting the child or of displacing equipment; under such circumstances, the nurse should identify small tasks that parents can perform comfortably.

Some parents will be unable to stay at the hospital because of responsibilities to jobs and other children. These parents are often very anxious about their child's welfare and may feel guilty that they cannot be present all the time. Allow these parents to verbalize their concerns, and provide emotional support. The health care provider needs to make a special attempt to give these children some extra time and attention. Often, parents of another child, who are able to stay in the hospital on a full-time basis, will informally "adopt" a child whose parents are unable to stay. This is frequently a very satisfactory arrangement. Parents of hospitalized children frequently develop informal support systems.

Health care providers need to reassure children that they are not to blame for the illness or hospitalization. Children of preschool age frequently believe that their behavior caused an event to occur; this reflects the egocentrism characteristic of the preschooler. Castration and mutilation fantasies are not uncommon in this age group and merit special attention. These fantasies may come into play especially with preschoolers who are to undergo surgery.

Play therapy is an extremely valuable tool with this age group. Most hospitals have dolls to which casts or bandages have been applied. Some children like to have their own doll or stuffed animal treated as they will be treated. It is helpful to have a play area with discarded hospital equipment such as an old lab coat,

nurse's cap, stethoscope, empty and clean syringes, and an empty IV bag and tubing. Children can be allowed to manipulate the equipment and act out their feelings and fears; some of these fears may be unfounded, and the health care providers can use this opportunity to correct misconceptions (Fig. 7-11). Many pediatric units now incorporate play therapy into preoperative teaching. Play is a natural part of the child's world, and learning in this way is appropriate to the child's cognitive level. Play therapy can be equally effective in an outpatient setting. Games and simulated experiences may be used to teach preschool children certain health practices and safety habits.

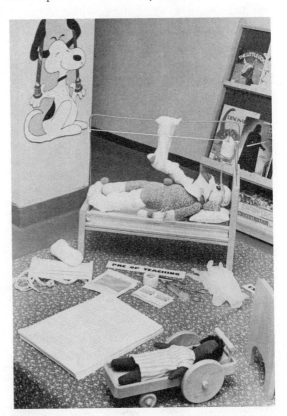

Figure 7-11. *Such props as dolls, dressings, syringes, surgical gloves, etc. are effective in preparing a preschooler for surgical procedures. (Courtesy Chapel Hill Newspaper.)*

In order for verbal communication to be effective, the health care provider must communicate with preschoolers on their level. The use of short, simple sentences is appropriate. At this age, children still use a great deal of nonverbal communication, and the health care provider needs to be alert to nonverbal clues to what the child is thinking and feeling. Children are also very much attuned to body language and the nonverbal communication of the health care provider.

Whenever possible it is important for the health care provider to encourage children to carry out self-care activities, since this fosters growth and development. Children need the opportunity to practice these skills, and their sense of accomplishment is threatened when they have to relinquish these responsibilities. Obviously, some children will be unable to care for themselves because of their illness, but they may still be allowed to carry out a small task such as brushing the teeth or combing the hair. It is helpful to ask children how they do these things at home and attempt to provide the care in the same way. This allows children to feel they have some control over the activity. Similarly, although it is very difficult to provide a consistent routine in the hospital setting, whenever possible, it is desirable to do so. For the child who is accustomed to bathing in the evening the bath time might be changed to the evening. If a bedtime ritual is followed at home it is helpful to try to follow this routine in the hospital setting. Such consistency aids the child in adapting to the new setting.

SUMMARY

The preschool years are a relatively calm period in the lifespan, characterized by slow but steady gains in both height and weight. During these years the body systems continue to grow and mature slowly. Most are functionally mature (one exception being the reproductive organs) and some, such as the immune system, reach adult levels of function near the end of the preschool years. Since the growth rate is slower the nutritional needs are not as great as in earlier years.

Psychosocial development during the preschool years is characterized by the development of social relationships with peers. Children progress from parallel play to cooperative play. The major developmental crisis of this period is, in Erikson's terms, the struggle for initiative versus guilt.

The preschooler remains in what Piaget calls the preoperational stage of cognitive development. During these years, there are tremendous gains in language development as the child learns to use longer and more complex sentences and increases his vocabulary rapidly.

The refinement of motor skills allows children to gain increased control over their bodies. As motor skills and cognitive ability improve, the child becomes more able to perform self-care activities. The many changes of the period also prepare the child for formal education.

REFERENCES

American Academy of Pediatrics. Vision screening of preschool children. *Pediatrics*, 1972, *50,* 966–967.

Ames, L. The development of the sense of time in the young child. *Journal of Genetic Psychology*, 1946, *68,* 97–125.

Baer, M. J. *Growth and maturation: an introduction to physical development.* Cambridge, Mass.: Howard Drake, 1973, pp. 49–76.

Chapman, A. H. *Harry Stack Sullivan, his life and his work.* New York: G. P. Putnam & Sons, 1976.

Chinn, P. *Child health maintenance* (2nd ed.). St. Louis: Mosby, 1979.

Cohen, M. *Enuresis. Pediatric Clinics of North America* 1975, *22,* 545–549.

Driekurs, R., and Soltz, V. *Children: the challenge,* New York: Hawthorn Books, 1964.

Elkind, P. *A sympathetic understanding of the child six to sixteen,* Boston: Allyn and Bacon, 1974.

Erikson, E. *Assessment and management of developmental changes in children.* St. Louis: Mosby, 1976.

Erikson, E. *Childhood and society* (2nd ed.). New York: Norton, 1963.

Furth, H. G., and Wachs, H. *Thinking goes to school: Piaget's theory in practice.* New York: Oxford, 1975.

Frankenburg, W., and Drumwright, A.: *The Denver articulation screening exam.* distributed by Mead-Johnson Laboratories Evansville, Ill., 1973.

Freud, A. *Normality and pathology in childhood: assessments of development.* New York: International Universities Press, 1965.

Freud, S. *Beyond the pleasure principle.* New York: Bantam, 1959.

Garvey, C. *Play.* Cambridge, Mass.: Harvard University Press, 1977.

Ginsberg, H., and Opper S. *Piaget's theory of intellectual development.* Englewood Cliffs, N.J.: Prentice-Hall, 1969.

Glenzen, W. P., and Denny, F. W. Epidemiology of acute lower respiratory infections in children. *New England Journal of Medicine*, 1973, *288*, 498–505.

Gordon, I. *Human development: a transactional perspective*, New York: Harper and Row, 1975.

Helms, D., and Turner, J. *Exploring child behavior: basic principles.* Philadelphia: Saunders, 1978.

Ilg, F., and Ames, L. *Child behavior.* New York: Harper and Row, 1955.

Inglis, S. The noctural frustration of sleep disturbance. *American Journal of Maternal-Child Nursing*, 1976 September/October, *1*(5), 280–287.

Inhelder, B. Some aspects of Piaget's genetic approach in cognition. In Kessen, W., and Kuhlman, C. (Eds.), *Thought in the young child. Monograph of the Society for Research in Child Development*, 1962, *27*, 19–33.

Inhelder, B., and Sinclair, H. Learning cognitive structures. In Mussen, P., Langer, J., and Covington, M. (Eds.), *Trends and issues in development psychology.* New York: Holt, Rinehart and Winston, 1969.

Johnson, J., Moore, W., and Jeffries, J. *Children are different: development physiology* (2nd ed.). Columbus, Ohio: Ross Laboratories, 1978.

Kastenbaum, R. *Humans developing.* Boston: Allyn and Bacon, 1979.

Kohlberg, L. *Stages in the development of moral thought and action.* New York: Holt, Rinehart and Winston, 1969.

Lewin, E. B. Middle ear disease. In Hoekelman, B. A. (Ed.): *Principles of pediatrics: health care of the young.* New York: McGraw-Hill, 1978, pp. 1730–1738.

Lewis, M. Language development, cognitive development and personality: a synthesis. *Journal of American Academy of Child Psychiatry*, 1971, *16*, 646–61.

Londe, S. and Goldring, D. High blood pressure in children: problems and guidelines for evaluation and treatment. *American Journal of Cardiology*, 1976, 37, 650–657.

Lowenberg, M. The development of food patterns in young children. In Pipes, P., *Nutrition in infancy and childhood.* St. Louis: Mosby, 1977.

Luria, A. R. *The role of speech in the regulation of normal and abnormal behavior.* New York: Pergamon Press, 1961.

Meredith, H. *Human body growth in the first ten years of life.* Columbia, S.C.: The State Printing Company, 1977.

Murray, R., and Zentner, J. *Nursing assessment and health promotion throughout the life span.* Englewood Cliffs, N.J.: Prentice-Hall, 1979.

Nelson, K. Aspects of language acquisition and use from age 2 to 20. *Journal of American Academy of Child Psychiatry*, 1977, *16*, 584–607.

Piaget, J. *The child's concept of numbers.* New York: Humanities Press, 1952.

——. *Judgement and reasoning in the child* (M. Gabain, trans.). New York: Harcourt, Brace and Jovanovich, 1932.

——. *The child and reality.* New York: Grossman, 1973.

Rosen, C. The effects of sociodramatic play on problem-solving behavior among culturally disadvantaged pre-school children. *Child Development.* 1975, *45*, 920–927.

Schuster, C., and Ashburn, S. *The process of human development.* Boston: Little, Brown, 1980.

Smith, D., and Bierman, E. *The biological ages of man*, Philadelphia: Saunders, 1978.

Skinner, B. F. *Walden two.* New York: Macmillan, 1948.

——. *Beyond freedom and dignity.* New York: Knopf, 1971.

Stangler, S., Huber, C. J., and Routh, D. K. *Screening growth and development of preschool children.* New York: McGraw-Hill, 1980.

Sullivan, H. S. *The interpersonal theory of psychiatry.* New York: Norton, 1953.

Springer, D. Development in young children of an understanding of time and the clock. *Journal of Genetic Psychology*, 1952, *80*, 83–96.

Valadian, I., and Porter, D. *Physical growth and development from conception to maturity.* Boston: Little, Brown, 1977, pp. 193–224.

Vaughan, V. C. III: Growth and development. In Vaughan, V. C. III, and McRay, R. J. (Eds.), *Nelson's textbooks of pediatrics* (10th ed.). Philadelphia: Saunders, 1975.

Wheatley, G. Childhood accidents. *Pediatric Annals*, 1977, *6*, 12–26.

Whipple, D. *Dynamics of development: euthenic pediatrics.* New York: McGraw-Hill, 1966.

8

The School-Age Child

PATRICIA HUMPHREY

The school-age years, or middle childhood (ages 6 through 12), is characterized by gradual physical growth in almost all body parts. Coordination and muscle control increase, and there are many opportunities to develop and practice skills. Considerable emotional and social development occurs as the child begins to interact with people outside the family in school and community activities (Fig. 8-1).

PHYSIOLOGICAL GROWTH

By 6 years of age, the overall appearance of the child is different from that of the preschooler. The school-age child is generally taller and thin-

ner than he was as a preschooler. Growth in height and weight during these years is relatively steady. The average gain in weight is about 3 kg (6.5 lb) and that in height, about 6 cm (2.5 in) per year (Johnson, Moore, and Jeffries, 1978). Although at this age boys differ little from girls in height and weight, boys are generally slightly taller and heavier, while girls generally retain more fatty tissue. The preadolescent increase in growth occurs in girls around 10 years and in boys around 12 years of age. Once this "growth spurt" begins, girls will generally be larger than boys of the same age for several years. Normal children of the same age exhibit a wide range in height. Meredith (1977) found that children in upper-class families are generally taller and heavier than

Figure 8-1. *The school-age years comprise an important period in physical, emotional, and social growth and development. These children, at work on "friendship necklaces," are obviously enjoying the social interaction and are also using fine motor skills that younger children do not possess. (Courtesy Chapel Hill Newspaper.)*

those in lower-class families. These findings were consistent in various parts of the world; the tallest and heaviest children come from parts of the world where food is abundant, where infectious diseases are under control, where the size of the average family has decreased and where good personal and community hygiene are practiced (Meredith, 1977). In addition, in the more prosperous parts of the world, children reach physical maturity at an earlier age. This is primarily attributable to better nutrition. Children continue to be taller and heavier than they were in the past (Fig. 8-2).

When plotted on a standard growth chart, the height and weight curves should correspond. An increase in height usually precedes a gain in weight, but the two are closely related.

The size of the head gradually increases; the average circumference is 20.5 in. at 6 years and 21.5 in. at 12 years. The face changes slightly, and the features increasingly resemble those of an adult. The mandible becomes more prominent, giving the chin a more well-defined appearance.

If a parent or health care provider is concerned about an extremely short or tall child, a further assessment is called for. The initial step in the assessment is to observe and inquire about the heights of close family members. If both parents are very short or very tall it is likely that the child's adult height will be comparable. It is helpful to know when the parents had their growth spurt, since the child will probably have a similar pattern.

If there is still concern about the child's height, x-rays can be used to evaluate bone growth. X-rays of the hand and wrist permit an evaluation of ossification centers, which reveal how much cartilage has turned to bone, and how closely the ends of each bone are fused to another; the more advanced the skeletal age, the greater the extent of fusion. The child's x-ray is compared to a set of standard x-rays; the x-ray that most closely matches the child's x-ray indicates the bone age (skeletal age) of the child. Skeletal age is then compared to chro-

nological age to determine if the child is an early, average, or late developer. In early developers, the skeletal age is advanced one year or more beyond chronological age. In the average developer, the skeletal age is within one year of the chronological age, and in the late developer, the skeletal age is one year or more behind chronological age. The earlier developer has a shorter future course of growth. Children destined to be tall usually reach their full growth sooner, while shorter people take longer to reach full growth.

Short stature is sometimes caused by a deficiency of thyroid hormone, which can be identified by thyroid function studies. The injection of human growth hormone at regular intervals before puberty can sometimes accelerate development in height.

During childhood the child's body is constantly forming new bony tissue, which ac-

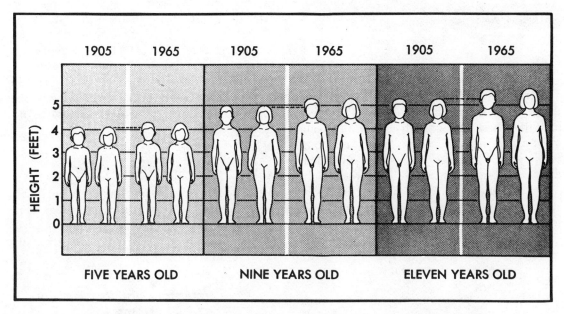

Figure 8-2. *Increases in the size of children of average socioeconomic backgrounds, 1905 to 1965. The figures are based on measurements taken in the United States and Europe. (Reproduced with permission from Tanner, J. M., "Earlier Maturation in Man," Scientific American 1968, 218 (1). Copyright © 1968 by Scientific American, Inc. All rights reserved.)*

counts for the rapid healing of fractures sustained at this time. Bones lengthen and become harder. Skeletal growth may sometimes be more rapid than the growth of muscles and ligaments, which may result in a swaybacked posture and a gangling appearance.

The muscle aches that schoolchildren sometimes complain of may be caused by the muscles trying to catch up with skeletal growth. By 12 years, the bone contours are generally established and will change only slightly thereafter. Good posture should be encouraged during these growing years. The child's gait should be even; the feet should be straight and the arms should swing slightly at the child's sides. The hips should be parallel to the floor in both the sitting and standing positions. A slight flaring of the ribs in the lower costal margin may still be present in the school-age child, but this should disappear during puberty (Chinn, 1979). In later childhood the curvature and function of the spine, particularly of girls, should be checked for scoliosis.

The Genitourinary System

The urinary system is functionally mature when the child enters middle childhood. The kidneys have the capacity for growth and for increasing cell number in response to insult; therefore, renal disease is less likely to result in permanent damage. From 6 to 11 years of age the weight of each kidney increases from 112 g to 164 g, and the length from 8.9 cm to 10.81 cm (Johnson, Moore, and Jeffries, 1978). As the child grows older the kidneys are better able to concentrate urine, and the glomerular filtration rate decreases from 134 ml/min at 6 years to 123 ml/min at 11 years.

The Gastrointestinal System

The organs and tissues of the gastrointestinal system reach adult maturity in late childhood. The stomach has a "fish-hook" shape between 7 and 9 years, and assumes the adult "bagpipe" shape at around 10 to 12 years, by which time children should be able to digest almost any food.

The Neurological System

The nervous system continues to develop gradually. The brain achieves 90 percent of its ultimate size by 6 years, and almost 100 percent by 12 years. The growth of the brain consists primarily of developing interconnections in the form of myelinated pathways through the brain and the spinal cord (Swainman, 1978). The cells multiply and orient themselves in the proper layer within the brain cortex which allows their fibers to make connections with the fibers of other nerve cells in a specific and predetermined manner and in a specific place at a specific time. Myelinization and the development of the sulci of the cortex proceed as intellectual function expands. These changes in the neurological system result in improved memory and the ability to conceptualize. Full voluntary control of gross and fine motor function becomes possible because of neurophysiological changes. Gains in the size and strength of the small muscles that control fine motor activity contribute to increasing control of such movements.

The Endocrine System

All endocrine system functions, except those regulating reproductive functions, reach adult capacity before puberty. Endocrine control of homeostasis and response to stress gradually increases during the school-age years.

The Circulatory and Respiratory Systems

Circulatory and respiratory functions reach adult capacity during the school-age years. By the seventh year the apex beat of the heart gradually shifts from the fourth to the fifth left intercostal space at the midclavicular line. The mean heart rate from 6 to 12 years is 95 beats per minute (Johnson, Moore, and Jeffries, 1978).

As in younger children, "functional," or "innocent," heart murmurs may be present in as many as 50 percent of school-age children. Functional murmurs, characterized by soft, blowing heart sounds caused by blood passing through normal heart valves, are of no significance.

Any child of 6 to 9 years with a blood pressure reading above 120/75 mm Hg on three separate occasions should be referred for medical evaluation. Children 10 to 12 years old with readings above 130/80 on three separate occasions should also be referred (Londe, 1976). In most instances, hypertension in childhood is a secondary manifestation of an underlying disease and warrants immediate attention. Some children may have *essential hypertension* (i.e., that in which the problem is primary and has no underlying cause). (See Table 8.1 for mean systolic and diastolic blood pressure readings for 6- to 12-year olds.)

The respiratory rate gradually declines during earlier childhood; by school age, it should be about 19 to 21 breaths per minute. The vital capacity of the lungs continues to increase as the lungs grow. At 6 years the vital capacity of the lungs is 1,450 ml for boys and 1,220 ml for girls; by 12 years it is 2,800 ml for boys and 2,620 ml for girls (Johnson, Moore, and Jeffries, 1978). Since the respiratory structures and the immunological system are fully matured the body is better able to localize infection in the respiratory system.

The Immunological System

The immunological system's main function is the detection and elimination of substances that are foreign to the body. Lymphoid tissues, which are a major part of the immunological system, reach the height of their development during the early school years and are present in amounts greater than are found in adults. The greater lymphoid tissue growth and increased vulnerability of the mucous membranes to congestion and inflammation make sore throats, upper respiratory infections, and ear infections common occurrences. Any contagious disease will spread rapidly because of the close physical contact among school-age children. Children should be properly immunized before entrance to school, and should receive the appropriate booster immunizations thereafter. Complications rising from minor infections are less severe in middle childhood because the body has the ability to localize infections (Chinn, 1979).

The Sense Organs

The senses of taste and smell are fully mature before the school years, but develop further during these years, allowing for greater discrimination of tastes and smells. By school age, children should have the ability to identify common solid objects by touch alone, and the ability to identify and locate the application of heat, cold, and pinprick to various body surfaces.

The sense of hearing matures in late infancy; by school age, children should be able to discriminate differences in pitch. The child's hearing should be tested before or during the first year in school. Any child who is not doing well in school or who displays signs of hearing difficulties should be tested every six to 12 months, depending on the severity of the problem.

Table 8–1. Mean Systolic and Diastolic Blood Presure, Ages 6–12

Ages	Mean Systolic Pressure	Mean Diastolic Pressure
6–8	105	60
8–10	110	60
10–12	110	60

SOURCE: Lowrey, G. H.: *Growth and development of children* (6th ed.) Chicago, The Yearbook Medical Publishers, 1975.

Central visual acuity is established by 6 years; normal visual acuity approaches 20/20. By the time children enter school they have binocular vision and can focus clearly. Visual ability increases as the child uses the eyes for the many tasks of childhood. In some children the *hyperopia* (farsightedness) of the preschool years is replaced by *myopia* (nearsightedness), which may gradually become worse during childhood. The child with myopia should be referred for correction of the problem as early as possible. Vision screening using the Snellen eye chart is done yearly in many public school systems, and many children are referred for a complete eye examination by an ophthalmologist as a result, since the screening alone does not enable an examiner to identify all eye problems. Careful attention should be given to children who complain of frequent headaches, are constantly blinking or rubbing their eyes, or have frequent drainage from their eyes. These signs may be indicative of eye problems and should be promptly evaluated and treated.

Dental Development

Permanent teeth begin to erupt at about age 6. The final emergence of the tooth is only one stage in a process of development that begins earlier and continues afterwards. The crowns of all teeth, except those of the third molars (wisdom teeth), are completely calcified by eight years of age (Johnson, Moore, and Jeffries, 1978).

Most permanent teeth are present by 12 years except for the third molars, which erupt between 17 and 21 years. Each primary (deciduous) tooth is shed at about the time its permanent successor erupts. Several months pass between the age at which a tooth pierces the alveolar bone and when it pierces the gingival tissue. One study found the average interval to be eight months for the permanent lower first premolar and 11 months for the permanent lower canine. Another several months pass before the tooth has emerged enough to make contact with the teeth of the opposite jaw (Meredith, 1977). Figure 8-3 shows the chronology of the emergence of the permanent teeth. A review of studies done since 1940 indicates a long-term trend toward earlier emergence of the permanent first molars and incisors (Meredith, 1977).

Nutritional Requirements

The 7- to 10-year-old requires 80 calories per kilogram or 35 calories per pound. This requirement stays the same, for both sexes, from 7 to 10 years; from age 11 through adolescence, males require more calories while females need more iron and protein. Prior to the adolescent growth spurt there is a gradual increase in the nutrient requirements as the child is storing nutrients in preparation for the growth spurt. Table 8-2 shows the recommended dietary allowances for school-age and early adolescent children.

Factors Influencing Nutritional Intake

Once the child starts school the family has less influence on eating habits. School-age children

Table 8–2. Recommended Dietary Allowances for 7- to 10-year-olds and 11- to 14-year-olds

	7–10 yr, (male and female)	11–14 yr male	11–14 yr female
kcal.	2,400	2,700	2,200
Protein (g)	34	45	46
Calcium (mg)	800	1,200	1,200
Iron (mg)	10	18	19
Vitamin D (IU)	400	400	400
Vitamin C (mg)	45	50	50

SOURCE: Recommended Dietary Allowances, Revised 1979. Food and Nutrition Board, National Academy of Sciences, National Research Council, Washington, D.C.

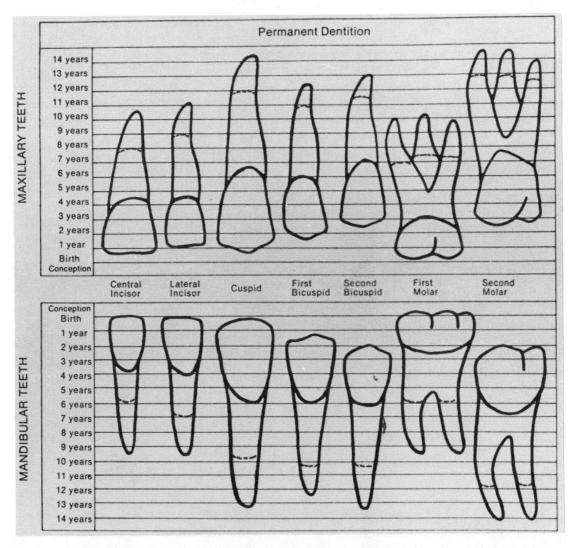

Figure 8-3. *Chronology of the development of the permanent teeth. (From Johnson, T. R., Moore, W. M., and Jeffries, J. E., eds., Children are different: developmental physiology, Copyright © 1978 by Ross Laboratories. Reproduced with permission.)*

may be less punctual for meals and often gobble down food as they are in a hurry to get back to what they view as more important activities. School-age children have somewhat more control over what they eat in that they usually have some spending money of their own and have access to food.

School-age children should be encouraged to try new and different foods. Food preferences established during the preschool years often change. However, eating habits become firmly established during later childhood, and care should be exercised to help the child learn sound eating habits.

Table Manners

Many adults are disturbed by the school-age child's terrible table manners. Children at this age often eat very fast and talk constantly while eating. Such behaviors are readily observable if one visits a school cafeteria. Parents can be reassured that eating behaviors will improve with time. It is also helpful to remind them that, from the child's point of view, eating is not as important as socializing. Manual dexterity is necessary for the skillful manipulation of eating utensils, and this is dependent upon fine motor coordination, which increases during the school years. By 9 years, most children are able to manipulate eating utensils skillfully, and by 12 years most children possess acceptable table manners.

Language Development

Tremendous growth in the ability to use words occurs during middle childhood. Average children expand their first-grade vocabulary of 20,000 to 24,000 words to 50,000 words by the sixth grade (McNeill, 1970). Of course, there is more to language than vocabulary; children must learn to put words into grammatical constructions. According to McNeill (1970), it is not until around 8 years that children recall complete grammatical sentences rather than strings of words, and that they make grammatically consistent word associations.

The use of faulty syntax declines during middle childhood while the use of compound and complex sentences increases. Children improve in their ability to apply new rules of grammar. There is also an increase in the use and understanding of prepositions. Goodglass, Gleason, and Hyde (1970) found that 97 percent of the 10-year-olds in their sample possessed full comprehension of prepositions. The use of adjectives and pronouns increases sharply during the school years.

Motor Development

By 6 years of age the child has achieved good control of his body. Strength and resistance to fatigue increase steadily, and neuromuscular maturation allows for improved manual dexterity.

Play now serves a variety of functions. Participation in strenuous physical activity, which requires large-muscle movement, helps strengthen and further develop the muscles. Improvement in motor skills during the school years contributes to the child's growing sense of competence.

Six-year-olds enjoy roller skating and jumping rope, and are beginning to ride a bicycle; their athletic skills are rudimentary—bouncing and throwing a ball, and running. The 6-year-old has not yet achieved the cognitive ability necessary for cooperation and organized games. Seven-year-olds have improved muscular skills and are beginning to perfect the movements necessary to throw, catch, and hit a soft-ball. Eight-year-olds are ready for cooperative games with rules such as soccer, basketball, and baseball. In the following years children work on improving previously acquired skills and may add new skills.

At the start of school, girls are slightly superior in activities requiring fine muscular coordination; boys are superior in strength and endurance in the latter half of middle childhood. This sex difference is attributable to the increased muscle mass that boys develop. As is true of all facets of growth and development, the timing of the acquisition of motor skills varies greatly among children of the same age. By the end of middle childhood many children will have attained a nearly adult level of proficiency in certain motor skills.

As myelination of the central nervous system increases, fine motor skills improve. Balance and hand-eye coordination gradually improve. Six-year-olds are able to hammer, paste, tie shoes, and fasten clothes. Seven-year-olds prefer a pencil (rather than a crayon or pen) for printing. Coordination develops sufficiently to allow the child of 8 to 10 years to write cursively rather than print. Fine motor skills are sufficient to allow children of this age to sew, paint, build models, or play a musical instrument. (Table 8-3 summarizes the motor skills and appropriate activities of school-age.)

PSYCHOSOCIAL DEVELOPMENT

Psychosocial development is centered around "three great outward pushes." Socially, the child moves out of the family environment and into the world of peers; physically, the child begins to participate in games and activities requiring increased skill and coordination; mentally, the child begins school and begins to acquire a grasp of concepts, symbols, communication, and logic (Fig. 8-4).

The school is the one agency in the United States that by force of law reaches all children who are physically and mentally able to attend. School is instrumental not only in the teaching of certain subject matter, but in the promotion of health and physical fitness, the development of character, and preparation for citizenship. In many ways school is the child's "job"; the attitudes a child develops toward school have very important implications for later life (Fig. 8-5). By the time children enter school they should have developed confidence in their own worth and have a concept of who and what they are; this enables them to turn their attention outside themselves.

With entry into school, the child begins to look elsewhere for goals and standards of behavior as well as for new role models. During this time teachers, coaches, scout leaders, and other adults may serve as role models and become important in the child's life. Children begin to compare their parents with other adults and especially with the parents of their special friends. Parents are sometimes disturbed by the child's constant comparing them with another adult; parents need to be assured that this is normal behavior, that it is necessary for psy-

Table 8-3. Development of Motor Skills and Appropriate Activities for School-Age Children

	Level of Motor Development	*Appropriate Activities*
6 years	Rudimentary coordination	Roller skating, jumping rope, running, bouncing and throwing ball, bicycle riding (usually not proficient), pasting, cutting, hammering
7 years	Improved motor skills; perfection of throwing, catching, hitting ball	More proficient at skating, jumping rope, dancing, riding bicycle, and playing ball; interest in games with identified goal.
8 years	Improved coordination (and cognitive ability) enables participation in organized sports	Soccer, basketball, baseball, sewing, painting, model-building, playing musical instruments
9 years	Hand-eye coordination greatly improved	Crafts
10–11 years	Continued refinement of skills; beginning of attempts to perfect skills	Team sports, crafts, artistic activities

Figure 8-4. *These older school-age boys are being coached in the fundamentals of soccer, which calls for somewhat more physical coordination and teamwork than the games that younger children play. (Courtesy Chapel Hill Newspaper.)*

chological growth, and that it does not mean that the child loves or needs them less.

Play is an opportunity for association with the peer group that assumes great importance during the school years. Peer groups arising during these years tend to have some common characteristics: similarity in age, sex, race, and social status appear to be important factors in the selection of members, sameness of sex being the most important variable in the early school years.

The child intuitively strives to be independent, but independence also brings apprehension. New situations, such as school, create fear. Association with the peer group helps the child deal with these fears. Peer groups provide the child with an opportunity to interact in the absence of adult supervision or control. The child develops new roles and more independence, and learns to cooperate, compromise, and compete during peer play.

At around 9 to 10 years the *chum stage* occurs, and affection is focused on the chum rather than the group (Fig. 8-6). A relationship with an agemate of the same sex is an important step in the child's social development; it leads ultimately to the ability to relate to a mate (Sullivan, 1953). The love the child develops for the friend is the child's first loving attachment outside the family and is a mark of increased maturity. The friend becomes as important as the self, and *I* becomes *we*. The chum relationship includes discussions of shared gripes, irritations, and frustrations. Children discuss their parents, siblings, and homes, and are able to compare idiosyncrasies of the family with those of the chum's family. Brothers and sisters as well as parents are rejected by the school-age child, since they represent ties he is trying to break. At this stage loyalty to the chum may seem to be greater than loyalty to the family (Sutterly and Donnelly, 1973). While children may seem to reject family, home, and parents temporarily, they will ultimately incorporate that which they find good in the parents' values into their own value system.

Figure 8-5. *Health habits, character, and a sense of citizenship are shaped in the classroom; the attitudes a child develops toward school are invariably reflected in later life. (Photo by John Elkins.)*

Figure 8-6. *Chums share everything from ice cream cones to discussions of dreams, frustrations, and their family lives. The chum relationship is the first loving attachment outside the family and is prerequisite to later, intimate relationships with members of the opposite sex. (Photo by John Elkins.)*

While homosexual overtones are present in the chum stage, they are not indicative of homosexuality. The foundation for later intimacy with people of the opposite sex is established during the chum stage, and without the chum relationship the child has difficulty with adolescent heterosexuality and intimacy. Homosexuality can result if the child becomes fixated at the chum stage of development (Sullivan, 1953).

According to Freud's psychoanalytic theory, the school-age child is in the *latency period* (Freud, 1959). Freud saw latency as a relatively stable period between the turbulent preschool years of infantile sexuality and the storminess of adolescence, during which the child first experiences feelings of adult (genital) sexuality. School-age children masturbate, sometimes as a nonspecific form of tension-reducing behavior. At this stage children do not associate masturbation with a sexual urge. They are, however, interested in sexual matters (including sexual anatomy); all too often, the attitudes of adults convey the notion that sex is "bad," and the child's curiosity is then tempered with guilt, leading, of course, to an inner sense of conflict. A large part of the child's libidinal energy appears to be channeled into generalized intellectual curiosity and the development of mental abilities and physical skills (Vaughan, McKay, and Buchman, 1979).

The child's superego (conscience) becomes more firmly established and serves in aiding adaptation to the external world as well as in controlling and redirecting instinctual drives. The developing ego (sense of self) consolidates the technique of sublimation, which is important in learning and in the development of the capacity for thinking (Vaughan, McKay, and Buchman, 1979). During the middle childhood years the self-concept is largely derived from parents and peers. Feelings of worthlessness or inferiority may develop; one may hear the child say, "I wish I were like Tom" or "I wish I were pretty like Sue."

The developing ego is threatened by the loss of the protective mantle of home and early childhood. The child develops *defense mechanisms* that help him cope with fears, doubts, unacceptable impulses, and unfamiliar experiences. The use of defense mechanisms is normal, but the excessive use of them can result in an immature personality. Commonly used defense mechanisms (which may persist throughout adult life) include regression, repression, sublimation, projection, reaction formation, ritualistic behavior, and rationalization.

Regression, the return to a behavior of an earlier age, may be used as a defense against anxiety. For example, the child who feels that a younger sibling is getting undue attention may revert to thumb-sucking; the child whose parents have separated may regress to bed-wetting. Regression is the child's attempt to recapture the security of an earlier time.

Repression, the means by which unpleasant thoughts or feelings are thrust from the conscious mind to the unconscious, may be employed in anxiety-producing situations.

Sublimation is the channeling of sexual or aggressive impulses into socially acceptable activities. Sublimation is frequently used during the school years, since the child is often made uncomfortable by his or her emerging sexuality.

Projection allows the child to deal with unacceptable thoughts and motives by attributing them to another. For example, one child may say that another child does not like him when it is really the first who dislikes the other.

The defense mechanism of *reaction formation* is characterized by acts or attitudes that reflect the opposite of what one really feels. For example, the child who hates a new sibling may adopt the appearance of being the loving, caring sibling because such a hostile impulse is unacceptable to the superego.

Ritualistic behavior is often employed to deal with imagined harm or anxiety and allows the child to feel in control. Commonly observed examples of ritualistic behavior include avoiding the cracks in sidewalks, putting school supplies in the same place each day, or eating the foods of one meal in a certain order.

Rationalization is the process of devising a seemingly plausible justification for ideas, acts, or beliefs. For example, a child who gets a low grade on a test may say that it was not an important test.

The reactions of others are very important to the development of the self-concept. If children are loved by their parents, they learn to love and to accept themselves, while parental rejection leads to a negative self-concept and feelings of shame or inferiority. The reactions of peers and adults other than parents during the school years contributes further to the development of the self-concept. Children with negative self-images often feel defensive and adjust poorly to school (Murray and Zentner, 1979).

While minority-group children from families of lower socioeconomic standing sometimes have poorer self-concepts, two studies indicate that the child's intimate family situation probably exerts more influence on self-image than the broader social context. In a study that compared the self-perceptions of poor, mostly black and Puerto Rican children with white, mostly middle-class children, Soares and Soares (1969) found that the children from disadvantaged homes had higher self-perceptions. Males from poorer homes had higher scores than females, while middle-class females thought more of themselves than middle-class males. Trowbridge and Trowbridge (1972) compared low-income and middle-class third- through eighth-graders and found that the lower-income children had higher self-esteem. The low-income children also saw themselves as more successful in school. The middle-class children who did poorly in school tended to blame themselves, while the low-income children tended to blame the teachers or the school.

According to Erikson's theory of personality development, *industry*, the wish and will to achieve mastery of whatever one is doing, is the central task of the child from seven to 12 years (Erikson, 1963). Children apply themselves to tasks and to the acquisition of skills, and receive satisfaction from their accomplishments. The ability to produce things is important in the school-age stage. The child is no longer satisfied by play for its own sake; instead, he directs play to some meaningful end.

Failure to achieve a sense of industry results in a sense of *inferiority*, feelings of inadequacy and defeat; this in turn may result in a child (and later, an adult) who does not like to work or try new tasks. Children who feel inadequate compared to their peers may revert "back to the more isolated, less tool-conscious familial rivalry of the Oedipal time" (Erikson, 1963, p. 260). Children who immerse themselves in tasks in order to prove their worth and gain attention may become bossy, aggressive, and overly competitive. Such children may neglect peer relationships. According to Erikson the adult who is totally immersed in business and rarely sees family and friends may have behaved similarly during the school years.

To contribute to the sense of industry, parents should encourage peer-group activities and give recognition for the child's accomplishments and talents (Fig. 8-7). Expectations for the child should be realistic, since unrealistic expectations can contribute to feelings of inferiority.

Cognitive Development

By 6 years of age most children are able to understand abstract concepts, enabling them to begin learning to read, write, and do arithmetic. As the child's intellect develops, he shifts gradually from a tendency toward thoughtless activity to a capacity for thought without activity.

The 6-year-old remains in what Piaget called the *preoperational stage* (discussed in the previous chapter). Between the ages of 5 and 7 years, the child moves into what Piaget characterized as the *operational stage*, in which the child gradually begins to understand abstract symbols and is able to use them to carry out mental operations (Piaget, 1932).

Piaget's stage of *concrete operations*, characterized by relatively logical thinking about tangible or familiar situations, spans the years from 7 to 11. Classification, nesting, seriation, multiplication, reversibility, and conservation are mental operations characteristic of this period. The child begins to understand the relationship between classes and sizes of objects.

Children of this age are capable of *classification*, i.e., they are able to sort objects into groups according to attributes such as color, size, or shape. *Nesting* is a special form of classification permitting the association of a concept with a larger concept, for example, "Ford is a car, a rose is a flower, and a table is a piece of furniture." *Seriation* is the ordering of objects according to increasing or decreasing size. *Multiplication* is the ability to classify according to two variables such as classification and seriation; the child can now group objects according to color and size simultaneously. School-age chil-

Figure 8-7. *Activities such as this, which involve the child with peers and with the community, contribute to the development of a sense of industry. (Courtesy Chapel Hill Newspaper.)*

dren are also capable of *reversibility*, that is, they can reverse mental operations and return to the starting point; for example, the child can add numbers and then reverse the operation by subtracting. *Conservation* allows the child to see that objects remain the same although outside appearances may change; for example, the child would be able to recognize that equal amounts of water in two glasses of identical size remain equal when poured into two glasses of unequal size. Conservation of substance is present at age 6 or 7, conservation of weight at 9 or 10, and conservation of volume at 11 or 12.

Realism, or a sort of confusion that exists between psychological events and objective reality, is still present in some degree. From ages seven to nine, children are in a period of *immediate realism* characterized by the ability to realize that thoughts originate in one's mind. Thus, the child is able to realize that dreams result from his thinking about something, but he is still convinced that his dreams actually become real while he is dreaming them. By 10 to 11 years, the child is able to distinguish between psychological and objective reality.

The grasp of time concepts improves during the early school years. The average child will have a fair grasp of clock time, making an occasional mistake, by entrance into school. In later childhood the child develops comprehension of time in terms of years and dates.

Cognitive development in the middle school years (9 to 12 years) is marked by a great improvement in the ability to do more complex intellectual tasks. At this age children are able to answer questions more sensibly, to see significant details in a situation as well as to detect absurdities, to use words correctly as well as defining abstract words more precisely, to derive generalizations from verbal and mathematical relationships, and to draw on a larger fund of general knowledge.

Moral Development

Moral development parallels cognitive development. Children must achieve a certain level of cognitive maturity and shed egocentric thinking before they can make moral judgments. Moral development occurs when thought processes become more logical. Thus, the level of moral reasoning depends on age and maturity.

Piaget proposes two stages in the acquisition of morality, the morality of constraint and the morality of cooperation. *Morality of constraint* is characterized by rigidity while *morality of cooperation* is characterized by flexibility. In the former, the child views every act as being either right or wrong, and because the child's thinking is egocentric, he believes that everyone shares his view. During this stage, children obey rules because they fear the consequences of disobedience. Children favor severe punishment and often believe that physical accidents that occur after a misdeed are a form of punishment willed by God (Piaget, 1948).

Children appreciate the morality of cooperation at a later stage, when they are able to put themselves in the place of others. They are less egocentric and able to see the possibility of more than one point of view. The child now judges an act by intentions and not by the conse-

quences. Children recognize that rules can be changed because rules are made by people.

Lawrence Kohlberg's research on the moral development of children in a variety of cultures confirmed Piaget's earlier findings. Kohlberg (1969) redefined Piaget's theory by defining different types of moral reasoning and dividing them into four levels. The child entering school is at what Kohlberg calls Stage 3, characterized by striving to please others and gain approval. Most children progress to Stage 4 at around age 10 or 12; at this stage the child is concerned with maintaining order and following rules, and his concept of right is influenced by religious beliefs. Kohlberg's theory of moral development is summarized in Table 8-4.

Spiritual Development

During the early school years children learn the basic tenets of their religion, from which they develop a religious philosophy. By 6 years of age children understand the concept of God as the creator of the world. The 6-year-old views God as a powerful father-figure who is responsible for many things. Six-year-olds expect their prayers to be answered. Seven-year-olds ask more appropriate questions about God and religious teachings. As children grow older they may show skepticism about religious teachings as they distinguish what they are learning from what they were told in the past. Questions may arise about the existence of a god who is not visible or tangible. Ten-year-olds often think of God as a kind of invisible man and will seek God's help for protection when fearful. By 11 years, children may view God as a spirit, and may begin to feel that what happens to them is influenced more by their own actions (Kaluger and Kaluger, 1979).

Sex-Role Development

During the school years children socialize mainly with members of their own sex. During this

Table 8–4. Kohlberg's Scheme of Moral Development

Levels		*Developmental Stages*
I. Premoral Stage	Stage 0:	The child is unable to understand rules or to judge good or bad in terms of rules or authority; guided only by what he can do and wants to do. Good is pleasant; bad is painful.
II. Preconventional Level	Stage 1:	Punishment-and-obedience orientation. The consequences of an action determine its goodness or badness. Avoidance of punishment and unquestioning deference to power are valued.
	Stage 2:	Egoistic orientation: right actions are those that satisfy one's own needs and occasionally the needs of others. Instrumental reciprocity; "you do something for me and I'll do something for you."
III. Conventional Level	Stage 3:	Good-girl/nice-boy orientation: behavior is aimed at pleasing others and gaining approval. Behavior is now judged by intent.
	Stage 4:	Law-and-order orientation: authority, fixed rules, and maintenance of the social order are important.
IV Post-Conventional Level	Stage 5:	Social-contract, legalistic orientation: right actions defined in terms of individual rights and standards agreed upon by society. Aside from the agreed upon standards right is a matter of personal values and opinions. Conscience begins to play a role, but is not yet based on well-thought-out principles.
	Stage 6:	Universal ethical principle orientation: universal principles of justice, or reciprocity and equality of human rights, and respect for the dignity of human beings as individuals.

SOURCE: Adapted from Kohlberg, L, Turiel, E.: Moral development and moral education. In Lesser, G. S. (ed.), *Psychological and educational Practice*. Glenview, Ill.: Scott, Foresman and Co., 1971.

time they begin to assume the mannerisms, attitudes, values, and recreational patterns of adults of the same sex (Cohen, 1972). Sex roles begin to be perceived in near-adult fashion between 8 and 11 years. Children whose mothers are not employed outside the home are more likely to associate certain activities with one sex or the other than children who have working mothers. Intelligence may be a factor in sex-role stereotyping during the middle years; one group of researchers found that on the whole, brighter children exhibited more advanced sex-role identity than average children (Kohlberg and Zigler, 1967). Sex-role identity, like morality and other facets of growth, appears to develop parallel to cognitive development.

The role of women in society has changed greatly, especially since the late 1960s, and children have had greater opportunities to view women in a wider variety of roles. Has this resulted in a change in the sex-role concepts among children? Greenberg surveyed 1,600 suburban fourth, sixth, and seventh graders from various social classes. While some stereotyped ways of thinking existed, Greenberg found girls to be consistently more egalitarian than boys, and older students more egalitarian than younger students. Many students were willing to "grant women greater participation in the social, economic, and political spheres" (Greenberg, 1972, p. 10).

The Impact of Television on Development.

Television is in almost every American home and has become part of our way of life. According to one study, television was turned on more than six hours per day in homes with children (Liebert, Neale, and Davidson, 1973). What influence does television have on the child's socialization, behavior, and communication patterns? Some studies contend that unmonitored viewing of television can stimulate aggressive behavior; excessive watching of television by 8-year-olds can cause aggressive behavior as much as 10 years later (Liebert, Neale, and Davidson, 1973). However, other researchers contend that the viewing of television violence provides an outlet for violent feelings rather than stimulating the acting-out of such feelings (Feshbach and Singer, 1971). Excessive television viewing can stifle creativity and initiative in children and can interfere with normal personality development. According to one study, teachers report more children entering school with decreased capacities for imaginative play and creativity, displaying lower frustration levels, less persistence, and decreased attention spans. They also reported children confusing reality and fantasy (Cohen, 1974).

In the mid-1970s the American Medical Association, the National Education Association, and the Parent-Teacher Association united in opposition against programming containing violent and sexual excesses. While there have been some improvements in programming during the earlier evening hours, and warnings advising parental discretion, there are still a great many programs that, because of excessive sexual or violent content, are not appropriate for the young child. Parents should be encouraged to monitor their children's television viewing and to help them interpret what they view. The unmonitored television, used as a "babysitter," may have a negative influence on the child's development.

Family Relationships

The family is still a major influence on the socialization of the child between the ages of 6 and 12. Children continue to learn their social roles through identification. Although peers and school influence the child's socialization, the family continues to be the primary influence. The availability of parents for interaction and discussion of problems is an important factor in the development and maintenance of self-esteem. Without the concern and acceptance of the family or family substitute, it is unlikely that the child will develop a high degree of self-esteem.

Parental attitudes contribute greatly to the child's success. When parents value education and place emphasis on learning, children are likely to adopt these attitudes and will be highly motivated to learn. Such families are likely to support the child's efforts and provide opportunities for learning, such as through educational games, trips to museums and zoos, and involving the child in household tasks.

By gradually increasing the child's responsibilities and encouraging success, the family promotes the development of a sense of industry, which in turn leads to a sense of independence. By 7 years of age most children have some responsibilities around the house, such as setting the table for meals or making their own beds. As children grow older their responsibilities should be increased. Success in these relatively simple chores develops self-confidence and lays the foundation for mastery of more complex tasks and responsibilities in later years. Being responsible for certain chores also makes children feel they are contributing to the family.

Most families begin to provide the child with

an allowance at the beginning of the school years. In some families the allowance is contingent on the successful completion of assigned chores. An allowance gives children the opportunity to make judgments and teaches them money management. Even though children occasionally make bad decisions concerning the use of their allowances, such decisions and their consequences can be instructive.

The school-age child will continue to have frequent conflicts with siblings. While sibling rivalry is less acute in school-age children than in preschool children, it may still exist. In particular, school-age children are often jealous of the relative freedom and additional privileges granted the older sibling. While siblings may disagree between themselves, they usually have a deep affection for each other and are quick to stick up for one another in disputes outside the home.

Common Health and Developmental Problems of the School-Age Child

The middle childhood years are for many people the healthiest time of life. The level of physiological maturity achieved by this age allows the system to effectively fight off most infections. At this stage the child has the capacity to recover from injury or infection rapidly. Since middle childhood is a relatively calm and stable period the risk of emotional disturbance is minimal.

While these generalizations apply to most children of this period, there will of course be children who have physical and emotional difficulties. Careful health assessment is a must for the identification and correction of health problems; uncorrected health problems may cause further injuries and illness as well as difficulties with academic and social development.

Accidents and Injuries

Accidents are the leading cause of deaths among school-age children; they account for more deaths in this age group than the six next most commonly reported causes combined. It is estimated that 40,000 to 50,000 children are permanently injured each year, and a minimum of one million seek medical care because of injuries from accidents (Mofenson and Greensher, 1977). Among the most common accidents are those involving motor vehicles, those involving firearms, and drownings.

In analyzing how and why accidents occur, it is helpful to look at three factors: the *host* (the person injured), the *agent* (the object that is the cause) and the *environment* (the conditions under which the accident occurs). Some factors that have been found to be associated with the occurrence of accidents include: time of year, time of day, illness, hunger or fatigue, hyperactivity, relocation to a new neighborhood, and frequent accidents among other family members. Boys are generally more prone to accidents than girls. The child who has one poisoning episode is more likely to have a second. According to Arena (1971), accident-prone children are easily distracted and may be antisocial, aggressive, nonconforming, intolerant, impulsive, and inclined to take risks; their accidents may be a means of attracting attention.

The dramatic increase in the popularity of bicycle riding since the late 1970s has been accompanied by a parallel increase in bicycle accidents. According to the U.S. Consumer Product Safety Commission (n.d.) there are more than 1.25 million bicycle accidents annually in the United States. Horseplay, riding double, and ill-fitting bicycles are frequent causes of bicycle accidents. Children should be taught the rules of safe bicycling just as adults are taught to drive cars safely. Another important factor in bicycle safety is adequate maintenance and repair of the bicycle (Waller, 1971).

Accidents connected with other kinds of play and athletics are common. Of the estimated 100,000 people who receive emergency room treatment for accidents associated with public and home playground equipment every year, most are children between 5 and 10 years old. Fractures and lacerations occur most frequently. In almost half of the injuries the face and the head are involved.

Since accidents in and around school are common, parents and health care providers should urge school officials to offer programs focusing on the prevention of accidents. Periodic evaluations of the school environment for health hazards should be part of the program. Often the health care professional can be instrumental in getting such programs started.

Any child with two or more significant accidents in a twelve-month period warrants further evaluation and follow-up by the health care provider.

School-Related Problems

Danger signs during the school years manifest themselves in many ways. The child whose behavior is "too good" is a cause for concern. The child who is unable to establish relationships with his or her age-mates may turn toward adults and be labeled as the teacher's pet, mommy's boy, or the goody-goody. While the child's behavior may cause little disruption, it can be cause for concern since it indicates the child is failing to establish the necessary separation from his parents and is failing to establish friendships with his peers.

A major developmental task for middle childhood is adjustment to school. Since school takes up many hours of the child's day it affects and is affected by every aspect of his growth. Most children at least initially view learning as fun and actively participate in new experiences. The teacher is an especially important person during the early years, being a role-model who influences the child's self-concept. An ideal teacher is kind, sympathetic, considerate, consistent in discipline, interested, and enthusiastic. The ideal teacher explains things so that children can learn, and leads the child into learning rather than pushing the child.

There has been much debate and criticism about American public schools. A variety of approaches to education have been tried—open versus closed classrooms, report cards versus no report cards, new math or old math, special enrichment programs versus the "three Rs." There has been much change in our public schools since the early 1970s and as yet the debate over which educational methods are best is unsettled. Parents need to be involved in schools and be aware of what is being taught and how, since the school is a major influence on the child. Increasingly, schools are assuming responsibility for safety and sex education as well as preparation for family living. The responsibility for education in these areas should not rest solely with the schools. Parents can help children interpret what is taught in school and supplement this information as needed. Often the child has questions that are not answered in school, and parents are especially needed in these instances.

Learning Difficulties

Learning difficulties may surface even in children of normal or above-average intelligence. Problems with speech, hearing, or vision (which are of course not "learning disabilities" per se) can cause the child to have difficulty with language and reading. Effective screening programs with adequate resources for referral and treatment should be part of the school health program.

As many as 10 percent of all school-age children suffer from a variety of "specific learning disabilities" (Brutten, Richardson, and Mangel, 1973). These children have difficulty learning to write, read, or work with numbers. They also have difficulty *processing* what they learn, i.e., they may "know it in their heads" but cannot write or say what they mean. Learning-disabled children need to be diagnosed early so that special teaching techniques can be employed to help correct the difficulty. With special attention these children may overcome their difficulties and some even go on to college. The learning-disabled child who is not diagnosed and treated will grow up feeling inadequate.

School Phobia

School phobia occurs in some young children. Among kindergarten through fourth-grade children, it is equally common in both sexes and in all socioeconomic strata. Usually it is separation from the mother and not what is happening at school that makes the child anxious.

School phobia often goes unrecognized for a time as the child presents organic symptoms, which parents and school officials accept. All organic causes must be ruled out before school phobia is suspected. The most common complaints associated with school phobia are abdominal pain and headaches. Affected children are often described as "good students" and are doing average or above-average school work. Often the family situation contributes to the problem; the child's parents may see him as vulnerable for some reason and may overprotect him. Often the mother is overly involved with the child.

The most appropriate treatment for school phobia is to get the child back to school as soon as possible. This is not always easy since the child often cries and has temper tantrums. Parents need much support during such times. If children come home during the day, they should be returned to school quickly. If the child has no other emotional problem, insistence on school attendance usually results in prompt resolution of the problem. The parents, child, and school must all be involved if the problem is to be managed.

When the child returns to school attempts may be made to decrease the child's anxiety and improve his self-image. The teacher can give special attention to the child such as the giving him or her tasks to do and rewarding the child.

When the child does complain of a headache or stomachache the child can be allowed to rest instead of being sent home. If the phobia continues for a prolonged period or if the child has other emotional problems, some form of psychotherapy is indicated.

Getting Along with Other Children

Establishing relationships with age-mates is another developmental task of the school years and contributes to the child's developing self-concept (Fig. 8-8). Feelings of loneliness, feelings of being disliked by other children, excessive anger or aggression, or refusal to conform to behavioral norms are indications of social incompetence. Children who display such feelings and behaviors for a prolonged time should be referred for evaluation and help.

Often, children who have appeared normal in development will display a change in behavior when a crisis occurs in their family such as the birth of another child, divorce, remarriage, or death. Such children need help in expressing their feelings in healthy ways that do not interfere with social development.

Behavior Problems

Since the school-age years are a trying time for the child in that they present new tasks and situations to master, even the child who is growing and developing normally will occasionally have behavior problems. For most children such problems prove to be transitory if they are handled correctly.

Children sometimes employ aggressive, negative, or disobedient behavior in an attempt to feel important and to control others. Other forms of aggression, such as lying and stealing, usually constitute an attempt to gain recognition. In effect, the child is saying "look at me, pay attention to me." Sometimes lying is indulged in just to see if one can "pull it off." At other times school-agers may resort to lying to

Figure 8-8. *Getting along with age-mates, as in this game of four squares, is a developmental task of the school years. (Photo by John Elkins.)*

avoid punishment or to impress their peers. When the child is caught in a lie it is useful for an adult to help him to analyze his behavior and to understand why he acted in such a way. The same is true in cases of extreme aggression, which it usually physical, involving fighting or vandalism. Often extreme aggression is caused by the child's reaction to either severe punishment, parental overindulgence, or the overprotection associated with lack of emotional depth on the parents' part. Such behavior should be immediately stopped; if it goes unchecked, the child may unconsciously fear that there is no limit to his destructive ability. Implementing a reward system for the desired behavior rather than punishing the undesired behavior is the more effective course.

Respiratory Tract Infections

In the close confines of classrooms many minor respiratory infections spread rapidly. Since the structures of the respiratory tract are more mature, school-age children are better able to localize infection. The incidence of otitis media declines because of changes in the anatomy of the ear, which attains adultlike proportions during the school years. The average child has three to four colds a year.

Communicable Diseases

Thanks to immunizations, the incidence of communicable diseases has greatly decreased. Most states require basic immunizations before the child is allowed to enter school. The immunization level of all children should be assessed and immunizations should be provided for those not protected. Much concern has been aroused in the United States because of the failure of some people to get immunizations for themselves or their children. Small outbreaks of communicable diseases such as rubella have occurred in some areas of the country. In 1978, the U.S. Public Health Service launched a massive program to increase immunization levels; in the previous year, the National League for Nursing (NLN) sponsored immunization clinics in many areas of the United States. School systems have become much stricter in requirement for proof of proper immunizations.

Streptococcal infections are a major health problem among school children. Since transmission occurs primarily by direct contact with infected persons, the school is an excellent vehicle for infection. The incidence is highest among 6- to 16-year-olds in areas where the weather is temperate, especially during the winter months. Prevention and control of the spread of streptococcal infection among children is important because serious sequelae such as rheumatic heart disease can result. Children suspected of having "strep throat" should be referred for throat culture and appropriate treatment.

Nutritional Problems

While there is most compelling evidence to indicate that obesity tends to be familial (Gain and Clark, 1976), children need to be given sound nutritional information in any case (Gain and Clark, 1976). It is quite probable that, in families that are inclined toward obesity, learned attitudes toward eating and physical exercise constitute part of the cause.

Overweight and obesity are serious health problems in the adult population in the United States, and there is considerable evidence that obesity begins in childhood for many people. One group of investigators found 27 percent of children aged 5 to 10 to be obese or overweight (Humphrey, 1979). Children should be introduced to the basic food groups and taught the importance of selecting foods from each group. Making children aware of the hazards of "junk-foods," especially those high in sugar and carbohydrates, is important for proper nutrition as well as the prevention of dental caries, another major health problem for school-age children. In addition to the hazards of obesity itself, the child who is overweight may not be receiving the proper nutrients needed for healthy growth and development.

Children who are extremely thin for their height also merit further investigation. Most frequently the thinness is attributable to a genetic predisposition, but in some cases extreme thinness is a symptom of a more serious underlying chronic disease. A small percentage of very thin children will be found to lack proper diets because of poor eating habits and/or inadequate financial resources on the family's part. Children who do not receive the proper nutrients during the growing years are not likely to reach their full potential for growth and may suffer from chronic nutritional diseases throughout life.

Skeletal Injuries

Since school-age children are increasingly mobile and active as well as more daring, injuries to the skin and skeletal system are more common. Dislocated joints, muscle and ligament injuries, and broken bones become more prevalent. Careful assessment of the injury and the prevention of bleeding into the joint is the first priority. Care should be exercised to protect the injured areas from further insult. Treatment should be secured from a specialist in skeletal trauma care if possible.

Some children suffer repeated skin and skeletal injuries and may be accident-prone. Such children may need to be referred for evaluation.

Common Orthopedic Conditions

The child who complains of pain in a lower extremity should be carefully evaluated. Transient *synovitis* is more common in boys and is sometimes preceded by trauma. The onset is acute, causing pain in the hip, thigh, or knee; usually the child refuses to bear weight on the joint. The child has no fever and otherwise appears well. Synovitis is a self-limited condition and usually resolves with several days of bed rest. If the child has a fever or other signs of systemic illness, more serious conditions of the hip such as septic arthritis, acute osteomyletis of the proximal femur, and, occasionally, tuberculosis of the hip must be considered.

Legg-Calvé-Perthes disease (osteochondrosis of the capital femoral epiphysis) generally begins with an insidious limp and pain on motion. It usually runs a prolonged course and may result in a flattening of the head of the femur. A child with prolonged hip pain and associated limp warrants evaluation. Legg-Calvé-Perthes is treated with bed rest, with or without traction. Once the acute inflammation has subsided, some form of long-term splinting to keep the hip in abduction and internal rotation is employed.

Structural *scoliosis* is a defect in the bones and supporting structures of the spine. Idiopathic scoliosis, the most prevalent type of structural scoliosis, commonly presents between the ages of 8 and 15. Idiopathic scoliosis may be inherited as a dominant trait with incomplete penetrance and occurs five to six times more often in females than in males. The onset can be quite insidious; one shoulder or one hip being higher

than the other may be the first symptom. Since the late 1970s, screening programs in public schools have contributed greatly to early identification of children with abnormal curvatures of the spine. Such children should be referred for x-rays of the spine, which are necessary to confirm the diagnosis. Many patients do not show progression of a mild curve but observation and periodic x-rays are necessary to evaluate the stability of the curve. If the curvature progresses, treatment is desirable for functional and cosmetic reasons. Severe scoliosis may impair pulmonary function and cause back pain in later life. Treatment consists of either bracing or spinal fusion.

Urinary Tract Infection

Urinary tract infection (UTI) results from bacterial invasion of any of the urinary tract tissues. *Cystitis* refers to inflammation of the urinary bladder; *pyelonephritis* is an infection of the kidney and renal pelvis. The most common symptoms of UTIs in the school-age child are urinary frequency and urgency; fever and dysuria may also be present. When pyelonephritis is present, costovertebral angle (CVA) tenderness accompanies the other symptoms. UTI may be asymptomatic. Diagnosis of UTI is confirmed by the presence of pathogenic bacteria in the urine. The most common organisms include the *Enterobacteriaceae, Pseudomonas,* and *Staphylococcus epidermidis.* Childhood UTI is more common in females than in males. Kunin (1972), in a 10-year comprehensive follow-up study of school-age children, found that approximately 5 percent of females acquired at least one UTI; in contrast, the incidence in males is only 0.003 percent. Kunin's study demonstrated that once a female had a UTI, one or two recurrences were likely. Any female with more than three UTIs should be referred to a urologist for a complete work-up to rule out functional problems in the urinary tract. Uncomplicated UTIs are treated with an antibiotic given in high enough doses to kill the organisms causing the infection.

Allergies

According to household surveys conducted as part of the United States National Health Survey of 1959–1961, asthma was the most common chronic health problem of children. Twenty-five percent of the days lost from school because of chronic health problems involved allergic conditions (Schiffer and Hunt, 1963). It is generally thought that a generalized predisposition to allergy (rather than a specific allergy) is inherited. The most common allergens include feathers, fur, pollen, dust, smoke, plants, chemicals, and foods.

There are two general types of allergic responses, immediate and delayed. The immediate reaction occurs suddenly after exposure to an antigen to which the host is sensitized, and is a result of the antigen-antibody complex formation. With the immediate reaction the child may develop urticaria, allergic rhinitis, eczema, or asthma. The delayed reaction occurs hours or days after exposure, and there are no demonstrable circulating antibodies. Fungus allergy, allergic contact dermatitis, and bacterial allergy are examples of allergies typified by delayed reaction (Schuster and Ashburn, 1980).

When the allergy is caused by one or two known allergens, treatment is simple: the child avoids these substances. Some children are allergic to many substances and it may be difficult to identify them, in these cases the children may need to be referred for allergy testing. In children who are allergic to substances that cannot easily be removed from the environment, desensitization programs may help to alleviate symptoms.

Bronchial asthma, a common manifestation of allergy in school-age children, can be aggravated by emotional factors. During an asthma attack, children frequently wheeze severely and may have dyspnea (difficulty in breathing).

Dyspnea is frightening to persons of any age but is especially frightening for the school-age child. Unfortunately, the fear itself may intensify the attack. Bronchodilators such as theophylline may be used. One of the most important aspects of the treatment of asthma is helping the affected child to understand the cause of the attack and how he or she can assist in the treatment. Many health care providers begin teaching children how to recognize the early symptoms of an asthma attack and initiate treatment when they appear. The asthmatic child should not be treated differently since this often creates an emotional state that is conducive to more attacks.

Symptoms of *food allergies* may include hyperactivity, tension, fatigue, irritability, restlessness, mental confusion, and excessive sweating. When such signs appear and a cause cannot be identified, the possibility of a food allergy should be considered. When a food allergy is the cause, the child will show a dramatic improvement if the offending foods can be identified and eliminated from the diet.

Dental Problems

Since most of the permanent teeth erupt between 6 and 12 years of age, it is recommended that children have dental checkups every six months during these years. There is a high probability that children of this age will develop one or more cavities (Holm and Wiltz, 1973). Prompt attention must be given to caries in order to prevent loss of teeth and systemic infection secondary to toxins produced by infected teeth. Some children will also need the care of an orthodontist.

By first grade most children are responsible for their own oral hygiene. However, most children will need frequent parental supervision, since they are not yet proficient at brushing their teeth. Children should also be taught to floss their teeth but, again, they will require parental supervision, since they lack the fine motor coordination required to do this task well. Good oral hygiene habits should be learned early in the school years. By 10 years of age most children will have sufficient fine motor coordination to allow them to be responsible for their own oral hygiene.

ANTICIPATORY GUIDANCE

The assessment of the child by a nurse or other health care worker provides the foundation for anticipatory guidance. It is helpful to find out how children feel about themselves with regard to their bodies, how the development of motor skills is progressing, and how well they relate to their families and peers.

Accident Prevention

Since accidents account for the greatest number of injuries and deaths among school-age children, it is always appropriate to discuss the matter of accident prevention with both the parents and the child. The nurse may begin by assessing the environment in which the child lives to identify what risk factors are present. Most school-age children ride bicycles; the nurse can provide information about bicycle maintenance and safe bicycling. Regarding unsupervised play, it is helpful for parents to devise a plan of action should an accident occur; what would the child do, for example, if he or another child was injured while playing in the park (Fig. 8-9)? Many parents have never taught their children what to do in case of an accident. If the family has such a plan, the nurse should review it and make suggestions for improvement. Attempts to teach basic first-aid should take into account the child's cognitive level. Children should know how to apply pressure to a cut to stop bleeding, not to try to move themselves or another child if a broken limb is

Figure 8-9. *Since unsupervised play is far more common among school-age children, parents should make sure their children have some idea of what to do if an accident occurs. (Courtesy Chapel Hill Newspaper.)*

suspected, and who to contact for help in case of an emergency.

School-Related Problems

The nurse should assess the child's progress in school. For the child who is having difficulty it is helpful to assess the parents' attitudes toward the school problem. Do they feel it is the child's fault, or the teacher's? What contact have they had with the school? Is the child receiving assistance?

Some parents who feel their children are not doing well are hesitant to bother teachers and principals. These parents may benefit from encouragement and suggestions for contacting school officials. Other parents are not aware of special resources for children who need help; for children with learning disabilities, the nurse can assist the family in finding appropriate community resources.

Sex Education

Sex education should begin in the early school years. Information about sexual anatomy and physiology should be provided. Most parents experience some anxiety when they discuss sex with their children. The nurse can help by acknowledging that it may be difficult for parents to discuss sex with their child, and by providing educational materials that parents can share with their children.

Some elementary schools are now introducing basic sex education as part of the curriculum. Parents are usually invited to preview the materials used and are required to give permission for their children to participate. It is imperative that parents, teachers, and nurses who are involved in sex education know the facts and answer each question honestly. It is helpful to find out what children already "know"; often children have some information, accurate or not. The information should be aimed at the child's immediate interest. The school-age child is interested more in physical aspects and is not ready to deal with the affective component of sex. Fourth- and fifth-graders are much more interested in conception and birth than in sex per se. They are often fascinated by the unusual and will want to know, for example, why and how twins occur and what causes Siamese twins.

Fifth- and sixth-graders are very aware of the changes occurring in themselves and are keenly interested in learning about their bodies and what is going on. It is important that they learn not only what is occurring in their bodies but also what changes are taking place in children of the opposite sex.

Since it has been documented that some children engage in sexual experimentation at an early age, information about birth control should be provided. This is best handled on an individual basis since many children are not ready for this information.

Respiratory Illness

The nurse may encounter children who suffer from repeated respiratory infections or who seem to have runny noses constantly. These children should be referred to a physician for careful evaluation to rule out the possibility of a more serious underlying illness. Most of these children will have some form of allergic sinusitis, which responds well to treatment. It is helpful to review with parents the treatment of minor respiratory illnesses as well as the more serious symptoms, which require medical attention. Since sore throats are a common complaint of children, it is useful to explain the danger associated with streptococcal infections and why a throat culture is useful. Parents should be encouraged to seek medical care if the child has a lingering sore throat, a low grade fever that persists after the sore throat has cleared, or complaints of joint pains, since these are all possible symptoms of rheumatic fever.

Behavior Problems

It is helpful to discuss behavior problems with parents and to assist them in viewing their child's problems in the proper perspective. If the undesired behavior is only occasionally displayed the nurse can suggest ways the parents may cope with the undesired behavior as well as ways to reinforce desired behaviors. When the undesirable behavior is excessive or totally unacceptable, the health care worker may provide information about behavior modification and assist the family in setting up a behavior modification program designed to eliminate the undesired behavior. When working with a family in which a behavior problem exists it is necessary to assess the total family dynamics to identify the factors that contribute to the problem. It is often helpful to ask the parents to keep a diary for one or two weeks in which they record the number of times the behavior occurs as well as the time of day and the event surrounding the behavior, and the parent's response.

Occasionally behavior problems in children will be indicative of more serious underlying emotional problems in the child and/or the family. In these cases the nurse can assist the family in locating the health care provider who is best qualified to treat the problem and provide emotional support for the family.

Health Maintenance

Nurses should assess the immunization status of school-age children and assist families in obtaining immunizations. It is helpful to review the times when booster immunizations are necessary and the reasons why these are needed.

The nurse should address both the parents' and the child's concerns about physical growth and development. Even when children are growing and developing normally, they may have concerns about their bodies. School-age children who are advanced in size for age often need support and reassurance that they are normal and that their peers will eventually catch up with them. Children with extremely advanced growth should be referred for further evaluation. Children who are over- or undernourished merit further assessment and intervention. The undernourished child may require counseling regarding the selection of nutritious foods that contribute to growth. Sometimes, as with the overnourished child, the problem is simply an imbalance between the amount of food taken in and the energy expended. Intervention should be effected as quickly as possible

with children who are gaining weight too rapidly; just as with adults, it is much easier to lose 10 pounds than it is to shed 20 or 30 pounds. The nurse can assist the parents and the child in a careful diet assessment and make recommendation for dietary alterations. The health care provider may also need to make suggestions on ways in which children can increase their activity level and thus increase their energy expenditure.

Allergies

When working with children and parents of children with allergies the nurse should assess how the child and the family are managing the allergy. Does the allergy restrict the child's activity; does it interfere with his or her progress in school? Since it is known that asthma attacks can be precipitated by emotional factors, it is helpful to identify when asthmatic children have attacks, what was happening in the child's life at the time, and the parents' response to the attack. Having the parents keep a written record is useful for a child who has frequent attacks. The nurse can use the record to review the situation with the family and make suggestions for altering the emotional environment. Asthma attacks are also frequently precipitated by infection. Both the parents and the child should be reminded of the need for early and aggressive treatment of even minor respiratory infections. The frequency of asthma attacks often diminishes when the child reaches late childhood. Some parents will have been overprotective of the child with asthma, and as the attacks decrease or subside, such parents will need encouragement from the health care provider to allow the child to become more involved in physical activities.

The Transition From Childhood to Adolescence

During middle childhood the body is preparing for the changes that will occur during adolescence. Just as the body prepares physically for these changes, so should health care providers assist children and their families in preparing for these changes emotionally. The health care provider may introduce the topic by discussing the physical changes that will occur in the next few years and then progress to emotional and social changes. If possible, the health care provider should talk with the parents in private and allow them to express their concerns about the child. Just as some parents have concerns about letting go of the preschooler, some parents are hesitant of letting go of their child. Many parents are fearful of what will happen to their child during adolescence. The nurse can aid the family in identifying ways to gradually increase the child's responsibilities and decision-making ability. It is much less traumatic for both the child and the family if such issues can be discussed and planned in advance rather than confronted only when the child demands increased freedom. For example, the parents and the child may discuss and agree upon a curfew when the child enters junior high school and begins to attend nighttime group activities. It is also most appropriate for the family to discuss smoking, drugs, and alcohol. The health care provider can encourage parents to discuss such topics with their children and serve as a resource person for the family.

Hints on Working with the School-Age Child

School-age children are generally less fearful of nurses and other health care providers than preschool children. They are much more aware of and responsible for their own bodies and are thus interested in what the nurse does and why. It is important to acknowledge the child's cognitive level; when collecting data for the history, allow the child to provide as much information as he or she can. Such data can be validated with the parents.

Whenever the nurse is doing a procedure, she should explain it and the rationale for it. Older school-age children will be studying the body in science and health classes and often have numerous questions. By answering them, the nurse can teach the child and build rapport.

Whenever possible, the same staff member should work with the same child since this allows for good rapport. However, school-age children can handle interactions with more than one provider relatively well. Hospitalization, if necessary, is easier for children of this age than for younger children since they are now capable of reason and can understand the concept of time. Separation anxiety is usually minimal; school-agers develop friendships with other patients and staff, and can enjoy the social aspects of hospitalization if they receive adequate preparation and emotional support. The occasional immature school-age child may react to hospitalization by regressing. The likelihood of regression is reduced if the child is prepared for treatment and given the opportunity to participate in his own care.

It is important to foster growth and development when children are hospitalized for a prolonged period of time. Attempts should be made to develop routines for these children that foster social, academic, and motor development. School-agers should be introduced to other children of similar age and some opportunity should be provided for group activities. This may be done through playroom activities or informal gatherings on the ward.

It is imperative that school-age children keep up with their school work if at all possible. Most larger hospitals employ teachers who work with hospitalized children. If such a service is not available, the nurse can contact the child's teacher and make arrangements to receive homework assignments. Nurses may need to help children plan times during the day when they will do school work. Children who are hospitalized for a prolonged time often become discouraged and need encouragement and positive feedback from the staff.

It is also important for the school-age child to maintain contact with peers. In many hospitals children under 12 years are not allowed to visit although this rule is often "bent" for children who are hospitalized for a long period. If visiting is not possible, children can correspond with their classmates via telephone or mail.

School-age children do not usually require the constant presence of their parents but need regular contact with the family for reassurance. If the child is acutely ill or regressed he may benefit from the parents' rooming-in.

School-age children should be given as much responsibility for self-care as possible. When preparing the school-age child for surgery or other hospital procedures, it is helpful to use diagrams, drawings, or models. Children are eager for knowledge and are interested in learning the correct terminology.

SUMMARY

The school-age years (middle childhood), the years from 6 to 12, are characterized by gradual but steady change. All of the organ systems except the reproductive and endocrine systems are functionally mature at the onset of middle childhood. During the school years, children gain greater coordination and muscle control.

The school-age years are hallmarked by many advances in psychosocial growth. As children enter school they spend a large part of the day outside the home and are greatly influenced by people outside the family. As a result, children of this age often question the goals and standards of behavior established by their parents. Such behavior, while sometimes frustrating to parents, is a normal phase of psychosocial growth.

School-age children are very involved with their peers. At the time children enter school they play readily with children of both sexes, but there is a gradual shift to play groups of the same sex. The chum stage, occurring at around 9 to 10 years, is the child's first love attachment

outside the family. The friend, an age-mate of the same sex, becomes as important to the child as the self.

Freud saw the school-age child as being in the latency period, a relatively stable period of development characterized by intellectual curiosity and the development of mental abilities. According to Erikson the major conflict during this period is the development of industry versus inferiority. In Piaget's terms, 6-year-olds are in the preoperational period of cognitive development, while the years from seven to 11 constitute the stage of concrete operations, i.e., those that involve systematic reasoning about situations that are familiar to the child.

The most frequently encountered illnesses of middle childhood are respiratory infections, usually the common cold. Accidents remain the leading cause of death for children of this age and one of the leading causes of morbidity. Many minor injuries are the result of unsupervised play, including bicycling. Behavior problems such as aggression, lying, and stealing may be present in some children.

A small percentage of children will have school-related problems. Brutten, Richardson, and Mangel (1973) estimated that 10 percent of school children have some form of learning disability. A small percentage of children will have difficulty getting along with other children, and a very small number of children will experience school phobia.

Since school-age children are very interested in learning about their bodies, health care providers can capitalize on this interest and help children to develop good health habits. When hospitalization is necessary it is usually less traumatic to the school-age child, since children of this age often enjoy the social aspect of the hospital experience.

REFERENCES

Arena, J. M. *Dangers to children and youth.* Durham, N.C.: Moore, 1971.

Bernard, H. W. *Human development in western culture* (2nd ed.). Boston: Allyn and Bacon, 1975.

Bigley, N. J. *Immunologic fundamentals*, Chicago: Year Book, 1975.

Bower, E. W., and Nash, C. L., Jr. Evaluating growth and posture in school-age children. *Nursing 79*, 1979, *9*(4), 58–63.

Brutten, N., Richardson, S., and Mangel, C. *Something's wrong with my child: a parent's book about children with learning disabilities.* New York: Harcourt Brace Jovanovich, 1973.

Chinn, P. *Child health maintenance* (2nd ed.). St. Louis: Mosby, 1979.

Cohen, D. Is television a pied piper? *Young Children*, 1974, *30*, 4–14.

Erikson, E. *Childhood and society* (2nd ed.). New York: Norton, 1963.

Feingold, B. F.: *Why your child is hyperactive.* New York: Random House, 1975.

Feshbach, S., and Singer, R. D. *Television and aggression.* San Francisco: Jossey-Bass, 1971.

Fomon, S. *Nutritional disorders of children: prevention, screening, and follow-up.* Washington, D.C.: U.S. Government Printing Office, DHEW Publication No. (HSA) 76-5612.

Frazier, C. *Coping with food allergy.* New York: Quadrangle, 1974.

Freiburg, K. *Human development.* North Scituate, Mass.: Duxbury Press, 1979.

Freud, S. *Beyond the pleasure principle.* New York: Bantam, 1959.

Garn, S., and Clark, D. Ten state nutrition survey of the United States. In *Report on the second Wyeth symposium.* New York: Wyeth Laboratories, 1976.

Ginsburg, H., and Opper, S. *Piaget's theory of intellectual development.* Englewood Cliffs, N.J.: Prentice-Hall, 1969.

Goodglass, H., Gleason, J. G., and Hyde, M. R. Some dimensions of auditory language comprehension in aphasia. *Journal Speech Hearing Research*, 1970, *13*, 595–606.

Gordon, I. *Human development.* New York: Harper and Row, 1975.

Greenberg, S. F. Attitudes toward increased social, economic and political participation by women as reported by elementary and secondary students. Paper presented to AERA Convention, Chicago, 1972.

Guyton, A. C. *Textbook of medical physiology* (5th ed.). Philadelphia: Saunders, 1976.

Hartup, W. W. Peer interaction and the behavioral development of the individual child. In Schopler, E., and Reichler, R. J. (Eds.), *Psychopathology and child development.* New York: Plenum, 1976.

Helms, D., and Turner, J. *Exploring child behavior: basic principles.* Philadelphia: W. B. Saunders, 1978.

Hoffman, M. L. Moral development. In Mussen, P. (Ed.), *Carmichael's manual of child psychology* (3rd ed.). New York: John Wiley & Sons, 1970.

Holm, V., and Wiltz, N. Childhood. In Smith, D., and E. Bierman (Eds.), *The biologic ages of man from conception through old age.* Philadelphia: Saunders, 1978.

Humphrey, P. Weight/height disproportion in elementary school children. *The Journal of School Health* 1979, *49*, 25–29.

Hurlock, E. *Child development* (3rd ed.). New York: McGraw-Hill, 1956.

Johnson, T. R., Moore, W. M., and Jefferies, J. E. *Children are different: developmental physiology* (2nd ed.). Columbus, Ohio: Ross Laboratories, 1978.

Kaluger, G., and Kaluger, M. *Human development: the span of life* (2nd ed.). St. Louis: Mosby, 1979.

Kastenbaum, R. *Humans developing.* Boston: Allyn and Bacon, 1979.

Kohlberg, L. *Stages in the development of moral thought and action.* New York: Holt, Rinehart and Winston, 1969.

Kohlberg, L., and Zigler, E. The impact of cognitive maturity on the development of the sex-role attitudes in the years 4 to 8. *Genetic Psychological Monographs* 1967, *75*, 84–165.

Kunin, C. M. *Detection, prevention and management of urinary tract infections.* Philadelphia: Lea & Febiger, 1972.

Kravitz, H. Preventing injuries from bicycle spokes. *Pediatric Annals*, 1977, *6*, 713–716.

Liebert, R. M., Neale, J. M., and Davidson, E. S.: *The early window, effects of television on children and youth.* New York: Pergamon Press, 1973.

Liebert, R. M., and Neale, J. M. Television violence and child aggression. *Psychology Today*, 1972, *5*, 38–40.

Londe, S., and Goldring, D. High blood pressure in children: problems and guidelines for evaluation and treatment. *American Journal of Cardiology*, 1976, *37*, 650–657.

Lowery, G. H. *Growth and development of children* (6th ed.). Chicago: Yearbook Medical Publishers, 1975.

McNeill, D. The development of language. In Mussen, P. (ed.). *Carmichael's manual of child psychology* (3rd ed.). New York: John Wiley & Sons, 1970.

Meredith, H. *Human body growth in the first ten years of life.* Columbia, S.C.: The State Printing Co., 1977.

Mofenson, H., and Greensher, J. Childhood accidents. In Hoekelman, R., et al. (Eds.), *Principles of pediatrics.* New York: McGraw-Hill, 1978.

Murray, R., and Zentner, J. *Nursing assessment and health promotion through the life span.* Englewood Cliffs, N.J.: Prentice-Hall, 1979.

Neumann, L. Obesity in the preschool and school-age child. *Pediatric Clinics of North America*, 1977, *24*, 117–122.

Papalia, D., and Olds, S. *Human development.* New York: McGraw-Hill, 1978.

Petrillo, M., and Sanger, S. *Emotional care of hospitalized children.* Philadelphia: Lippincott, 1980.

Piaget, J. *Judgment and reasoning in the child* (M. Gabain, trans.). New York: Harcourt, Brace, and Jovanovich, 1932.

—. *The moral judgment of the child* (M. Gabain, trans.). Glencoe, Ill.: Free Press, 1948.

—. *The child and reality*. New York: Grossman, 1973.

Schiffer, C., and Hunt, E. *Illness among children*. U.S. Children's Bureau Publication, No. 405. Washington, D.C.: U.S. Government Printing Office, 1963.

Schuster, C. S., and Ashburn, S. S. *The process of human development*. Boston: Little, Brown, 1980.

Salkurt, E. E. (Ed.). *Physiology* (4th ed.). Boston: Little, Brown, 1976.

Soares, A., and Soares, L. Self-perceptions of culturally disadvantaged children. *American Educational Research Journal*, 1969, *6,* 31–45.

Sullivan, H. S. *Conceptions of modern psychiatry* (2nd ed.). New York: Norton, 1953.

Sutterly, D., and Donnelly, G. *Perspectives in human development*. Philadelphia: Lippincott, 1973.

Swainman, K. F. Brain development in the middle childhood years. *Journal of School Health*, 1978, *48,* 289–292.

Tanner, J. M. *Foetus into man: physical growth from conception to maturity*. London: Open Books, 1978.

Trowbridge, H., and Trowbridge, L. Self-concept and socioeconomic status. *Child Study Journal*, 1972, *2,* 123–149.

Turner, J., and Helms, D. *Life span development*. Philadelphia: Saunders, 1979.

U.S. Consumer Product Safety Commission. *Bicycles*. Fact Sheet No. 10. Washington, D.C.: n.d.

Valadian, I., and Porter, D. *Physical growth and development from conception to maturity*. Boston: Little, Brown, 1977.

Vaughan, V. C. III, McKay, R. J., and Buchman, R. E. (Eds.): *Nelson's textbook of pediatrics* (11th ed.). Philadelphia: Saunders, 1979.

Waller, J. Bicycle ownership: use and injury patterns among elementary school children. *Pediatrics*, 1971, *47,* 1042–1050.

Wheatley, G. Childhood accidents. *Pediatric Annals*, 1977, *6,* 12–26.

Education U.S.A. *Bulletin of University of Southern California*, October 1973.

9

The Adolescent

PATRICIA HUMPHREY

Adolescence is characterized by change—most obviously, by physical changes, but also by changes in relationships and changes in values. Adolescence begins with *pubescence,* or *puberty,* which is marked by rapid physiological growth, maturation of the sex organs and reproductive functions, and the appearance of the secondary sex characteristics. Pubescence usually covers a two-year span.

It is difficult to establish when adolescence begins since its timing is very individualized. Females generally enter pubescence at around 10 or 11 years, while males enter this period at 11 to 12 years. It is equally difficult to identify the end of adolescence as it also varies greatly, depending not only on the individual but also on cultural definition. In American society,

adulthood may be said to begin when one is self-supporting, has chosen a career, and has established one's own home. Individuals attain psychological adulthood when they become independent from their parents, develop their own system of values, and are able to form mature friendships and loving relationships. Thus, the length of adolescence can vary greatly depending on the individual. It must be recognized that some people never leave adolescence regardless of their chronological age.

Not until the twentieth century, after Darwin's presentation of the theory of evolution, was adolescence considered a stage in human development. As a developmental phenomenon adolescence is unique to the higher primates; the exclusively human maturational pattern of a delay in the attainment of full growth and sexual maturity appears essential to human development and therefore suggests that adolescence is an important evolutionary trait.

Even though adolescence covers a span of less than ten years, there are great differences in children at different periods during adolescence. It is helpful to divide the period into early, middle, and late adolescence.

Early adolescence, or preadolescence, is the period from 11 to 13 years that includes the prepubertal "growth spurt," which marks the onset of puberty. Early adolescence is characterized by instability and emotional upheaval. Early adolescents frequently have great difficulty relating to their families, and parents often find coping with the child extremely difficult. Fritz Redl (1975) states "During preadolescence the well-knit pattern of a child's personality is broken up or loosened, so that adolescent changes can be built into it and so that it can be modified into the personality of an adult." p. 265. According to this theory, some of the repressed and forgotten impulses of earlier childhood resurface, which explains the silly antics and irritating behavior displayed at this time. Previously developed values and standards are no longer operative; thus the child of this age has less self-control and may appear highly disorganized. Redl also states "During preadolescence it is normal for youngsters to drop their identification with adult society and establish a strong identification with a group of their peers" (Redl, 1975 p. 266). At no other age do youngsters show such a need for group formation and identification with the peer group (Fig. 9-1). Until this period in life the youngster has lived by adult codes and values; now the youngster must establish his own values and codes. Group identification and support aid the child in developing a personal value system.

Middle adolescence (14 to 16 years) is characterized by attraction to the opposite sex. Conformities to peer group standards and ideologies

Figure 9-1. *The need for identification with a peer group is strongest during adolescence. (Photo by John Elkins.)*

are still dominant. Development of a personal value system is an important task of middle adolescence. The middle adolescent is often ambivalent and prone to mood swings.

Late adolescence (17 to 20 years) is a relatively calm period by which time physical growth has slowed and the individual has adapted to a new body image. The individual is on the way to knowing "who am I" and is relatively secure in relationships with peers of both sexes. Relationships with parents are smoother and the adolescent can see the parents' point of view and accept their help in planning for the future. The individual learns rapidly during this period, refining interpersonal skills. The late adolescent has acquired more mastery over drives and emotions. The major developmental task is preparation for adulthood.

Adolescents who find the transition into adulthood especially difficult fall into four major categories: (1) adolescents who must enter the work world early and are forced to become adults almost overnight; (2) students who must depend on others for economic sup-

Table 9-1. Sex Maturity Ratings (Tanner Ratings), Male

Stage	Pubic Hair	Penis	Testes
1	None	Preadolescent	Preadolescent
2	Scanty, long, slightly pigmented	Slight enlargement	Enlarged scrotum, pink, texture changed
3	Small amount, darker, begins to curl	Increases in length	Increases in size
4	Begins to resemble adult type, but still scanty; coarse, curly	Becomes larger	Becomes larger, scrotum becomes darker
5	Adult distribution, spread to medial surface of thighs	Adult characteristics	Adult characteristics

SOURCE: Reproduced with permission from Daniel, W. A., Jr. *Adolescents in health and disease*. St. Louis: C. V. Mosby, 1977; adapted from Tanner, J. M. *Growth at adolescence* (2d. ed.). Oxford: Blackwell Scientific Publications, 1962.

port until the early years of adulthood; (3) females whose parents have encouraged them to be dependent; and, (4) children who have not mastered the developmental tasks of childhood and are ill prepared at adolescence to make the changes necessary for a smooth transition into adulthood (Hurlock, 1973).

PHYSIOLOGICAL GROWTH AND DEVELOPMENT

During puberty the child secretes growth hormones while awake as well as while asleep and there are larger and more numerous peaks of secretion as puberty advances (Johnson, Moore, and Jeffries, 1978). The progress of physical change during adolescence can be assessed with J. M. Tanner's Sex Maturity Ratings (SMRs) (Tables 9-1 and 9-2). All normal adolescents proceed through the sequence of changes described in the table, although the age at which the events occur varies. Using the Sex Maturity Ratings allow the health care provider to identify where the client is in the process of physical development, and to answer questions the adolescent or the parent may have regarding the growth process.

Table 9-2. Sex Maturity Ratings (Tanner Ratings), Female

Stage	Pubic Hair	Breasts
1	Preadolescent	Preadolescent
2	Sparse, slightly pigmented, straight, at medial border of labia	Breast and papilla elevated as small mound; areolar diameter increased
3	Darker, increased amount and beginning to curl	Breast and areola enlarged, without contour separation
4	Coarse, curly, abundant but less than in adult	Areola and papilla form secondary mound
5	Adult "feminine triangle," spread to medial surface	Mature; nipple projects; areola part of general breast contour

SOURCE: Reproduced with permission from Daniel, W. A., Jr. *Adolescents in health and disease*. St. Louis: C. V. Mosby, 1977; adapted from Tanner, J. M. *Growth at adolescence* (2d. ed.). Oxford: Blackwell Scientific Publications, 1962.

Changes in Height and Weight

Skeletal growth is regulated by pituitary growth hormones, but thyroid hormones must also be present for the process to occur. Skeletal maturation occurs through a combination of lengthening of the bones and changes in the ossification centers.

Before the adolescent growth spurt children will have reached 75 to 80 percent of their adult height and approximately 50 percent of adult weight. The female growth spurt begins approximately 2 years earlier than the male, is less intense, and is of shorter duration. It is thought that this earlier and less dramatic growth spurt in females may account for their smaller stature, since they are virtually equal in stature or larger before the growth spurt. In females the growth spurt usually occurs late in SMR Stage 2 or early in Stage 3. Girls are initially taller than boys (between 11 and 13 years); and reach a plateau of height by age 17 (Tanner, 1970).

In males the growth spurt occurs late in SMR Stage 3 or early in Stage 4. Between the ages of 14.5 to 15 years most males enter a period of rapid development by the end of which they surpass girls in height. By age 21, most males have reached their adult height (Roache and Davike, 1972).

The adult height of a child who experiences the growth spurt at an early age is likely to be less than that of the child whose growth spurt occurs later in adolescence. The gain in weight in adolescence follows the same pattern as the acquisition of height.

Cardiovascular Development

The heart continues to increase in size until age 17 or 18. The EKG pattern should resemble that of an adult. From 12 years on, boys have larger transverse cardiac diameters. In females the increase in cardiac size occurs around the time of menarche (Smith and Bierman, 1978). The pulse rate continues to decrease in both sexes,

but is greater in males. The mean pulse rate per minute during adolescence is 82 (Johnson, Moore, and Jeffries, 1978). Systolic blood pressure rises in both sexes, although females experience the rise earlier. Diastolic pressures show little change during adolescence and no sex difference. The mean blood pressure during adolescence is 120/65 mm Hg.; 130/90 should be considered the upper limit of normal between 12 and 18 years. Any client with a reading above 130/90 on three or more occasions should be referred for further evaluation.

Pulmonary Development

There is a gradual increase in lung weight in both sexes during adolescence; adult lung weight is achieved by about 17 years. Vital capacity increases rapidly during adolescence. The vital capacity of the lungs is greater in males even when corrected for differences in body size and surface area. This sex difference increases rapidly during the adolescent years (Smith and Bierman, 1978). The respiratory rate decreases steadily throughout childhood and continues to do so throughout puberty. The respiratory rate averages 17 to 22 respirations per minute during early adolescence and reaches adult levels of 15 to 20 per minute between 15 and 19 years, (Johnson, Moore, and Jeffries, 1978).

Digestive System Development

The organs of the digestive system undergo rapid growth and reach their mature size and shape during late adolescence. The stomach becomes longer and less tubular, and increases in capacity. The intestines increase in both length and circumference. The smooth muscles of the intestinal walls and the stomach become thicker and stronger, resulting in stronger peristaltic movement (Hurlock, 1973). During the adolescent years the liver attains adult size, location, and function.

The basal metabolism rate (BMR) declines

during adolescence. The decline is less in males, probably because of the greater muscular development, which requires more oxygen consumption. Hormonal differences may also play a part in the sex differences.

The Nervous System

Growth of the structures of the nervous system is completed before adolescence and is therefore not involved in the growth spurt. Myelination of the greater cerebral commissures and the reticular formation of the central nervous system continues. The electrical activity of the brain of an awake 12-year-old, as recorded on an electroencephalogram, is dominated by low-voltage, fast activity in the frontal and temporal leads. By 16 years the pattern is similar to that of an adult.

Musculoskeletal Development

Bones increase in length, weight, and density; and grow more rapidly than the supporting muscles. Large muscles develop faster than small muscles. Skeletal weight for both males and females increases throughout puberty, but the increase appears to be greater in males (Katchadourian, 1977). Muscle growth in males continues during late adolescence because of androgen production, while muscle growth in females is less obvious and is proportional to the growth to other tissues. The overall increase in strength and exercise tolerance is greater in males. Females appear to experience peak growth in muscle strength at about 12.5 years while males reach this peak much later (Anyan, 1978). However, many outstanding female athletes appear to have capacities equal to those of their male counterparts. As more communities and schools provide more exercise and sports programs for females, we may see more equal development of exercise tolerance and strength in the next generation.

Ossification proceeds in a fairly orderly sequence. Leg length accelerates first, followed by hip width, chest width, shoulder width, trunk length, and chest depth. Data suggest that growth in the peripheral parts of the limbs is more advanced than growth in the proximal parts. Two constituents of stature, the peaks of total leg length and total trunk length, are separated by approximately one year. The ratio of trunk length to leg length always increases during adolescence; the increase in trunk length appears to be more responsibile for the spurt in height than the increase in leg length. The increase in hip width in females is caused mainly by changes in the pelvic bones.

Estrogen influences ossification and early fusion of the epiphyses with the shafts of long bones, resulting in shorter stature in females. Ossification occurs later in males, accounting for their arms and legs being longer in proportion to trunk size, and for the greater shoulder width.

These rapid and unequal changes in the musculoskeletal system may at times produce an unusual appearance; for example, the rapid growth of facial bones may produce a temporary thickening and coarsening of features, and the rapid growth of the hands and feet may at one point result in a disproportional appearance. Increased skeletal growth and the simultaneous lag in muscle growth may result in poor posture and decreased coordination.

Hormonal Changes

The biological changes associated with puberty are initiated by the liberation of hormonal secretions in response to hypothalamic stimuli. At this time the pituitary glands send out messages to certain parts of the body, resulting in the secretion of thyroid-stimulating hormone (TSH), adrenocorticotropic hormone (ACTH), gonadotropic hormone, and growth hormone. These substances then stimulate other endocrine glands to produce their own growth-related hormones (Figure 9-2).

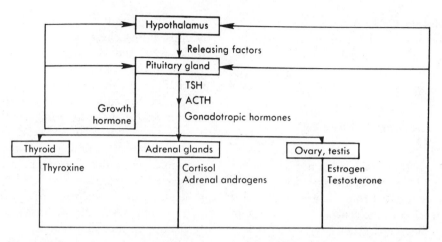

Figure 9-2. *Endocrine control at adolescence. Activating hormones from the pituitary gland stimulate other endocrine glands, activating their own growth-related hormones. These hormones in turn provide for orderly and progressive physical and psychological changes in adolescence. (Reproduced with permission from Humphrey, P., and Hewitt, D. Adaptation throughout the Life Cycle. In Phipps, W. J., Long, B. C., and Woods, N. F. (Eds.). Medical-surgical Nursing. St. Louis: C. V. V. Mosby, 1979).*

The sex hormones regulate the structure and function of sex organs as well as the appearance of secondary sex characteristics (Figs. 9-3 and 9-4). Androgens are hormones that promote the development and maintenance of male secondary sex characteristics and structures; testosterone is the principal androgen. Estrogens, found principally in the ovaries and also in the placenta, stimulate the accessory sex structures and the secondary sex characteristics in females. Progesterone is the female hormone that prepares the uterus to accept a fetus and maintain pregnancy. The cessation of the secretion of progesterone at the end of the menstrual cycle largely determines the time of onset of menstruation. Both sexes have certain amounts of all three sex hormones; males produce greater amounts of androgens while females produce greater amounts of estrogen and progesterone. Urinary estrogens show a gentle increase in premenarche, a plateau for 18 months postmenarche, then a further and steeper increase in late puberty. In females serum progesterone levels are low during premenarche, rise slowly two years postmenarche, and then show a sudden increase to adult levels in SMR Stage 5; this reflects the maturation of the ovaries, which by State 5 are ovulating regularly in response to increased levels of luteinizing hormone. Longitudinal studies in males have shown increasing serum testosterone levels in SMR Stages 2 to 4 (Johnson, Moore, and Jeffries, 1978).

In both sexes the sequence of sexual development is fairly predictable. In females this process starts with the development of breast buds; at the same time the reproductive organs are maturing. This is followed by the appearance of downy pubic hair over the mons pubis, the onset of menstruation, the continued development of the breasts and pubic hair, and the characteristic widening of the hips (Pierson and D. Antonio, 1974). Breast development generally begins between 10 and 13 years; mature size is generally reached over a span of three years (Katchadourian, 1977).

Menarche, the onset of menstruation, generally occurs following the initial growth spurt, during a cessation of growth, provided the fe-

sterility lasting from 1 year to 18 months after menarche. Females are generally ovulatory at SMR 5.

Since 1830, the age of menarche has decreased steadily (Fig. 9-5). This change in menarcheal age is observed in many countries and in many races. One explanation, offered by Frisch, is that since critical body weight triggers menarche, and children are bigger than their counterparts of the past, girls achieve critical weight sooner (Frisch, 1974).

In males the growth of the testes begins to

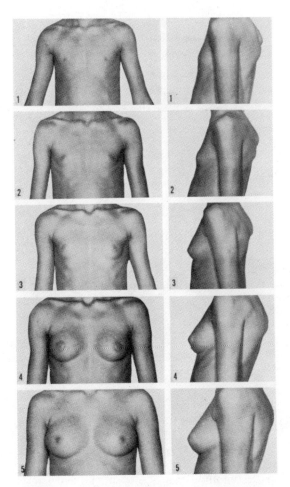

Figure 9-3. *Development of the breasts, front and side views; the numbers correspond to the Sex Maturity Ratings (SMRs) shown in Table 9–2. (Reproduced with permission from Tanner, J. M., Growth at adolescence, 2d. ed., Oxford: Blackwell Scientific Publications, 1962.)*

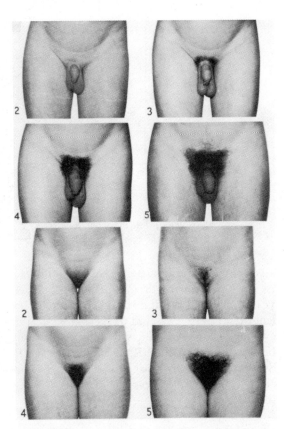

Figure 9-4. *Development of the male and female genitals and pubic hair; the numbers correspond to the Sex Maturity Ratings (SMRs) shown in Tables 9–1 and 9–2. (Reproduced with permission from Tanner, J. M., Growth at adolescence, 2d. ed., Oxford: Blackwell Scientific Publications, 1962.)*

male's body is 24 percent fat, and that she weighs no less than 94 lb; this must be achieved in order for regular menses to occur (Martin, 1978). Most American girls experience menarche by 12.8 years; the normal range is 10 to 15 years (Zacharias, Rand, and Wurtman, 1976). The early menstrual cycles in some girls are more irregular than later ones and are often anovulatory. There is often a period of adolescent

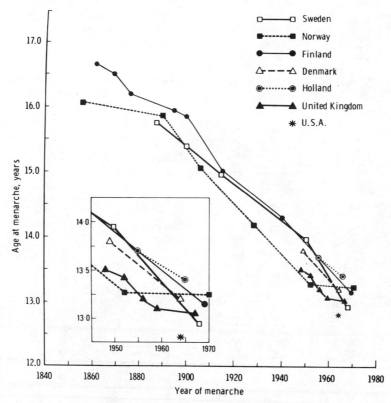

Figure 9-5. *Secular trend in age of menarche, selected countries, 1830 to 1980. (Reproduced with permission from Tanner, J. M., Growth at adolescence, 2d ed., Oxford: Blackwell Scientific Publications, 1962.)*

accelerate by approximately 11 years, followed by accelerated growth of the penis by 12.5 years. On the average the penis doubles in length between 12.5 and 17 years, while the average volume of the testes increases tenfold between ages 12 and 19 (Rogers, 1972). At SMR Stage 3 the male is capable of ejaculation, and at SMR Stage 4 he is capable of impregnating the female.

Vision

Visual development is complete before adolescence. It is important to continue routine vision screening in order to identify and treat any problems. Because the adolescent is very con-

cerned with body image, those with eye defects may not wish to wear glasses. They need to have their feelings recognized and respected by adults, who should try to encourage them to wear glasses. Many adolescents may wish to switch to contact lenses. Of course, these are often expensive and are easily lost. It may be helpful to assist the adolescent in planning ways to earn money and save up for contact lenses.

Skin

The apocrine sweat glands (specialized sweat glands found mainly around the nipples, armpits, anus, and genitals) begin to be active between ages 8 and 10. These glands become more active in response to emotional, mechanical, and pharmacological stimuli during adolescence. Eccrine glands may be less numerous than during the school years but are fully developed and respond to emotional stimuli. These glands tend to be more active in males. Sebaceous glands increase in size and also become more active.

Total Body Water

Body water decreases gradually after infancy and reaches adult levels at between 17 and 18 years. In females 50 percent of the total body weight is fluid; in males the figure is 60 percent fluid (Chinn; 1979). This difference is caused by the greater percentage of muscle mass in the male and the greater percentage of fat in the female.

Racial Differences in Physiological Development

The health care provider needs to be attuned to racial differences in physical development. Black children attain a greater proportion of their adult stature earlier, although in the United States adult stature for blacks and whites is about the same. Skeletal mass is greater in blacks; therefore, the use of norms for whites could result in failure to detect bone loss when assessing blacks (Garn and Clark, 1976). Black females tend to develop secondary sexual characteristics earlier than their white counterparts, and reach the menarche earlier. The normal value for hemoglobin concentration is one gram less for blacks (Garn and Clark, 1976).

Nutritional Needs

The rate of physical growth during adolescence is second only to that of infancy. Thus, there is a greatly accelerated demand for calories and nutrients.

Growth during adolescence is highly individualized; the health care provider will therefore need to adjust the nutrient allowances for each client. The SMR scale also provides a useful guide to nutritional needs since the highest nutrient and energy demands occur at the peak of growth, which is at SMR Stage 3 for the female and Stage 4 for the male. Males tends to deposit more lean body mass and therefore have an increased need for protein. Age or weight alone is not a useful predictor of energy needs. If one must use a single measure, height best expresses energy requirements since it correlates well with physiologic development. Table 9-3 shows the recommended daily dietary allowances for adolescents.

Rebellion against parents in the process of striving for independence may lead to fad diets, irregular eating, and skipping meals. With increased independence the adolescent is often out of the home for more time and makes more decisions regarding his or her diet. Peer pressure often influences the selection of foods. Snacking up to 4 or 5 times a day becomes a part of life and may provide up to $\frac{1}{3}$ of the daily intake.

Because of the adolescent's concern with body image, exploited by advertising and the mass media, females often want to lose weight and may resort to fad diets which, if adhered to for long, can compromise growth. Males often want to gain weight and are especially concerned with having larger muscles, and may resort to high-protein diets and pills in an attempt to increase muscle mass.

Adolescents are susceptible to food advertising and new ideas. In their attempts to try new things and to be accepted they may espouse vegetarianism or other fad diets. In order to be nutritionally sound, such a diet must be carefully planned and varied, and most adolescents are not motivated to do this.

Other activities and behaviors during adolescence also affect food intake. Smoking, alcohol, and drugs affect the appetite and metabolism. Females taking oral contraceptives probably need increased amounts of folic acid and vitamin

Table 9-3. Recommended Daily Dietary Allowances

			Males				
Age	Weight (kg/lb)	Height (cm/in)	Energy Needs (kcal)	Protein (gm)	Calcium (mg)	Iron (mg)	Vitamin D (μg)*
11–14	45/99	157/62	2,000–3,700	45	1,200	18	10
15–18	66/145	176/69	2,100–3,900	56	1,200	18	10
19–22	70/154	177/70	2,500–3,300	56	800	10	7.5
			Females				
11–14	46/101	157/62	1,500–3,000	46	1,200	18	10
15–18	55/120	163/64	1,200–3,000	46	1,200	18	10
19–22	55/120	163/64	1,700–2,500	44	800	18	7.5

* As cholecalciferol; 10 μg cholecalciferol = 400 I.U. Vitamin D.

SOURCE: Recommended dietary allowances. Washington, D.C.: Food and Nutrition Board, National Academy of Sciences–National Research Council, Revised 1979.

B_{12}, but as yet there is no multiple vitamin complex that supplies the nutrients needed by women taking the pill.

Dental Care

Most permanent teeth are present by 12 years of age, except for the third molars, which erupt between 17 and 21 years. During the school years children should be taught to care for their teeth and should enter adolescence with good dental hygiene habits. During adolescence the client enters another peak in the appearance of dental caries. An increase in dental caries begins around 10 years, peaks at 16 years, and gradually declines in early adulthood.

PSYCHOSOCIAL DEVELOPMENT

Although adolescents reach physical maturity at different times, all proceed through the phases of physical development in an orderly,

predictable sequence. The same cannot be said of psychosocial development, which is characterized by many mood swings and is usually a stormy period for even the most normal child.

C. Stanley Hall is considered the father of a psychology of adolescence; he introduced his work in 1916. Hall explained that the difficulties of adolescence are the result of biological maturation and are therefore inevitable. Anna Freud (1965) and Peter Blos (1970), influenced by Freudian theory, recognized that psychological changes during adolescence are as important as biological changes. Freud and Blos view typical adolescent behavior patterns as reactions to both the physical changes that the adolescent undergoes and residual unresolved psychological problems of childhood. The viewpoints of these early theorists hold that adolescence is a stage that all youngsters pass through and is a universal developmental phenomenon.

The acknowledgment of the role of cultural and societal factors by more contemporary theorists adds to an understanding of adolescence. Cultural anthropologists have docummented the differences in adolescence in various cultures; in more primitive cultures in which the transition from childhood to adulthood occurs in a short time, individuals do not appear to have such traumatic psychological conflicts. The prolonged period of dependence in modern society may be a factor in these conflicts.

From a sociological point of view, adolescence can be seen as a transition from dependence to independence. In order to become an independent person the adolescent must complete four major developmental tasks: (1) achieve a stable identity, (2) establish independence from the family unit, (3) assume adult sex roles and (4) select a vocation or career.

Identify Formation

According to Erikson (1963), the developmental crisis of adolescence is one of *identity formation versus identity diffusion*. In order to establish one's identity as a separate person, there must be some breaking away from the family. Identity formation, however, can be enhanced by supportive parents and other adults who have stable identities and therefore serve as role models. Too much parental control can stifle identity formation, while too little can have equally negative effects. The peer group also provides support for adolescents in their search for identity (Fig. 9-6).

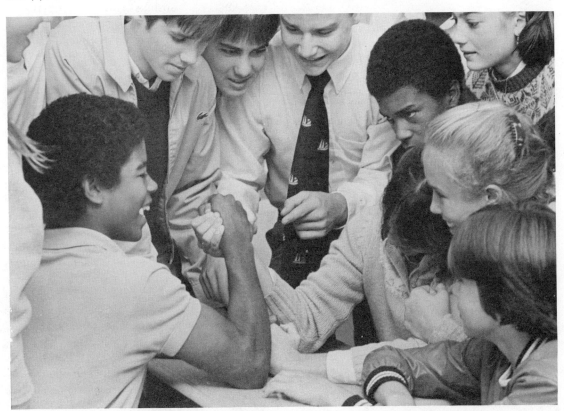

Figure 9-6. *The peer group provides support for the adolescent during the quest for identity, the primary task of psychosocial development in this stage. (Courtesy Chapel Hill Newspaper.)*

There is a restaging of each of the previous developmental crises, and the extent to which the adolescent has successfully resolved earlier developmental tasks influences the success in finding identity. Values, beliefs, and attitudes now become internalized and the adolescent decides what is and is not acceptable to him or her.

If the crisis of identity is not successfully resolved, identity diffusion results. Adolescents who do not develop a sense of identity will feel self-conscious and will have doubts about themselves and their roles in life. Delinquent behavior may be one result of failing to establish one's identity.

Establishing Independence

During the school years, children begin to move into their own social world and the family becomes less important. During adolescence this shift is accelerated as the adolescent strives to establish independence from the family. Whereas in earlier years the family was viewed as the source of security and status, during adolescence the youth looks to the peer group.

The peer group provides the support the adolescent needs to separate from the family, and provides the arena for trying out new ideas. However, during early adolescence, group standards are still tempered by previously learned standards. There are occasions when parents need to set limits, as adolescents do not possess the cognitive ability or experience necessary to make decisions independently in certain areas. The youth expects the parent to set some limits and is often more comfortable when they do. While youths wants to be independent and make decisions, they still need a certain amount of stability in order to feel some degree of security in this relatively unstable period of development.

The period during which the youth breaks away from the family and establishes ties with the peer group is a time of emotional upheaval.

Adolescents often do not understand why they are having so much trouble getting along with their siblings and parents. Even while trying to break away, they need love and the security of acceptance within the family unit.

Peer group relationships provide the basis for later adult roles and relationships. The focus of the peer relationships varies with the age of the adolescent; in early adolescence it is a friend of the same sex, and this relationship serves to increase the individual's self-esteem. Later in adolescence the individual is involved with a whole group of peers of both sexes rather than with one individual. Next, couples form within the group. In late adolescence, one-to-one heterosexual relationships come into play, although some couples will spend a significant amount of time with other couples.

Sexual Identity

Along with establishing identity and independence, the adolescent begins to assume adult sex roles. The sexual identity that was established in early childhood becomes more firmly established in adolescence. Parents, other adults, the mass media, and the peer group all influence sexual identification.

As with other aspects of adolescent development, the acquisition of adult sex roles is facilitated by the experimentation that occurs during adolescence. The adolescent often experiments with clothes, makeup, and various aspects of the adult sex role in the process of acquiring an identity. The female often selects clothes and makeup that may seem to her parents to be inappropriate for her age; while her selection may indeed be outlandish at times, this is part of the process of trying out adult roles. There is also a fierce insistence on dressing like the rest of the peer group. As the adolescent grows older and has some experience with various clothing styles, he or she will learn to select clothes that are appropriate for his or her own self-image.

The telephone is a safe way to experiment with intimacy, as it provides some safety in distance. Much time during adolescence is spent on the telephone, often with a peer of the same sex during early adolescence, while in later adolescence it may be with a member of the opposite sex, although, even in older adolescence, girls may spend a great deal of time talking with girlfriends (Fig. 9-7). Long letters are often written to close friends; written communication, like the telephone, is a safe way to say intimate things that may be difficult to say in a face-to-face encounter.

Early dating tends to be group-centered, with little emphasis on selecting a special person. Dating activities often center around school and sports activities. The selection of a date is frequently based on who is popular and accepted by the peer group. Some youths are not ready for dating during adolescence. Inasmuch as it appears that early dating is often encouraged by adults, as evidenced by preteen dances and other such social events, the youth needs to be reassured that it is okay if he or she does not feel ready for a heterosexual relationship.

According to Harry Stack Sullivan (1953), sexuality first becomes a significant force in personality development at the beginning of adolescence. With adolescence comes the "final integrating tendency," lust, which Sullivan defines as "the felt aspects of genital sexual urges." (p. 62). Lust draws the individual into closer relationships with persons of the opposite sex, and genital sexual activity is the implicit goal. The early adolescent spends 3 to 4 years developing interpersonal skills and becoming comfortable with persons of the opposite sex before achieving *intimacy*, which Sullivan defines as a state in which the well-being of another person is as important to one as one's own well-being (Sullivan, 1953).

In later adolescence (ages 17 to 20), true one-to-one heterosexual relationships are developed. The adolescent is now concerned with finding that "one special person." Dates are selected on the basis of mutual respect and enjoyment.

One reason for the amount of time adolescents spend on hygiene, grooming, and clothing is that the time spent in front of the mirror helps the adolescent to integrate body image changes. Since the body is so important to adolescents, they may see it as the reason for rejection or acceptance by others.

Figure 9-7. *The inseparability of the teenager and her telephone is not merely a staple of situation comedies but a fact of life. The telephone allows the adolescent to experiment with intimacy but also affords the safety of distance. (Photo by John Elkins.)*

Body Image

During the time youths are struggling with the question of "who am I" and with taking on adult sex roles, they are also experiencing a great deal of change in *body image*, the mental picture one has of one's body. Since adolescence is marked by rapid changes in the body, adolescents must alter their mental pictures of themselves in order to function. Of course, the adolescent will have experienced body image changes earlier in life; if earlier experiences were positive, the adolescent should feel good about his or her body (Fig. 9-8). If a bad feeling about one's body has already been established, adolescence is likely to be a difficult period.

Many studies have found adolescents' self-images to be unrealistic when compared with

Figure 9-8. *Primping before the mirror, a time-honored activity of adolescents, is more than a manifestation of vanity; it allows the adolescent to adjust his or her body image in accordance with the rapid and dramatic physical changes of the period. (Photo by John Elkins.)*

actual measurements (Simmons, et al., 1973; Learner, et al., 1976). Girls appear to see themselves as fat and want smaller waistlines, hips, or thighs, and larger breasts. Young males are often concerned with having larger biceps, chest, shoulders, and forearms.

The individual who has some physical defect or disability may have an especially difficult time during adolescence. The adolescent's physical ideal is greatly influenced by the mass media and their current heroes and heroines. Since standards for acceptability are very rigid in early adolescence, this may create problems for some. The adolescent who develops a negative self-image may have problems in later life and may display deviant social behaviors.

Sex is of great concern to the adolescent. The adolescent is experiencing hormonal and physiological changes and is thus preoccupied with developing sexuality. Sexual activity during adolescence may include masturbation, heterosexual experimentation, and sometimes, homosexual experience. Parents are often very disturbed when they realize their adolescent is sexually active. The adolescent must make a decision about whether to engage in sexual activity; this decision will be influenced by past experience, one's personal moral code, the society, and the media.

For the young adolescent who is still operating at a concrete level (i.e., experiencing things strictly in terms of past experiences), the awakening of sexual feeling may be very difficult to handle. Such a young person may not be mature enough to consider alternatives and possible outcomes of actions.

Masturbation

Masturbation is a normal activity and is one way the individual learns about sexual feelings. Peter Blos (1970) suggests that adolescent masturbation is a phase-specific developmental task that must be mastered if the individual is to achieve adult genitality. He believes that total abstinence from masturbation during adolescence indicates an inability to deal with sexual drives. Unfortunately, even with the so-called "sexual revolution" in the United States, masturbation is still a somewhat taboo topic. Myths regarding the harmful consequences of masturbation still have some currency. Gallagher and Harris (1976) found college students who were guilt-ridden and deficient in overall effectiveness because of their struggle to stop masturbating. Adolescents need reassurance that masturbation is a normal activity and is a response to developing sexual feelings.

Early Sexual Activity

The young male has greater desire for coitus because of greater localized genital sensation. The male reaches his peak of sexual desire at a younger age than the female; this often occurs in mid or late adolescence. Males are more likely to become sexually experienced during adolescence because of the higher sexual drive and also because sexual activity for the male is still more socially acceptable. In addition, males are less likely to be closely supervised and thus have more opportunity than females.

It is often difficult for adults, including health care providers, to come to grips with sexually active adolescents. Research data indicate that more adolescents are sexually active, and many are sexually active at an earlier age than was the case in the past; Hunt (1974) collected information about sexual behavior from a sample of 2,000 white adults in 24 cities around the United States in 1972. He found that over half of the men who had gone to college reported having had premarital intercourse by 17 years, and 3 out of 4 noncollege men experienced coitus by this age. One out of five married women interviewed reported having had sexual intercourse by age 17, while 1 in 3 single women reported having sexual intercourse by age 17.

In contrast with Kinsey's data, reported in 1953, the percentage of women in their late teens and twenties who have had sexual intercourse before marriage has doubled (Hunt, 1974). Females are now experiencing first intercourse at an earlier age than when Kinsey reported his interviews with women. Zelnik and Kantner (1972) found that 11 percent of white females were no longer virgins by age 15, and 40 percent had had intercourse by age 19. According to Sorensen (1973), 56 percent of all nonvirgin girls and 71 percent of all nonvirgin boys surveyed had their first sexual intercourse by age 15.

Adolescents may become sexually active for a variety of reasons, including the desire to form a significant relationship, but they may also feel guilty about having sex. Making a decision about sexual activity is often one of the most difficult decisions an adolescent has to make. Despite our "enlightened" sexual attitudes, young people still receive mixed messages. The mass media portray sex as glamorous, the "in" activity. Most parents, while striving to help their children see that sex is part of meaningful relationships, still discourage sexual activity before marriage.

In many areas of the country, particularly on college campuses, being sexually active is the norm. The male or female who is not sexually active often receives a great deal of peer pressure to engage in sexual activity. Many adolescents first engage in sexual activity out of curiosity and in response to peer pressure, and experience great disappointment when they do not find the experience satisfactory. They are often filled with self-doubt and the feeling that there must be something wrong with them. Often they are afraid to share these feelings with their peers and unable to admit to their parents that they have been sexually active. Such feelings are often revealed by females who seek advice on contraception from health care providers. If the health care provider takes the time to ask how the female feels about her sexuality, the client will often reveal her ambivalence. The health care provider can allow the client to express her feelings and provide reassurance that there is nothing wrong with her and that it is okay not to be sexually active. If she has a steady boyfriend, she should be encouraged to discuss her feelings with him. Many young men have similar feelings but are more reluctant to seek out someone to talk with.

Planning A Career

The last major developmental task of adolescence is the selection of a vocation or career. In other cultures, and in earlier times in the United States, the child's career was chosen by the parents. In more recent times, the adolescent has been allowed to make this decision. The pressures of society unfortunately require teenagers to make decisions earlier; during the last two years of high school, the young person must decide whether to pursue a vocation or to go on to college, where to live after graduation from high school, and what to do with his or her life. Many factors influence this decision, including the family, the school system, the cultural and moral attitudes of the community, the job market, and the interests of the individual. The decision is a very difficult one to make and is confusing to many adolescents (Fig. 9-9). High school guidance counselors and vocational testing centers may aid the adolescent.

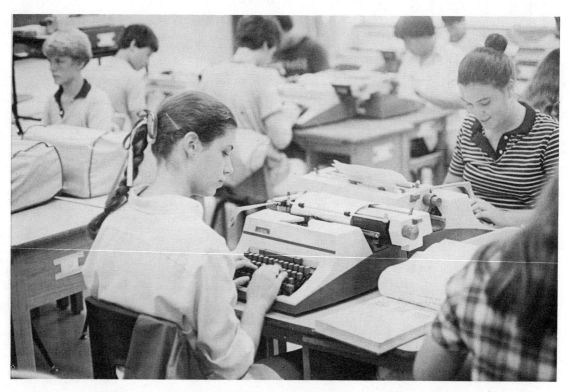

Figure 9-9. *By its very nature, adolescence is often a bad time for important decisions, but it is during this period that one must make plans for a career, including a decision as to whether to attend college or to attempt to establish oneself in a vocation on the strength of skills already acquired. (Courtesy Chapel Hill Newspaper.)*

COGNITIVE DEVELOPMENT

According to Piaget's scheme of cognitive development, the adolescent should be entering the period of *formal operations*, which is characterized by abstract thought and the ability to conceptualize (Piaget and Inhelder, 1969). Thus the adolescent becomes capable of logical thought and is able to make inferences. Adolescents who have entered the formal operations stage are able to use both inductive and deductive reasoning; they no longer learn solely from experience.

Cognitive development is achieved on a continuum; the older adolescent usually has greater cognitive ability than the younger counterpart. Physical and cognitive development do not necessarily occur at the same time, so the health care provider must assess adolescent clients carefully in order to avoid expecting too much of one youth or talking down to another. Not all people progress to formal operations; some adolescents will remain at the level of *concrete operations*, at which the individual is not capable of abstract thought, and logical reasoning is restricted to the concrete, trial-and-error level. The individual can acknowledge the reality of only those things that he or she has had experience with. This is why some adolescents, especially young ones, continue to deny certain things. One frequently encountered example is the denial of pregnancy by the adolescent female. If she is still in the stage of concrete operations, she cannot accept the fact that she is pregnant, having had no prior experience with pregnancy.

MORAL DEVELOPMENT

The adolescent subscribes to ideas of right and wrong that are close to the ideas held by adults. Moral development is closely related to cognitive development. According to Piaget there is a transition from a morality based on laws made by others to a more mature morality based on one's own judgment and convictions (Piaget, 1932). In other words, a shift from moral realism to moral relativism occurs; whereas the moral realist follows the letter of the law, the moral relativist takes account of intentions as well as consequences.

Peer interaction is important in making the transition from moral realism to moral relativism. Through peer interaction the individual learns to share in decision-making. The individual also learns to take the role of another, which in turn leads to being able to see things from another perspective. Peer interaction allows the individuals to gain an understanding of the inner states that underlie other people's acts.

RELIGION AND ADOLESCENCE

In many religions adolescence is regarded as an important period, a time when the person takes full responsibility as a member of his or her faith. In some faiths the transition from childhood to adulthood is recognized with rites of confirmation, bar mitzvah or bas mitzvah, and other forms of introduction into adult religious practices and obligations.

Many young people experience doubts about religion during adolescence. In their quest for religious identity they ask, "who am I?" "Why am I here?" "What is my purpose in life?" (Rice, 1975). Questioning one's religion, if it has previously played an important role in one's life, is tantamount to questioning an important foundation of life. One is in a sense questioning one's capacity for understanding as well as the relationship with those in whom one has placed one's trust. Questioning one's religion can be a frightening and disturbing experience for the adolescent. It can also be an enriching experience, as such inquiry can lead to resolution of inconsistencies in belief and a fuller and richer understanding of the meaning of religious beliefs.

COMMON HEALTH AND DEVELOPMENTAL PROBLEMS OF THE ADOLESCENT

The special needs of the adolescent did not gain recognition until the 1950s, when Dr. Rosewell Gallagher advocated special services and support for the adolescent client. Before this development, adolescence had been viewed as a relatively healthy period of life with few true "medical" illnesses and diseases. Although it is true that few illnesses are characteristic of this period in the life cycle, the adolescent has a great need for health supervision and counseling. In accordance with Gallagher's suggestions, several health care agencies opened adolescent outpatient clinics designed to deal with health care needs specific to adolescents. As a result of the success of these clinics, the specialty of adolescent medicine was born.

In many parts of the United States, services for adolescents are still inadequate, and the majority of adolescents therefore receive inadequate and inappropriate health care. However, even if special facilities are not available, the health care provider can make the existing facilities more acceptable by being aware of the unique needs of the adolescent and the health concerns common to this age group.

Faulty Nutrition

Faulty nutrition is a common problem during adolescence. Since the 1950s the dietary patterns of adolescents have changed dramatically. From 1955 to 1965 there was a decline in the nutritive intake of adolescents, and the pattern has continued (Toore, 1977). Along with the increased intake of "fast foods" has come a decreased intake in fresh vegetables and fruits. Most adolescents are not deficient in protein and calories, but their diets often lack sufficient quantities of iron, vitamin A, and calcium. The adolescent diet is also likely, to a lesser extent, to include insufficient vitamin C, riboflavin, and thiamin, among other nutrients.

Teenagers often restrict their diets to a small number of favorite foods, often "fast foods," which contain more calories than needed and can contribute to obesity. These foods lack vitamins A, D, and C, and folic acid. One fast-food meal per day can be acceptable if the diet is augmented with two glasses of milk, green leafy vegetables, and a glass of fruit juice at other meals. The practice of skipping breakfast has also contributed to the nutritional deficiencies of adolescents, since breakfast foods tend to constitute the major source of vitamin C in the diet.

Adolescent athletes, especially those involved in rigorous sports such as football, are at risk for nutritional deficiencies. They require 2,300 to 5,000 calories per day depending on the energy expenditure; 25 to 35 percent of the calories should come from fats, 10 to 15 percent from protein, and the rest from carbohydrates (Toore, 1977). These young people also require an increased intake of vitamin C and the B complex.

Weight Problems

Underweight during adolescence may be caused by a rapidly increasing basal metabolism rate. Underweight young people may not be taking in enough food to meet all the body's requirements during the adolescent period of rapid growth. Faulty nutritional patterns only compound the problem. One group of researchers found approximately 20 percent of their teenage population to be underweight (Berenberg, 1974).

Of even greater concern is the problem of *overweight*, or overnourishment. Adolescents may be considered overweight if their weight falls more than two standard deviations away from the mean on a standardized growth chart. *Obesity* refers to an excess of body fat; males with 20 to 25 percent excess body fat are classified as obese while females with 25–30 percent excess body fat are classified as obese. (Berenberg, 1974). The prevalence of obesity among adolescents in the United States is estimated to be fairly high and its incidence correlates inversely to the socioeconomic status (Meyer and Neumann, 1977; Ross Roundtable, 1971). If one or both parents are obese the adolescent is at an increased risk for being obese. According to several studies the prevalence of obesity in high school students ranges from 11 to 30 percent (Eden, 1977; Meyer and Neumann, 1977). Research indicates that up to 80 percent of obese children remain obese as adults (Knittle, 1972; Meyer and Neumann, 1977).

The earlier the obese child can be identified and treatment begun, the better the chance of establishing a healthy diet and a return to normal weight. Treatment during adolescence is difficult, since high nutritive demands are placed on the body during this period of rapid growth. A combination of moderate dietary restriction and exercise is the safest treatment. The incorporation of behavior modification techniques appears to be useful in helping the adolescent learn to change eating behaviors.

Iron-Deficiency Anemia

Yet another problem that may arise from faulty nutrition is *iron-deficiency anemia*. The adolescent female is more likely to be anemic because of a combination of factors: poor dietary intake, a lack of testosterone to stimulate erythropoiesis, and blood loss caused by menstruation. Anemia may be further complicated in the adolescent female who uses an intrauterine device (IUD) since this increases menstrual flow and depletes iron levels. Such adolescents may require an iron supplement.

Since normal hematocrit and hemoglobin levels for adolescents differ somewhat from adult levels, the health care provider should be familiar with the normal range. The lowest limits of normal for both sexes are as follows: from 13 to 16 years, a hemoglobin of 13 gm and a hematocrit of 39; from 17 years and older, a hemoglobin of 14 g and a hematocrit of 42 (Toore, 1977).

Pregnancy

Pregnancy during adolescence creates a nutritional crisis for the female. Whereas pregnancy at any age creates a demand for increased nutrients and calories, pregnancy that occurs before the female completes her growth spurt may result in a situation in which the gravida and the fetus are competing for nutrients. If the nutrients are not supplied in sufficient amounts, both the gravida and her fetus will be at increased risk. Mothers below the age of 15 deliver 2.2 times the number of premature infants that mothers in the 20- to 24-year-old range deliver (*Eleven Million Teenagers*, 1976) Even when the adolescent mother carries a pregnancy to term she is more likely to have a smaller baby. Since anemia is common among adolescent females, the young pregnant adolescent is placed in double jeopardy of iron-deficiency anemia. Pregnant adolescents will require iron supplements and will need careful counseling regarding the importance of iron.

While the overall rates of childbearing among teenage mothers declined in the early 1970s, the increasing teenage population resulted in more infants being born to teenage mothers than ever before (Baldwin, 1976). In 1975, 20.3 million 10- to 19-year-olds gave birth to 594,880 infants (Eddinger and Forbrush, 1977). In one-third of teenage marriages the female is pregnant before the marriage, and half of all teenage marriages end in divorce. Pregnancy during adolescence carries physiological risks to the gravida as well as increased risks to the fetus.

Adolescence is considered a normal developmental crisis; when pregnancy, a situational crisis, is superimposed, the psychological stress is compounded. Pregnancy is seldom a planned event during adolescence and thus may result in conflict with other personal goals. Pregnancy often means the interruption of school, sports and other school-related events, and the usual dating patterns. Depression may result from these losses; the depression may be so severe that it leads to a suicide attempt.

The teenage mother is at high risk for having additional children while still an adolescent, thus increasing the physiological risks for herself and her fetus (Oppel and Royston, 1971). Graves and Bradshaw (1975) found that young mothers who married after giving birth had a higher pregnancy rate within one year than those who remained single. Special programs with intensive follow-up and support for the young mothers seem to help reduce the incidence of repeated pregnancies (Currie et al., 1972).

Pursuing one's education while pregnant is difficult, and without an education the adolescent is more likely to be dependent on others. Furstenberg (1976) found that only half of pregnant teenagers finished high school while 90 percent of a comparative group who were not pregnant completed high school.

The pregnant adolescent who marries is more likely to get divorced and is faced with the task of assuming the role of wife at a time when she is ill prepared to assume this role. Adolescent males who marry are equally ill prepared to assume the roles of husband and father, since they will not have completed their education and will probably be poorly equipped to support themselves, let alone a wife and child. Approximately 50 to 75 percent of all teenage marriages are "shotgun weddings" (Gordon, 1973) and about half of all teenage marriages end in divorce (Wagner, 1970).

Since the adolescent is physiologically, anatomically, and immunologically different from the mature female, she is at greater risk during pregnancy for a variety of complications including preeclampsia or eclampsia, premature rupture of the membranes, uterine dystocia, cephalopelvic disproportion, infections, first and/or third trimester bleeding, and fetal distress (Daniel, 1977). The younger the girl, the greater the risk. Teenage girls probably need prenatal care even more than older pregnant women but usually obtain care later in the pregnancy and on a less regular basis.

Aiman (1976) found that the pelvis of the average pregnant adolescent 16 years old or younger is smaller than that of the average woman older than 16 years. His findings suggest that growth of the pelvis is not complete by age 16, which accounts for the high incidence of cephalopelvic disproportion in adolescent pregnancies.

Toxemia of pregnancy is one of the most feared of all complications of pregnancy; reports of rates observed in teenagers range from 4.3 percent to 23.5 percent (Mercer, 1979). The high incidence of pregnancy-induced hypertension in teenagers has been observed worldwide (*Population Reports*, 1977).

It has been suggested that the uterus of the adolescent is structurally and functionally less proficient since it has fewer cycles of exposure to ovarian hormones (Zlatnik and Burmeister, 1977). The incomplete development of the myometrium may contribute to premature delivery.

Hatcher (1972) offers a picture of the typical pregnant adolescent. The young adolescent (11 to 13 years) who becomes pregnant lacks information or understanding about contraception and almost never uses contraceptive devices. Distortion and denial characterize the young adolescents who become pregnant. The girl refuses to accept responsbility for becoming pregnant and usually blames and despises the boy because of the pregnancy. Young adolescents may continue to deny the pregnancy even into the last trimester and thus will not readily accept instructions regarding nutritional advice and care of the body. Those who continue to deny the pregnancy are unable to make plans for themselves or the baby and require a more intensive follow-up by health care providers.

Midadolescent females (14 to 16 years) are aware that pregnancy is likely to result from sexual intercourse and are aware that contraceptive agents can be obtained. Hatcher found, however, that most midadolescents refused to use contraceptive measures and always blamed someone else for the pregnancy—parents, doctors, or the boyfriend. At this age pregnancy may be used to establish independence or to punish parents. Feelings about the pregnancy are ambivalent; guilt and pride may alternate. Midadolescent girls often romanticize about being pregnant but later become depressed at the thought of having to provide long-term care for an infant. Rebellion and antagonistic reactions characterize most midadolescents who become pregnant. Parent-child conflict is greater as well as conflict with doctors or nurses. Health care providers have to work harder to establish trust with the pregnant midadolescent.

The late adolescent female (17 to 20 years) knows that contraceptive agents can be easily obtained and is generally knowledgeable about the various methods. Pregnancy is likely to be the result of not using contraceptives, using them irregularly, or from coitus occurring during alcohol intoxication. The late adolescent female usually accepts responsibility for pregnancy and blames herself. Late adolescent girls are more likely to hope for marriage and a home, and worry about the future. They are also ambivalent about actually giving birth to a child.

Abortion

Abortion is becoming increasingly common among pregnant teenagers. In 1974, 45.2 percent of all pregnancies in females under 15 years of age and 27.4 percent of all pregnancies in females 15 to 19 years ended in abortion (*Eleven Million Teenagers*, 1976). The single white teenager chooses abortion more frequently than does the single black teenager (Zelnik and Kantner, 1974). Abortion remains a controversial issue. While many parents have been opposed to abortions in the past because of the fears associated with illegal procedures more parents are now seeking abortions for their pregnant adolescents as they see abortion as a safe alternative to an unplanned pregnancy.

Hatcher (1972) found that young adolescents are often in favor of abortions, since they deny the pregnancy and have no qualms about a procedure they consider to be unrelated to it. Midadolescents were found to have difficulty making the decision to have an abortion; they seemed to want someone else to make the decision for them. Late adolescents were more realistic about their pregnancies but often hoped for marriage and therefore sometimes waited too long before seeking abortion. Since these girls were more aware of the developing fetus, the decision to have an abortion was reported to be more difficult. The late adolescent was more likely to have given serious consideration to several alternatives.

In abortion, as with all aspects of adolescent care, the legal aspects must be considered. A Supreme Court decision in 1976 affirmed the right of an adolescent to have an abortion without parental consent. Neither should a minor submit to an abortion only because her parents insist; she has the same right of self-determination as an adult woman.

Since abortion carries with it certain physiological and psychological risks, careful counseling and teaching is important. Barglow (1976) found that the developmental immaturity of teenage women contributed to ambivalence about abortion, and that most of the young women experienced the procedure as frightening, punitive, and overwhelming. Those who could be contacted for follow-up three months later reported they had all experienced symptoms of grief but now felt that the procedure had been positive and beneficial. Follow-up care after an abortion is desirable since some young women experience difficulty adjusting after an abortion.

Prevention of Adolescent Pregnancy

Many adults, including health care providers, often have difficulty understanding why adolescents do not use contraceptives. The reasons are varied; to begin with contraception may not be easily available or acceptable. Frequent reports of the dangers of oral contraceptives and intrauterine devices have caused many adolescents to be afraid of these methods. Ignorance is another factor; adolescents often lack information on reproduction and contraception. Finally, there is the previously mentioned possibility that the young female, still in what Piaget calls the concrete operations stage, cannot preceive herself as capable of reproduction.

Insufficient motivation often leads to noncompliance among those teenagers who do have access to contraceptives. Often a teenage girl will stop taking her pills after a fight with her boyfriend and will vow never to see him again; she does not stop to consider the possibility of reconciliation in the next few weeks and the renewed possibility of pregnancy. The use of a diaphragm or an IUD requires a visit to a physician, which involves admitting one's sexuality and planning ahead; many adolescents are unable to do this. Using condoms requires enough motivation to have them available when coitus is likely to occur, i.e., it also involves planning ahead. Many teenagers feel this "destroys spontaneity."

Venereal Disease

Venereal disease (VD) has reached epidemic proportions in the United States and is especially prevalent among adolescents and young adults. Venereal disease is highly communicable and is usually transmitted through sexual in-tercourse, hence it can be referred to as a *sexually transmitted disease* (STD). The incidence of VD among all ages has soared since the early 1960s. Some of the reasons for the increase are increased sexual activity stemming from changes in sexual attitudes and mores, the increased use of oral contraceptives, which do not protect against VD as the condom does, the misunderstanding that VD can be easily cured, the thought that "it can't happen to me," and the willingness of teenagers to take risks. (see Table 9-4). According to Sorensen (1973) most teenagers know the basic facts about VD but are often reluctant to seek help because of embarrassment or fear that their parents will find out. Parental consent is not necessary for treatment of adolescents in most states.

Breast Masses and Menstrual Dysfunctions

The adolescent female often has concerns about her breasts and menstruation. According to the Adolescent Clinic at the University of Alabama in Birmingham, most breast masses seen in their adolescent population were fibroadenomas, which are firm and movable but not hard. Although fibroadenomas are not malignant, they can cause discomfort and concern (Daniel, Brown, and Garrison, 1979). Many physicians recommend surgical excision if the fibroadenomas persist or enlarge after several menstrual cycles.

Menstrual concerns most frequently involve delayed or irregular menstruation or abnormal uterine bleeding. The absence of menses by age 17 is termed *primary amenorrhea*. If a thorough history and physical exam reveal no abnormal anatomical findings and if the girl is at a sex maturity rating of 3 or 4, she should be reassured that her periods should start. A five-day trial regimen of progesterone may be prescribed; bleeding in the week following cessation of treatment indicates that the girl can menstruate and that she will do so when her

Table 9-4. Common Venereal Diseases

Disease	Causative Organism		Symptoms	Treatment
Gonorrhea	*Neisseria gonorrhoeae*	Male:	Asymptomatic at times; urethritis; thin, watery, white discharge that becomes yellowish, purulent; dysuria.	Aqueous penicillin
		Female:	Asymptomatic in up to 80 percent of cases; dysuria and purulent discharge; often accompanied by cervicitis and/or salpingitis.	
Syphilis	*Treponema pallidum*	Male and Female:	Primary: chancre; painless ulceration on genitalia, mouth, nipples, anus 10 to 28 days after sexual contact; lesions heal in 4 to 6 weeks, but treponemes multiply rapidly. Secondary: Rash covering skin, mouth, genitalia; hair loss, inflammation of eyes and ears; low grade fever, lymphadenitis, sore throat; pain with bone movement; positive blood test after five weeks. Latent: Symptoms disappear in two to six weeks and absent for months to lifetime. Positive blood test but person noninfectious except for the female, who can have child with congenital disease. Tertiary: Complications may occur after 3 to 30 years; Affect CNS, heart, blood vessels, bones, skin; positive blood test.	Benzathine penicillin G. or Procaine penicillin

Table 9-4. (*Continued*)

Disease	Causative Organism		Symptoms	Treatment
Moniliasis	*Candida albicans*	Females:	White, cottage-cheese-like vaginal discharge; intense vulval itching and/or burning.	Monostat vaginal cream for one week
		Male:	Often asymptomatic, but may have penile discharge and itching.	
Trichomonal vaginitis	*Trichomonas*	Female:	Greenish-yellow, foamy, foul-smelling vaginal discharge; vaginal irritation and itching.	Flagyl tablets; partner should be treated also
		Male:	Usually asymptomatic but may develop urethritis or prostatitis.	
Herpes progenitalis	Type 2 herpesvirus		Usually affects skin and mucous membranes below the umbilicus; indurated papules or vesicles surrounded by erythema located in genital area; dysuria and tenderness of infected area; fever, headache, malaise.	As of this writing, no effective treatment, although there is the prospect of an effective drug in the near future. Treatment is primarily symptomatic.
Chlamydia	*Chlamydia*		Dysuria, frequency; milky discharge on arising. This condition may mimic gonorrhea.	500 mg Tetracycline every six hours for seven days

hormone production is adequate. Should bleeding not occur, further endocrine studies are indicated.

The absence of menses for over 150 days in a female who has previously menstruated regularly is termed *secondary amenorrhea*. Pregnancy is the most common cause of secondary amenorrhea; sudden weight loss is the next most common cause. Careful attention should be given to the girl's history, which may reveal emotional, nutritional, or environmental stress.

Excessive or irregular uterine bleeding can be quite distressing to the adolescent. Dysfunctional uterine bleeding is most frequently associated with anovulation. Other causes include blood dyscrasias and hypothyroidism. In an adolescent whose mother took diethylstilbestrol (DES) during pregnancy, this excessive bleeding may signal clear-cell adenocarcinoma or adenosis of the vagina or cervix. If no pathologic condition exists, the excessive bleeding may be controlled with a progestational compound such as norethindrone (Goldfarb, 1976). In adolescents with dysfunctional uterine bleeding the hematocrit should be monitored for the presence of anemia, and replacement therapy instituted as indicated.

Dysmenorrhea (painful menstruation) is a frequent complaint of adolescent females; it may be psychological or physiological in origin. A careful assessment of the client will enable the health care provider to determine when the pain occurs as well as what attitudes and expectations the client has about menstruation. If she has been conditioned to expect pain with menstruation, the dysmenorrhea may be psychological. If the pain has a physiological basis it will occur just prior to and early in menstruation. The pain is caused by the hormone prostaglandin, which in excessive amounts produces smooth muscle cramping, diarrhea, and flushing. If a thorough physical exam (including a pelvic examination) reveals no organic dysfunction, a prostaglandin inhibitor such as aspirin or ibuprofen (Motrin) may be prescribed. Used 2 to 3 days before the period, these provide relief in most cases.

Delayed Development in Males

Males who have not began their sexual development by age 13 or 14 may be as concerned as the female with delayed menarche. Frequently, these males are short in stature and the family history reveals a delay in bone maturation. Reassurance that all is normal and that development will occur is all that is needed in most cases. If sexual development has not occurred by age 14 or 15, endocrine studies are indicated. If the endocrine studies are normal, a six-month trial of monthly testosterone injections may be instituted. This induces some sexual development and may restore the client's self-confidence (Daniel, 1977).

Skeletal Problems

Problems with the skeletal system are common in adolescence since it is a period of rapid growth. The incidence of idiopathic *scoliosis*, a lateral curvature of the spine, increases dramatically at adolescence and is seven times more common in females than in males. Not all scoliosis is idiopathic; it may be secondary to spinal lesions, or muscle or neurological problems.

Scoliosis (which was also discussed in Chaper 8) may not be evident until the adolescent growth spurt. The onset is frequent in girls at SMR 2 and in boys at SMR 3 (Daniel, 1977). Since treatment either with a Milwaukee brace or with spinal fusion followed by an extensive period in a body cast makes the adolescent client different from his or her peers, considerable adjustment on the part of the adolescent is required. During the treatment the adolescent is unable to participate in certain activities and may find less acceptance by the peer group. The onset of scoliosis during the adolescent period may interfere with the adolescent's ability to establish and accept his or her body image. Accomplishing the tasks of adolescence will also be more difficult for the youth with scoliosis. Treatment requires that the adolescent be dependent on parents and siblings.

Health care providers working with adolescents with scoliosis need to be aware of these hinderances to normal growth and development. It is helpful to acknowledge the problems and to allow both the adolescent and the parents to express their feelings. Once the problems are recognized many parents and adolescents are able to identify ways in which some of the usual activities can be continued.

Osgood-Schlatter disease, a partial avulsion of the tibial tubercle caused by trauma to the immature epiphysis, is frequently seen in active, athletic adolescents who overuse the quadriceps femoris. Pain, point tenderness, and swelling at the site of insertion of the tibial tendon are the characteristic symptoms. Bed rest decreases the inflammation and the youth can be allowed to gradually resume activity. Once ephiphyseal closure is achieved the condition disappears.

The adolescent who presents with hip pain and a limp should be evaluated for slipped capital femoral epiphysis (SCFE), which is more common in obese males than in females. SCFE is an acquired disease in which the upper femoral epiphysis translocates away from its normal location on the neck of the femur; the dislocation is usually in a posterior and medial direction. SCFE is an orthopedic emergency and demands immediate treatment.

Sports Injuries

Many adolescents participate in competitive sports and they are susceptible to a variety of musculoskeletal injuries. Adolescents who were previously inactive often begin to participate in sports without the proper conditioning and training, thus increasing their risk of injury. Since there is a wide variation in bone growth among adolescents of the same age, competition is often more dangerous to those whose growth is at a less mature stage. Any adolescent who is going to participate in competitive sports should have a thorough physical examination to ensure that he or she is in good physical condition and has no organic problems that might be aggravated by participation in sports.

Garrick (1977), in a two-year study of athletic injuries at four high schools, found that the 3,049 students participating in 19 sports sustained 1,197 injuries—or 39 injuries per 100 partcipants. Nearly two-thirds of the injuries were strains and sprains; over half of the injuries were to the lower extremities, with the thigh, knee, and ankle being involved equally.

The two major categories of orthopedic problems in sports are acute injuries and overuse syndromes. Acute injuries are most often sprains or strains of the lower extremeties. Less common acute injuries associated with sports include costochondral separation and subluxation (separation and incomplete dislocation of the ribs), quadriceps contusion (bruising of the large extensor muscles of the thigh), and myositis ossificans (inflammation of muscle with formation of bone). Overuse syndromes are insidious in onset and are initially benign; they are

related to specific athletic activities such as excessive running, pitching, or swimming. Usually, stopping the activity alleviates the pain and the athlete is likely to seek treatment only when the pain interferes with his or her performance. When dealt with early, overuse syndromes rarely progress to permanent disability. Treatment of overuse syndromes involves resting the involved parts. Resumption of athletic activity must be gradual.

Automobile Accidents

Automobile accidents are the leading cause of adolescent mortality and morbidity. Motor-vehicle-related injuries to the head, spinal cord, and skeleton, as well as abrasions and burns, are common. In 1975, more than 24,000 motor vehicle accidents involving 15- to 24-year-olds occurred (Monthly Vital Statistics, 1976). Males were involved four times more often than females. While poor coordination, carelessness, alcohol and drug abuse, and increased risk-taking are the cause of many motor vehicle accidents among adolescents, there is also great concern that many automobile accidents may be the result of suicide attempts.

Suicide

The escalating rate of adolescent suicide is a growing concern. The suicide rate for adolescents increased by approximately 300 percent from 1950 to 1979. Suicide ranks as the second leading cause of death for the 12 to 24 age group in the United States.

Some behavioral scientists theorize that the roots of suicidal behavior are established in infancy. Rohn (1977), theorizes that the developmental task of establishing trust is never accomplished because of death, abandonment, or even severe depression on the part of the mother; thus the seed is planted for later suicide.

A crisis or death in the family, an abortion, breaking up with a boyfriend or girlfriend, or the diagnosis of a fatal illness may precede the suicide attempt. The depressed adolescent may exhibit a variety of symptoms, including acting-out behavior, to relieve periods of depression. Once the depressed adolescent is identified, the health care provider should take a careful history and assess various aspects of the adolescent's life. The source of the problem may relate back to earlier childhood. It is appropriate to ask the adolescent if he is having thoughts about hurting himself; often this is a source of relief and allows the individual to vent feelings. If the adolescent is having suicidal ideas it is important to refer the individual for immediate psychiatric help. Without intervention it is highly probably that suicide may occur. Two-thirds of the adolescents who commit suicide express their feelings to someone a few hours to a few weeks before the suicidal act (Prophit, 1979). Signs that may indicate a tendency to suicide include an unusual degree of withdrawal; a drop in grades and loss of interest in school; despondency and talk of death; an increase in the use of drugs and alcohol; and the giving away of personal belongings. If an adolescent who has been depressed and withdrawn suddenly becomes very cheerful and happy, this also should be taken as a warning signal; often the individual who has made the decision to die is much calmer and happier.

Many adolescents have transient suicidal preoccupations. Sometimes impulsive acts are committed in an attempt to gain parents' attention and are pleas for help. Often the troubled adolescent reflects a troubled family, and the whole family needs treatment. It is very important to have the family involved in the care of the adolescent; without family involvement the prognosis for the adolescent is not very good.

Drug Abuse, Alcoholism, and Cigarette Smoking

Since the late 1960s drug abuse among adolescents has been a major concern. It has often been pointed out that Americans comprise a drug-using society, relying on drugs to relieve constipation, headaches, anxiety, insomnia, arthritis, indigestion, and other ailments. The family habits and societal attitudes of which this dependency is a reflection undoubtedly influence adolescent attitudes about drugs. Other reasons for drug use include anxiety, curiosity, peer pressure and the desire for social acceptance, the need for rebellion, the desire to escape from reality, and the wish to be sophisticated and more "mature." Adolescents seldom seek care because of a particular drug habit or the impairment of a specific organ but rather because of a symptom complex that is causing concern. Often the adolescent will present an unrelated symptom or request a routine examination, desiring reassurance that the drugs have not created permanent damage (Schonberg et al., 1978). Since the character of illicit drug use constantly changes, health care providers should keep informed about what drugs are being abused in their part of the country. One of the primary considerations in the evaluation of drug abuse in adolescents is whether or not the use of the drug is a symptom of a more serious underlying problem. Most adolescents who habitually abuse drugs are unable to cope successfully with the problems of daily living. Other adolescents cope well before experimenting with drugs, and the drug use itself is the problem. General signs of drug abuse include loss of interest in school, difficulties in peer relationships, personality changes, short temper, loss of interest in physical appearance, and decreased coordination (Fig. 9-10).

Treatment is generally highly directive and much time is spent in helping the individual identify ways to change or modify the social environment. Most adolescents are treated in community-based programs but hospitalization is almost always indicated for adolescents presenting with drug intoxication, as this may signal a suicide attempt or the loss of ability to control use.

The use of alcohol among the adolescent population has increased dramatically since the early

Marijuana	• Dilated pupils, bloodshot eyes
	• Loss of critical judgment, distorted time perception
	• Tachycardia, transient elevated blood pressure
	• Sore throat and bronchitis
	• Long-term use produces decreased testosterone levels, decreased sperm counts, decreased sexual functioning, and gynecomastia in males
Amphetamines	• Increased pulse rate, increased alertness and activity level
	• Dilated pupils; sluggish pupillary response
	• Weight loss closely mimicking hyperthyroidism
Barbiturates	• Pulse, respirations, and blood pressure decreased
	• Depression
	• Symptoms of alcohol intoxication without alcohol on breath
Heroin	• Constricted pupils; absence of pupillary response
	• Decreased visual acuity
	• Subcutaneous fat necrosis and cutaneous scars ("tracks") among those who inject the drug
	• Constipation, urinary retention
	• Central nervous system depression
	• Severe abdominal pain, anorexia, vomiting, acute gastritis and pancreatitis, and GI hemorrhage may accompany use of heroin combined with acute ingestion of excessive alcohol
PCP("angel dust")	• Central nervous system depression
	• Depression, possible suicidal thoughts, violent behavior

Figure 9-10. *Physiological effects of commonly abused drugs.*

1970s. A national survey conducted among almost 14,000 American seventh- through twelfth-graders in 1975 revealed that most adolescents had some experience with alcoholic beverages (Rachal et al., 1975). About one in three adolescents can be classified as an infrequent or light drinker, but almost one in four is a moderately heavy or heavy drinker. Well over a million Americans aged 10 to 19 are addicted to alcoholic beverages (The new youth, 1977). According to the national survey, boys drink more heavily than girls; beer was the beverage most frequently preferred, followed by wine and hard liquor (Rachal et al., 1975). As with other drug use patterns, children tend to follow the drinking patterns of parents. An adolescent with at least one parent who drinks regularly appears to be about twice as likely to be a moderate to heavy drinker (Rachal et al., 1975).

Adolescents who drink regularly usually start at an early age; by late adolescence they may well be alcoholics (The new youth, 1977). This has resulted in the formation of groups for the teenager with a drinking problem, sponsored by Alcoholics Anonymous.

The 1964 U.S. Surgeon General's report on the effects of cigarette smoking, indicating a relationship between smoking an lung cancer, heart disease, and emphysema, has spurred national concern about smoking. While many adults are quitting or struggling to quit, many teenagers are taking up smoking. Most adolescents are aware of the hazards of smoking but feel they won't be harmed and can quit easily.

One out of four teenagers smokes. According to a 1972 survey adolescent boys were less likely to smoke in 1972 than in 1970, but girls were taking up the habit at a much more rapid rate (National Clearinghouse for Smoking and Health, 1972). Most teenagers smoke because they feel it makes them appear more mature. As with other forms of drug abuse, the adolescent whose parents and peers smoke is more likely to take up the habit.

Immunizations

Immunizations should be completed by adolescence; however, many clients may have incomplete or undocumented immunizations and will require immunization. Adolescents who have never received immunizations or are incompletely immunized should receive two tetanus/diphtheria vaccine doses and two doses of trivalent oral poliomyelitis vaccine eight weeks apart, followed by a booster of each in one year. The completely immunized adolescent should receive a tetanus/diphtheria booster every 10 years.

Since the rubella vaccination is given primarily to prevent the transmission of rubella to a fetus, girls should be vaccinated prior to the menarche. In the postmenarcheal female, pregnancy must be ruled out and the client must agree not to conceive for three months after immunization. A single dose of rubella vaccine is given.

It is important to verify the age at which the adolescent received measles vaccine. A single dose of live attenuated measles vaccine should be given to any adolescent who was vaccinated with killed vaccine, or vaccinated with inactive vaccine. Adolescents who have never had mumps or the mumps vaccine should be given a single dose of mumps vaccine.

Common Skin Disorders

Almost all youths have some evidence of acne or other visible skin disorders during adolescence. Pimples are of great significance to adolescents as they are convinced that anything that detracts from their appearance is threatening to their acceptance by other adolescents.

Acne usually reaches a peak in late adolescence and then decreases. It is often more severe in males. Most adolescents will have a mild form of acne at some times; the majority will be helped by good personal hygiene and one of the many over-the-counter preparations available.

Acne is a complex disease; its cause is only partially understood. It is thought that hyperkeratinization occurs in the wall of sebaceous follicles resulting in the formation of free fatty acids from sebum at the skin's surface. Both the free fatty acids and the sebum are irritants when in contact with the dermis. Hyperkeratinization plugs the follicles and gives rise to comedones (blackheads). If the follicular wall ruptures permitting sebum to spread within the skin, inflammed lesions (papules, pustules, and nodules) develop. *Corynebacterium acnes* is found deep in the follicle and is thought to be lipolytic thus contributing to the inflammatory reaction (Daniel, 1977).

Commonly used over-the-counter preparations contain either sulfur, resorcinol, salicylic acid, or benzoyl peroxide, either as single agents or in combination. Most of these preparations produce a slight irritation and exfoliation of the skin.

Vitamin A acid is also being used by physicians in the treatment of acne. The liquid or cream is applied to the skin at night and irritates the skin. Clients must be warned that the face may temporarily look worse. Vitamin A acid is used only at bedtime and should be washed off thoroughly in the morning. It is recommended that the use of vitamin A acid be discontinued during the summer months since the combination of vitamin A acid and sun is suspected of being carcinogenic.

Systemic treatment is recommended for pustular acne. Tetracycline is the drug of choice as it is thought to decrease Corynebacterium acnes in the follicles thus lowering the production of fatty acids. The client receives 250 mg of tetracycline twice a day initially and the same amount once a day after a few weeks.

The use of washing agents and astringents to remove surface oil from the skin has not proven effective in the treatment of acne. Excessive washing should be discouraged. Dietary management is of little use in the treatment of acne; adolescents need not restrict their diets unless they are aware that particular foods cause flare-ups.

Many adolescent boys have tinea pedis ("athlete's foot") or tinea cruris ("jock itch"). It is exacerbated by heat and perspiration. Tinea is easily treated with tolnaftate (Tinactin). Instructing the male to wear a clean athletic supporter and clean gym trunks can help decrease the likelihood of recurrence of tinea cruris.

Sexual Abuse

Incest is sexual intercourse or other sexual activity between persons too closely related to marry legally. The incidence of incest in the United States is unknown since people are reluctant to admit to it, although it is indicated that incest is more common than was once generally suspected. Incest occurs in all segments of society and is not limited to the lower socioeconomic levels although it is more frequently reported in this stratum.

Incest commonly occurs between brother and sister, father and daughter, stepfather and daughter, and, occasionally, mother and son. Intercourse is more likely to occur in the first three relationships, while in the last, seduction and sexual manipulation are more common (Masters and Johnson, 1976).

All of these relationships can result in emotional trauma to the adolescent involved, who will frequently have difficulty establishing heterosexual relationships with peers. Sexually abused females especially may feel guilt and shame, and may be unable to trust males of their own age. These feelings may be repressed and may take the form of psychosomatic complaints unassociated with sexual matters (Daniel, 1977). In any event, adolescents involved in incestous relationships will need psychiatric therapy.

Rape

There has been an increase in the number of rapes reported through the 1970s but law enforcement agencies believe that the reported cases represent only a small portion of the total number of rapes. Most hospitals have a standard protocol for the treatment of rape victims. This involves a careful physical examination including a pelvic examination. Documentation of the extent of physical injury is necessary and will be used as evidence in court. Photographs may be taken if there are bruises or lacerations. Specimens are taken when the vagina is examined. Most physicians will prescribe mediation to prevent pregnancy and will treat the patient prophylactically for gonorrhea and syphilis.

The emotional trauma associated with rape is usually more severe and of longer duration than the actual physical abuse. Burgess and Holmstrom (1974) found both immediate and long-term reactions in rape victims. Immediate reactions were either controlled (i.e., seemingly calm and remote) or expressed in frequent crying, extreme anxiety, and tension, hostility, and anger in response to questions. Physical symptoms such as difficulty in sleeping and loss of appetite were common in both groups in the next few days after the attack.

The most common sign in rape victims is fearfulness; feelings such as anger, guilt, desire for revenge, and lack of self-worth may also be present. Mood swings are common in the weeks that follow. Most rape victims experience some degree of emotional and/or physical withdrawal from activities and friends. The adolescent girl may delay returning to school and may not want to talk with her friends. Since adolescent girls are frequently raped by a male student or a gang of boys from school, the girl is often afraid to return to school. She may also be fearful that other students know what happened and are talking about her. It is not unusual for the adolescent girl to have difficulty with heterosexual relations following rape and to be fearful of sexual contact.

Counseling should be initiated shortly after the rape occurs. The initial aim is to get the girl to return to school and back into her usual activities as quickly as possible. Rape victims who deny the problem and refuse to admit that they are having difficulties have more problems resolving their emotional trauma. Rape crisis centers located in many cities offer trained counselors who can provide valuable help to rape victims.

ANTICIPATORY GUIDANCE

As with other age groups, the health care provider can play a vital role in the care of adolescents by providing anticipatory guidance. At times this guidance is best shared in a family conference, and at other times it is more appropriate that it be given separately to the parents and the adolescent.

The first step in anticipatory guidance is to prepare the family for the events that occur in adolescence. Both parents and adolescents should be given information regarding the physical and psychological changes that take place. Since adolescence does not occur at a set age, decisions must be made on an individual basis. The Sex Maturity Ratings may be a useful guide for the health care provider. The SMRs enable the health care provider to assess whether the individual has begun to develop secondary sex characteristics and where the individual is in the growth process, and allows a prediction of the

time of the growth spurt and the anticipated onset of menses. The SMR also serves as a guide to when to discuss birth control with the youth. Since the male at SMR 3 is capable of ejaculation, and is capable of impregnating a female by SMR 4, the health care provider should start counseling on birth control at SMR 2. Since the growth spurt in females occurs at late SMR 2 or early SMR 3, followed closely by the onset of the menses, it is appropriate to discuss birth control with the female who is at SMR 2. Not all adolescents will be sexually active, but since there is no way to predict when they will be, it is prudent to introduce the subject of birth control, preferably when alone with the adolescent. The health care provider should furnish information about birth control and let the client know that contraceptives are available. Such information should be given in a nonjudgmental way and the youth should be given an opportunity to ask questions. Establishing rapport with the adolescent before the onset of sexual activity will increase the likelihood that the individual will feel comfortable enough to use a health care facility if the need arises.

Adolescents need information about the changes occurring in their lives. They often have difficulty understanding why they are frequently at odds with their parents and siblings; often they feel guilty about such happenings. The health care provider can help the adolescent to understand that these occurrences are normal and are a part of establishing independence. Parents also have difficulty understanding what is happening within their family. It is important to explore the parents' feelings about "letting go" of their child. Often parents still view the youth as a child and have difficulty perceiving him or her as moving toward adulthood. It is useful to explore the parents' reaction to the adolescent's push for independence. Do they allow the adolescent to make some decisions? In what area? Do they allow the adolescent to express ideas different from theirs? It is equally important to get the adolescent's view. How does he or she feel about attaining independence? Most youths, while demanding independence, are sometimes nonetheless frightened by the responsbility that goes with independence. While they may not want to reveal their fears to their parents, the health care provider can help them to express these feelings. Adolescents who are gradually given responsibility tend to have less difficulty than adolescents who are severely restricted or who are suddenly given total responsibility for all decisions (Fig. 9-11).

The adolescent often has difficulty making decisions, which can be frustrating to both the adolescent and to parents. Decision-making is a learned technique and can be taught. The health care provider can teach a decision-making framework, i.e., a process for developing a planned course of action toward some goal (Taylor, 1976). First, the goal is identified; next,

Figure 9-11. *Most adolescents seek independence from their parents but are simultaneously fearful of independence and the responsibilities that go with it. Adolescents who have gradually been given responsibility for tasks such as household chores are the most successful in making the transition to independence. (Photo by John Elkins.)*

various courses of action are considered, and finally, the individual selects the alternative that carries the greatest possibility for achieving the desired goal. In teaching the technique, it is best to use some nonthreatening decision, such as the purchase of an article of clothing, for example, which pair of jeans to buy. Once the decision to buy a new pair of jeans is made, the adolescent must then decide where to buy them, which style to buy, and how much to spend on the jeans.

With this technique the adolescent can be asked to imagine that he or she is at a store and has narrowed the selection down to three or four pairs of jeans. What factors might affect his or her decision as to which jeans to buy? Appropriate responses could include cost, style, or quality. The teenager is then asked to find alternatives for each of the factors mentioned. Through this process it can be illustrated that decision-making is a part of a person's everyday functioning. The same process can be used to help youths make decisions about drinking, smoking, and sexual and other activities.

Another important area of anticipatory guidance is health care. Many females will experience their first pelvic exam during adolescence. They should be prepared for this experience and know what to expect. Many health care providers use a mirror during the initial pelvic exam so that the client can see what goes on. Most adolescents respond positively to this experience and are extremely pleased to learn about their own bodies. Since breast masses do occur in adolescents, females should also be taught breast self-exams. The young male should be taught to do testicular self-examinations since the incidence of cancer of the testicles is highest in 15- to 25-year-olds.

Dental counseling is also important during adolescence. Poor diet and poor oral hygiene place adolescents at risk for caries. The nurse can capitalize on the adolescents' concerns about self-image to help them eat sensibly, practice good oral hygiene, and seek care when needed.

HINTS ON WORKING WITH ADOLESCENTS

The key to working with adolescents is *to remember to accept and understand each adolescent as an individual*. It is often tempting to lump all adolescents together, but this presents great dangers since people differ from one another more during adolescence than at any other stage of the life span.

It is most important to assess the cognitive level of each adolescent client and to acknowledge the level of maturity. How one relates to the individual and the type of teaching one does are directly related to the client's cognitive level. It is important to provide thorough explanations of procedures and to allow the adolescent to ask questions. Encounters with health care professionals during adolescence do much to influence the individual's attitudes toward the health care system in later life. Thoughtless action or a lack of thought and care on the part of the health care provider can turn a person away from the traditional health care system.

Adolescents respond better if they are treated as adults and not as children. When treated as mature individuals they will generally respond in a like manner. Allow adolescents as much independence as possible and let them make choices whenever possible. When rules and restrictions are necessary, as in a hospital, they should be clearly identified from the onset. Make the adolescent aware of such rules and regulations when he or she is first admitted. Try to provide flexibility whenever possible. Many hospitals still do not have special adolescent units, and adolescents are admitted to the pediatric unit. Often bedtime on the ward is 9 PM. It is unrealistic to expect an adolescent who may be accustomed to an 11 PM bedtime to go to sleep at 9 PM. Certainly the adolescent can't be allowed to disturb other patients, but it may be possible to provide another area where reading or watching TV is permissible. Since peer relationships are so much a part of adolescence, it is important to facilitate contact with close friends and classmates.

The health care professional must be alert to adolescents' concerns about their bodies. It is extremely important to provide privacy for adolescent clients; many are very shy about their bodies and will respond better to examination and care by a member of the same sex. Since this is not always possible it is important for the health care provider to respond to the adolescent in a mature and appropriate manner. A male adolescent will respond well to care given by a female student nurse who is very close to his age if the nurse herself acts maturely and professionally.

It is equally important to be open and unafraid to introduce subjects for discussion with adolescent clients. Adolescents are often hesitant to ask for information, especially about sexual matters. It is the duty of the health care provider to let them know it is all right to talk about these subjects and to create an atmosphere conducive to discussion. Don't be fooled by adolescents who appear blasé and uninterested; they may actually be very concerned and may be absorbing a great deal of information.

The stresses of normal adolescence may be compounded by a life-threatening or terminal illness, or the death of a parent or close friend. Death is still a very uncomfortable subject for most people in American society to talk about, and health care providers themselves may have difficulty caring for adolescents who face death or who have experienced the death of a loved one. Yet adolescents are likely to need more understanding, knowledge, empathy, and concerned care than individuals in any other phase of the life span. To deal successfully with adolescents and death, the health care provider must examine his or her own attitudes toward life and death and become knowledgeable about the ways adolescents view death. (For a more detailed discussion of grief, loss, and dying, see Chapter 14.)

It is helpful if there is a separate time and

place where the adolescent can be seen. If this is not possible it is best to see the adolescent client first, alone, and to talk with the parents later. Some parents will resist this approach and such situations must be dealt with individually. Be honest with adolescents at all times, since they are very skeptical and are testing adults. Don't make promises to the adolescent that cannot be kept, for example, regarding confidentiality. Be familiar with your state laws, know which diseases must be reported, and be honest with the adolescent. In most states individuals 14 years of age and older can be treated without parental consent.

It is vitally important for health care providers to examine their own feelings and values before electing to work with adolescent clients. They must ask themselves: How do I view adolescents? Am I able to relate to them on an individual basis or am I stereotyping them as a group? Am I flexible enough to allow the adolescent to make decisions or am I guilty of imposing my values and beliefs on them? How do I feel about premarital sex, drugs, teenage drinking?

SUMMARY

Adolescence is a period characterized by rapid changes in all aspects of growth and development. The physical changes are very obvious and quite dramatic, but equally dramatic are the psychosocial changes that occur in the individual. Adolescents become sexually mature, and many adolescents will become sexually active.

Adolescence is a stormy period characterized by mood swings with extreme highs and lows. During early and midadolescence, family relationships are often placed under stress. The nurse can help to ease the stress by preparing adolescents and their families for the changes that will occur during this period. Health care providers need to be alert for signs of depression

in adolescents and to intervene appropriately, since suicide is one of the leading causes of death in this age group. Another major concern with the adolescent population is the prevention of accidents, the leading cause of death among those between 12 and 19. Other health concerns in the adolescent population include drug abuse, venereal diseases, nutritional problems, skin disorders, and sports injuries.

Adolescence is a very exciting period of transition. The role of the health care provider is to facilitate healthy growth and development, and to promote as easy a transition into adulthood as possible. By being knowledgeable in normal growth and development the health care provider can identify deviations early and intervene to prevent negative long-term consequences.

REFERENCES

Aiman, J. X-ray pelvimetry of the pregnant adolescent, pelvic size and the frequency of contraction. *Obstetrics and Gynecology*, 1976, *48*(3), 281–286.

Anyan, W. R. *Adolescent medicine in primary care*, New York: John Wiley & Sons, 1978.

Baldwin, W. H. Adolescent pregnancy and childbearing— growing concerns for Americans. *Population Bulletin*, 1976, *31*(2), 2–34.

Barglow, P. D. Abortion in 1975: The psychiatric perspective with a discussion of abortion and contraception in adolescence. *Journal of Obstetric, Gynecologic and Neonatal Nursing*, 1976, *5*(1), 41–48.

Berenberg, S. R. (ed.). *Puberty: Biological and psychosocial components*. Leiden: H. E. Stenfert Kroese, 1974.

Blos, P. *The young adolescent*. New York: Free Press, 1970.

Burgess, A. W., and Holmstrom, L. L. *Rape: Victims of crisis*. Bowie, Md.: Brady, 1974.

Chapman, A. H. *Harry Stack Sullivan, his life and his work*. New York: Putnam's, 1976.

Chinn, P. *Child health maintenance* (2nd ed.). St. Louis: C. V. Mosby, 1979.

Currie, J. B., et al. Subsequent pregnancies among teenage mothers enrolled in a special program. *American Journal of Public Health*, 1972, 62(12), 1606–1611.

Daniel, W. A., Jr. *Adolescents in health and disease*. St. Louis: C. V. Mosby, 1977.

Daniel, W. A., Jr., Brown, R. T., and Garrison, C. L.

Adolescence: The clinical encounter and common health problems. In Mercer, R. T. *Perspectives on adolescent health care*. Philadelphia: J. B. Lippincott, 1979.

Eddinger, L., and Forbrush, J. *School-age pregnancy and parenthood in the United States*. Washington, D.C.: The National Alliance Concerned With School-Age Parents, 1977.

Eden, A. How to "fat-proof" your child. *Family Weekly Magazine*, March 13, 1977.

Eleven million teenagers: what can be done about the epidemic of adolescent pregnancies in the United States. New York: Allan Guttmacher Institute, 1976.

Erikson, E. *Childhood and society* (2nd ed.). New York: Norton, 1963.

Faigel, H. C. Hematocrits in suburban adolescents: A search for anemia. *Clinical Pediatrics*, 1973, *12*, 494–496.

Freud, A. *Normality and pathology in childhood: Assessments of development*. New York: International Universities Press, 1965.

Freud, S. *Beyond the pleasure principle*. New York: Bantam, 1959.

Frisch, R. E. Critical weight and menarche: Initiation of the adolescent growth spurt, and control of puberty. In Rumbach, M. M., Graves, G. D., and Mayer, F. E. (Eds.), *Control of onset of puberty*. New York: John Wiley & Sons, 1974.

Furstenberg, F. F., Jr.: The social consequences of teenage parenthood, *Family Planning Perspectives*, 1976, *8*(4), 148–164.

Gabrielson, I. W., et al. Suicide attempts in a population pregnant as teenagers. *American Journal of Public Health*, 1970, *60*(12), 2289–2301.

Gallagher, J. R., Heald, F. P., and Garrell, D. C. *Medical care of the adolescent* (3rd ed.). New York: Appleton-Century-Crofts, 1976.

Gallagher, R., and Harris, H. I. *Emotional problems of adolescents*. New York: Oxford University Press, 1976.

Garn, S., and Clark, D. Problems in nutritional assessment of black individuals, *American Journal of Public Health*, 1976, *66*(3), 262–267.

Garrick, J. G. Sports medicine. *Pediatric Clinics of North America*, 1977, *21*(4), 737–747.

Goldfarb, A. Balancing hormonal imbalance. In *Clinician: Adolescent gynecology*. Chicago: Searle and Co. Medcom, Inc., 1976.

Gordon, S. *The sexual adolescent*. North Scituate, Mass.: Duxbury, 1973.

Graves, W. L., and Bradshaw, B. R. Early reconception and contraceptive use among black teenage girls after an illegitimate birth. *American Journal of Public Health*, 1975, *65*(7): 735–740.

Hatcher, S. *The adolescent experience of pregnancy and abortion: A developmental analysis*. Ann Arbor, Mich.: University of Michigan, 1972.

Hunt, M. *Sexual behavior in the 1970's*. New York: Dell, 1974.

Hurlock, E. *Adolescent development* (4th ed.). New York: McGraw-Hill, 1973.

Johnson, T. R., Moore, W. M., and Jeffries, J. E. *Children are different: Developmental physiology* (2nd ed.). Columbus, Ohio: Ross Laboratories, 1978.

Katchadourian, H. *The biology of adolescents*. San Francisco: W. H. Freeman, 1977.

Keasey, C. B. The lack of sex differences in the moral judgement of preadolescents. *Journal of Social Psychology*, 1972, *86*, 157–158.

Knittle, J. Obesity in childhood: A problem in adipose cellular development. *Journal of Pediatrics*, 1972, *81*, 1048–1059.

Kohlberg, L. Stage and sequence. In Goslin, D. A. (ed.), *Handbook of socialization theory and research*. Chicago: Rand-McNally, 1969.

Learner, R. M., et al. Physical attractiveness, physical effectiveness and self-concept in late adolescents. *Adolescence*, 1976, *11*, 313–326.

Maccoby, E. E., and Jacklin, C. N. *The psychology of sex differences*. Stanford, Calif.: Stanford University Press, 1974.

Manela, R., and Hillebrand, W.: *EPSDT: Child health*. Washington, D.C.: United States Department of Health, Education and Welfare, Health Care Financing Administration (HCFA), 24528, 1977.

Martin, L. M. *Health care of women*. Philadelphia: J. B. Lippincott, 1978.

Masters, W. H., and Johnson, V. E.: Incest: The ultimate sexual taboo. *Redbook*, March 1976, *146*, 54.

Mercer, R. T. *Perspectives on adolescent health care*. Philadelphia: J. B. Lippincott, 1979.

Meyer, E., and Neumann, C. Management of the obese adolescent. *Pediatric Clinics of North America*, 1977, *24*, 123–132.

Millar, H. E. C. *Approaches to adolescent health care in the 1970's*, Rockville, Md.: U.S. Department of Health, Education, and Welfare, Pub. No. (H.S.A.) 75-5014, 1975.

Monthly vital statistics report. National Center for Health Statistics: Washington, D.C. Vol. 25, No. 10, Dec. 1976.

Monthly vital statistics report. *Advanced report, final mortality statistics 1977*. Washington, D.C.: National Center for Health Statistics, DHEW Pub. No. (PHS) 79-1120, Vol. 28, No. 1, Suppl. May, 1979.

National Clearinghouse for Smoking and Health. *Patterns and prevalence of teenage cigarette smoking*: 1968, 1970, and 1972. Washington, D.C.: US Department of Health, Education, and Welfare, 1972.

Okie, S. Death rates for teens continue to rise. Durham, N.C.: *Durham Morning Herald*, November 18, 1979.

Oppel, W. C. and Royston, A. B. Teen-age births: Some social, psychological and physical sequelae. *American Journal of Public Health*, 1971, *61*(4), 751–756.

Piaget, J., and Inhelder, B.: *The psychology of the child.* New York: Basic Books, 1969.

Piaget, J. *The moral judgement of the child.* New York: Harcourt, 1932.

Pierson, E. C., and D'Antonio, W. C. *Female and male: Dimensions of human sexuality.* Philadelphia: J. B. Lippincott, 1974.

Population reports. Washington, D.C.: George Washington University Medical Center, Series J, No. 14, 1977.

Prophit, P. The enigma of adolescent suicide. In Mercer, R. T. (ed.), *Perspectives on adolescent health care.* Philadelphia: J. B. Lippincott, 1979.

Rachal, J. V., et al. *A national survey of adolescent drinking behavior, attitudes and correlates.* Washington, D.C.: National Institute of Alcohol Abuse and Alcoholism, US Department of Health, Education, and Welfare, Contract No. HSM 42-73-80 (NIA), April, 1975.

Recommended dietary allowances. Washington, D.C.: Food and Nutrition Board, National Academy of Sciences, National Research Council, Revised 1979.

Redl, F. Pre-adolescents—what makes them tick? In Sze, W. (ed.), *Human life cycle.* New York: Jason Aronson, 1975.

Rice, F. P. *The adolescent: Development, relationships and culture,* Boston: Allyn, Bacon, 1975.

Roache, A. F., and Davike, G. H. Late adolescent growth in stature. *Pediatrics,* 1972, *50,* 874–880.

Rohn, R. D. Adolescents who attempt suicide. *Journal of Pediatrics,* 1977, *90,* 636–638.

Rogers, D. *The psychology of adolescence.* New York: Appleton-Century-Crofts, 1972.

Ross Roundtable. *Obesity in pediatric practice.* Columbus, Ohio: Ross Laboratories, 1971.

Saltzstein, H. D., Diamond, R. M., and Belenkly, M. Moral judgement level and conformity behavior. *Developmental Psychology,* 1972, 7, 327–336.

Schonberg, K., et al. Drugs, alcohol and tobacco abuse. In Hoekelman, R., et al (Eds.), *Principles of pediatrics: health care of the young.* New York: McGraw-Hill, 1978.

Simmons, R. G., et al.: Disturbances in self-image at adolescence. *American Sociological Review,* 1973, *38,* 553–568.

Smith, D., and Bierman, D. *The biological ages of man.* Philadelphia: W. B. Saunders, 1978.

Sorensen, R. C. *Adolescent sexuality in contemporary America.* New York: World Publishing Co., 1973.

STD Fact Sheet (Ed. 34). Washington, D.C.: US Department of Health, Education, and Welfare, Public Health Service, Center for Disease Control, HEW Publication No. (CDC) 79-8195.

Sullivan, H. S. *The interpersonal theory of psychiatry.* New York: Norton, 1953.

Sze, W. (ed.): *Human life cycle.* New York: Jason Aronson, 1975.

Tanner, J. M. Physical growth. In Mussen, P. (ed.), *Carmichael's manual of child psychology* (3rd ed.). New York: John Wiley & Sons, 1970.

Taylor, D. A new way to teach teens about contraceptives. *American Journal of Maternal Child Nursing,* 1976, *1*(6), 378–383.

Toore, C. T. Nutritional needs of adolescents. *American Journal of Maternal Child Nursing,* 1977, *2,* 118–127.

The new youth, *Life,* Fall, 1977, 12.

Wagner, N. N. Adolescent sexual behavior. In E. D. Evans (ed.), *Adolescents: Readings in behavior and development.* Hinsdale, Ill.: Dryden Press, 1970.

Weisbroth, S. P. Moral judgements, sex and parental identification in adults. *Developmental Psychology,* 1970, *2,* 369–402.

Zacharias, L., Rand, W. M., and Wurtman, R. J. A prospective study of sexual development and growth in American girls: The statistics of menarche. *Obstetrical Gynecological Survey* (Supp.), 1976, *31,* 325–327.

Zelnik, M., and Kantner, J. *Survey of female sexual behavior.* Conducted for the Commission on Population, Washington, D.C., 1972.

Zelnik, M., and Kantner, J.: The resolution of teenage first pregnancies. *Family Planning Perspectives,* 1974, *6*(2), 74–80.

Zlatnik, F. G., and Burmeister, L. F. Low gynecological age: an obstetric risk factor. *American Journal of Obstetrics and Gynecology,* 1977, *128*(2), 183–186.

10

The Young Adult

BERNICE D. OWEN

WHEN DOES YOUNG ADULTHOOD BEGIN?

In some societies, adult role expectations are placed upon young people when they reach a certain age. A number of cultures inaugurate the youth into young adulthood through such ceremonies as puberty rites, thereby clearly marking the boundary between the stages of youth and adulthood. The individual is then expected to act like an adult, and to assume the responsibilities of an adult.

In modern American society, however, there is no such demarcation between adolescence and young adulthood. Of course, legal sanctions confer some rights and responsbilities at certain ages; upon reaching 18 the individual is granted the legal right to vote and to hold some civic offices; the military accepts, and under some circumstances may draft, males when they reach the age of 18; and in some states, the 18-year-old has the right to purchase and drink alcoholic beverages. Age, however, is not a good criterion for determining when young adulthood begins. Some young people are young adults while they are still teenagers, and others do not attain young adulthood until their middle or late twenties, as it is not until then that they are able to cope with the responsibilities of adulthood.

Papalia and Olds (1978) state that many authors feel that the individual could be considered a young adult when financial independence is achieved, but many personality theorists, such as Freud, Erikson, and Fromm, cite maturity as the major criterion of young adulthood. Maturity, as these theorists see it, lies primarily in the ability to accept and cope successfully with certain developmental tasks. Starr and Goldstein (1975) cite personality characteristics that suggest a standard for maturity, Table 10-1 describes these personality traits.

Lugo and Hershey (1979, p. 182) list several more behaviors that reflect maturity:

Freedom to feel: the mature person is able to acknowledge feelings and express them; there is congruence between feelings and behavior.

Appropriateness of emotional expression: the mature individual is able to appropriately express various feelings as they are experienced.

Authenticity of being: all aspects of the mature person's life reflect the person as he really is.

Spontaneity: the mature person can respond without rigidity or compulsiveness to feelings at the moment of occurrence.

Sense of humor: maturity is reflected in the ability to look at oneself and others and smile.

Ability to handle stress: the mature person realizes that stress is a part of daily living and learns how to cope with it effectively.

There does not seem to be a single best criterion for the determination of when young adulthood begins, since individuals experience and cope with growth and development differently and at different chronological ages. There are, however, certain patterns that tend to correlate within certain time spans in relation to physical growth, cognitive development, and psychosocial development; and, hence, there are some common health and developmental problems. The patterns that tend to occur between the ages of about 20 to 40 will be considered in this chapter on the young adult.

Table 10-1 Personality Characteristics and Behaviors that Evince Maturity

Personality Characteristic	Behavioral Manifestations
A sense of responsibility	The person can be relied upon to carry out a defined role.
	The person can function independently of external supervision or coercion.
Adequate impulse control	Impulse can be channeled into constructive behavior.
	Expressed anger is appropriate according to the standards of society.
Tolerance of frustration	Can sacrifice immediate pleasure for long-term gain.
	Confident of ability to deal with stress.
Ability to plan	Able to look to the future.
	Able to set goals and implement plans.
Ability to accept differences in others	Has tolerance for diversity.
	Has respect for individuality.
Capacity for intimacy	Has the ability to trust another human being.
	Can share self with others.
Movement toward development of own potential	Has confidence in self.
	Can reach out and take risks.

SOURCE: Adapted from Starr, B., and Goldstein, H. *Human development and behavior: psychology in nursing*, pp. 70–73. Copyright © 1975 by Springer Publishing Company, Inc., New York. Used with permission.

The Emergence of Developmental Research in Young Adulthood

For years, the growth and development field was dominated by research in child development. Not until after World War II was there a shift in focus to the growth and development of the elderly. By the mid 1960s, emphasis was placed on the needs of the middle-aged adult, and by the 1970s, research on the young adult finally began to appear.

Elias, Elias, and Elias (1977) credit Sidney Pressey as one of the first authors to deal with some developmental needs of the young adult (in 1939). Levinson et al. (1978) consider Carl G. Jung to be the father of modern research in adult development. Jung recognized young adulthood as a time of coping with the demands of emotional involvements in family, work, and community; he also found that the individual experienced a time of turmoil around age 40. His writings on young adulthood appeared in the 1930s. Also in the 1930s, Charlotte Bühler developed a theoretical model of the life span that included young adulthood. She postulated that the life span comprises five phases and that the major tasks within these phases are those of goal establishment and goal attainment. The second and third phases occur during young adulthood. By young adulthood, the individual is able to establish goals that are much more specific and definitive. The experiences that the individual has play a major part in the person's ability to establish and maintain personal goals; successful experiences help the person in this respect, while failure can stymie growth (Bühler and Massarik, 1968).

Erik Erikson's theoretical model of the life span is also based on the concept of development by stages. These stages envelop the total life span, although Erikson emphasizes the crises of childhood. According to Erikson's theory, the major crises of young adulthood can be expressed as paired opposites: intimacy versus isolation and generativity versus stagnation.

In the 1970s, Levinson et al. (1978) and Sheehy (1976) postulated that there were major transitional periods in the lives of every young adult. These periods are generally devoted to particular tasks and fall within ranges of several years each; for example, Levinson cites the "leaving the family" period (ages 16 to 24), and Sheehy speaks of "pulling up roots" (ages 18 to 22).

It is ironic that research in human development, and especially in young adult development, has not advanced nearly as dramatically in any 20-year period as does the human being in the same span of time. Lugo and Hershey address the need for more research into young adulthood emphatically: "Although adulthood forms the central stage for the acting out of the human experience, it is the least researched and least understood of all the life-stages. Thus, there is less data available about its development than for infants, children, adolescents and the aged . . . " (1979, p. 658).

The impact of this lack of research upon the young adult is hinted at by Havighurst: "Early adulthood is the most individualistic period of life and the loneliest one, in the sense that the individual, or, at the most, two individuals, must proceed with a minimum of social attention and assistance to tackle the most important tasks of life" (1972, p. 83). As more research emerges and is absorbed by the general public, these young adults may well receive increased social attention and more will be known about how to help them with the successful completion of the tasks of young adulthood.

Societal Changes and The Transition from Adolescence to Young Adulthood

The time needed to make the transition from late adolescence to early adulthood has increased quite significantly since the 1960s. This has been especially true for middle- and upper-class individuals as more opportunities seem to be available to this group and they can generally defer becoming financially independent for a longer period of time. In comparing the young adults of the middle and upper class with those of the working class, it seems that the latter tend to complete their formal education earlier, assume full-time jobs at an earlier age, marry on the average of two years earlier (Levinson et al. 1978), and have children earlier in their marriages.

Whereas our age-based, mandatory educational system imposes role expectations and structure on the life of the adolescent, the end of secondary schooling brings with it an end to uniform societal expectations. In fact, great emphasis has been placed on the individual's freedom of choice, and these choices have become much more difficult to make.

Several changes of the 1970s provided young Americans with more alternatives regarding their futures. The federal government increased educational funding and programs, creating more educational opportunities for minority members. The withdrawal from Vietnam ended compulsory military service. Technological advances created new jobs and careers (and made some existing jobs obsolete). Some jobs that had been traditionally attained through promotion within the seniority system, such as supervisor or foreman, became attainable with the credentials of a formal education. Legislation designed to combat racist and sexist discrimination has changed the job market.

Sheehy (1976) states that society is generally more tolerant of diversity and there are therefore more choices in life-style open to the individual. She feels there is more acceptance of communal living, cohabitation, singlehood, couples who desire to remain childless, and experimentation with homosexuality and bisexuality.

Society also appears to grant more approval to those individuals who seek more formal education, thereby further postponing their adult responsibilities (Fig. 10-1). Society may even be more tolerant to those who need a time of moratorium during the difficult transition from adolescence to young adulthood; these moratoria may take the form of dropping out of school for a period or taking a number of different short-term jobs.

There appears to be less societal pressure on young adults to marry early. During the 1970s the average age for first marriages in the United States climbed into the early twenties for both men and women; also, the marriage rate per 1,000 dropped from 11.1 in 1972 to 10.0 in 1976 (National Center for Health Statistics, 1978) but by the end of 1979 the rate had climbed up past 10.5 (National Center for Health Statistics, 1980). The trend may eventually indicate a decrease in the number of people who are getting married. Levinson et al. (1978) found that society placed pressure on the male who was not married by his late twenties.

The increased number of choices open to many young adults has created in some a feeling of disorganization and a sense of lack of direction. Havighurst (1972) feels that it takes several years in this transitional phase to attain a somewhat secure feeling of direction in the adult world. Sheehy (1976) found that this period of transition is needed so that one can identify the things one wants to pursue in adulthood.

This period of transition may last a number of years for some young adults and may be nonexistent for others; the wide range of choices mentioned earlier is not available to all. For instance, some may not have the money or the ability to pursue studies in higher education; others may not be able to choose military service because of family responsibilities. However, a narrowing of one's range of choices does not necessarily arise from a lack of financial resources; affluence can have the same effect. For example, Yankelovich (1974) found that 72 percent of those who went to college (rather than a job) after high school came from families in which it was taken for granted that they would go on to college.

Kaluger and Kaluger (1979, p. 343) point out that the young adult's lack of experience is a significant limitation on good decision-making.

Figure 10-1. *Society is increasingly tolerant of the quest for additional formal education. (Reproduced with permission from University Publications, The Ohio State University.)*

According to them, "what is not recognized or fully realized by this age group is that their capacity to make good, valid judgments is restricted because they are limited in the effective use of their reasoning powers by their limited experiences, depth and breadth of knowledge, and inadequately developed foresight. Their idealism further influences their decision-making processes. As a result, their perspective concerning total reality is extremely narrow, restricted, and sometimes even distorted. . . . They do have a glimpse of the truth of reality but they are unprepared to see the whole picture." The lengthening of the transition period between adolescence and young adulthood will undoubtedly allow more young people to have more of the experiences that Kaluger and Ka-

luger feel are important in the preparation for the competent decision-making needed in adulthood.

Physical Growth of the Young Adult

Young adulthood is the time when the physical capacity of every system of the body peaks. Most physiologic functions reach maximum levels during the middle twenties; thereafter, in most adults, decline is so gradual as to be difficult to detect until middle age. Some of this decline may be caused by disuse rather than the normal aging process; if this is so, the process is probably reversible (de Vries, 1975).

In most people, skeletal development is completed in the early twenties. The skeletal differences between men and women are more pronounced during this range of years than at any other time during the life span. The young man is generally taller and heavier than the young woman; he has broader shoulders, a larger chest, and narrower hips, his feet and hands are larger, and his fingers and toes are heavier. The height that one achieves during these early adult years begins slowly to decrease at around 25 years of age (Schell, 1975) but the decline is generally unnoticeable until old age.

Both muscle tone and muscle strength appear to peak at between 20 and 30 years of age (Timiras, 1972); after this age there is a gradual decrease in the power and speed of muscular contractions and a reduced capacity for continued muscular effort. Chaffin (1975) studied 400 people who were employed in jobs requiring various degrees of lifting strength; he found that women had a mean strength of about 58 percent of the men's mean strength. Snook and Ciriello (1974) found that a woman's strength may average as low as 43 percent of a man's strength, depending on the woman's level of physical activity. This difference in physical strength may be partially attributable to the sex-role stereotyping that was prevalent until the early 1970s; boys and young men were encouraged to engage in activities that increased strength and endurance, while girls and young women engaged in activities that highlighted coordination. Men consequently developed strength while women did not.

Muscle fibers continue to grow thicker until age 25. One's level of physical activity exerts a considerable influence on muscle fiber thickness. Muscle fiber thickness has been found to increase even later into young adulthood as a result of strenuous exercise; however, it is felt that this increase is caused by an increase in the diameter of the fiber rather than in an increase in the number of muscle fiber cells (Timiras, 1972).

In most people, muscle mass is replaced by approximately one pound of body fat per year from ages 20 through 55 (de Vries, 1975). This means that even if the individual maintained his or her body weight at young adult values, that person would nevertheless be getting flabbier because of the decrease in muscle mass and the increase in fat tissue.

In healthy young adults the cardiovascular system is functioning at its peak level. De Vries (1975) indicates there are researchers who have found that there is a loss of strength of the myocardium of approximately 1 percent per year after the age of 20. Blood pressure readings remain stable during this period; systolic pressure of less than 140 and a diastolic pressure of less than 90 are considered normal. The pulse stabilizes at around 72 beats per minute. Evidence has been found, though, of significant changes in the lumen of the coronary arteries even in the young adult. Autopsies on 200 American battle casualties of the Korean war (mean age = 22.1 years) indicated that fatty streaks and early atherosclerotic plaques were present in 77.3 percent, and 15 percent had a narrowing of at least one coronary artery to 50 percent of its original size (Enos, Homes, Beyer, 1953). The American diet, which has been high in cholesterol and triglycerides, is a key suspect in these findings.

Development of the brain continues into the twenties; studies on myelination indicate that this process is completed by about 25 years of age (Timiras, 1972). Brain waves assume adult patterns in all cortical areas in the early twenties. As with other organs and systems, brain function generally declines at such a very slow pace that the decline may go unnoticed throughout one's life span.

According to Troll (1975), manual dexterity declines predictably during the young adult years. After about age 33, hand and finger movements become slower and progressively more clumsy. Troll also mentions that brain size begins to decrease after age 30, as there is an increase in moisture in the cerebral cortex as the solid material atrophies. However, no link between this change and a change in behavior has been established.

Simple tasks requiring movement take more effort after the age of 30, although, once again, the change takes place very slowly (Bromley, 1974). Reaction time improves until approximately age 19; it then remains constant through the age of 26, and then declines gradually and progressively (Schell, 1975).

Visual acuity is generally at its peak at around 20 years of age, and remains steady for most young adults. However, there are many exceptions to this generality, as visual acuity is one of the most variable human senses (Atchley, 1972). The rate of accommodation of the eye begins to diminish in young adulthood; this diminution continues throughout the life span.

Hearing loss, which begins in late adolescence, continues throughout young adulthood, although the loss is gradual. As one's age increases, higher frequencies and very soft sounds become harder to hear (Knox, 1977).

The reproductive systems of both men and women are fully mature by young adulthood. According to Kinsey (1948, 1953), the male libido peaks at somewhere between 15 and 20 years and the female libido peak is between 26 and 40 years of age. By young adulthood, the woman should be experiencing a regular pattern of menstruation.

According to de Vries (1975), researchers are questioning the extent to which physiological decline is directly attributable to aging, and to what extent it can be seen as a result of the sedentary way of life that is so common in America. His study indicated that with "vigorous" exercise, cardiovascular and respiratory functioning were significantly improved. Exercise improves strength and circulation in the cardiac muscle and decreases the severity of arteriosclerotic lesions, and exercise of the leg muscles is effective in facilitating the blood flow to the heart and therefore indirectly improving the functioning of all systems.

Cognitive Development of the Young Adult

Lugo and Hershey (1979) mention that a number of researchers, Piaget and Flavell among them, think that cognitive development may continue during adulthood. Such development may relate to such aspects as judgment, attitudes, and beliefs rather than the skills used in logical or mathematical operations.

Many of the early intelligence tests demonstrated a decrease in intellectual ability beginning at around age 25; however, these tests were usually based on content only (as opposed to tests requiring reasoning, experience, and evaluative ability), and the content was much more appropriate for the younger young adult than for people closer to the age of 40, for whom norms had not been established. Current researchers do not seem to be in agreement among themselves that intellectual ability decreases with age, nor is there a consensus regarding a working definition of "intelligence."

Wechsler saw intelligence as much more than the ability to deal with and arrange content; he defined it as "the aggregate or global capacity of the individual to act purposefully, to think rationally, to deal effectively with his environment . . . " (1950, p. 7). Horn (1972) deals with this dichotomy by proposing that there are two types of intelligence. *Fluid intelligence* is relatively independent of education and experience. It involves perceiving relationships, reasoning inductively, doing abstract thinking, and forming concepts. It also relies on speed of thinking, dexterity, and short-term memory. This type of intelligence is tested by using common tasks such as remembering a series of numbers. Because this kind of intelligence is dependent upon neurological development, it begins to decrease in the middle twenties. In contrast, *crystallized intelligence* evolves from experience. It is dependent upon the learning of specific tasks, the use of reasoning to solve problems of everyday living, and experiential evaluation. It can be measured by tests of verbal comprehension and general knowledge. A person's crystallized intelligence can continue to develop throughout most of the life span. Since IQ tests measure a mixture of fluid and crystallized intelligence, the IQ can remain quite stable throughout the adult years. As one's fluid intelligence decreases, one's crystallized intelligence increases and, therefore, the overall quotient remains essentially the same.

Kohn found in his sample of 687 men that the complexity of a job was the most important criterion in determining how much potential for intellectual growth the job offers. He states "if two men of equivalent intellectual ability were to start their careers in jobs differing in substantive complexity, the man in the more complex job would be likely to outstrip the other in further intellectual growth. This, in time, might lead to his attaining jobs of greater complexity, further affecting his intellectual growth. Meantime, the man in the less complex job would develop intellectually at a slower pace, perhaps not at all, and in the extreme case might even decline in his intellectual functioning. As a result, small differences in the substantive complexity of early jobs might lead to increasing differences in intellectual development" (1979, p. 71).

It is generally agreed that creativity and productivity peak during young adulthood. Studies of creativity indicate the peak is generally in the twenties for a wide variety of fields (Troll, 1975). The peak of productivity (that is, output) usually occurs around the middle or end of the thirties.

PSYCHOSOCIAL DEVELOPMENT OF THE YOUNG ADULT

For most people, young adulthood is a time when many difficult decisions need to be made. These decisions are essentially made quite in-

dependently, and many of them will affect the person for the rest of his or her life. This is the time to complete the breaking away from the family in order to establish emotional and financial independence. Intimate relationships must be established. One must commit oneself to a job or career. One's personal values must be studied, tested, and established as part of oneself. One must decide whether to marry and whether to have children. Other developmental tasks include forming friendships, choosing leisure activities, and taking on civic responsbilities (Fig. 10-2).

Breaking Away from the Family

For most people the process of breaking away from the family is not really one of separation,

Figure 10-2. *Deciding whether to marry is one of the most important developmental tasks of young adulthood.*

but of changing relationships; one continues to be a daughter or son, brother or sister.

Levinson et al. (1978) studied a group of 40 men between the ages of 35 and 45 who were biologists, novelists, executives, or salaried workers. Seventy percent of their sample were college educated. Even though these men had accomplished the task of breaking away, their responses are revealing in regard to this task. Levinson et al. found that the changes the individuals had to make were both internal and external. External changes included becoming financially independent and evolving new roles and living arrangements. The internal changes were those of increasing the psychological distance from the family, escalating the differentiation between self and parents, and reducing emotional dependence on parental support and authority (Levinson et al., 1978). Most of these men formed a life in this period that was quite different from that of their parents. Both Diekelmann (1977) and Kaluger and Kaluger (1979) report the same type of changes occuring within family relationships during the young adult years.

Emotional independence (which Levinson et al. consider an internal change) is the most difficult change to achieve. The individual must learn not to become unduly influenced by the emotional responses of parents. One must be able to make decisions in accordance with one's own judgment and problem-solving abilities. *Economic independence* is more than just receiving a few pay checks and making ends meet for a couple of months; it manifests itself as the ability to support oneself completely in all aspects of living. This type of independence generally implies growth in money management knowledge, knowledge of credit buying, purchasing skills, and an adequate income. *Physical independence*, of course, is easier to achieve if the individual lives away from the parents' home. If one does live with the family, it is essential for continued growth that the young adult be granted some freedom for decision making and that the adult role be expected.

Establishing Intimate Relationships

The capacity for sexual intimacy begins during adolescence, but at that time the person is generally not capable of entering into a fully intimate relationship until the identity crisis of adolescence is approaching resolution. One must have a fairly firm sense of who one is before one can consolidate one's identity with that of another human being in an intimate relationship. Once the young adult has achieved a sense of personal identity, there is a normal tendency to risk a relationship with another person.

It will be recalled that Erikson characterized young adulthood as the stage of intimacy versus isolation. He defines intimacy as "the capacity to commit [oneself] to concrete affiliations and partnerships and to develop the ethical strength to abide by such commitments, even though they may call for significant sacrifices and compromises" (1963, p. 263). Therefore, the young adult who attains intimacy becomes less egocentric. Inability to resolve conflicts in intimate relationships results in the *isolation* of the young adult, which is characterized by competitiveness, distance, egocentricity, and avoidance behaviors. Growth through intimacy, on the other hand, is not only important in itself but may also lay the groundwork for the selection of a mate.

Many factors influence the choice of a mate. It has been observed that in most couples, the partners are of similar socioeconomic, religious, and national background, are of the same race, and are close in age. According to Starr and Goldstein (1975) there is a higher divorce rate among couples in which the spouses are from different backgrounds. Other characteristics that are important to a lasting marriage are likeness in values, role expectations, and interests. Knox (1977) feels these characteristics are important but also adds effectiveness in interpersonal relationships, a satisfactory fit between one's actual mate and one's ideal, and having parents who were well-adjusted and good role models in their marriage.

Other significant factors in the choice of a partner are geographic and social proximity. Obviously, two people must have an opportunity to come in contact with one another before an attraction can develop. Some settings in which this can occur are one's neighborhood, place of employment, classes, place of worship, social organizations, etc.

Making a Commitment to a Career

Working is part of adulthood; both Freud (1930) and Erikson (1963) include the ability to work in their criteria of maturity and mental health. According to Freud, a good person is one who can love and work; Erikson's concepts of intimacy and generativity also embrace these capacities.

In all societies, work is a significant part of the life of every person within the social structure. One's occupation is the major determinant of one's level of prestige and amount of income. One's job influences where one lives, what kind of clothes one wears, who one's friends are, the kind of recreation and social life one has, the organizations one is involved in, and even one's health. For many, work provides a sense of personal worth, a level of recognition by others, a sense of influence on the course of one's own life, and a feeling of individuality, growth, and security. Therefore, one's job plays a major role in shaping one's self-concept.

Personality factors influence vocational choice, but other factors are essential to the process also. Parental encouragement is an important factor; this is especially true if the parent experiences job satisfaction and talks about job responsbilities. The socioeconomic status of the family also tends to influence vocational choice in several ways. It affects the level of the family's expectations regarding the child's career, and the family's ability to pay for schooling. Also, the child may be motivated to choose a career that will guarantee his ability to remain at the same socioeconomic level as his parents.

Abilities and interests demonstrated in formal education may have an impact on one's vocational choice. Also, a teacher that one has had or a specific course may become the stimulus for a vocational choice. Sometimes, too, a special incident has a great impact on the young person and eventually inspires a career choice; for instance, a positive experience as a patient in a hospital might cause an adolescent to consider a career in the health field. Part-time jobs may also help in suggesting a direction for the adolescent; for example, experience in baby sitting might suggest a career that involves working with children, such as teaching or pediatric nursing.

Unfortunately, many young adults must make vocational choices before they have much knowledge about the job. In some jobs the young adults must acquire the necessary skill and credentials before they are socialized enough to the job to make a good judgment about the wisdom of entering the field. The majority of young adults, however, do have some time to experiment with various types of jobs, since most enter the work force after graduation from high school.

Troll (1975, p. 135–136) divides the process of choosing and establishing oneself in a career into five stages:

Crystallization occurs between ages 14 and 18 and is marked by a developing awareness of various career choices.

Specification envelops the ages of 18 to 21 and involves the training of the individual for the job; some of this may occur in formal educational or trade programs, and some takes place in on-the-job training.

Implementation occurs between 21 and 24; at this stage one enters a beginning-level position.

Stabilization occurs between 25 and 33 years of age; at this stage one becomes established in one's field, having developed one's skills. This is often a time of great competition for many young adults who want to achieve greater status in the job.

Consolidation occurs from 35 years of age onward; at this stage one is very secure in one's job and may become a mentor to a younger person.

Levinson et al. (1978) stressed the importance of a *mentor* during the stages of specification and implementation; a mentor is one who takes a personal interest in a younger person, giving support, advice, and guidance concerning the job, perhaps within the work setting. A mentor could be a senior person on the production line who shields a new worker from the hostility directed at him for slowing down the line because of inexperience or the mentor might be an established professor of research who is willing to support and guide a younger faculty member in writing proposals or in seeking research funds. According to Levinson et al. (1978, p. 98), the mentor serves as a teacher, a sponsor, a host and guide who welcomes the young adult into the occupational and social world and who acquaints this person with the values, customs, resources, and other workers in the work setting, an exemplar to be emulated by the young adult, and a counselor who can provide moral support at times of stress and crisis. To put it simply, the most important function of a mentor is to provide support. The mentor's role is a mixture of parent and peer. As the young adult grows older and develops his skills the relationship between him and his mentor becomes more like that of peers, and there is more of a balance between giving and receiving. Levinson et al. found that these relationships often end in conflict when the young adult no longer needs a mentor.

Job Satisfaction

Yankelovich (1974) interviewed 3,522 men and women between the ages of 16 and 25 (with and without college educations) and found that blue-collar workers were less satisfied with their present jobs than either white-collar, professional, or managerial young people.

However, 78 percent of both the blue-collar and the white-collar workers were "very satisfied" or "somewhat satisfied" with their jobs. Yankelovich also found that only one out of three blue-collar workers feels there is a future in the job; these blue-collar workers recognize that their lack of education is a career barrier, and these same workers desire more training and would even take a cut in pay in order to acquire it.

In this same study, the blue-collar workers differed very little in their listing of things they wanted in a job. Both the blue-collar and white-collar groups selected the same top five priorities: friendly coworkers, interesting work, a chance to use your mind, work results that you can see, and a chance to develop skills and abilities. Good pay was the sixth priority for the blue-collar worker and the ninth for the white-collar worker.

When asked about specific sources of satisfaction in the present job, 50 percent of the total sample named friendly coworkers, good pay, and tangible work results. Only 50 percent of the white-collar workers felt that their work was interesting. Seventy-three percent of the total population said they would continue to work even if they had enough money to live comfortably without working (Yankelovich, 1974).

Caplan et al. (1975), under contract from the National Institute of Occupational Safety and Health, found that workers who reported higher job stresses were those who complained of underutilization of their abilities, little participation in decision making, minimally complex work, and little responsibility for others. These were factory assembly workers and relief workers on machine-paced lines who professed to high levels of boredom and who were dissatisfied with their work loads. The most satisfied workers in this study were university professors, family physicians, white-collar supervisors, police officers, and air traffic controllers in small airports.

According to Troll (1975), job satisfaction

among blacks and other minority groups has been consistently lower than among whites. Also, those young people who started college but were unable to complete it and those who wanted to go to graduate school and did not were less satisfied in their jobs than those who had less educational experience.

Women and Work

Making a commitment to a career is as important to the young woman as to the young man. According to the United States Department of Labor (1976), 56 percent of minority women and 54 percent of white women who were 18 to 64 years of age were in the labor force in 1975. During the 1970s it became more common for young mothers to return to work after the birth of a baby instead of waiting until the youngest child began school. That it has become more common for mothers to work is evidenced by such conveniences as day care centers.

More young mothers are also returning to formal education after the birth of the last expected child. Many are preparing themselves for better jobs. Some are starting or continuing a college education which was prevented or interrupted by marriage and child rearing. Others who were not initially interested or able to attend college are now doing so. Educational institutions throughout the country have devised special programs for these older students, who have different needs than the typical young college student.

Rapaport and Rapaport (1976) have found that over one-third of their sample of college-educated women were involved in continuous work orientations and were combining or expecting to combine the roles of wife, mother, and worker. Sheehy (1976) has labeled this life pattern as that of the "integrator." She warns, though, that a woman is rarely successful in this integrator role if she strives to integrate marriage, career, and motherhood while in her twenties; the combination is more feasible after 30 years of age.

As of this writing, women are seeking the same types of jobs that they have sought in the past. Forty percent of all professional and technical workers are women (United States Department of Labor, 1976) but these continue to be primarily in the typical female careers of teaching (below college level) and health care. Rapaport and Rapaport (1976) mention a number of reasons why women are not entering the traditional male fields as quickly as expected: lags in motivation among women to make the extra effort necessary to compete for the jobs; a cultural lag in the redistribution of domestic responsibilities that would allow working mothers to spend less time on child rearing and housework; and the efforts of men to defend their strongholds of power. Pressure from family members and peers may also be hindering factors.

Work, then, has different meanings for different people. It can be a source of prestige and recognition; it provides a basis for self-respect and a sense of worth; it may give enjoyment and may allow self-expression. Work provides an opportunity for social participation and a way of being of service to others. And, of course, it is a way of earning a living.

Establishing a Set of Values

In establishing one's own set of values, one must analyze and evaluate the set of values that has been internalized through childhood and adolescence under the direction of adults. The young person must develop his own philosophy, which will guide his actions throughout life.

Most young people are confused about values. They are idealistic but they find so many contradictory values in society. Many young people have been taught to respect the law, the lives and property of others, and the virtue of honesty; yet they see cheating and stealing, corruption in government, and countries at war with one another. Too many older adults exemplify the "do as I say, not as I do" attitude, which further confuses the young adult. In addition to this, there is a societal emphasis on personal freedom, on "doing your own thing"; consequently there are fewer guidelines that may be helpful as a starting point for formulating one's own values.

Even though many young adults are not in agreement with the policies and values of government and other institutions, the majority feel confident about their personal lives. They are in favor of social and political reform but they are also able to reconcile themselves to the present society.

A number of young adults are deeply dissatisfied with the values of society and the impact that these values have on themselves and others. In the 1960s and the early 1970s it was these young people who protested against these values. Most were generally very intelligent, from upper class families, and placed little value on conformity (Mussen et al., 1979). Many young adults remain dissatisfied with the values and practices of society; but with the termination of the 1970s, one saw few protests and little violence (Fig. 10-3). Many young adults seemed to be turning to groups such as Young Christian Fellowship groups to provide the organizing principles for guidance in establishing a personal set of values.

According to Kaluger and Kaluger young adults throughout the world are confused and agitated about the values of their own society and those of the world. "There appears to be a universal tendency for young people to want to change or reject the lifestyle imposed upon them by their elders" (1979, p. 346). Their method of expressing this confusion and agitation varies from country to country depending upon the degree of permissiveness in each society.

Kohlberg (1976) conducted a longitudinal study with a group of 50 boys aged 10 to 15;

every three years he questioned them about the resolution of a set of eleven moral dilemmas. He found that changes in moral thinking proceeded step by step and can progress through six stages. He found that the stages, and the order of the stages, were the same with subjects from urban middle and lower classes and subjects from other countries. Kohlberg found that the moral reasoning of many young adults was at the *conventional* level, in which the expectations of an individual, family, group, or nation are to be adhered to and upheld regardless of the intrinsic validity of those expectations. However, he also found that some young adults were able to progress to the higher level of moral reasoning which he called the *post-conventional*, autonomous, or principled level. At this level, the person makes an effort to evolve moral values that are valid and applicable regardless of who is in authority and regardless of the degree to which the person identifies with the authority. This level of reasoning is rooted in a belief in the rights of the individual, the principles of justice and equality, and the dignity of human beings.

Maturity of moral judgment is not highly correlated with intelligence but the development of logical reasoning is a necessary prerequisite for moral development. Advancing to the upper stages of moral reasoning requires opportunities for problem solving and reasoning in the personal, social, and moral realms of life. These experiences help the young adult develop judgment about what is right and fair. Kohlberg (1976) feels that very few adults reach the higher levels of moral reasonings. It should be noted, however, that the research done on moral reasoning has been studied essentially in hypothesized situations. There is as yet little evidence to indicate that the reasoning expressed in these hypothetical situations carries over into real situations.

In his study of college-educated and noncollege-educated young adults, Yankelovich (1974) found that both groups expressed satisfaction

Figure 10-3. *The peaceful demonstration of support for social and political reform is a significant part of life for many young adults. (Courtesy Chapel Hill Newspaper.)*

with their lives and both were optimistic about the future. In his comparison of the Vietnam War era to the post-Vietnam War era of the 1970s, he found that the values of the noncollege-educated and college-educated youth were much closer after the war. The noncollege-educated youth had become more like the college-educated youth in that they were more liberal in their sexual mores, expressed a rejection of unquestioning obedience to authority, looked to religion less for guidance in moral behavior,

and felt less patriotic. There was also less emphasis on the work ethic and more on self-fulfillment. About 15 percent of Yankelovich's sample were nonwhite, noncollege-educated youth; these minority members were more pessimistic about their immediate situations, were more critical of society, and placed much more importance on the value of money. In 1977, Yankelovich found a somewhat differing trend concerning sexual mores; premarital sex and extramarital affairs were approved by fewer young adults, and more of these young adults disapproved of having children out of wedlock (Kaluger and Kaluger, 1979).

Religious values typically undergo a change during young adulthood. Knox (1977) found a reduction in attendance and involvement in organized worship in young adulthood, which he attributes to the necessity of considering five major dimensions of religious commitment before a pattern of religious stability can evolve. After the individual has evaluated the five major dimensions—experimental, ideological, ritualistic, intellectual, and consequential—there is often a return to membership in a religion by the late twenties (Fig. 10-4).

Choosing an Adult Life-Style

Another imperative task in the growth and development of the young adult is that of choosing an adult life-style. Since the 1970s, young adults seem to be less pressured to live only in a tra-

Figure 10-4. *Sounding the shofar, the ancient ram's-horn trumpet used during the Jewish High Holidays. A renewed or heightened religious commitment often occurs in early adulthood. (Courtesy Chapel Hill Newspaper.)*

ditional marriage. The life-styles of singlehood, cohabitation, mate swapping, communal living, contract marriage, group marriage, and marriage without children appear to be more acceptable. Although these life-styles are not new in our society, there is more societal tolerance for them. Much of this tolerance is a direct result of the emphasis that has been placed upon individualism. The woman's liberation and gay liberation movements may have had an impact on societal tolerance also. Although it seems that more people are exploring these nontraditional life-styles, most couples still select the traditional marriage.

Singlehood

A trend toward singlehood began in the early 1970s; Kaluger and Kaluger (1979) cite statistics indicating that the number of young adults between the ages of 25 and 34 years of age who never married increased by 50 percent between 1970 and 1975. Also, the number of unmarried women 20 to 24 years of age increased from 28 percent to 40 percent in this same period. Kaluger and Kaluger feel that more young adults are remaining single because of less societal pressure to marry, because the media have made more people aware of marital conflicts and divorce, and because there are more opportunities for women in careers and they can therefore be their own breadwinners. However, it is too early to tell if this is really a trend toward singlehood; it may be merely the beginning of a trend toward marriage at a later age.

There are many reasons for selecting singlehood. Some people may not have mastered the establishment of intimate relationships, some may have painful memories of unhappiness in their parents' marriage, some may be impeded by mental or physical health, certain career goals may be incompatible with marriage, some people have not met a suitable mate, some may not desire the responsibilities of a marriage partner or parent, and others may simply not want to get married or may not feel ready for marriage. It is commonly thought that if a woman is not married by the age of 30, she probably will not marry as she is usually building up economic independence, is exhibiting investment in her work and has developed a value system that allows her sources of personal and social satisfaction in areas other than marriage (Papalia and Olds, 1978).

A number of stresses are common to those who select singlehood. Perhaps the greatest stress comes from the prejudices of society, family, friends, religious and professional organizations, and employers. As friends marry, they often associate more with married couples and they may see their single friends on special occasions only. Some religious groups tend to place emphasis on married couples and family through such things as family Bible study and mother-daughter or father-son banquets. Many professional organizations have activities built into their convention meetings for the spouse, who is expected to accompany the professional. Businesses may prefer to hire a married male executive since it is assumed that his wife will manage entertainment related to business affairs. Ibrahim (1980) points out that a number of studies have indicated that men and women who are single do not enjoy as high a level of physical health as their married counterparts. In addition to that, single men do not evidence as high a level of mental health as married men or single women (Gove, 1973). Married women are also low in mental health unless they are gainfully employed (Gove, 1977).

Marriage

Marriage is one of the few institutions that exists in all human societies. According to Abernathy: "the orderliness of society and the security of the individual are enhanced to the degree that a culture recognizes the legitimacy of some forms of social organization that clar-ifies the sexual and economic rights and obligations of men and women, and, through reproduction, provides for the continuity of the social group" (1976, p. 33). Marriage, then, is a socially sanctioned union between a man and woman, governed by laws, attitudes, and traditions of the society. Approximately 95 percent of all Americans marry sometime in their lives (Kaluger and Kaluger, 1979).

In American society it is expected that marriage is characterized by living together, having sexual relations, providing mutual support in times of crisis, and the expectation that the union is permanent (Starr and Goldstein, 1975). The permanency of this union is important since it necessitates a deeper level of personal involvement and commitment. Permanency is also desired for the optimal growth and development of the children born within the union; 90 percent of American families have children (Schell, 1975).

There are other reasons for marriage besides those of companionship, sexual relationships, love, support in crises, and procreation. Other reasons include economic security, desire for a home, parents' wishes, and prestige. Sheehy (1976) concluded from her interviews with 115 people that people's first marriages were prompted by others' expectations that they would marry; in other words, the reason for these marriages was to fulfill the expectations of others. Besides feeling that they "should" marry, these young adults also felt a need for safety (i.e., they wanted someone to care for them), they wanted someone to fill some emptiness in themselves, they wanted to get away from home, and some felt that marriage would provide prestige.

Preparation for Marriage

In our society there is no mandatory preparation for marriage other than the purchase of a marriage license within a certain number of days before the wedding date. It is usually the couple who decide when they are ready for marriage.

Levinson et al. (1978) feel that partners are never fully prepared for marriage, no matter how long and how well they have known each other. Some churches require that the couple complete a marriage preparation program before they can be married within the church; most of these programs take the form of small group discussions covering such aspects as communication, managing finances, sexual relationships, and adjustment to new roles.

Rapaport (1963) concluded that there were three major tasks in personal preparation for marriage: preparing oneself for the role of husband or wife, disengaging oneself from any exceptionally close relationships that might interfere with one's marital relationship, and accommodating the patterns of gratification that one had in single life to those of married life. Rapaport proposed that one could determine mastery of these three tasks by considering whether one is willing to be an exclusive sexual partner, is realistic about the economic problems that marriage entails, and has a realistic understanding of the other person's characteristics.

It is imperative that the young adult feel a high level of commitment to marriage, because very soon after the marriage ceremony, many developmental tasks face the couple as husband and wife and as a beginning family unit. At this same time, each of these young persons is expected to continue in the mastery of growth and developmental tasks that are common to all young adults.

Tasks of the Developing Family

Duvall (1977) has outlined eight basic tasks that are essential for the survival and continuity of the family (see Fig. 10-5). She warns that if the family does not fulfill these functions, society

1. Provide such aspects as shelter, food, clothing, and health care to family members.
2. Meet family costs and allocate such resources as time, space, and facilities according to the needs of each member.
3. Determine who does what in the support, management, and care of the home and its members.
4. Assure the socialization of each member through the internalization of increasingly mature roles in the family and outside the family.
5. Establish ways of interacting, communicating, expressing affection and sexuality, and exhibiting aggression within the limits acceptable to society.
6. Conceive (or adopt) and raise children; incorporate and release family members appropriately.
7. Relate to school, religion, work and community; establish policies for in-laws, relatives and guests.
8. Maintain morale and motivation; reward achievement; meet personal and family crises; set attainable goals; and develop family loyalties and values.

Figure 10-5. *Basic Family Tasks. (Adapted from Duvall, E. M. Basic family tasks, pp. 176–177 in Marriage and Family Development (5th ed.), [J. B. Lippincott]. Copyright © 1977 by J. B. Lippincott Company. Reprinted by permission of Harper & Row, Publishers, Inc.)*

will protect the family members by performing the functions for them.

The young adult may have already developed some skills before marriage that may be quite helpful in working through these basic family tasks with a mate. For instance, some may have been responsible for their own shelter, food, clothing, and health care; many have been financially independent so they have some skill in meeting their financial obligations and allocating resources. Some have already handled the responsibilities of managing their own apartments; and most have tested out various behaviors so they have some grasp of the limits that are acceptable to society.

Even though they may have developed these skills, the couple must decide who has the power for making the decisions needed in order to accomplish the tasks, what process will be usual for making these decisions, and which methods will be used when conflict arises in relation to these decisions. The courtship and engagement period allows the couple time to gain some experience with this decision-making process. Examples of some of the decisions generally made during this time are where the couple will live, what the relationships and responsibilities to in-laws will be, who will work to gain the income necessary to meet financial obligations, and whether children are desired immediately.

Some couples will experience difficulty with this decision-making process. Duvall (1977) cites a study reported in 1975 by the United States Office of Education in which it was found that 28 percent of the U.S. population were incompetent in problem solving, and another 23.4 percent were found to be able to "just get by." In addition, this study also found that only 58 percent of the U.S. population were proficient in writing, 46 percent in reading, 51 percent in ability to use community resources, and 37 percent proficient in managing a family budget.

It will take time for the couple to learn to live as a couple rather than as singles, to achieve a mutually satisfying sexual relationship, to integrate each of their styles of living and personality traits, to become aware of the realities of married life, and to plan realistically the short-term and long-term goals that they wish to achieve in their marriage.

The determination of the roles of husband and wife seems to be more the couple's decision than that of society. The traditional father-dominated family seems to be on the decline, and a democratic relationship within marriage on the rise. Yankelovich, Skelly, and White (1975) found that 61 percent of the spouses in their study felt they were partners in decision making within the marriage; the other 39 percent felt it was important that the husband be the main provider and decision maker and that wives should not work unless it was absolutely necessary. In his study of 1974, Yankelovich found that the majority of noncollege-educated married young women were devoting their lives to the roles of housewife and mother. Among the minority population (primarily black) 56 percent felt "the woman's place is in the home" and 54 percent felt that a man should make the decisions about family matters. Forty percent of the noncollege-educated women (of all races) felt a man should make the decisions, while only 20 percent of the college-educated stated this (Fig. 10-6).

Safilios-Rothschild (1976) claims that egalitarian relationships in marriages are merely given lip service; the husbands are still expected to be (and are rewarded for being) dominant, and they are ridiculed for yielding power to their wives. She found that husbands made the major decisions; for example, the husband decides how much money will be spent on food, and the wife decides how that money will be distributed. If, however, the wife works in a career that is not significantly below that of her husband's in terms of income and prestige, then she is more likely to share in making major decisions. Many authors agree that there is less role differentiation in young middle-class families as evidenced by greater sharing in house-

Figure 10-6. *The roles of husband and wife continue to evolve, partly in response to changes in the larger social order, but couples themselves ultimately determine whether their relationship as partners and parents will follow the older (male-dominated) pattern or a more democratic one. A shift to the latter is apparent among many couples, as evidenced, for example, by the increased involvement of many fathers in the tasks of parenting.*

hold chores and more joint responsibilities for the developmental needs of the children.

Sex in Marriage

It is essential that the couple strieve for a mutually satisfying sexual relationship; sex occupies an important place in the marital relationship. Masters and Johnson (1966) feel that sex is more important in marriage than it was during the Kinsey era (i.e., the 1940s); they found that intercourse was more frequent, sexual activities were more varied, and that both men and women were deriving more pleasure from sex. They felt this was because of greater societal emphasis on sex as a normal, healthy, and pleasurable aspect of marriage, a more liberal attitude toward sex, and the greater availability of professionals in the field of sexuality to provide clinical help and to make more information available.

According to Giele (1976), marital sexuality serves not only a biological function but is also an aspect of abstract moral and ethical issues of commitment. Sexual activity is symbolic of that commitment but the symbolism varies in different subcultures. In some subcultures the body is viewed as the most important thing a person has to give; therefore, sexual intercourse becomes the symbol of the deepest and most far-reaching commitment that a person can make, and this should be strictly limited to a one-pair bond. In other subcultures, sexual activity may be seen as a merely physical expression that does not involve the whole personality and consists primarily of the pleasure of the moment; in such a subculture sexual expression may be viewed more permissively.

Sexual adjustment in marriage depends upon a number of factors such as the strength of the partners' affection for each other and the meaning that sexual intercourse has for each. Also, the attitudes about sex that the young adult developed in childhood are important in this sexual adjustment. If one learned that sexuality is a normal, acceptable, and pleasurable aspect of life, then one has a better chance of developing a positive attitude about sex, and a mutually satisfying sexual relationship. If, however, the individual learned that sex is disgusting and dirty, then the negative attitudes developed may make a satisfying adjustment much more difficult to achieve. Sometimes an attitude change occurs when the young adults discover that they no longer need to feel guilty about having sexual intercourse or having sexual thoughts as they now have societal approval for sexual freedom within the marriage. It is also important that, if children are not desired, the young adults feel secure about whichever birth control method they are using.

Sexual adjustment requires an openness of communication and a willingness to learn. Each spouse must be able to let the partner know what is satisfying and not satisfying; this involves a mutual respect for the other's likes and dislikes as well as a willingness to accommodate the variations that each has in sexual needs.

Sheehy (1976) discussed the differences in the sexual needs of young men and women. She stated that men and women are most alike at birth, at 18, and past 60; therefore, between the ages of 18 and 60 they move toward opposite needs and reach the extreme of opposition at around age 40. Sheehy proposed that this progression from sameness to differentness and back again could be viewed as a diamond. Men and women start out quite alike and in their twenties begin to move apart in sexual capacity. Then by the late thirties or early forties the distance across the diamond is at its greatest as it is at this time that men and women are most strikingly dissimilar in their sexual capacities. Sheehy feels that around the age of 18 the sexual capacities of men and women may be quite similiar. She points out that the sexual capacity of women of this age is as yet unrecorded, but her data led her to believe that societies have always suspected a height of sexual capacity in young women, but took efforts to suppress these urges. Women are now beginning to admit the sexual intensity they felt when they were young, but many could not acknowledge or express it at the time because of society's double standard and religious prohibitions against premarital sexual activity. The young woman reaches another peak of sexual intensity in her mid-thirties and is capable of multiple orgasms; by this age, the man has lost his capacity for multiple orgasms and requires more time between erections. These changes in sexual capacity and sexual need indicate a need for dynamic sexual adjustment throughout young adulthood.

Sexual dysfunction, according to Starr and Goldstein (1975), is generally an expression of tensions, basic disagreements, and unresolved conflicts in nonsexual areas of daily living. They feel that the 1970s saw more couples seeking help for sexual dysfunction since sexual matters began to be discussed more openly. As was mentioned earlier, there are more professionals in the clinical practice of treating sexual dysfunction, and treatment has shown a high degree of success.

Marital Conflicts and Adjustments

A crucial ingredient in successful marriage is the willingness and ability of the partners to learn about one another and to grow in their ability to achieve important goals established as a couple. Knox (1977) found that good adjustment was more likely to occur when the husband and wife individually were happy people, when each was satisfied with the role expected of each, when each felt a mature love for the other, when the couple had a satisfying sexual relationship, and when they had shared interests that related to recreation and the care of the home. He also found that blue-collar workers found marital adjustment more difficult to cope with since they had less money and less occupational stability.

Most young couples do not live together before marriage; therefore, they must not only adjust to all the social roles associated with marriage, but they also must adjust to living with another person. However, the success of a marriage ultimately depends on the desire of the young adults to make their marriage a success and the appropriateness of their expectations of what the marriage will be.

Conflicts in marriage are inevitable. How these conflicts are resolved depends to a great extent on the communication pattern and problem-solving abilities developed by the couple to deal with differences between them. These conflicts generally arise from differences in the values and expectations held by the partners. Sheehy (1976) points out another potential source of conflict: she states that if couples are near the same age their developmental tempos are not synchronized and therefore conflicts arise easily. For instance, she found that men gain confidence in their twenties but women are losing confidence; men in their thirties want to settle down but women are becoming restless; a 40-year-old man may feel his strength and power decreasing while a 40-year-old woman may be brimming with ambition and ready to start a new life. Whatever the conflict, compromises need to be reached. Early in the relationship, rules need to be established on how conflicts will be resolved.

Childlessness

Not all married couples have children. Reasons that have been given by couples for childlessness include feeling incapable of raising children, not wanting their careers affected by added responsibilities, financial inadequacy, not wanting to increase the world's population, wanting to avoid transmitting a hereditary defect, not wanting to risk the wife's health, dislike of children, or inability to conceive or successfully carry to term a live infant. The childless marriage is of course more likely to be successful if both husband and wife desire no children, and if the wife has major interests to give her life meaning and direction.

Couples that experience problems with infertility often need special attention from health professionals. According to Moore, "A couple is considered infertile when conception does not occur within a year during which intercourse is frequent and contraception is not used" Moore, 1978, p. 149). Approximately 15 percent of couples who desire children will experience problems of infertility (Clausen, Flook, and Ford, 1977).

Infertility is found in all cultures, all economic groups, and all social classes. The ages of the man and woman, the frequency of intercourse, and the length of exposure (number of months of intercourse without the use of contraception) are all important factors affecting fertility. Moore (1978) found that male and female fertility is at its highest at approximately 24 to 25 years of age; intercourse at least 4 times a week enhances fertility; and 90 percent of couples will conceive within 18 months if they are using no contraception during that time.

Causes of infertility are often multiple; these causes may be nutritional, hormonal, emotional, coital, immunological, or physiological, i.e., abnormalities in the reproductive system of the man, woman, or both. Behrman and Kistner (1975) found that 30 to 35 percent of cases of infertility were caused by male factors such as low sperm count or poor motility of the spermatozoa.

Nurses need to be very sensitive to the needs of couples experiencing infertility problems. Support and encouragement are needed as the couple attempts to alleviate the infertility problem. Alternatives such as adoption may be considered if a partner is found to be sterile or if therapy is not successful.

Parenthood

The birth of a first child marks a major transition for the parents. Up until this event their major responsibilities have been to themselves as individuals and as a couple. With the entrance of a child into this relationship, there arises a need for the growth and development of the *family*.

According to Abernathy (1976), children are looked at as the desired outcome of marriage in all countries. However, Sheehy (1976) found in her interviews that whereas many couples in America expressed no desire for children early in their marriage, after age 30, many women decided they wanted children. Yankelovich (1974) interviewed young adults between 16 and 25 years of age, some enrolled in college and some not, and found that three out of four looked forward to having children; almost half of the noncollege youth felt it very important

to have children, while less than one-third of the college group felt that having children was very important.

Statistics indicate that the period between marriage and the birth of the first child is lengthening (Troll, 1975). Starr and Goldstein (1975) found that the more education the wife has, the longer the time before having children. They felt the reasons for this were that education generally means more job satisfaction as well as greater knowledge and use of birth control measures. Possible reasons for lengthening the time between marriage and having children are summarized in Figure 10-7.

The average size of the American family is also changing. In families with children the average number of children was 2.29 in 1970; in 1977 the average number of children was 2.0 (Grossman, 1977). Schell (1975) reported that 90 percent of American women have at least one child.

As to why couples desire to have children, some of course have no choice because of the lack of or ineffectiveness of a birth control method, and/or the religious or ethical views of the couple. Rising (1979) found that some

1. Desire to achieve a stable, intimate relationship before trying to accommodate the needs and demands of another human being.
2. Desire to become more financially solvent before incurring the expenses of having a baby.
3. Desire to wait until ready for the developmental tasks of parenthood.
4. Desire to complete education before having a child.
5. Desire to postpone being "tied down" until attainment of certain goals, such as travel.
6. Inability to decide whether children are desired.
7. Desire to permit the wife to get started in a career before bearing a child.

Figure 10-7. *Potential reasons for lengthening the time between marriage and having children.*

Figure 10-8. *The desire to have children is for many a powerful motivating force of adulthood. Erikson theorized that the drive to raise a family was a natural outgrowth of development. (Photo by Sam Gray, M. Photog., Cr.)*

couples desired children in order to fill a gap in their lives or in their relationship, to prove masculinity or femininity, or because they genuinely enjoyed children.

Erikson (1963) theorized that the desire to have children may arise from normal developmental needs (Fig. 10-8). After successful resolution of the intimacy conflict, Erikson states, the individual needs to share self with others; this can be accomplished by conceiving and rearing one's own children or by giving nurturance to young people such as scouts or students. Creative contributions to society may also fulfill the need for generativity. Erikson feels that lack of growth toward generativity results in self-absorption, and often in a sense of stagnation and impoverishment.

Parenthood as a Crisis Period

The birth of the first child is thought to be a major crisis for most young adults. Troll (1975) found that many women claimed that the greatest change in their lives took place at the time of the birth of their first child and that the crisis was caused by an abrupt shift in role and status for which they were inadequately prepared. She found that new mothers complained of months of sleeplessness, chronic exhaustion, and nervousness, and were frustrated by the lack of social activities, long hours of work, and their physical appearance. They also felt "tied down," and many longed for adult companionship; at the same time they felt ashamed of their negative feelings. Mussen et al. (1979) also cite the reduction in social activities, the increase in household duties, the loss of income from the wife's job at a time when expenses were very high, the fatigue of the mother, and the father's feelings of being neglected. Rising (1979) found that research did not consistently demonstrate that childbearing was a crisis; however, she found that parenthood results in changes or periods of adjustment and that the state of organization in the family at the time of the entrance of the new member is critical.

Most of the literature written for parents describes effective parenting techniques, teaches parents about the growth and developmental needs of the child, and discusses common problems that parents have with children. However, there is little written that helps parents through their own transition period so that they may effectively cope with parenthood.

Needless to say, there are joys and satisfactions in having children too. Campbell (1975) found that marital satisfaction was at a peak before and with the advent of the first child; however, during the transition period of adjustment to parenthood, marital satisfaction falls dramatically and does not again peak until the children have left home and are married. In addition to this, Troll (1975) found that most studies indicated that much of the responsibility for the children rests with the mother and that the father's greatest involvement was during the pregnancy and immediately following the birth of the baby. There is evidence, however, that fathers are becoming more involved with the care of the children, but this trend is less evident in lower-class families.

Preparation for Parenthood

Preparation for parenthood is an essential component of becoming a parent. Natural parents, as well as adoptive parents, have months to prepare for the entrance of the new member into the family. Most of these expectant parents have had little first-hand experience with the responsibilities that will be expected of them as new parents. The Group for the Advancement of Psychiatry (1973) proposes that there are four phases of parenthood. The first phase is that of *anticipation*; during this phase plans are made for the baby and the couple begins preparation of parenting skills. The second phase, the "honeymoon," is the time of adjustment of roles and the learning of new skills as the new member is incorporated into the family. The *plateau* phase covers infancy through the teenage years; during this phase the parents must continually adapt their behavior and expectations to the level of the child. The last phase of parenthood is *disengagement*; during this phase the parents must relinquish their active parental roles to permit complete independence of the child.

Some parents begin their anticipation phase by recalling their own experiences as children and making decisions about the extent to which they would like to model themselves, as parents, after their own parents. They may begin to read some of the many books that are available on the growth and developmental needs, as well as the care, of infants and children. The involvement of the expectant couple in groups, such as prenatal classes and natural childbirth groups, can offer the couples opportunities to share ideas, experiences, and anxieties with other expectant couples. These groups also help them to begin thinking about themselves as parents as well as to begin their preparation in certain practical facets of parenting such as planning for the clothing of the newborn.

Rising (1979) stresses that one of the most important tasks for the couple during the anticipation phase is the reassessment of their relationship. Together they need to determine which activities are important and enjoyable to them both. These activities need to be incorporated into the couple's routine in order to facilitate their continued growth as individuals, as a married couple, and as parents.

The expectant couple needs to assess their financial position and plan their fiscal strategies so that their needs and desires can be met within their budget. If the wife contributes income and can no longer do so, the couple needs to plan how they can maintain their standard of living at this time of added expense. It may be that the wife will return to work shortly after the birth of the child, or maybe the husband will take on a second job. Espenshade (1977) stated that the direct costs of raising a child from birth through college ranged from $44,200 for those with a net annual income of $10,500 to $13,500, to $64,200 for those with an income of $16,500 to $20,000. It seems that more families are opting for the wife's return to work after the birth of the baby. Grossman (1977) found that in two-parent families in 1977, the percentage of mothers in the work force was as follows: 37 percent if the children were under six; 48 percent if the children were between 6 and 13; and up to 55 percent if the children were 14 to 17 years of age. If the child and mother lived without the father, the proportion was substantially greater.

Growth as a Family

Havighurst (1972) proposes that there are two major developmental tasks that occur within the honeymoon and plateau phases of parenthood, namely, starting a family and rearing children. Some of the facets that he includes as part of these developmental tasks are learning to become parents, learning to meet the physical and emotional needs of children, adapting the schedule and pattern of living to the family, and assisting teenage children in becoming responsible and happy adults.

Duvall lists the following as developmental tasks that appear to be common within any stage of the family cycle:

1. learning the growth and developmental needs of the children and learning how to help fulfill these needs in order to enhance the growth of the children;
2. developing routines and rituals for the smooth functioning of the household and for highlighting special events within the family;

3. learning and adapting to the changing needs of individual family members in such areas as privacy, morale, and support;
4. learning and enhancing communication patterns that include all family members and that are open and honest;
5. planning for changing space needs for the family;
6. developing and maintaining the ability to provide financially for the needs of the family;
7. maintaining ties outside the nuclear family with friends and relatives;
8. defining and implementing roles and responsibilities in order to enhance the growth of the family and the individuals (Duvall, 1977).

Some developmental tasks that appear to be common for the married couple within any stage of the family's development are planning for future children, maintaining satisfying relationships with one another as husband and wife, and supporting and accepting each other in child rearing (Duvall, 1977).

In years gone by, it was the accepted role of the extended family to help the young and maturing nuclear family in the identification and mastery of these developmental tasks. However, with the high level of geographical mobility in our society and the increased emphasis on the independence and rights of the individual, the young family is now more isolated from the influence of the extended family. According to Yankelovich (1974), this is especially true of white-collar families.

Some communities, congregations, and mass media programs have become more responsive to the growth and developmental needs of modern-day married couples and families. Some communities have developed such resources as adult education courses in family living, family planning programs, libraries, and supportive network systems such as mental health centers. Religious institutions have also experienced a renewed interest in family growth and development as evidenced by the creation of marriage encounters groups and family living programs. Some junior high and high schools are developing classes for parents to help give them an understanding of what their children are learning. Some hospitals, clinics, and community education centers are offering health education programs on topics like drug dependency and the needs of the adolescent. Obviously, many of these programs serve needs that were formerly met by the extended family.

Parents who have similar and specific needs have banded together to establish national support groups such as Parents of Twins, Parents Anonymous, and Mothershare (a support group for mothers of children up to 3 months of age who help each other to ease the transition into motherhood). Parents have also established baby-sitting pools to enable members to take turns in fulfilling their needs to socialize or be alone.

Adoptive parents go through parenthood in essentially the same way as natural parents. However, in addition to the common developmental tasks of the couple and the family, these parents experience the additional stressors of accepting their own inability to have children (if this is the situation), acknowledging that the child is not their "own flesh and blood," and anticipating the time when they will reveal the fact of the adoption to the child (Papalia and Olds, 1978).

One-parent Families

Parenting in one-parent families carries with it a number of stressors also. Most parents in these families find it extremely difficult to try to meet the needs of their children alone. The majority of one-parent situations consist of mother and child (or children). In most of these families, the mother assumes sole responsibility for earning the family income in addition to caring for the children and the home. Grossman (1977) found that 11 percent of all children in the United States lived in one-parent families in 1970; however, by 1977 this figure had risen to 17 percent. She found that the majority of these children lived with their mothers and that black children were far more likely to be in one-parent families than white children. Duvall (1977) stated that 35.2 percent of all black children under the age of three were living with their mothers only in 1974.

High quality day-care centers and "after school programs" have been helpful to the working parent in these one-parent families. The Tax Reform Act of 1976, which granted tax credits for child care expenses, also was beneficial to these working parents. Fathers who are raising their children alone experience most of the same problems that confront mothers without partners.

Alternate Life-styles

Alternate life-styles to marriage are not new in our society. It is difficult to know if these alternate styles of living have actually become more popular; most authors feel there has been little or no increase, only that society has been more open about acknowledging such arrangements. In all societies there are a few people who develop their own life-styles while the majority remain unaffected. Troll (1975) found that the frequency of alternate life-styles was low and their duration was short. However, it is too early to tell if these styles of living truly herald any new directions.

Constantine and Constantine (1976) indicated that there has been a surge of interest and involvement in sodalities and extended groups because of the isolation of the nuclear family from the extended family. Some theorists in marriage and family life feel that drastic revisions of the monogamous relationship are coming, at least with regard to sex. It is also felt that marriage may be less stable for some couples as both partners become independent of one another (Giele, 1976).

Any life-style that deviates from those sanctioned by our changing and arbitrary societal norms could be defined as an alternate life-style. Examples of some of the present alternate life-styles follow.

Cohabitation

In this life-style a man and a woman live together without being legally married. Children may or may not be a part of this relationship. Theorists are not in agreement as to how the cohabitants view the relationship; some have found that they look at it as a trial marriage and others have found that the relationship is not viewed as having any potential for permanence. Schell (1975) found that many of the couples who live together do eventually marry. Kaluger and Kaluger (1979) state that unmarried couples living together make up 1 percent of the U.S. population.

"Swinging" (Mate Swapping)

In this situation, married couples exchange partners for limited encounters. Generally the relationship is sexual only. Troll (1975) states that most of these swingers are between 20 and 30 years of age, have been married for about 5 years, and are parents. They are predominately middle-class suburban couples who keep their swinging hidden from their neighbors. Troll states that a survey found that about 1 percent of American couples swing.

Intimate Network

In this alternate life-style, one or both of the spouses have an extramarital lover. There is emphasis on a close, open, and honest emotional relationship as well as on sexual intimacy. The married couple retains their own residence; their extramarital intimacies may take place in their own homes or away from home.

Group Marriage

In this arrangement three or more people commit themselves to one another and live together in a common residence. Generally, there is sharing of income and responsibilities. The total family life is affected, since the participants live together all the time. A common theme among participants is personal growth and security, along with sexual variety. These "marriages" are generally under male leadership. Communes may or may not be thought of as a form of group marriage; in some communes there is free sex among members, but other communes are characterized by serial monogamy. Yankelovich (1974), in his interviews with college men and women, found that in 1972, 36 percent of his sample were interested in spending some time in a communal or collective living situation; however, in 1974 only 30 percent were interested.

Constantine and Constantine (1976) determined that interpersonal tensions and pressures were inevitable concomitants of intimate groups. Members tended to have second thoughts about their involvement with a radical family life-style and about possible long-term consequences of such a relationship (especially with regard to children); there were also feelings of isolation from the mainstream of society. Constantine and Constantine's research, however, found no significant psychological problems among "group marrieds" as a class.

Important Transitional Periods in Young Adulthood

Gould (1975), Levinson et al. (1978) and Sheehy (1976) are among the researchers who have found that young adults experience significant transitional periods that often eventuate into crises during the course of normal growth and development. The first major transitional period generally occurs during the entrance into the adult world. For many it is difficult to pull up roots from the family and to try to establish oneself as an independent person. It may be hard to terminate or modify the long-term relationships developed during childhood and adolescence, and it certainly takes some sacrifice to support oneself on a somewhat limited income. The young adult may be chagrined to find that the support system of a peer group is quite different from that of the family. Independent decision making, and the acceptance of the consequences of those decisions, may be very difficult for some young adults. The number of changes occurring at this time, together with one's own and others' expectations of high-level achievements, creates the potential for crisis during this transitional period.

The early and mid-twenties also frequently present the young adult another transitional period with the necessity of making difficult and important decisions concerning marriage and occupation. Levinson et al. (1978) found that this transitional period created a moderate to severe crisis for 70 percent of their subjects. Those with no spouse, those having problems with a spouse, those who had yet to decide on an occupation, and those with no home base experienced the most difficulty.

The late twenties through early thirties comprise a crucial period of transition for young adults. This is a time for evaluating decisions made about career, marriage, values, children, life-style, peer relationships, goals, and dreams for the future. Dreams may be converted into more realistic goals, careers or jobs may be changed, marriages may be dissolved, extramarital "flings" may be started, the decision may be made to have a child, and the single adult may feel growing pressure to marry, if only because he or she perceives the pool of eligible partners to be shrinking. Levinson et al. (1978) stated that 62 percent of their sample went through a moderate to severe crisis during this transitional period; 18 percent acknowledged no crisis during this time. With or without turmoil, many young adults are ultimately satisfied with decisions made earlier, and experience a sense of confirmation and enrichment during this period.

Both Sheehy (1976) and Levinson et al. (1978) place the "midlife crisis" period between the mid-thirties and 40 years of age. Sheehy reports that the transition into middle life is as critical as the crisis of adolescence and may be even more harrowing. It is during this transitional period that the woman surveys her roles and options and then decides on the direction she needs to take for the second half of her life. She may now decide that staying home is too narrow for her and that she must return to work or go to school before aging closes her off from these opportunities. If she is working, she may decide it is time to attempt advancement or a change in careers. She may decide that this is the time for episodes of infidelity in her marriage, since she is reaching her sexual peak and she may fear that her attractiveness may soon begin to decline with age. Levinson et al. (1978) found that some of their sample of men also experienced turmoil in this period. However, more than half found they were happy with their work, family, community involvements, interests, and friendships; they then were able to refine and advance their plans for continued goal achievement. On the other hand, 20 percent felt they had failed or declined in the fulfillment of their goals.

Young adults go through many struggles in the quest for self-mastery and growth. Maladaptation is evidenced in such problems as continued narrowness and self-indulgence, isolation, stagnation, depression, dependency on alcohol or drugs, and divorce.

Establishing a Social Network

Friends are important for a number of reasons; they are sources of emotional support, stabilizers of self-image, and facilitators of integration with society. Researchers in stress and crisis intervention such as Selye (1956), Aquilera (1978), Rueveni (1979) and Mechanic (1978) all emphasize the imperativeness of a social support system to help the individual adapt to stress in times of crisis. Cobb (1976) has defined a social support system as a mechanism that assures the individual that he is cared for and loved, that he is esteemed and has value, and that he indeed belongs to a network of communication and mutual obligation.

Troll (1975) found that most married young adults generally have a large number of acquaintances and just three or four close friends. Close friends are those who share intimate feelings, see each other frequently, have a bond of love and deep respect between one another, and who remain friends for a long time. According to Knox (1977) there are five important dimensions of friendship: similarity of experience, reciprocity, compatibility, structural proximity, and contribution (see Fig. 10-9).

The well-educated young adult is generally as mobile as opportunities for advancement require him to be; therefore, his friendships are not necessarily long-lasting. However, less-educated people are usually not mobile and are thus more likely to have longer-lasting friendships that revolve mainly around job, neighbors, high school friends, and even elementary school friends.

DEVELOPING SATISFYING LEISURE ACTIVITIES

Leisure activities can help the young adult to meet needs for relaxation, creativity, self-expression, and exercise. The increased interest in physical fitness that arose in the late 1970s has prompted many young adults to participate in more active leisure activities such as jogging, running, swimming, and bicycling.

Yankelovich, Skelly, and White (1979) found that the adults who exercised regularly tended to be between 18 and 34 years of age, were more affluent, and were suburban family members; also, women who worked outside the home did more regular exercising than women who were not employed outside the home. However, only 36 percent of the 1,918 adults interviewed got some planned physical exercise at least several times a week. The adults in this study were asked what they liked to do best with their leisure time and were given three possible choices; 51 percent preferred to sit around and relax, 24 percent chose physical exercise, and going out to a movie, bar, or restaurant appealed to 25 percent.

It has often been pointed out that many jobs

Similarity of experience: Friends share experiences, enjoy the same activities, have similar interests, and communicate easily with one another.

Reciprocity: Friends are mutually supportive, accepting, trustworthy, and confidential.

Compatibility: Friends like one another and get along well.

Structural proximity: Friends are geographically close; it is convenient for them to see one another.

Contribution: Friends are role models for one another, help one another, and each can serve as the other's mentor.

Figure 10-9. *Important Dimensions of Friendship. (From Knox, A. B. Adult Development and Learning. San Francisco: Jossey-Bass, 1977, p. 229. Adapted with permission.)*

in modern America are routine and unfulfilling, and give the individual little opportunity for creativity, physical activity, or self-expression. Therefore, it is important to invest time and energy in the development of satisfying leisure activities. Starr and Goldstein (1975) feel that the de-emphasis of the work ethic permits people to enjoy leisure without feeling guilty.

Duvall (1977) suggests that leisure activities that can be enjoyed by a couple are especially important in the first few years of marriage and then again when the couple is adjusted to the departure of the last grown child. She reports that individual activities requiring no communication tend to be negatively correlated to marital satisfaction; parallel activities providing some interaction are somewhat positively correlated to husband-wife satisfaction; but a high degree of interaction and role exchange are positively correlated to marital satisfaction.

Most young adults do not use all the leisure time at play. Some go back to school, work in the garden, spend time practicing music, or develop some other skill such as sewing, cooking, or woodworking. Knox (1977) states that each year about one out of four adults engages on a part-time basis in educational activities provided by schools, colleges, libraries, churches, employers, associations, and community agencies. Also, in any five year period, half of the adults in the United States participate in at least one continuing educational activity. Knox also related that a national Gallup poll found that two thirds of those questioned read a newspaper every day, half read one or more magazines each week, a third had used a library during the year (another third had never done so), and those with more education tended to engage in more of these activities and watched little television.

Taking on Civic Responsibilities

Havighurst (1972) and Erikson (1963) are among the theorists who believe that it is important for the young adult to contribute to the community in order to continue to grow toward greater maturity. Havighurst (1972) found that young adults generally did not sense a need for developing any civic responsibility until they had property to care for, had children growing up in the community, or were paying local taxes.

Knox (1977) points out that there is generally a difference between white-collar and blue-collar adults in their assumption of civic responsibilities. The life-style of the blue-collar adult most often embraces work, home, and informal activities with friends and family; he is therefore usually neighborhood-oriented, or perhaps his associations extend as far as religious, union, and some fraternal organizations. White-collar adults, on the other hand, are generally more oriented toward events, activities, and ideas in the larger community. Community organizations tend to reflect the white-collar orientation toward formal procedures and written communications; these seem to discourage blue-collar adults.

Role changes (such as those brought about by the birth of a child) and frequent mobility have a significant impact on the young adult's involvement in civic affairs. Young adults who are new to the community may find it difficult to become involved in neighborhood or community affairs until they are better informed about such aspects as the values of the people, the issues and problems of the community, political factors, and informal support groups. Young adults undergoing role changes because of a new family member or a new job generally have less time for participation in civic affairs.

COMMON HEALTH AND DEVELOPMENTAL PROBLEMS OF YOUNG ADULTS

Young adulthood should be a time of good health and physical stamina, and indeed it is for most young people. Stress, however, is a major culprit in many of the health and developmental problems of young adulthood. Problems such as chemical dependency, family and community violence, suicide, hypertension, accidents, and many other physical and mental problems are manifestations of maladaptive responses to stress.

Many of these problems are found in the young adult years. The leading causes of death in this age group are accidents, suicide, malignant neoplasms, diseases of the heart, and homicide. Studies have shown that chemical dependency plays a significant role in some of these leading causes of death; in the *Forward Plan for Health* (1975), the U.S. Department of Health, Education and Welfare estimates there are at least 100 million American drinkers, and of these at least 9 million are classified as alcohol abusers. Alcohol plays a significant part in half of the nation's highway fatalities, half of the homicides, and one third of all suicides. Alcohol, along with drugs, has been implicated in 70 percent of all cases of domestic violence (Messolonghites, 1979). The national Institute on Alcohol Abuse and Alcoholism (1975) states that research has been limited thus far concerning the role of alcohol in crime; but, studies do indicate that alcohol consumption is relatively frequently involved in robbery, rape, and assault. The National Institute on Alcohol Abuse and Alcoholism also determined that the peak ages for problem drinking in men are 18 to 20 years of age, and again at the time of midlife crisis (37 to 40 years of age), which is also the peak problem drinking time for women.

Inadequate management of stress is responsible for at least half of all illness in society (Smith and Selye, 1979). Reubin (1979) found that stress—either acute stress or a stressor of any magnitude placed upon one who is already in a chronic state of stress—is generally the trigger for a suicide attempt.

Divorce is another problem that many young adults (and children) must deal with; it may also be a result, and certainly a cause, of stress. The

National Center for Health Statistics (1980) estimated that 1,170,000 divorces were granted in the United States during 1979; this represents an increase of 2 percent over the divorce rate for 1978. Most authors agree that the dissolution of a marriage is accompanied by great stress before and after the divorce; some even feel that the stress created is greater than that caused by the death of a spouse because of the stress, tension, and bitterness that occur before as well as after the divorce.

Family Violence

Family violence includes wife beating, husband beating, and child abuse. All forms of family violence are maladaptive responses to psychological tensions and external stresses, which affect all families at all socioeconomic levels. Gelles (1979), in citing a study that found marital violence in all social strata and in couples of all ages, remarked that families that experienced greater stress were more violent; also, in families where the husband and wife shared the decision making, the rate of violence was the lowest.

Davidson (1978), in his history of wife abuse, found reports that wives have been raped, choked, stabbed, shot, beaten, have had their jaws broken, and have been beaten with whips, pokers, bats, and bicycle chains for as long as records of family life have been kept. He also found that law enforcement officers estimated conjugal crime to affect about half of American families sometime during the marriage and that the safety of a significant percentage of women and their children was seriously threatened. Straus, Gelles, and Steinmetz (1979) estimated that 1.7 million of the 47 million couples living together in the United States experienced violence involving the use of a gun or knife, and that the beating of one spouse by another has occurred in over 2 million couples.

Gelles emphasizes that "People are more likely to be hit, beat up, physically injured, or even killed in their own homes by another family member than anywhere else, and by anyone else, in our society" (1979, p. 11). He interviewed 2,143 American families on how they dealt with marital conflict. He listed a number of responses, including acts of physical violence such as punching, biting, kicking, and the use of a knife or gun. Sixteen percent of the couples had used at least one of these forms of violence within the previous year.

In the early 1960s child abuse was recognized as a significant community and national problem. Then, in the early 1970s, legal and public attention (in addition to research) began to be directed toward wife beating, and since 1975, attention has also been focused on husband beating. By the late 1970s, attention began to be called to the beatings of elderly parents by their children.

There does not appear to be much societal concern about spouse beatings; this is probably because the spouse is not looked upon as an innocent, helpless victim, as children are. All 50 states and the District of Columbia have child abuse laws, but very few states have domestic violence laws applicable to spouse beating.

Authors generally seem to feel that husband beating is the least significant family violence problem. Gelles (1979) reported that as many husbands kill wives as vice versa, but wives are seven times more likely to kill in self-defense. There seem to be two problems in looking at husband abuse; the first is that women "abuse" primarily in self-defense, and the second is that women generally do not inflict as serious an injury upon the husband as the husband does upon the wife. Gelles also speculates that women are more vulnerable to violence than men because they are socially at a disadvantage; violence is most common with young adult families in which there are young children, the wife does not work outside the home, she has no income, and she is not prepared to leave home in order to protect herself.

Parker and Schumacher (1977) concluded

from interviews with 50 wives, 20 of whom were battered, that if the mother in the wife's family of origin was a victim of beatings, there was a significant probability that the wife would be battered by her husband. There was no significant correlation between the wife's victimization as a child and her being beaten as an adult. Another finding was that women who did not observe violence in their families of origin found wife battering inconsistent with their role and were able to cope with and avoid any further violence, either by leaving, threatening to leave (and following through on the threat if necessary), etc.

Gelles (1979) interviewed 80 families, 41 of whom were violent families. He found that if the abused wife had been abused by her parents, she was more inclined to stay with her husband. He also found that women who do not seek outside help are less likely to have completed high school and are likely to be unemployed. Those women who did seek help from police, other agencies, or a court generally did so only once because the initial attempt to get help was fruitless. In ten of Gelles's families, the violence occurred while the wife was pregnant. Gelles proposes five factors that may contribute to violence during pregnancy: sexual frustration, added stress to a family already under much pressure (such as from inadequate income or unemployment), unanticipated changes in the pregnant woman such as irritability and egocentrism, a desire to terminate the pregnancy or damage the unborn child, and the defenselessness of the wife because of the pregnancy. It appears, then, that violence during pregnancy may grow out of other stresses compounded by the stress of a child entering the family.

Eisenberg and Micklow (1977) conducted a study of 20 middle-class families in which the husbands' average educational attainment was 13.2 years; in half of the cases, violence occurred between the parents of the husband. These husbands attacked their wives' heads most frequently (and their faces second most frequently). The weapons used, in addition to feet and fists, were guns, knives, brooms, leather belts, a brush, a pillow (to smother), a hot iron, lighted cigarettes, and a piece of railroad track. Verbal threats to kill the wife and children were also common.

Although it has been widely assumed that alcohol is the exclusive cause of marital violence, there is growing evidence that the cause may be the family's inability to handle stress, and the husband's feelings of helplessness, hopelessness, loss or threatened loss of self-esteem, fear, and his inability to understand his feelings (Davidson, 1978). Valenti (1979) points out that the problem of wife beating is essentially a relationship problem and that most helping professionals can help families deal with relationship problems. The professional must be aware of the dynamics involved in the perpetuation of unhealthy relationships (Fig. 10-10).

Wife beating has a great impact on the children in the family. Some have been found to defend themselves psychologically by becoming apathetic; others take on a "battered personality," fearing every authority figure and expecting bad treatment as a part of life. Some identify with the power of the abuser and become aggressive toward strangers, loved ones, and even the mother. A very few may escape any psychological impairments, but research indicates that many of these children will comprise the next generation of wife beaters and physically abused wives (Davidson, 1978).

---

1. Both the abused and abusing spouse may have been abused as children or witnessed parents abuse each other.

2. Although violence is not acceptable to the victim, he or she has been exposed to violence so much that it is difficult to leave a violent family situation; the victim fears winding up in a similar situation again.

3. Low self-concept is often seen in both husband and wife.

4. The woman is socialized to be a "good wife," "good mother," and is therefore motivated to "keep the family together."

5. Many abused women feel financially dependent and are therefore less likely to leave a violent husband.

6. Many victims fear reprisal and do not know where to get help.

7. Those who fear loneliness may prefer an unhealthy relationship to none at all.

8. Some hope that things will get better.

9. Some believe they deserve physical abuse.

10. Alcohol may be involved; abuser and victim may both blame the alcohol for the violence.

Figure 10-10. *Dynamics of the Perpetuation of Unhealthy Relationships (From Valenti, C. Working with the physically abused woman. In Kjervik, D. K., and Martinson, I. M. (Eds.). Woman in stress: a nursing perspective. New York: Appleton-Century-Crofts, 1979, pp. 188–190. Adapted with permission.)*

The organized movement against wife beating started in England in 1971. In 1972 a group in Minnesota founded Women's Advocates, Inc. Members of this organization began a telephone information and referral service and even accepted physically abused women and their children into their own homes. Shelters were then established by 1974 (Davidson, 1978). Progress appears to be slow in the establishment of programs and shelters for battered women. By the late 1970s there were fewer than 100 shelters and most of these have been funded by private contributions only. In 1978, Congress defeated the Domestic Violence Assistance Act, which would have provided $16 million for research and services for domestic violence.

Sexually Transmitted Diseases (STD)

Syphilis and gonorrhea have long been major health problems primarily affecting adolescents and young adults. Other sexually transmitted diseases, such as trichomoniasis, nongonococcal urethritis, and genital herpes, have more recently become more common; probably since the early 1970's.

Trichomoniasis

According to the marketing data for medications sold for the treatment of trichomoniasis, G.D. Searle and Company claims that 2.8 million cases of trichomoniasis were treated during 1978; Hume (1979) estimates an increase of 500,000 cases in a one-year period.

Trichomoniasis is caused by a parasite of the genus *Trichomonas* that can live and multiply in the urinary or genital tract. According to Shapiro (1977) the man is usually asymptomatic but is a carrier of the disease. The woman's symptoms are usually itching and irritation of the vulva and a foul-smelling yellowish-green vaginal discharge. Trichomoniasis is especially common in men and women between the ages of 16 and 35. Hume (1979) states that 1 in 5

American woman may suffer from trichomoniasis during her lifetime.

The most effective treatment available is metronidazole (Flagyl) taken orally by both (or all) sexual partners. This treatment is effective, but a great obstacle to eliminating the problem is that asymptomatic men do not seek treatment. Topical creams or douches may eliminate the itching that the woman experiences, but they will not kill the parasite.

Nongonococcal Urethritis

About 2.5 million cases of nongonococcal urethritis occur in the United States each year (Hume, 1979). It is sometimes caused by Chlamydia trachomatis bacteria, but usually the causative agent is not identifiable. It is not caused by the gonococcus bacteria, although the signs and symptoms (discharge, urethritis, and cystitis) in the male are similiar to those of gonorrhea. Eye infection and pneumonia in infants born to infected mothers and pelvic inflammatory disease in women are some of the known consequences of infection by Chlamydia.

Genital Herpes

Genital herpes (herpes simplex, Type II) is caused by a virus; the signs and symptoms appear in men and women. Lesions appear on the penis or urethra in the infected man and on the external genitals, cervix, or vagina of the infected woman. Genital herpes is a known cause of neonatal death and is a prime suspect in cancer of the cervix. Califano (1979) points out that genital herpes can cause severe neurological damage in a newborn who is infected while passing through the birth canal; this occurs in about half of the infants delivered vaginally by infected morthers. For this reason, active lesions in a pregnant women indicate the need for Cesarean section. Approximately 500,000 cases of genital herpes occur annually in the United States (Califano, 1979).

Syphilis and Gonorrhea

Syphilis is caused by the spirochete Treponema pallidum and if untreated progresses through three stages. Signs and symptoms are experienced by both infected men and women. During the primary stage, painless chancres may be present at the site of the infection such as the lips, tongue, tonsils, vulva or penis; this stage lasts 2 to 10 weeks. If the individual remains untreated, there is progression to the next stage. This secondary stage of 3 to 14 weeks involves a generalized maculopapular rash and lymphadenopathy, malaise and fever. Without treatment, the third stage is spread over five years and there is involvement in the skin, bones, liver, cardiovascular system and the central nervous system. Penicillin continues to be the drug of choice in the treatment of syphilis.

Califano (1979) believes that syphilis has almost been eliminated in the general population of the United States with the exception of certain high-risk groups such as homosexuals, migrant workers, and the poor.

During the first quarter of 1979 the number of reported cases of syphilis increased by 19.2 percent over the number reported in the first quarter of 1978. When comparing only the number of cases reported in the months of March 1978 and March 1979, one finds an increase of 23.5 percent (*Morbidity and Mortality Weekly Report*, May 25, 1979). Public health officials believe the increase came after several years of a plateau because public attention has been shifted from the control of syphilis to that of gonorrhea. By 1973 all state and local health departments had established gonorrhea prevention programs with federal assistance. Within these programs therapy recommendations were made, clinical services were expanded, educational programs were offered, a culture-testing program was started for women in high-risk disease groups, and contact-tracing was emphasized in order to reach infected sexual partners. Consequently, attention has been directed toward gonorrhea, with quite successful results (*Morbidity and Mortality Weekly Report*, November 16, 1979).

From 1965 to 1975 the number of reported cases of gonorrhea in the United States increased from 324,925 to 999,937 cases. Since 1975 the number of cases leveled off, and by 1979 there had been only a 1 percent increase over the 1975 figure. This small increase in reported cases seems even more negligible when one considers that the high-risk age group (15 to 24 years of age) increased by 3.4 percent from 1975 to 1979. The federally assisted prevention programs have been credited for this success (*Morbidity and Mortality Weekly Report*, November 16, 1979).

Gonorrhea is caused by *Neisseria gonorrhoeae*. The disease is easy to treat but difficult to control because 70 to 80 percent of affected women and 10 to 20 percent of affected men are asymptomatic but are capable of spreading the disease (Ramaswamy and Smith, 1977).

The symptoms in the man are the same as those of urethritis: local burning and dysuria followed by a purulent urethral discharge. If untreated, urethral damage, evidenced by fibrous stricture and its complications, occurs. Often the infected woman is asymptomatic, but pelvic inflammatory disease and arthritis can occur. Untreated gonorrhea frequently leads to sterility in both men and women.

In 1979 the Center for Disease Control established new recommendations for treatment schedules for gonorrhea; there are now 4 drug regimens that can be used for the treatment of uncomplicated gonorrhea (*Morbidity and Mortality Weekly Report*, January 19, 1979).

Young adults must know about sexually transmitted diseases, the early symptoms of them, the kind of sexual activity that increases the risk, the necessity of seeking medical attention if they suspect they have such a disease, and how and where to secure such treatment. Men with several sexual partners should be advised to use a condom, since it offers some protection against STDs.

Periodontal Disease

Periodontal disease is a rather common dental health problem of young adulthood. Most dentists agree that the cause of this disease is multifaceted, the local cause being the various microorganisms that normally reside in the oral cavity; these microorganisms attach to the teeth, and if left undisturbed, they multiply to form colonies. These colonies, called bacterial plaque, contribute to the inflammatory disease process of the periodontium, resulting in gingivitis and periodontitis (Naval Graduate Dental School, 1975).

Systemic factors are also felt to be important in the origin of periodontal disease. Studies indicate that a number of systemic disorders affect the periodontium but no systemic disorder has been found that will cause chronic destructive periodontitis in the absence of bacterial plaque (Naval Graduate Dental School). Faulty nutrition and stress, which reduce resistance, may well contribute to periodontal disease. Stress can create an endocrine imbalance that can encourage the periodontal disease process.

Periodontal disease is preventable. The major factors in the control of this problem are plaque removal by a dental professional at least once a year, daily plaque control by the individual, good nutrition, and successful management of stress.

Although prophylactic visits to the dentist at least once every year for plaque removal are recommended, Bauer, Pierson, and House (1978) found that only half of the American population visit a dentist in any given year. Those who have higher incomes and those with dental insurance are most likely to go. Those with higher incomes come most often for preventive and restorative dentistry, while those with lower incomes come mainly for extractions and relief of pain.

Plaque control by the individual is the most important aspect of prevention. As yet there is no agent available that will totally prevent plaque formation or remove bacterial colonies once they have formed. Every surface of every tooth must be cleaned; this can be accomplished through proper tooth brushing and flossing techniques. Young adults must have the knowledge about how and why to do this, must be motivated to do it routinely and effectively, and must have the manual dexterity needed to perform the tasks and reach the tooth surfaces.

A number of methods of tooth brushing have been recommended for the control of plaque. The goal of all these techniques is the proper cleansing of the dentogingival junction, which is the critical area in periodontal disease. In effective brushing, the bristles of a soft brush are placed at the dentogingival junction and the brush is moved back and forth in very short, vibratory movements. The lingual, facial, and occlusal surfaces of the teeth are brushed in this manner (Naval Graduate Dental School, 1975). It is recommended that the tongue also be brushed in an effort to decrease the colonies of microorganisms that can contribute to re-formation of plaque on the surfaces of the teeth. The toothbrush should be replaced when the bristles lose their shape.

The proximal surfaces of the teeth must be flossed in order to remove plaque from between the teeth. Floss-holding devices should be recommended for those who have difficulty in flossing. Those with plaque problems should be encouraged to use a disclosing medium, which pinpoints the exact location of plaque in the mouth.

Cancer

Breast cancer is the most common malignancy among American women; one out of every 13 women in the United States will develop breast cancer at some time during her life (American Cancer Society, 1979). Although breast cancer occurs most frequently among women between the ages of 35 and 55, it is important that the young woman learn and practice breast self-examination as early as possible. This skill can help the young woman to detect any abnormality at the earliest possible time, and early detection and prompt treatment are the best defenses against cancer. Sadowsky (1979) states that 85 percent of symtomatic lesions are discovered by the patient. Only about 10 percent of all breast masses prove to be malignant (Burger, 1979).

The incidence of breast cancer is higher in white women, and especially Jewish women in higher economic brackets (Strax, 1974). The risk of breast cancer increases eightfold if one's mother or sister had premenopausal breast cancer. Also, there is some evidence that a diet high in animal fat may contribute in some way to cancer of the breast. On the other hand, a woman seems less likely to develop breast cancer if she delivers her first full-term baby before the age of 20, or if artificial menopause (removal of the ovaries) occurred before the age of 35 (Marchant, 1979).

Young women should examine their breasts every month, after the cessation of the menstrual flow. The breast self-examination includes visual inspection and palpation of the breasts. Several techniques are available for teaching the young adult to do breast self-examination. Caine (1979) has found it easy to teach patients to palpate properly by having them imagine the breast divided into four quadrants—central, superior, inferior, and medial, plus the axilla; the whole hand is used in palpation. Burger (1979) teaches the palpation technique recommended by the American Cancer Society, which involves palpating with the fat pads of the fingers in a circular fashion from the outer aspect of the breast to the nipple.

Until simpler and safer tools for the detection of early breast cancer are available, nurses must take every opportunity to help young women understand and implement breast self-examination on a routine monthly basis. Many free teaching materials, including easy-to-read pamphlets, are available from the American Cancer Society and from some pharmaceutical companies.

Men as well as women should be encouraged to do breast self-examinations. Even though cancer of the breast is not common in men, when it does occur, it metastasizes due to the scant amount of breast tissue present.

Testicular cancer is the most common solid tumor among American men in the 20-to-40 age group and is responsible for at least 12 percent of all cancer deaths among young men (American Cancer Society, 1978). Testicular cancer is most common in white men and is rarely found in blacks or Asians. The American Cancer Society (1978) states that if testicular cancer is caught in the early stages, it is one of the most curable cancers; if not, it is one of the most deadly cancers. Tumors of the testicle are usually malignant and they generally tend to metastasize early to the periaortic lymph nodes and lungs.

Altman and Malament (1967) point out that a malignancy is 48 times more likely to occur in an undescended testicle than in a normal descended testicle. Even testes that have been brought down surgically into the scrotal sac at an early age may become malignant. However, testes that are left in the abdomen are four times more likely to become malignant than testes that have been moved surgically. Other high-risk factors are family history of testicular cancer, recurrent injury, incomplete testicular development, and damage secondary to mumps and infection.

The testicular self-examination is easy to perform and takes only a few minutes; it should be done monthly (See Figure 10-11 for important pointers on testicular self-examination.) Any lumps that are found should be examined immediately by a physician.

Conklin et al. (1978) interviewed male college students about their knowledge of testicular

1. The best time to do the testicular self-examination is immediately after a warm bath or shower, when the skin, tissue, and muscles will be relaxed.

2. Examine each testicle separately.

3. Hold the scrotum in the palms of the hands and then slowly and gently palpate each testicle; palpate by rolling the testicle between the thumb and the fingers of both hands. The total surface area of each testicle should be palpated.

4. The examination should be done monthly; this will enable the examiner to establish a baseline for the consistency and size of his testicles.

5. The surface area should be smooth, with no palpable lumps, but the epididymis should not be mistaken for a lump; it is located on top of each testicle and extends downward behind it.

6. Males should examine themselves beginning at around 15 years of age and continuing throughout life.

7. The commonest tumor sites are the lateral and anterior surfaces of the testicles.

Figure 10-11. *Testicular Self-examination.*

self-examination; 58 percent had had a health-education course within the past two years, yet 75 percent had never heard of cancer of the testicules, and none knew how to examine their testicles correctly. These young men felt that they could and would do a monthly testicular self-examination; they mentioned that written instructions would be helpful.

Back Injury

Back injury has been found to be second only to the common cold as a cause of absenteeism in industry (Rowe, 1971). Back injuries account for the greatest amount of compensation paid by worker's compensation insurance (Taugher, 1973). It has been estimated that 70 to 80 percent of the American population will sometime suffer a back injury that will curtail usual daily activities for at least several days (Hipp, 1976).

The young adult years encompass the modal age group for the onset of back injury. Researchers, however, are not in agreement about the cause of back injury, but certain facts and theories can be brought to bear on an attempt to minimize the likelihood of back injury. Armstrong (1965) and other researchers have found, through cadavar studies, that minor but repeated stress to the vertical cartilage causes disc degeneration over time; therefore, frequent twisting movements of the back should be kept to a minimum. Lind and Petrofsky (1974) cited fatigued muscles caused by constant lifting as a contributing factor in low back injury. Therefore, the young adult should be counseled to take rest periods and to watch carefully for muscle fatigue.

Brown (1973) and Jones (1973) indicate that much more research is needed to determine the proper techniques for lifting. However, researchers are in agreement that *the straight-back-and-acutely-flexed-knees technique is not a safe position for lifting* (Chaffin, 1975). In this position most individuals cannot exert the upward thrust needed for lifting the load (and the weight of the body) and still retain their balance. Jones (1973) postulates that the best position from which to lift is one that employs a flexed back and flexed (but not acutely flexed) knees, with the feet spread to about shoulder width. He states that this position provides for a synchronized effort of the lumbar, abdominal, and thigh muslces, and that it provides also for maintained balance of the lifter.

Snook et al. (1978) determined that 49 percent of the back injuries in their sample were caused by lifting. Sixty-one percent of the worker's compensation claims for back injury studied by Taugher (1973) were also caused by lifting. A twisting movement during lifting is one of the commonest causes of back injury (International Labour Office, 1971).

Chaffin (1975) concluded, not surprisingly, that the weight of the load to be lifted is important. Lifting a load of more than 35 lb held close to the body, or 20 lb held about 50 to 75 cm in front of the body, may be potentially dangerous for some people. Also, if the arms are extended horizontally away from the body (either in front or to the sides of the torso), lifting loads smaller than 35 lb can produce equally high and potentially injurious forces in the low back. Tichauer, Miller, and Nathan (1973) agree that even light loads produced greater stress on the lumbar muscle groups if the load was bulky.

Chaffin (1975) and Schultz (1979) found that one should bring the torso as close to the load's center of gravity as possible before lifting it. Hyperflexion of the torso is potentially dangerous, as it places a greater stress on the posterior portions of the annulus of the discs. However, moderate flexion of the torso provides for more effective assistance from abdominal pres-

sure during lifting and therefore decreases the stress on the back musculature. Chaffin (1975) agrees that this "stooped over" position results in about one third less compressive stress on the low back than the squatting type of lift. This same researcher also found that the shoulder joint cannot withstand high forces when flexed or abducted; therefore, lifting from the chest, head, or above-the-head levels, in addition to the sides of the torso, is potentially injurious to the lumbar muscle group. Tichauer (1966) found that unanticipated motions during the lifting process, such as in trying to catch falling objects or regaining one's grip or foothold, contributed to back injuries.

Smith (1979) and Krames (1977) are adamant about the need for proper back care throughout life in order to decrease stress to the back; this includes sleeping on a firm mattress in order to keep the back in correct anatomical alignment. It is important, too, that one sleep in a position that assures correct alignment, such as on either side but not on the stomach. Good posture and low-heeled shoes that offer good support are important to back hygiene too.

The frequency with which one lifts heavy loads has also been found to be important. Chaffin (1975) concluded that lifting heavy loads more than 150 times a day, or less than 50 times a day, eventuated in a greater incidence of back injury; those whose daily routine necessitates lifting heavy loads 150 times a day or more are at risk for accelerated tissue degeneration, muscle fatigue, and uncoordinated muscle action during a lift. Those whose routines require 50 or less lifts a day, on the other hand, may be less accustomed to lifting and perhaps not in shape for it.

Preliminary findings indicate that people with family histories of back problems succumbed to back injury at an earlier age then those who had no family history of back problems (Owen, 1980). People with family histories of back problems should not hold jobs that

require frequent lifting, should know and practice good back care, and should use proper lifting techniques. They should be encouraged to
do back-strengthening exercises such as the pelvic tilt and sit-ups with flexed knees.

ANTICIPATORY GUIDANCE

Regarding anticipatory guidance for the young
adult, Havighurst (1972, p. 83) states: "of all
the periods of life, early adulthood is the fullest
of teachable moments and the emptiest of efforts to teach. It is a time of special sensitivity
and unusual readiness of the person to learn.
Early adulthood, the period from eighteen to
thirty, usually contains marriage, the first pregnancy, the first serious full-time job, the first
illnesses of children, the first experience of fur-
nishing or buying or building a house, and the
first venturing of the child off to school. If ever
people are motivated to learn and to learn
quickly, it is at times such as these." He goes
on to reiterate that society makes little effort to
teach those in this age group about early adulthood; even those who go to college receive little
formal education about the growth and development of the young adult.

Young adults may need help and support as
they struggle with attempts at independent decision making. Health care professionals can be
helpful by guiding them in establishing good
patterns for decision making. The individual
should establish goals, determine the alternatives
for attaining the goals, weigh the pros and cons
of each, select the best choice, plan a strategy
for implementation, and determine how one
will evaluate the decision. Using this delibera-
tive process can enable the young adult to make
good decisions, and this in turn can have an
impact on how the individual resolves conflicts
throughout life.

Premarital counseling is an excellent idea but
is generally not feasible because there are too
few trained professionals in this field of practice

for the number of couples getting married every
year. Therefore, the nurse should take every
opportunity to discuss potential problems in
family development with couples who are planning to get married. Special attention should be
given to establishing good communication patterns, and to planning ahead for strategies to be
used in conflict resolution. Nurses in schools,
industrial settings, and family planning clinics
may have greater opportunities to get involved
in these aspects.

Nurses need to be astute listeners when talking with clients. Often, the client may express
tensions relating to sexual dysfunction, which
may often be the result of unresolved conflicts
in other areas of daily living. Support, problem
solving, and, possibly, referral are important
strategies to use in averting a crisis.

Anticipatory guidance during pregnancy is
important for the expectant family. Health care
professionals need to anticipate, with the couple, important changes during pregnancy. It is
essential that both the man and the woman understand the normal physiologic and emotional
changes that occur during pregnancy. For instance, if they anticipate the fatigue usually associated with the first trimester, they can plan
for a reduction in the amount of work the
woman can accomplish, and plan sexual activity
for a time other than at the end of a tired day.
Classes and groups have been very helpful to
many expectant couples in preparing them to
plan for the many changes that pregnancy
brings.

The onset of parenthood is another time of
much information-seeking and learning. Parents should anticipate the incorporation of the
new child into the family, detecting and dealing
with sibling rivalry if there are other children
in the family, and caring for the needs of the
infant such as feeding, bathing, elimination,
clothing, sleeping, crying, safety, and toys.
Many couples may need help with learning how
to schedule and share domestic tasks so that the
whole family can progress in normal growth

and development. The family may need help and support with fulfilling the developmental tasks of the family as the children progress in age. Some communities and neighborhoods have groups and classes in family living; health care professionals should encourage family involvement in these if at all possible.

The nurse, then, can have an impact on the young adult's ability to anticipate and plan for normal growth and developmental needs, and on his or her ability to establish good patterns for decision making, communication, and use of a social support system. Also, the nurse can have an influence on the establishment of good habits of everyday living. Patterns developed in young adulthood for stress management, nutrition, exercise, immunizations, and birth control have a significant impact on health later on in life.

Stress

Stress is an essential component of daily life. However, prolonged or excessive stress can easily exceed the body's ability to adapt. Most people are confronted daily with stressful situations such as running behind schedule; while it is impossible to avoid stress (and sometimes an overload of it), it is not only possible but necessary to learn to live with it and keep the damaging side effects of prolonged or excessive stress to a minimum.

Selye (1956) defines stress as a state that manifests itself in what he calls the *general adaptation syndrome* (GAS). It is through this GAS that the internal organs, especially the endocrine and nervous systems, help one to adjust to the constant change that characterizes life. The three stages of Selye's GAS are the *alarm reaction* stage (the flight or fight reaction), the *resistance* stage (steadfastness/adaptation), and the stage of *exhaustion* (altered psychophysiological functioning). The stages of alarm reaction and resistance recur continually over a lifetime; exhaustion occurs when the individual can no longer offer resistance.

Any stressor can activate the GAS, including noise, overcrowding, poor nutrition, boredom, one's behavioral patterns, or infection. Cox (1970), in her sample of 63 college-educated young adults, found that frequent stressors were shattered love affairs, careers that were developing too slowly or in a disappointing way, divorce or conflict in one's parents' marriage, inability to determine if one had made a good vocational choice, financial dependency upon parents for too long, and relationships with mothers that were too close emotionally, too dominated by her, or too distant. The stressors most often mentioned by subjects in in-tact marriages were insufficiency of husband's income and dissatisfaction with sexual life.

The body's response to the GAS includes changes in the body chemistry such as increases in blood levels of lactate, epinephrine, and norepinephrine. These changes result in increased blood pressure, increased pulse rate, increased blood clotting time, irritability, and nervousness. In addition to these responses, O'Flynn-Comiskey (1979) reported on research findings that also indicated a rise in the serum cholesterol level in response to stress.

Besides these common body responses, there are others that seem to affect only some individuals, such as muscle tightness in the face, neck, shoulder, or back; a tight feeling in the stomach, an "upset stomach," or even vomiting; listlessness, apathy, fatigue, restlessness, depression, or insomnia; headaches, itching, diarrhea, anorexia, or a feeling of wanting to eat all the time; or a pervasive feeling of anxiety and tension. Factors that influence the resistance stage of the GAS must be identified. According to Smith and Selye (1979) there are three categories of variables that influence the individual's ability to adapt to the stressor. The first variable is the *perception of the individual*; the individual must recognize the stressor and must believe that he can adapt to the situation. The second factor is *conditioning*, which includes such aspects as ability and motivation to adapt, adequate nutritional status, appropriate genetic make-up, and sufficient self-esteem to allow the individual to react to, influence, and control the situation to which he is exposed. The general health status of the individual is important; one's prior experience with stressors is also important, as is the number of changes that are occurring at the same time. The third variable is *coping mechanisms*; the individual must have the ability, physiologically, emotionally, psychologically, and behaviorally, to maintain a state of equilibrium. As has been mentioned, the availability of a social support system in time of crisis is very important.

Inability to deal with stress can result in a number of physical and psychological problems. Smith and Selye (1979) report that between 50 and 80 percent of all diseases may be related to stress. Common stress-related diseases include ulcerative colitis, bronchial asthma, arthritis, myocardial infarction, hypertension, cancer, migraine headaches, cirrhosis, and much mental illness. Of course, such problems as alcoholism, drug abuse, violence, and suicide are also manifestations of maladaptation to stress.

Califano (1979) cited a study that revealed that pregnant women who were undergoing a great deal of stress and lacked a strong social support system were almost three times as likely to experience complications of pregnancy or delivery.

Holmes and Rahe (1967) hypothesized that any great change, even a pleasant one, produces stress, and if too many changes occur in too short a time, illness results. These researchers developed a scale assigning point values to common changes that frequently affect human beings such as death of a spouse, marriage, pregnancy, retirement, change in residence, and job change. (Of these examples, death of a spouse was assigned 100 points, and change of residence, 20 points.) They found that 80 percent of the subjects who exceeded 300 points within one year became depressed, had my-

ocardial infarctions, or developed some other serious ailment. Of those subjects who scored in the 250 to 300 range within one year, 53 percent succumbed to serious illness; 33 percent of those scoring up to 150 points were similarly affected.

According to Girdano and Everly (1979) one of the easiest and most effective techniques of stress management is to identify stress-promoting factors and develop a life-style that modifies or avoids these stressors; this falls under Smith and Selye's "perception" category. Many people need help in sorting out stressors and assigning priorities to those that can be changed. Some stressors may be easy to avoid, as in the case of driving to work via a different route in order to avoid the stress of freeway driving; other stressors may be very difficult to change, such as starting over in a new career or reducing the noise level at work. Smith and Selye (1979) suggest that if the stressor cannot be removed then the individual could be helped to lessen the intensity of the stressor by an approach such as changing goals.

Farquhar (1978) developed a self-scoring test for gauging stress and tension levels (see Figure 10.12). He recommends that the individual keep a daily stress and tension log that records the stress or tension felt, when it occurs, where, with whom, and the thoughts or feelings experienced. Farquhar found that this technique was helpful in identifying various stressors. Self-awareness groups may also be quite helpful to some individuals in identifying stressors.

Those aspects that Smith and Selye refer to as "conditioning factors" also need to be considered in the management of stress. Nutritional counseling may be essential for the young person whose usual diet is likely to contribute to stress. Caffeine, sugar, flour, salt, and a lack of certain vitamins have all been implicated as contributory to decreased stress management capability. Regarding vitamins, levels of vitamin C and the B complex vitamins need to be in-

BEHAVIOR	Often	A few times a week	Rarely
1. I feel tense, anxious, or have nervous indigestion.	2	1	0
2. People at work/home make me feel tense.	2	1	0
3. I eat/drink/smoke in response to tension.	2	1	0
4. I have tension or migraine headaches, or pain in the neck or shoulders, or insomnia.	2	1	0
5. I can't turn off my thoughts at night or on weekends long enough to feel relaxed and refreshed the next day.	2	1	0
6. I find it difficult to concentrate on what I'm doing because of worrying about other things.	2	1	0
7. I take tranquilizers (or other drugs) to relax.	2	1	0
8. I have difficulty finding enough time to relax.	2	1	0
9. Once I find the time, it is hard for me to relax.		Yes 1	No 0
10. My workday is made up of many deadlines.		Yes 1	No 0

Maximum total score = 18   My total score _____.

ZONE	SCORE	TENSION LEVEL
A	14–18	Considerably above average
B	10–13	Above average
C	6–9	Average
D	3–5	Below average
E	0–2	Considerably below average

Figure 10-12. *Simplified Self-scoring test for Gauging Stress and Tension Levels. (From Farquhar, J. W. The american way of life need not be hazardous to your health. New York: Norton. Copyright © 1978 by John W. Farquhar. Reproduced with permission.)*

creased during periods of stress in order to maintain the proper functioning of the endocrine and nervous systems. The B complex vitamins are also important because deficiencies in certain of these are responsible for anxiety reactions, depression, insomnia, and cardiovascular weakness; a deficit in certain others of the B complex group is known to cause stomach irritation and muscle weakness (Girdano and Everly, 1979).

The number of changes that are occurring simultaneously is another important conditioning factor in the control of stress. Many of the events that Holmes and Rahe included in their scale generally occur during young adulthood. These include marriage, marital separation, divorce, injury, dismissal from a job, pregnancy, sexual difficulties, birth of a child, change in financial state, taking out a mortgage, trouble with in-laws, beginning or ending school, and starting a new job. The young adult must anticipate such major events and try to ensure that not too many changes occur too quickly.

Coping with stress is a skill that must be learned, like any other skill. Learning how to manage stress effectively takes much patience and practice. Farquhar (1978) found in his experience in directing Stanford's Heart Disease Prevention Program that if people can learn and apply stress management techniques effectively, it is easier for them to make and maintain other life-style changes that are important to health such as in eating, smoking, and exercise habits.

Friedman and Rosenman (1974) have proposed that there are certain personality characteristics that tend to increase the risk of coronary artery disease; they term those who have such traits "Type A" persons. According to Friedman and Rosenman, the Type A person has a life-style that consistently involves a high level of stress. These individuals are highly competitive, constantly trying to move forward at a fast pace; they ignore any call for rest and relaxation, they get involved in many tasks at one time, and they often place unrealistic demands and deadlines on themselves. Some stress management techniques that might be recommended to such people are time management, the establishment of realistic goals, the development of strict limit-setting, and involvement in a relaxation training program. It is essential that these individuals recognize that their life-styles need changing, and that the goal of making the change must be assigned a higher priority than their everyday goals.

Girdano and Everly (1979) suggest that the problem of time restraints can be dealt with through the development of a time management system in which imperative tasks for a given period are listed, priorities are set, the time needed to complete each task is estimated; a 15-percent margin is added, and the tasks that cannot be accomplished are eliminated from the list. Because the Type A person has a problem with planning, it may be very helpful for him

or her to establish long-term and short-term goals. Relaxation training programs can also be helpful. The various methods of relaxing that have grown in popularity since the 1970s—yoga, meditation, progressive relaxation, hypnosis, and biofeedback—can all help people to learn how to relax.

Kezdi (1977), director of the Cox Heart Institute, suggests the use of *concentrative relaxation*, a technique akin to self-hypnosis in which one assumes a comfortable position in a relaxed atomsphere and concentrates on certain phrases that correspond to progressive levels of relaxation, e.g., "I'm completely relaxed," "my arm is very heavy," "my arm is warm," etc. Farquhar (1978) found the techniques of *deep muscle relaxation* and mental relaxation to be useful. The deep muscle relaxation technique is similar to the techniques proposed by Kezdi except that the individual tenses and then relaxes the muscles of the body. It is also very similar to the progressive relaxation techniques developed by Benson, who found (1975) that alternate relaxation and tension of various muscle groups for various periods of time enable the subject to relax. He also found, however, that when relaxation responses (lower heart rate, lower metabolism, and lower respiration rate) were elicited for more than 30 minutes twice daily, some subjects experienced symptoms such as insomnia and withdrawal behaviors (Benson, 1974). He warns that these exercises are meant only for relaxation and are not intended to provide an escape from everyday pressures.

In *mental relaxation* the individual practices imagining calm scenes and colors and repeats soothing words or sounds. When the individual has mastered the deep muscle and mental relaxation techniques, he can then combine them. Farquhar (1978) found that individuals who practiced the relaxation techniques regularly were able to reduce their blood pressure readings significantly.

Kezdi and Farquhar both warn that mastery of relaxation techniques takes frequent practice and much patience; it may take weeks of practice to produce even a remote feeling of relaxation. Also, several researchers suggest that the individual needs to enlist the support or active participation of a friend in order to increase the individual's likelihood of continuing with the exercises and of being successful at achieving a state of relaxation. Morris (1979), in her work with clinic patients, recorded the exercises and added background music to heighten the relaxing effects. The recording permitted the patients to concentrate on relaxation rather than on timing the periods of muscle tension and relaxation. Richter and Sloan (1979) report that one of the major effects of relaxation training is that of a feeling of self-reliance that stems from being able to make oneself relax instead of feeling helplessly controlled by tension. This, then, has a great impact on one's self-esteem.

Biofeedback is also popular in relaxation training. Biofeedback enables one to clearly distinguish between muscular tension and relaxation. The individual can therefore quite easily determine whether he is making progress in relaxation training. Some authors, however, feel that the electronic apparatus used in biofeedback is too expensive, and that one's subjective determination of progress is equally effective.

Blue-collar workers as a group also exhibit high levels of stress. Caplan et al. (1975) mentioned that since the mid 1960s much research has been devoted to decreasing stress in the workplace. However, these studies generally attempted to improve several aspects of the workplace at once and hence one cannot determine which factors made a difference; control groups were generally not used either.

Job rotation has been successful in some settings; this innovation involves scheduling the worker whose usual job is simple, boring, or repetitive to do other work for 4 hours out of an 8-hour shift, or during alternate months. Job expansion has also been effective in some companies; here the workers form teams, which become responsible for the total job, including the support functions of maintenance, quality control, and custodial and personnel functions. Some effort has been focused on changing the style of supervision so that workers can carry more responsibility for other workers, see job results, and hence gain more satisfaction from their work. Many companies are encouraging their workers to become involved in relaxation groups that often practice on company time.

The vital importance of a support system in times of crisis has already been discussed. Many authors indicate that one's capacity to handle the stresses of life depends on how effective the individual's social support system is. Rueveni (1979) points out that there are a number of different social support systems that may be available to the young adult, i.e., the family, one's friends and neighbors, professionals in therapeutic relationships with the individual, and fellow members of a club. Self-help groups such as Parents Anonymous and Mothershare can extend support to a parent in a time of stress. All of these different support groups can serve their function by helping the young adult to cope with stressful situations.

Involvement in hobbies and other diversions may be helpful for the young adult in times of stress. These activities may provide time for making decisions, reexamining problems, and deciding to seek out or make use of a support system. Care must be taken, though, that the individual does not turn to hobbies and diversions as a means of avoiding everyday pressures.

Routinely scheduled physical activity has also been found to be helpful in decreasing anxiety. Girdano and Everly (1979) found that their subjects reported a feeling of relaxation about 90 minutes after a bout of physical activity. Many who engage in regular physical activity cite benefits of enhanced self-esteem, an increased feeling of self-reliance, decreased anxiety, and relief from mild depression.

Since stressful events are an inevitable part of life, preparation for coping with them should begin early. Cobb stated "one cannot escape the conclusion that the world would be a healthier place if training in supportive behavior were built into the routines of our homes and schools, and support worker roles were institutionalized" (1979, p.103). Califano (1979) also emphasized the importance of training children and young adults in coping skills.

The nurse can play a vital role in helping young adults (and their children) to learn about stress and coping. In order to do this, the nurse must be knowledgeable about the causes and effects of stress, the assessment of factors that can help or impede adaptation, techniques that can be helpful in reducing stress, and how to help the individual in controlling stress. Of course, setting a good example is always part of the nurse's role. Because stress is so pervasive in our society, it is important that nurses consistently include stress assessment as a part of all patient encounters. Nurses must be astute in assessing tension and maladaptive responses to high levels of stress.

Prevention of the problems related to stress should be a primary goal of all health professionals. All nurses, regardless of the setting in which they work; should be helping patients strive toward effective stress management.

Nutrition

In 1977 the Senate Select Committee on Nutrition and Human Needs published a study noting that six of the ten leading causes of death in the United States are linked to diet: myocardial infarctions, cerebral vascular accidents, arteriosclerosis, cancer, cirrhosis of the liver, and diabetes. The committee advises Americans to:

1. decrease calories and increase energy expenditure if they need to lose weight; balance calorie intake with energy expenditure if they need to maintain weight;
2. increase intake of complex carbohydrates (whole grains and vegetables) and naturally occurring sugars (fruit) from 28 percent of intake to 48 percent;
3. reduce consumption of refined sugars by 45 percent so that it accounts for only 10 percent ot total intake;
4. reduce fat intake from 40 to 30 percent of intake;
5. reduce saturated fats to 10 percent of intake;
6. decrease cholesterol to 300 mg per day;
7. limit intake of salt to 5 gm per day (Singer, 1979).

Some interesting facts reported by Brewster and Jacobson (1978) can be helpful in understanding certain problems in the diet of the young adult. From 1946 to 1976, there was an almost 400-percent increase in the per capita consumption of soft drinks in the United States. In the same period, fruit consumption increased but processed fruit accounts for most of the increase; fresh fruit consumption declined by one third. Consumption of red meat, especially beef, increased every year. The intake of hydrogenated fats increased. (Fats arc hydogenated by a process that converts polyunsaturated fats into monounsaturated and saturated fats; saturated fats increase cholesterol levels, and monounsaturated fats have no effect on cholesterol level, but polyunsaturated lowers blood cholesterol levels.) Consumption of sugar and artificial sweetners increased every year; sugar is added to many processed foods such as catsup and hot dogs. These sweetners add no nutrition, and they contribute to obesity and tooth decay. Americans eat more vegetables than they used to, but most of them are canned and have salt added during the processing. Sodium consumption increased, mainly because of the addition of sodium in processed foods in the forms of sodium chloride, sodium nitrate, sodium benzoate, and monosodium glutamate. (The Food and Drug Administration does not yet require that salt or sugar additives to food be accounted for in nutritional information given on labels.) Caloric intake from alcoholic beverages increased by about 25 percent per day per person between 1961 and 1976. In addition the mean iron intake was 11.95 mg per person per day as discovered by the Health and Nutrition Examination Survey (HANES) of 1971–74 (Brewster and Jacobson, 1978). This intake is adequate for men but young adult women need 18 mg per day. This same study found that on the average young adults weighed six to seven pounds more in 1971 through 1974 then they did in 1960 through 1962.

In order to help young adults to develop good nutritional habits, a valid baseline must be established. An *honest* log of one week of usual intake, including all snacks and beverages, should be obtained; data from several 24 hour recalls can suffice if more data is not available. Data should also be gathered on the method of preparation of foods and additives to the food; for instance, was the meat broiled, boiled, roasted, baked, or fried? Were margarine, butter, salt, or sugar added to the food? Evaluation of this data will help the nurse to assess dietary adequacy.

The goal is to help the young adult develop healthy eating habits, and that means moderation. Therefore, caloric intake should balance energy expenditure, unless the patient is overweight, in which case caloric intake must be decreased and energy expenditure, increased. Red meats (pork and beef) should be eaten only several times per week; poultry and fish (not shellfish) have less saturated fats than red meats. Cholesterol intake should be lowered, so the young adult should limit egg yolks, organ meats, shellfish, and dairy products, which are high in fat content. Skim milk, or milk containing only 1 or 2 percent butterfat, should be recommended. Meats should be prepared by methods other than frying. Salt and sugar (including refined sugar) should be used sparingly if at all, and prepared and processed foods should be kept to a minimum. Fresh fruits and vegetables, whole grains cereals and breads, and low-calorie snacks and beverages should be encouraged. Women whose dietary sources of iron do not supply enough should be encouraged to take iron supplements.

Exercise

If a pattern of physical exercise is established in young adulthood it is more likely to become part of one's life-style for the rest of one's life. Thomas (1979) cited a 1977 Gallup Poll finding that almost half of all American adults exercise regularly to keep in shape, but Yankelovich, Skelly, and White (1979) found that two thirds of adults get almost no regular exercise.

The sedentary life-style of many Americans has contributed significantly to obesity and coronary heart disease. As discussed earlier, de Vries (1975) showed that exercise increased cardiac and pulmonary outputs and therefore a decrease in these functions were not due to normal aging but rather were due to the deconditioning of sedentary life.

Vigorous physical exercise (20 to 30 minutes, three times a week) has been found to decrease the risk of heart disease by reducing serum triglycerides and cholesterol, decreasing the resting heart rate, increasing cardiac stroke volume, increasing maximum working capacity of the heart and body, decreasing clotting tendencies, and decreasing resistance in the blood vessels. It is thought that exercise may also cause an increase in collateral circulation. Aerobic, or isotonic, exercise (that requiring large amounts of oxygen for energy production) is most beneficial to the cardiovascular system; examples of aerobic exercise are swimming, biking, brisk walking; and running.

Surprisingly, exercise and fitness programs have not received much emphasis from health professionals. Califano (1979) states that 80 percent of patients in one study could not recall that their physicians had ever recommended exercise. Similarly, whereas cardiac rehabilitation programs that emphasize prevention of a recurrence have been popular since the 1970s, primary prevention has not received equal attention.

Many businesses have begun to offer exercise and fitness programs to their employees on a trial basis. While the benefits of exercise to the individual cannot be doubted, large businesses are hoping that their exercise programs will benefit the companies also by reducing absenteeism, low productivity, hospitalizations, and premature deaths. Some programs are set up for executives and managerial staff only while others are established for all employees. Most of the programs are set up on the companies' premises, take place before, during, or after the work day, and generally cost the employee nothing.

Young adults should be encouraged to establish a pattern of exercise that is tolerable (or better yet, enjoyable). They should be reminded that some of their routine daily activities can be modified so as to provide exercise; for instance, a bicycle could be used instead of a car for short trips, or one can choose more active recreational pursuits such as dancing or playing tennis. Farquhar (1978) points out that the one pound per year of weight gain experienced by the average adult from ages 20 to 50 can be prevented by as little as 10 extra minutes of brisk walking per day.

Cooper (1977) warns that, before embarking on a cardiovascular fitness program, young adults under 30 should have had a complete physical examination within the past year; if between 30 and 39, they should have a physical exam within 3 months of commencing exercises, and the exam should include an electrocardiogram taken at rest. (An electrocardiogram while exercising is suggested for adults 40 to 59.) The individual should do limbering-up exercises before vigorous exercises. Of course, one should begin an exercise regimen slowly, gradually increasing the amount of activity. Cooper gives extensive guidelines to the exerciser on how to choose an exercise program, how to take the resting and training pulse rates, how to monitor one's progress, and how to evaluate the effectiveness of one's program.

Immunizations

Diphtheria/tetanus (Td) boosters at 10-year intervals throughout the adult life span are recommended. However, if an injury of the type that may cause tetanus is sustained, a booster dose should be given if five years have elapsed since the last booster.

Fulginiti (1979) states that diphtheria toxoid does not provide 100 percent protection but research data suggest that protection is increased five- to tenfold in the immunized. Also, if the disease is contracted by the immunized individual, the severity of the illness is decreased and death is rare. Children very rarely have a sensitive or toxic reaction to the diphtheria toxoid, but with increasing age and exposure to the toxoid a hypersensitive reaction could occur.

The tetanus booster is the most effective antigen used in immunization therapy. Epidemiologists feel that protection may be as high as 100 percent after the complete primary series. However, boosters continue to be recommended every 10 years to ensure protective levels of antitoxin in the circulation.

If a young adult has never received a primary series of tetanus injections, it is recommended that the person have two doses of Td administered two months apart, a booster six to twelve months later, and boosters at ten year intervals.

Side effects as a result of Td have generally been limited to minor reactons. Middaugh (1979) did a follow-up study following a mass immunization program for Td and found that the most frequent reaction in the young adult was a sore arm; this affected about 50 percent of that age group. Less than 40 percent stated their arms were swollen at the site of injection;

less than 30 percent complained of itching. It was interesting to note that there were more reactions with the jet injector than with the needle.

Fulginiti (1979) warns that there are problems if Td is given too frequently. A local necrotic reaction occurs because of the high levels of serum antitoxin.

Primary *polio* vaccination is no longer necessary for Americans over 18. The *Morbidity and Mortality Weekly Report* (1979) indicates that most adults are already immune and have a very small risk of exposure to poliomyelitis. They do, however, recommend immunization for those at greater risk for polio, such as those traveling to countries in which polio is epidemic or endemic, members of communities in which disease has been caused by wild poliovirus, laboratory workers who handle poliovirus, and health workers who are in close contact with patients who may harbor poliovirus.

Adults at high risk who have not been vaccinated against polio should receive three doses of inactivated (Salk) polio vaccine at intervals of 1 or 2 months, and a fourth dose six to twelve months later. The inactivated vaccine is preferred because the risk of vaccine-associated paralysis following administration of oral (Sabin) polio vaccine is slightly higher in adults than in children. The *Morbidity and Mortality Weekly Report* (1979) also states that unvaccinated parents of infants who are to be given oral polio vaccine are at a very small risk of developing vaccine-associated paralysis. However, some health professionals elect to give at least two doses of inactivated polio vaccine a month apart to these unvaccinated adults before the children receive the oral polio vaccine.

There has been no documentation of any serious side effects of currently available inactivated polio vaccine. However, this vaccine does contain trace amounts of streptomycin and neomycin, so there is a possibility of hypersensitivity reactions to these antibiotics. Also, although there is no documentation of adverse effects of polio vaccine on pregnant women or developing fetuses the *Morbidity and Mortality Weekly Report* suggests that this vaccination be avoided by pregnant women if possible.

Before *rubella* vaccine was available, the disease was seen most often in school-age children; now, however, most cases occur in adolescents and young adults. According to the *Morbidity and Mortality Weekly Report* (1978), 70 percent of the cases in 1977 occurred in those 15 years of age and older.

The most important consequence of rubella is the occurrence of anomalies in the fetuses of women who contract the disease, especially those infected during the first trimester of pregnancy.

The 1978 recommendations of the Advisory Committee on Immunization Practice of the Public Health Service emphasize the need for more effective ways to get the young adult population immunized. They suggest that "educational and training institutions such as colleges, universities and military bases should seek proof of rubella immunity (a positive serologic test or documentation of previous rubella vaccination) from all female students and employees in the childbearing age. Non-pregnant females who lack proof of immunity should be vaccinated unless contraindication exists" (*Morbidity and Mortality Weekly Report*, 1978, p. 453). Also, susceptible men and women in hospitals and clinics should be immunized. The Committee further suggests that hospitals conduct antepartum screening of expectant mothers; those susceptible to rubella should be vaccinated before being discharged from the obstetrical unit.

Before a woman of childbearing age can be immunized, a serum antibody titer (from a reliable laboratory) must demonstrate that the individual has no rubella antibody; it must be established that the woman is not pregnant; she must be made to understand why she should not become pregnant for at least 3 months after the vaccination; she should take measures to prevent a pregnancy during this time if she is sexually active; and the woman should be aware of the possible side effects of the vaccine.

Side effects that are somewhat common in young women are arthralgia and transient arthritis, which begin two to ten weeks after immunization and generally last for one to three days without recurrence. However, Fulginiti (1979) mentions a peculiar syndrome that occurs infrequently. Affected women experience nonlocalized pain around the joints of the upper and lower extremities, which is thought to be caused by a neuropathy induced by the live virus. Recurrences may last for about 2 years. In addition to this, the young adult may infrequently experience thrombocytopenia, rash, or lymphadenopathy.

Every effort should be made to inform young adults of the need to keep their immunization status up to date throughout their lives. The areas of teaching include: the need for the immunizations, the benefits derived from this, the possible side effects, where to get the immunization, and the need for record-keeping. Industries should be encouraged to have a mass immunization program in order to reach a large number of the young adult population.

Birth Control

Lieberman (1976) found no single factor more important in contributing to the enhancement of marriage and the mental health of the family than family planning before the first pregnancy and each succeeding pregnancy. Yet, many young adults are not knowledgeable about birth control measures. Some use ineffective measures, some use effective measures incorrectly, and others do not use any birth control method at all. McElmurray (1979) found in her study of low-income women that those who were sexually active exhibited a severe lack of information regarding the use of contraceptives.

Also, the fact that one million women seek abortion every year (Weinstock et al., 1976) is some indication of a need for greater emphasis on the prevention of unwanted pregnancies.

Barnes (1978, p. 30) and the Committee on Maternal Health Care and Family Planning proposed the following guideline for the accessibility of family planning services: "Family planning services should be available without regard to age, parity, marital status, parental or spousal consent, income level, receipt of government funds or aid, sex, color, race, national origin, religion, residence, or source of referral. Where fees are charged, they should be flexible enough to ensure that no one is denied service because of inability to pay. Services and supply hours should be available at times convenient to the population served, being scheduled frequently enough to provide continuity of service and a choice of time for patients to attend. Services should provide protection and regard for the dignity, sensibilities, privacy, confidentiality, and self respect of the individual at all times and in all interactions with staff."

Generally, family planning is successful if the method is effective, if it is acceptable to the couple, and if both partners are involved in the choice of the method(s). Implications for health professionals in family planning are many. Some young adults or couples who want information about birth control do not feel comfortable in initiating a discussion of the subject. The nurse must be very perceptive in listening and should feel at ease in eliciting a desire for information. Some nurses are more comfortable in beginning with an indirect approach such as "Are you interested in having a large family like your mother did?" Other nurses prefer a more direct approach, such as "Are you interested in discussing any aspects of birth control?" The important thing is that the young adult be granted an opportunity to discuss any and all aspects of family planning.

Birth control pills are by far the most common form of contraceptive method used in the

United States. Shapiro (1977) estimated that 30 percent of the women in the 18 to 44 age group are on a birth control pill. This method is probably favored because it is effective, easy to use, and rather inexpensive. However, an increasing number of couples are returning to some of the earlier methods of birth control such as the condom, foam, and the diaphragm. This may be because of more awareness of the side effects of birth control pills, the desire of the couple to have increased joint responsibility in methods of birth control, or because of the desire to reap the high degree of effectiveness afforded by the use of any two of these earlier methods in combination.

In addition to their knowledge about various types of birth control methods, the couple must also be aware of the side effects and possible complications of these methods. For instance, if the pill is the chosen method, the couple must realize that birth control pills with estrogen have been found to enhance the formation of thrombi. Therefore, women with superficial phlebitis, deep phlebitis, previous embolus, extensive or painful varicose veins, and some even with a family history of phlebitis should not be placed on the pill. Black women with sickle cell anemia must not take the pill. Women with a normal glucose tolerance may show abnormal findings while on the pill. (This chemical change produces no symptoms and disappears when the pill is discontinued.) Shapiro (1977) states that long-term effects of the pill remain unknown. Birth control pills may also cause an increase in blood pressure. Fatal and nonfatal heart attacks are more common in young adults who use the pill than in those who do not; the risk is even higher in those who smoke, are diabetic, are obese, or have high blood pressure.

Studies have indicated that the estrogen in oral contraceptives prevents the body from absorbing certain important vitamins and minerals such as vitamin C, some of the B complex group, and zinc (Shapiro; 1977). The recommended daily allowance of B_6 is 2 mg for women not on the pill, but 25 mg for those on the pill (Shapiro, 1977); foods rich in B_6 are wheat germ, meat, liver, soy beans, milk, fish, whole grain cereals, peanuts, and bananas. A sufficient daily intake of folic acid is also important because folic acid deficiency in the first trimester has been found to be associated with birth anomalies. Foods rich in folic acid are green leafy vegetables, citrus fruit, lean meat, organ meat, and yeast.

The couple should understand that if they select the pill as their method of contraception, they will need to use a different form of contraception for three months after termination of the pill in order to avoid a pregnancy, which is necessary because pregnancies that occur within the first three months after discontinuing the pill are more likely to result in the birth of deformed babies.

HINTS ON WORKING WITH YOUNG ADULTS

The young adult woman generally has more contact with the health care system than the man. She commonly enters the system for information and treatment relating to birth control, pregnancy, gynecological problems, and, if she has children, she accompanies them on their appointments also. Every opportunity should be taken to assess as fully as possible her health needs, growth and developmental needs as well as whatever specific concerns she has. The young man's trips to a health care facility are primarily for acute care or for confirmation and treatment of symptoms that he observes.

Observation is a powerful tool in the assessment process. For instance, the nurse may be able to detect nervousness, irritability, distractability, or certain withdrawal behaviors that might provide a clue to potential depression. Obesity, physical development, and injuries that may provide clues to family violence can also be observed, and a smile that reveals plaque formation and red, edemetous gums may indicate a need to pursue the possibility of periodontal disease. Finally, the young adult's interaction with her children will give some indication of communication patterns and parenting skills.

In talking with the patient, the nurse should listen for indications of how many changes are occurring simultaneously in the patient's life, and for evidence of a good social support system. Conversation provides opportunities detecting problems, teaching, referral, and giving support.

Young adults (and others) commonly complain that health professionals are in too much of a hurry, or too busy to be "bothered" with their patients' questions or problems. Nurses in all settings need to become cognizant of the behavior that creates this impression, and to take the time to answer questions, explore areas of concern, and point out other areas of concern detected during assessment. Often, an open-ended, general question, asked in a nonthreatening, unhurried manner, such as "How are things going for you?" may prompt the young adult to discuss specific health or developmental concerns.

Knox (1977) found that white-collar adults, when faced with a major change or personal problem, seek information first from media such as books. If answers or solace are not found there, they then seek the advice of experts such as counselors or physicians. Their third source of information is friends or acquaintances. Blue-collar adults, however, were found to first seek information from friends and acquaintances, since they read little and are less likely to spend money on consultations with professionals. McElmurry (1979) found in her study of low-income women that 58 percent said they would be influenced by advice from a nurse, but they would weigh this advice against that of family members or friends. These findings are of sig-nificance to health professionals working with blue-collar young adults. Health professionals—health educators, social workers, physicians, and nurses—should work together with neighborhood workers to reach these young adults. It might also be advisable to encourage patients to bring a friend along on visits to a health professional.

Nurses should be aware of the need for self-help groups for young adults. It has been shown that some people who are undergoing a crisis, or who are at risk for a certain problem, benefit from the support of others who have similar needs. Nurses should be knowledgeable about the availability of such groups.

The nurse must be nonjudgmental and accepting of the individual. Young adults are especially quick to detect bias. If some aspect of a patient's cultural or ethnic background has an impact on health or developmental concerns, the nurse should attempt to understand that aspect, by eliciting information from the patient if necessary. The fact that the nurse can admit her ignorance of another's culture may encourage the patient to work with her.

SUMMARY

The stage of young adulthood generally begins with the exhibition of behaviors that reflect maturity: a sense of responsibility, a capacity for intimacy, the ability to plan, etc. The 1960s saw a longer transition period between adolescence and young adulthood; society is more tolerant of this. White-collar young adults have more opportunities to lengthen the transition period than blue-collar individuals have.

Physical growth of all body sysems is at its peak during early young adulthood. By the mid twenties most systems begin to decline, but this decline is so gradual that it is generally not observed during young adulthood. When decline is noticed, it is usually caused by the effects of a sedentary life-style rather than normal aging. Vigorous physical exercise has been found to restore the cardiovascular system to peak performance during young adulthood. In cognitive development, fluid intelligence, which is independent of education and experience, declines during young adulthood, but crystallized intelligence, which evolves from experience, increases.

The psychosocial development of the young adult includes the accomplishment of six major developmental tasks:

1. Breaking away from the family.
2. Establishing intimate relationships.
3. Making commitments to a career.
4. Establishing a personal set of values.
5. Forming an adult life of one's own.
6. Establishing a social network.

Common health and developmental problems of young adulthood include prolonged or excessive stress. Manifestations of maladaptive responses to stress include chemical dependency, family violence, suicide, and hypertension. Other health problems of young adulthood include sexually transmitted diseases, periodontal disease, breast and testicular cancer, and back injury.

In addition to providing anticipatory guidance regarding these common problems, health professionals can be instrumental in helping young adults establish good health habits, which maximize the likelihood of continued health throughout the life span. Other areas of anticipatory guidance include nutrition, exercise, immunizations, and birth contol.

REFERENCES

Abernathy, V. D. American marriage in cross-cultural perspective. In Grunebaum, H., and Christ, J. (Eds.), *contemporary marriage: structure, dynamics and therapy.* Boston: Little, Brown, 1976.

Aguilera, D. C., and Messick, J. M. *Crisis intervention* (3rd ed). St. Louis: C. V. Mosby, 1978.

Altman, B. L., and Malament, M. Carcinoma of the testis following orchiopexy. *Journal of Urology,* 1967, *97,* 498–504.

American Cancer Society. *Facts on testicular cancer.* New York: American Cancer Society, 1978.

American Cancer Society. *How to examine your testes.* New York: American Cancer Society, 1978.

American Cancer Society. *Cancer facts and figures.* New York: American Cancer Society, 1979.

Anderson, R. A. *Stress power! How to turn tension into energy.* New York: Human Sciences Press, 1978.

Armstrong, J. R. *Lumbar disc lesions.* Baltimore: The Williams & Wilkins Company, 1965.

Atchley, R. C. *The social forces in later life: an introduction to social gerontology.* Belmont, Calif.: Wadsworth, 1972.

Barnes, F. E. (ed.). *Ambulatory maternal health care and family planning services: politics, principles, practices.* Washington, D.C.: American Public Health Association, 1978.

Bauer, J. C., Pierson, A. P., and House, D. R. *Facts which affect the utilization of dental services.* Hyattsville, Md.: DHEW, 1978.

Behrman, S. J., and Kistner, R. W. (Eds.). *Progress in infertility* (2d ed.). Boston: Little, Brown, 1975.

Benson, H. The relaxation response. *Psychiatry,* 1974, *37,* 37–46.

———. *The relaxation response.* New York: Morrow, 1975.

Brewster, L., and Jacobson, M. F. *The changing american diet.* Washington, D.C.: Center for Science in the Public Interest, 1978.

Bromley, D. B. *The psychology of human ageing* (2nd ed.). New York: Penguin, 1974.

Brown, J. R. Lifting as an industrial hazard. *American Industrial Hygiene Association Journal, 1973, 34,* 292–294.

Bühler, C., and Massarik, F. (Eds.). *The course of human life.* New York, Springer, 1968.

Burger, D. Breast self-examination. *American Journal of Nursing,* 1979, *79,* 1088–1089.

Caine, P. A. Teaching breast self exam. In Marchant, D. & Nyirjesy, I. (Eds.), *Breast Cancer.* New York: Grune and Stratton, 1979.

Califano, J. A. *Healthy people: the surgeon general's report on health promotion and disease prevention.* Washington, D.C.: U.S. Government Printing Office, 1979.

Campbell, A. American way of mating: marriage si, children only maybe. *Psychology Today,* May 1975, 37–43.

Caplan, R. D., Cobb, S., French, J. R., Van Harrison, R., and Pinneau, S. R. *Job demands and worker health.* Cincinnati, Ohio: DHEW, NIOSH, 1975.

Chaffin, D. B. Biomechanics of manual materials handling and low-back pain. In Zenz, C. (ed.), *Occupational medicine: principles and practical application.* Chicago: Year Book Medical Publishers, 1975.

Clausen, M. P., Flook, M. H., and Ford, B. *Maternity nursing today* (2d ed.). New York: McGraw-Hill, 1977.

Cobb, S. Social support as a moderator of life stress. *Psychosomatic Medicine*, 1976, *38*, 300–314.

Cobb, S. Social support and health through the life course. In Riley, M. W. (ed.), *Aging from birth to death*. Boulder, Col.: Westview Press, 1979.

Conklin, M., Klint, K., Morway, A., Sawyer, J. R., and Shephard, R.: Should health teaching include self-examination of the testes? *American Journal of Nursing*, 1978, *8*, 2073–2074.

Constantine, L. L., and Constantine, J. M. Marital alternatives: extended groups in modern society. In Grunebaum, H., and Christ, J. (Eds.), *contemporary marriage: structure, dynamics and therapy*. Boston: Little, Brown, 1976.

Cooper, K. H. *The new aerobics*. New York: Bantam Books, 1977.

Cox, R. D. *Youth into maturity* New York: Mental Health Materials Center, 1970.

Davidson, T. *Conjugal crime: understanding and changing the wife-beating pattern*. New York: Hawthorn Books, 1978.

de Vries, H. A. Physiology of exercise and aging. In Woodruff, D. S., and Birren, J. E. (Eds.), *Aging: Scientific perspectives and social issues*. New York: D. Van Nostrand, 1975.

Diekelmann, N. *Primary health care of the well adult*. New York: McGraw-Hill, 1977.

Duvall, E. M. *Marriage and family development* (5th ed.). Philadelphia: J. B. Lippincott, 1977.

Eisenberg, S., and Micklow, P. The assaulted wife: 'catch-22 revisited'. From *Women's Rights Law Reporter*. Newark, N.J.: Rutgers University, Spring/Summer, 1977.

Elias, M. F., Elias, P. K., and Elias, J. W. *Basic processes in adult developmental psychology*. St. Louis: C. V. Mosby, 1977.

Enos, W. F., Homes, R. H., and Beyer, J. Coronary disease among United States soldiers killed in action in Korea. *Journal of the American Medical Association*, 1953, *152*, 1090–1093.

Erikson, E. H. *Childhood and society*. New York, Norton, 1963.

Erikson, E. H. (ed). *Adulthood*. New York: Norton, 1978.

Espenshade, T. J. The value and costs of children. *Population Bulletin*, 1977, *32*.

Farquhar, J. W. *The american way of life need not be hazardous to your health*. New York: Norton, 1978.

Forward plan for health. Washington, D.C.: DHEW, Public Health Service, 1975.

Freud, S. *Civilization and its discontents*. New York: Jonathan Cape and Harrison Smith, 1930.

Friedman, K. C.: The image of the battered woman. *American Journal of Public Health*, 1977, *67*, 722–723.

Friedman, M., and Rosenman, R. H. *Type A behavior and your heart*. New York: Alfred A. Knopf, 1974.

Fulginiti, V. A. Active immunization for infectious diseases. In Conn, H. F. (ed.), *1979 current therapy*. Philadelphia: W. B. Saunders, 1979.

Gelles, R. J. *Family Violence*. Beverly Hills, Ca: Sage, 1979.

Gelles, R. J., and Straus, M. Determinants of violence in the family: Toward a theoretical integration. In Burr, W. et al. (Eds.), *Contemporary theories about the family, Vol. I*. New York: Free Press, 1979.

Giele, J. Z. Changing sex roles and the future of marriage. In Grunebaum, H., and Christ, J. (Eds.), *contemporary marriage: structure, dynamics and therapy*. Boston: Little, Brown, 1976.

Girdano, D. A., and Everly, G. S. *Controlling stress and tension: A holistic approach*. Englewood Cliffs, N.J.: Prentice-Hall, 1979.

Gould, R. Adult life stages: Growth toward self tolerance. *Psychology Today*, 1975, *8*(9), 74–78.

Gove, W. R. Adult sex roles and mental illness. *American Journal of Sociology*, 1973, *78*(4), 812–835.

———. The effect of children and employment on the mental health of married men and women. *Social Forces*, 1977, *56*(1), 67–76.

Grossman, A. S. *Children of working mothers*. Washington D.C.: U.S. Department of Labor, Bureau of Labor Statistics, 1977.

Group for the Advancement of Psychiatry. *The joys and sorrows of parenthood*. New York: Scribner's, 1973.

Grunebaum, H., and Christ, J. (eds.). *Contemporary marriage: Structure, dynamics and therapy*. Boston: Little, Brown, 1976.

Havighurst, R. J.: *Developmental tasks and education* (3rd ed.). New York: David McKay, 1972.

Hazen, S. (ed.). *Diet, nutrition and periodontal disease*. Chicago: The American Society for Preventive Dentistry, 1975.

Hipp, L. L. A new look at back-injury prevention. *Occupation Health and Safety*, 1976 *45*, 6–18.

Holmes, T. H., and Rahe, R. H. The social readjustment rating scale. *Journal of Psychosomatic Research*, 1967, *11*, 213–218.

Horn, J. L. Intelligence: Why it grows, why it declines. In Hunt, J. M. (ed.), *Human intelligence*. New Brunswick, N.J.: Transaction, 1972.

Hume, J. Trichomoniasis and other sexually transmitted diseases. *Occupational Health Nursing*, 1979, *27*(8), 16–23.

Ibrahim, M. A. The changing health state of women. *American Journal of Public Health*, 1980, *70*, 120–121.

International Labour Office. *Encyclopaedia of occupational health and safety* (Vol I). St. Louis: McGraw-Hill, 1971.

Janis, I., and Mann, L. *Decision-making*. New York: Free Press, 1977.

Jones, D. F. *Human factors—Occupational safety*. Ontario, Canada: Ontario Ministry of Labour, 1973.

Kales, J. D. Aging and sleep. In Goldman R., and Rockstein, M. (Eds.), *The physiology and pathology of human aging*. New York: Academic Press, 1975.

Kaluger, G., and Kaluger, M. F. *Human development, the span of life* (2nd ed.). St. Louis: C. V. Mosby, 1979.

Kezdi, P. *You and your heart*. New York: Atheneum/SMI, 1977.

Kinsey, A. *Sexual behavior in the human male*. Philadelphia: W. B. Saunders, 1948.

———. *Sexual behavior in the human female*. Philadelphia: W. B. Saunders, 1953.

Knox, A. B. *Adult development and learning*. San Francisco: Jossey-Bass, 1977.

Kohlberg, L. The cognitive-developmental approach to moral education. In Purpel, D., and Ryan, K. (Eds.), *Moral education—It comes with the territory*. Berkeley, Ca.: McCutchan, 1976.

Kohn, M. L., and Schooler, C. The reciprocal affects of the substantive complexity of work and intellectual flexibility. In Riley, M. W. (ed.), *Aging from birth to death*. Boulder, Col.: Westview Press, 1979.

Krames, L. A. *Back owner's manual*. Daley City, Cal.: Physician Art Services, 1977.

Levinson, D. J., Darrow, D. N., Klein, E. B., Levinson, M. H., and McKee, B. *The seasons of a man's life*. New York: Alfred A. Knopf, 1978.

Lieberman, E. J. The prevention of marital problems. In Grunebaum, H, and Christ, J. (Eds.), *Contemporary marriage: Structure, dynamics and therapy*. Boston: Little, Brown, 1976.

Lugo, J. O., and Hershey, G. L. *Human development, a psychological, biological, and sociological approach to the life span* (2nd ed.). New York: McMillian, 1979.

Masters, W. H. and Johnson, V. E.: *Human Sexual Response*. Boston: Little, Brown, 1966.

McElmurry, B. L. Health appraisal of low-income women. In Kjervik, D. K., and Martinson, I. M. (Eds.), *Women in stress, a nursing perspective*. New York: Appleton-Century-Crofts, 1979.

Mechanic, D. *Students under stress*. Madison, Wi.: University of Wisconsin, 1978.

Messolonghites, L. *Multicultural perspective on drug abuse and its prevention, a resource book*. Rockville, Md.: DHEW, 1979.

Middaugh, J. P. Side effects of diptheria-tetanus toxoid in adults. *American Journal of Public Health, 1979, 69*, 246–249.

Moore, M. L. *Realities in childbearing*. Philadelphia: W. B. Saunders, 1978.

Morbidity and mortality weekly report. Atlanta, Ga.: DHEW, National Center for Disease Control November 17, 1978; January 19, 1979; May 25, 1979; November 2, 1979; and November 16, 1979.

Morris, C. I. Stress: relaxation therapy in a clinic. *American Journal of Nursing*, 1979, *79*, 1958–1959.

Mussen, P. H., Conger, J. J., Kagen, J, and Geiwitz, J. *Psychological development: A life-span approach*. New York: Harper & Row, 1979.

National Center for Health Statistics. *Monthly vital statistics reports: Annual summary for the United States, 1977*, Hyattsville, Md.: DHEW, December, 1978.

———. *Monthly vital statitics report: provisional statistics*, Hyattsville, Md.: DHEW, February, 1980.

———. *Monthly vital statistics report: provisional statistics*. Hyattsville, Md.: DHEW, March, 1980.

National Institute on Alcohol Abuse and Alcoholism,: *Alcohol and health*. Washington, D.C.: DHEW, 1975.

Naval Graduate Dental School,: *Periodontics Syllabus*, Washington, D. C.: U.S. Government Printing Office, 1975.

O'Flynn-Comiskey, A. I. Stress: the type A individual. *American Journal of Nursing*, 1979, 79 1956–1958.

Owen, B. D. Back-injured persons' perception of circumstances that may have led to back injury causation and injury prevention. Unpublished, 1980.

Papalia, D. E., and Olds, E. W. *Human development*. New York: McGraw-Hill, 1978.

Parker, B., and Schumacher, D. N. The battered wife syndrome and violence in the nuclear family of origin: A controlled pilot study. *American Journal of Public Health*, 1977, *67*, 760–761.

Ramaswamy, S. and Smith, T. *Practical Contraception*. Philadelphia: Stickley, 1977.

Rapoport, R. Normal crises, family structure, and mental health. *Family Process*, 1963, *2*, 68–80.

Rapoport, R., and Rapoport, R. N.: Marriage and career. In Grunebaum, H., and Christ, J. (Eds.), *Contemporary marriage: structure, dynamics and therapy*. Boston: Little, Brown, 1976.

Reubin, R. Spotting and stopping the suicide patient. *Nursing 79*, 1979, *9*(4), 82–85.

Richter, J. M., and Sloan, R. Stress: A relaxation technique. *American Journal of Nursing*, 1979, *79*, 1960–1964.

Rising, S. S. Childbearing, its dilemas. In Kjervik, D. K., and Martinson, I. M. (Eds.), *Women in stress, A nursing perspective. New York: Appleton-Century-Crofts, 1979.*

Rowe, L. M. Low-back disabilities in industry: updated position. Journal of Occupational Medicine, 1971, *13*, 476–478.

Rueveni, U. *Networking families in crisis*. New York: Human Sciences Press, 1979.

Sadowsky, N. L. Mammography in early detection of breast cancer. In Marchant, D. J., and Nyirjesy, I. (Eds.), *Breast disease*, New York: Grune and Stratton, 1979.

Safilios-Rothschild, C. The dimensions of power distribution in the Family. In Grunebaum, H., and Christ, J. (Eds.), *Contemporary Marriage: structure, dynamics and therapy*. Boston: Little, Brown, 1976.

Schell, R. E. (ed.). *Developmental psychology today* (2nd ed.). New York: Random House, 1975.

Schultz, A. B. Personal Communication. March 28, 1979.

Selye, H. *The stress of life*. New York: McGraw-Hill, 1956.

Shapiro, H. I. *The birth control book*. New York: St. Martin's, 1977.

Sheehy, G. *Passages: Predictable crises of adult life*. New York: Bantam Books, 1976.

Singer, J. M. Taking charge: current events nutrition for healthy consumers. In Lazes, P. M. (ed.), *The handbook of health education*. Germantown, Md.: Aspen Systems Corporation, 1979.

Skolnick, A., & Skolnick, J. (eds). *The family in transition* (2nd ed.). Boston: Little, Brown, 1977.

Smith, K. U. *Industrial back hygiene*. Madison, Wi.: Wisconsin Department of Industry, Labor and Human Relations, 1979.

Smith, M. J., and Selye, H. Stress: reducing the negative aspects of stress. *American Journal of Nursing*, 1979, *79*, 1953–1955.

Snook, S. H., Campanelli, R. A., and Hart, J. W. A study of three preventive approaches to low back injury. *Journal of Occupational Medicine*, 1978, *20*, 478–481.

Snook, S. H., and Ciriello, V. M. Maximum weights and work loads acceptable to female workers. *Journal of Occupational Medicine*, 1974, *16*, 527.

Starr, B. D., and Goldstein, H. S. *Human development and behavior, pyschology in nursing*. New York: Springer, 1975.

Straus, M., Gelles, R. J., and Steinmetz, S. *Behind closed doors: Violence in the American family*. New York: Doubleday/Anchor, 1979.

Strax, P. *Early detection: Breast cancer is curable*. New York: Harper and Row, 1974.

Taugher, N. J. Incidence of industrial back injuries and the significance to our community. *Workmen's compensation data, back injuries*, Wisconsin Department of Industry, Labor and Human Relations, 1973.

Thomas, G. S. Exercise your treatment options. *Medical Dimensions*, *40*, February/March, 1979.

Tichauer, E. R. The biomechanics of the arm-back aggregate under industrial working conditions. *American Society of Mechanical Engineers*, *65*, September, 1966.

Tichauer, E. R., Miller, M., and Nathan, I. M. Lordosimetry: A new technique for the measurement of postural response to materials handling. *American Hygiene Association Journal*, 1973, *34*, 1–5.

Timiras, P. S. *Developmental physiology and aging*. New York: Macmillan, 1972.

Troll, L. E. *Early and middle adulthood*. Monterey, Ca.: Brooks/Cole, 1975.

Valenti, C. Working with the physically abused woman. In kjervik, D. K., and Martinson, I. M. (Eds.), *Women in stress, a nursing perspective*. New York: Appleton-Century-Crofts, 1979.

Wechsler, D. *The measurement and appraisal of adult intelligence* (4th ed.). Baltimore: Williams & Wilkins, 1958.

Weinstock, E., Tietze, C., Jaffee, F. S., and Dryfoos, J. G. Abortion need and services in the United States. *Family Planning Perspectives*, 1976, (March–April), *8*, 58–69.

Yankelovich, D. *The new morality: A profile of American youth in the 1970's*. New York: McGraw-Hill, 1974.

Yankelovich D., Skelly, and White.: *The General Mills American family report*, Minneapolis, Minn.: General Mills, 1975.

————. *The American family report, 1978–1979*. Minneapolis, Minn.: General mills, 1979.

11

The Middle-Aged Adult

ELSIE WILLIAMS-WILSON

By middle adulthood (the years from 35–64), physical development is essentially complete and a very gradual decline in functioning begins. Middle adulthood is a time of personal growth; however, it is the least researched and the least understood of the life stages.

In 1900, the average life expectancy was approximately 49 years; in 1976, it was approximately 72 years (DHEW, 1978). Some scientists predict that life expectancy will reach 100 years by the end of the twentieth century. Because of this rapid lengthening of the life span, the middle-adult population continues to grow. For example, in 1940, 39 million of a total U.S. population of 131 million were in the age range of 35 to 64 years; in 1976, of a national population of 214 million, 66 million were in that age range. It is important to note that these statistics represent a composite without regard to differences in life expectancy specific to sex or race (see Table 11-1).

Chronological age is not always a meaningful indicator of middle adulthood because of the vast biophysical and psychosocial variations among people in the 35-to-64 age group. However, the use of chronological age as a criterion is consistent with many legal requirements—Social Security regulations, income tax laws, and many insurance policies.

In terms of events, middle adulthood is the time of launching children from the home, grandparenting, reaching the peak in one's career, and preparing for retirement. However, this view is also limiting, considering those who

Table 11-1. Average Remaining Lifetime in Years at Specified Ages, by Color and Sex: 1900–1902, 1969–1971, and 1976[a]

Age in years	1976	1969–1971	1900–1902[b]	1976	1969–1971	1900–1902[b]
	White, male			*White, female*		
0	69.7	67.94	48.23	77.3	75.49	51.08
1	69.8	68.33	54.61	77.2	75.66	56.39
5	66.0	64.55	54.43	73.4	71.86	56.03
10	61.1	59.69	50.59	68.5	66.97	52.15
15	56.2	54.83	46.25	63.6	62.07	47.79
20	51.6	50.22	42.19	58.7	57.24	43.77
25	47.1	45.70	38.52	53.9	52.42	40.05
30	42.4	41.07	34.88	49.1	47.60	36.42
35	37.7	36.43	31.29	44.2	42.82	32.82
40	33.1	31.87	27.74	39.5	38.12	29.17
45	28.7	27.48	24.21	34.9	33.54	25.51
50	24.4	23.34	20.76	30.4	29.11	21.89
55	20.5	19.51	17.42	26.1	24.85	18.43
60	16.9	16.07	14.35	22.0	20.79	15.23
65	13.7	13.02	11.51	18.1	16.93	12.23
70	10.9	10.38	9.03	14.4	13.37	9.59
75	8.5	8.06	6.84	11.2	10.21	7.33
80	6.6	6.18	5.10	8.5	7.59	5.50
85	5.1	4.63	3.81	6.4	5.54	4.10
	All other, male			*All other female*		
0	64.1	60.98	32.54	72.6	69.05	35.04
1	64.9	62.13	42.46	73.3	70.01	43.54
5	61.1	58.48	45.06	69.5	66.34	46.04
10	56.3	53.67	41.90	64.6	61.49	43.02
15	51.4	48.84	38.26	59.7	56.60	39.79
20	46.8	44.37	35.11	54.9	51.85	36.89
25	42.5	40.29	32.21	50.2	47.19	33.90
30	38.2	36.20	29.25	45.5	42.61	30.70
35	34.0	32.16	26.16	40.9	38.14	27.52
40	30.0	28.29	23.12	36.4	33.87	24.37
45	26.1	24.64	20.09	32.1	29.80	21.36
50	22.5	21.24	17.34	28.0	25.97	18.67
55	19.2	18.14	14.69	24.3	22.37	15.88
60	16.3	15.35	12.62	20.7	19.02	13.60
65	13.8	12.87	10.38	17.6	15.99	11.38
70	11.3	10.68	8.33	14.3	13.30	9.62
75	9.7	8.99	6.60	12.3	11.06	7.90

Table 11-1. (*Continued*)

Age in years	1976	1969–1971	1900–1902[b]	1976	1969–1971	1900–1902[b]
	All other, male			*All other female*		
80	8.6	7.57	5.12	10.9	9.01	6.48
85	7.2	6.04	4.04	9.1	7.07	5.10

[a] For 1900–1902, data are for 10 States and the District of Columbia: 1969–71 and 1976 data are for the United States.

[b] Figures for the "All other" group cover only blacks; however, the black population comprised 95 percent of the corresponding "All other" population.

SOURCE: U.S. Department of Health, Education, and Welfare, *Facts of Life and Death*, 1978.

chose not to have children, people of 30 whose careers have already peaked, and the many in their seventies and eighties who continue to work productively. Moreover, studies have shown that the socioeconomic status of an individual greatly influences the timing of major life events; in general, those in the middle and upper classes tend to delay these events. Therefore, to identify middle adulthood solely in terms of the events that often characterize it would be as erroneous as regarding chronological age as the only criterion.

PHYSIOLOGICAL CHANGES

By middle adulthood, physical growth and maturation are completed. With adequate nutrition, sufficient physical exercise and rest, and in the absence of disease, the decline in level of physical function is inappreciable until much later in life. Indeed, this absence of dramatic physical changes is undoubtedly part of the reason for the dearth of studies on middle adulthood. Unfortunately, much of what is known or assumed about biophysiological functioning in middle adulthood has been based on studies about late adulthood; therefore, many people incorrectly perceive middle age as a time when they will develop degenerative processes and chronic alterations of health, such as arthritis, diabetes mellitus, emphysema, and hypertension.

Many chronic conditions and degenerative diseases are more prevalent in people of older ages. There is a small increase in the incidence of heart diseases, hypertension, and arthritis between the ages of 18 and 34 years. Between the ages of 35 and 64 years, the incidence of these diseases increases markedly; beyond the age of 65 years, the rates begin to soar. Between the ages of 45 and 54 years, all of the previously mentioned chronic alterations of health except rheumatoid arthritis are more common among males. In the age group of 55 to 65 years, males have a higher rate of coronary heart disease, while all other conditions are more prevalent among females (see Table 11-2) (DHEW, 1978).

It is accepted as inevitable that chronological aging brings with it some decline in the efficiency of some physiological processes. The time of onset and rate of these changes vary among individuals. Insufficient data are available regarding the process of aging in association with racial, ethnic, psychosocial, and environmental factors, although these variables are recognized. Because of the vast differences among individuals, it is important for health professionals to recognize that chronological age is not a good indicator of biological age.

Biological aging occurs at various levels of

Table 11-2. Prevalence of Heart Disease, Hypertension, and Arthritis in Adults by Age and Sex: 1960–1962

Diagnosis and Sex	Total	Age Group in Years[a]						
		18–24	25–34	35–44	45–54	55–64	65–74	75–79
Definite Heart Disease, Total[b]								
Both sexes	13.2	1.2	2.4	6.7	13.2	25.3	39.9	42.3
Men	12.6	1.4	2.9	7.4	13.8	24.2	33.2	38.8
Women	13.7	1.1	2.0	6.1	12.5	26.2	45.2	45.8
Suspect Heart Disease, Total[b]								
Both sexes	11.7	4.0	4.9	8.8	15.3	19.4	20.7	25.2
Men	13.9	6.4	6.6	11.4	18.3	18.5	25.3	27.1
Women	9.7	2.0	3.3	6.4	12.4	20.1	17.1	23.3
Definite Hypertensive Heart Disease								
Both sexes	9.5	0.3	1.3	4.7	9.6	17.9	30.3	31.8
Men	7.7	0.4	1.4	5.2	9.7	13.6	18.9	24.6
Women	11.1	0.2	1.2	4.2	9.5	21.9	39.5	39.0
Suspect Hypertensive Heart Disease								
Both sexes	4.3	0.7	1.1	2.6	4.4	8.4	10.4	14.1
Men	5.1	1.5	1.7	4.2	5.0	7.8	12.8	16.1
Women	3.5	—	0.6	1.1	3.8	9.0	8.5	12.1
Definite Coronary Heart Disease								
Both sexes	2.8	—	0.3	0.7	2.5	7.1	9.5	6.8
Men	3.7	—	0.4	1.1	3.5	9.7	11.6	9.1
Women	2.0	—	0.2	0.5	1.6	4.7	7.9	4.5
Suspect Coronary Heart Disease								
Both sexes	2.2	—	0.1	0.9	3.0	4.8	5.9	5.7
Men	2.2	—	—	1.3	3.4	4.4	5.3	3.8
Women	2.2	—	0.2	0.5	2.5	5.2	6.4	7.5
Definite Hypertension								
Both sexes	15.3	1.4	3.9	10.9	18.2	26.9	38.5	38.8
Men	14.1	1.7	4.8	13.5	18.3	22.3	27.1	32.4
Women	16.4	1.2	3.1	8.5	18.2	31.2	47.6	45.1
Borderline Hypertension								
Both sexes	14.6	5.7	7.4	11.5	16.5	25.9	24.5	27.5
Men	17.2	10.9	11.9	14.2	17.7	27.5	24.8	26.7
Women	12.2	1.4	3.2	9.0	15.3	24.5	24.3	28.3
Rheumatoid Arthritis								
Both sexes	3.2	0.3	0.3	1.3	3.0	6.3	9.2	18.8
Men	1.7	0.2	—	0.5	1.5	4.2	3.1	14.1
Women	4.6	0.3	0.6	2.1	4.4	8.3	14.1	23.5

Table 11-2. (*Continued*)

Diagnosis and Sex	Total	Age Group in Years[a]						
		18–24	*25–34*	*35–44*	*45–54*	*55–64*	*65–74*	*75–79*
Osteoarthritis								
Both sexes	37.4	4.1	9.7	24.7	46.6	69.4	80.7	85.4
Men	37.4	7.2	13.6	30.2	47.0	63.2	75.8	80.9
Women	37.3	1.6	6.2	19.6	46.3	75.2	84.7	89.8

[a] Rates per 100 persons in specified groups.

[b] Includes persons with other types of heart disease not shown separately.

SOURCE: U.S. Department of Health, Education, and Welfare, *Facts of Life and Death* (November, 1978), p. 18.

body organization: organ systems, organs, tissues, cells, subcellular particles, and molecules. Changes related to aging that occur at the lower levels are only microscopically evident and become obvious only when organs and systems are affected; most of the normal changes in the biophysical subsystems are likely to be observed in the latter phase of middle adulthood rather than in the earlier phase. It should always be remembered that heredity, environment, and other unknown factors will influence the body's response to the process of aging. Several theories of cellular aging have been proposed; none is wholly acceptable as an explanation but each contributes to understanding biophysiological changes.

The first is a biological theory of aging that stresses hereditary factors. The theory purports that genetic inheritance determines one's susceptibility to changes in various organs. Another theory stresses that the individual's immune system releases defective antibodies that destroy healthy tissue, eventuating in specific diseases of aging (Woodruff and Birren, 1975). The third theory emphasizes that the accumulation of certain substances in body cells, especially lipofusin, interferes with nourishment of normal cells and leads to cellular death. A fourth theory holds that with aging comes a crosslinking of elastin and collagen, resulting in rigidity and decreased pliability of connective tissue. Without elasticity of supporting tissues in the body, alterations in the functioning of the body's organs and structures will occur.

In order to better appreciate the relative merits of these cellular theories of aging, it is important to understand cellular functioning. In the organization of the body, the cell is the structural and functional units of the living organism. Microscopic studies of the cell reveal that the nucleus, located in the center, contains chromatin networks and controls vital activities of the cell, including reproduction. The organelles, including the endoplasmic reticulum, the mitochondria, the Golgi apparatus, and lysosomes, are the functional substances essential for cellular life. Because various parts of the body have a specific function, their constituent cells are structurally different. Differences among cells arise from cell specialization or differentiation. Cells that are specialized aggregate into units of higher order and become tissue. Tissues are, in turn, aggregated into organs; the functions that different organs perform depend upon the tissue and specialization of the cells therein. Any changes or defects within the cell, or abnormal outputs of the cell's component parts, will impair the functional capacity of the entire cell and subsequently lead to alterations in functioning of the organ. These biophysical changes are extremely complex and do not uniformly result in decreased functioning, nor will the change occur in all individuals.

Integumentary Changes

Wrinkling and graying are the initial indications of middle adulthood for many. The age at which these occur seems genetically determined (Fig. 11-1). The skin and its appendages—hair, hair follicles, nails, sebaceous and sweat glands—comprise one of the largest and most important body systems. Its basic functions are to protect underlying structures, to regulate temperature, and to prevent drying of tissue.

Although these functions are maintained throughout middle adulthood, physiological changes do occur slowly, influenced and accelerated by such factors as poor nutrition, extensive exposure to sun and chemicals, and inactivity. Other integumentary changes include drying of the skin, changing complexion, thinning of the hair, increasing baldness in males, and toughening of fingernails and toenails.

These changes in the integumentary system occur mostly in the dermis (the intermediate layer of the skin) and in the subcutaneous tissues (the innermost layer). The epidermis (the out-

Figure 11-1. *For many people, graying of the hair is one of the first physiological signs of having entered middle adulthood; but, as with many developmental phenomena, the age of onset varies greatly.*

ermost layer) undergoes little change since it is composed of epithelial cells, which constantly desquamate and are replaced. The dermis, which is composed of connective tissue, shows significant changes. The amount of connective tissue ground substance and the number of cells decreases, while collagen fiber formation, strength, and insolubility all increase. Calcium parts are also deposited in connective tissue with age, causing calcification (Groer and Shekleton, 1979). This process leads to a decreased blood supply to the epidermis, atrophic changes in the sweat and sebaceous glands, and a decreased amount of pigment-producing cells. The decreasing blood supply results in thinning of head and body hair, and thickening of toenails and fingernails; drying of the skin occurs in proportion to atrophic changes in the sweat and sebaceous glands, whereas changes in complexion and hair color are the result of decreased numbers of pigment cells. Decreased elasticity results in wrinkling and folding of the dermis. Loss of subcutaneous fat accentuates the wrinkled appearance. Most of the integumentary changes of middle adulthood are not physically significant; however, these changes may have a tremendous psychological impact, especially on those whose selfimages are based largely on physical appearance.

Cardiovascular Changes

At various times in middle adulthood, many people find it very difficult to perform physical activities at the same pace and for the same duration as was once possible. This is true to a much greater extent among those with sedentary life-styles. The reduction of physical activity causes atrophy of the left ventricular chamber of the heart, which impedes the transportation of oxygenated blood to the body's tissues.

The peak of cardiovascular performance is reached at the end of the growth spurt of adolescence (Smith, Bierman, and Robinson,

1978). After the age of 20 years, cardiac muscle strength declines at a rate of approximately 1 percent yearly. Generally, this alteration does not significantly affect cardiac output in the middle years. Changes in the rate and rhythm of the heartbeat become more evident after age 45.

Biophysical changes in the arteries and, to a lesser extent, the veins affect the functioning of the cardiovascular system because they become less elastic and less flexible. Thickening of the arteries occurs as a consequence of the deposit of lipids and, possibly, fibrin in their intimal linings. Subsequent calcification may also occur. The lumen of the affected arteries narrows, causing partial occlusion. Why this occurs is not clearly understood; some theorists relate it to diet while others emphasize heredity. For example, the incidence of hypertension is greater among blacks, and dietary factors, heredity, and, more recently, stress, have been cited as causative factors. Elasticity of the arteries is a major factor in regulating blood pressure; inelasticity causes increased arterial resistance and blood pressure. However, high blood pressure in middle adulthood may be associated not with changes in the arteries, but with other systems involved in the regulation of blood pressure, including the endocrine, urinary, and nervous systems.

Musculoskeletal Changes

Aging brings a gradual loss of bone, which reduces skeletal mass. Bone growth reaches its peak at approximately 20 years of age and begins to diminish between the ages of 30 to 40 years. Bone loss begins at an earlier age and at a faster rate among females (Martin, 1978). This observation lends credibility to the theory that hormonal changes associated with the climacteric (cessation of the reproductive function in women) result in bone loss. Along with other functions, estrogen maintains a balance between bone formation and bone resorption; as estrogen levels decrease in the postmenopausal phase of the life cycle, there is an increased loss of bone substance. With loss of bone substance, especially calcium, the bone becomes demineralized and less dense. Rarefied bone (osteoporosis) decreases skeletal strength and is more likely to fracture when placed under stress. For example, those who suffer falls are more prone to fractures of the hip and wrist. Postmenopausal osteoporosis initially affects the weight-bearing vertebrae (those below the seventh thoracic vertebra) and various deformities of the spine may occur. Biconcavity is the most common deformity of the spine because of the pressure exerted by the intractable intervertebral discs on the brittle vertebrae. Progression of the biconcavity can result in compression fractures of the vertebrae, "dowager's hump" (thoracic kyphosis), and a decrease in height (Martin, 1978).

In the absence of osteoporosis, by the age of 60, males and females are approximately one-half inch shorter (Hershey, 1974). This occurs because the cartilages between bones become thinner and are not as distensible. Stiffness may occur as connective tissues increase in collagen and decrease in elastin. In addition, the calcium levels in tissues increase, producing more rigidity.

The changes in connective and circulatory tissues cause a decrease in the total mass of muscular tissue, which in turn reduces physical strength. The powers used for climbing stairs, cranking a wheel, and walking on a treadmill decline 30 percent from age 35 to age 70 (Hershey, 1974). Strength peaks at ages 23 to 27 and thereafter declines slowly until the age of 50 years, when it declines at a faster rate. The total decrement by the age of 60 is between 10 to 20 percent of the maximum (Woodruff and Birren, 1975). Those who remain active throughout middle age will probably notice very little loss of strength, since strength and muscle mass can be maintained through exercise.

Pulmonary Changes

With aging, the tissue in the alveoli, like tissue in other organs, increases in collagen formation and decreases in elasticity and permeability. Environmental factors, such as air pollution, cigarette smoking, and prolonged exposure to occupational dust, will cause pathological changes in the alveoli.

Because the respiratory reserve is so great, a considerable amount of lung tissue must be altered before there is significant interference with respiration. Given the absence of disease, optimal functioning of the thoracic and abdominal muscles, and maintenance of the thoracic skeletal structures, persons in middle adulthood will maintain pulmonary functioning; however, a gradual diminution of breathing capacity may be noted by age 55 to 60. Two common phenomena, shortness of breath upon exertion and easy fatigability, are associated with disease or inactivity rather than age-associated factors.

ENDOCRINE CHANGES

The endocrine system remains reasonably functional throughout middle adulthood. The functioning of each endocrine gland is highly dependent upon the central nervous system, and often on other glands. Because of this complex interrelationships, it has been difficult to clearly identify age-related changes in this system. Some findings about age-related changes suggest that thyroid functioning is maintained throughout middle adulthood, level of pituitary hormone remains fairly constant throughout adulthood, and the Islets of Langerhans in the pancreas do not decrease in numbers (Hershey, 1974).

The incidence of diabetes mellitus increases after the age of 40 years, especially among people who are obese or genetically susceptible. According to one study, 85 percent of people over 40 years of age were obese at the time of a diagnosis of diabetes mellitus (Luckmann and

Sorensen, 1980). While the exact cause of diabetes mellitus that has its onset in maturity is unknown, it is throught that there is an alteration in the pancreatic beta cells' glucoreceptors. Diabetes mellitus presenting in maturity is characterized by a decrease in the production of insulin and by an alteration in the release of insulin in response to an increased level of glucose in the blood. Because of their excessive adipose tissue and excessive intake of food, obese people require abnormally high amounts of insulin. The prolonged, continued, and increased secretory demands for insulin upon the Islets of Langerhans eventually exhaust the pancreatic beta cells' glucoreceptors and cells do not respond to the increased levels of blood glucose (Groer and Skekleton, 1979).

Gastrointestinal Changes

Little change in gastrointestinal function occurs in middle adulthood. Although the mucous membrane lining of the esophagus and stomach thins out, healthy middle aged people experience no significant changes on account of this. However, other changes that have a bearing on gastrointestinal function do occur. By the age of 65 years, approximately half of American adults have lost over half of their teeth as the result of poor dental hygiene. Also, a progressive loss of taste buds causes the sense of taste to diminish, especially after the age of 55. Severe and chronic emotional states such as grief, fear, or anger may inhibit gastric motility and glandular secretions of hydrochloric acid and pepsin, thereby interfering with the process of digestion.

Immunological Changes

During adulthood, usually late adulthood, the body sometimes loses the ability to differentiate between "self" and "nonself" substances, and consequently produces autoantibodies that destroy the body's own tissue. This process, called *autoimmunity*, can affect almost all tissues and organs of the body. Some immunological responses occur in specific organs, while others involve antigens that are widely distributed throughout the body. These autoimmunological processes can lead to the development of autoimmune connective tissue diseases (e.g., Hashimoto's thyroid disease, rheumatoid arthritis, and myasthenia gravis). The theories of autoimmunity support the hypotheses that the occurrence of autoimmunity is influenced by genetic factors, that the incidence of autoimmunity increases with age, and that autoimmunity is generally more prevalent among women (Jones, Dunbar, and Jirovec, 1978).

Another important theory is that among people over 60 years of age, a decline in immunological functioning results in the development of cancer. It is believed that cancerous cells that were previously eliminated by an efficient immune system remain in the body because the lymphocytes fail to recognize cancerous cells as foreign (Woodruff and Birren, 1975).

Neurological and Sensory Changes

The peak functioning of the central nervous system occurs at approximately 25 years of age; thereafter, there is a very gradual decline in functioning of this system. According to Woodruff and Birren (1975), the reflexes remain relatively unaffected from 20 to 80 years of age. Fine coordination tends to show some decline after the age 35. Among mentally active adults, intellectual skills and productivity are at a very high level during middle adulthood and are maintained until after the age of 60 years.

During middle adulthood, there are gradual changes in many of the sensory faculties. Hershey (1974) suggests that after the age of 37, smell is diminished because of a decrease in the numbers of nerve fibers. A gradual loss of hearing, especially of high frequencies, begins during adolescence but becomes apparent only after the age of 40 (Woodruff and Birren, 1975).

By the age of 65, over half of all people have a decrease in visual acuity and wear corrective lenses. By the age of 50, many people have difficulty in seeing at close distances; this is caused primarily by the loss of elasticity of the eye lens (presbyopia). This change can be corrected with eyeglasses or contact lenses.

Many people in their forties will have difficulty adapting to glare and to darkness. The diameter of the pupils when in the dark is smaller for many adults over 50; therefore, more light is needed in order to see objects clearly. It has been suggested that, beginning at 33, and every 13 years thereafter, light intensity must be doubled in order to enable one to clearly see an object (Hershey, 1974). Peripheral vision is significantly reduced after 55 years of age, and the affected person will begin to turn the head more in order to compensate. Color discrimination reaches its peak at 20 years of age and decreases gradually to a point at which the individual is unable to distinguish the blues, blue-greens, and purple colors. This alteration in color differentiation occurs because, as one ages, yellowing of the lens occurs, affecting color discrimination at the blue end of the spectrum.

Genitourinary Changes

Throughout middle adulthood, the urinary system continues to function normally. Normal age-related changes result in a gradual decrease in the number of nephrons, the functional unit of the kidney. The nephron consists of glomerular, tubular, and interstitial tissue, and, although these tissues perform vital functions, a gradual reduction of them will not lead to major problems because the body has a tremendous functional renal reserve. In addition to a decrease in nephrons, the process of aging affects renal flow, bladder capacity, and muscle tone in the urinary structures. In late middle adulthood, many males will develop enlargement of the prostate gland. The cause of this

common occurrence is not clearly understood. Enlargement of the prostate gland can lead to dysuria, a reduced urine output, and oliguria. Enlargement of the prostate gland accompanied by partial or complete obstruction requires medical intervention.

Reproductive Changes

In middle adulthood, changes in the production of reproductive cells and the secretion of hormones from the ovaries and the testes occur. This process, the *climacteric*, tends to begin after the age of 35 years and may span a period of 15 to 20 years, depending on the individual's rate of biophysical change (Mims and Swenson, 1980). During the climacteric, the ovaries produce decreasing amounts of estrogen and progesterone. Initially, the decrease in these hormones causes irregular monthly menstrual flow; the flow is generally diminished in amount.

Complete absence of the menstrual flow for one year is regarded as the sign that *menopause* has occurred. The age at which menopause occurs, believed to be genetically determined, has been reported at as young as 35 years and as late as 60 years (Martin, 1978). Once a woman reaches menopause, the ability to reproduce ceases. Other biophysiological changes that may occur with decreasing levels of estrogen and progesterone are atrophy of external genital tissues; decrease in pubic hair; atrophy of the glandular and fatty tissue of the breast; increased alkalinity of vaginal secretions; atrophy of the uterus, ovaries, and fallopian tubes; and decreased blood supply to the genital tract. Decreased estrogen can also affect protein anabolism and result in thinning of the skin and brittling of bones. Because the adrenal glands synthesize hormones that are transformed into female sex hormones, these physical changes in the female reproductive system are very gradual and do not affect the female as drastically as was once believed.

Sexual drive is not diminished; on the contrary; many women find that sexual drive increases. Many women may experience transitory "hot flashes." Hot flashes are associated with estrogen deficiency. The arteries of the skin dilate and create a feeling of warmness accompanied by profuse perspiration. Patches of redness may appear on the skin, beginning at the chest area and extending to the neck and face. Some women may experience emotional changes taking the form of depression, crying, and insomnia. Many of the biophysiological symptoms can be controlled or eliminated by the use of estrogen replacement. Because of the connection between hormones and the occurrence of cancer of the reproductive system, many physicians are taking a conservative approach to the use of estrogen replacement therapy. According to a report from the U.S. Department of Health, Education, and Welfare (1979), estrogens are known to affect some division of cells, and exogenous hormones are associated with increased risk of cancer of the endometrium. For these reasons, estrogen replacement therapy is used cautiously, and many physicians believe it is contraindicated for women who have a personal or family history of cancer of the reproductive system.

Unlike women, men in middle adulthood do not experience a significant decrease in sex hormones. After testosterone production peaks in young adulthood, it very slowly declines, with a plateau between the ages of 40 and 60 (Jones, Dunbar and Jirovec, 1978). In most men, the production of testosterone and pituitary gonadotropin is sufficient to maintain reproductive capacity beyond the age of 80 (Burnside, 1976). However, decreased secretion of testosterone can result in decreased force of ejaculation, decreasing numbers of sperm, and the need for a longer time to achieve an erection. Continued decline in the level of testosterone can cause decreased mucular strength, hot flashes, anxiety depression, and nightmares (Jones, Dunbar, and Jirovec, 1978).

COGNITIVE CHANGES

Middle adulthood can be, and is for many, a very dynamic and productive period in the life cycle. It is a time for the fullest use of one's knowledge and skills, a time for maximum control and influence over one's personal and social environment, a period when people increase their self-awareness, and a time to assume some new personal and social roles.

Longitudinal studies, in which the subjects were tested at different ages, suggest that general intelligence (the sum of fluid and crystallized intelligence, explained below) increased slightly until approximately age 50 and remained relatively unchanged until late adulthood (Birren and Schaie, 1977). Several longitudinal studies involving intellectually superior subjects showed improved intellectual functioning from early childhood or young adulthood through the early years of middle adulthood. Another longitudinal study of subjects whose initial scores were average showed that scores declined over a period of 13 years (Birren and Schaie, 1977). Other studies found that after 50 years of age, verbal skills increased and numerical skills declined.

Fluid intelligence affects the intellectual skills and abilities used for motor-visual coordination, intellectual speed, memory, and inductive reasoning. Development of these abilities and skills is dependent upon the functioning of the central nervous system. Fluid intelligence increases rapidly in adolescence, reaches its peak between the ages of 18 and 21 years, and begins to decline in middle adulthood. Much of the decline in general intelligence in middle adulthood is the result of slowing of performance speed. Therefore, the results of timed tests may not accurately reflect intellectual abilities in all adults. Tests in which speed is not a factor have yielded results that are consistent with the individual's ability (Woodruff and Birren, 1975).

Crystallized intelligence depends upon the acquisition of knowledge and skills through ex-

Figure 11-2. *Education is one of several factors that affect one's level of crystallized intelligence, which peaks during the middle years but continues to develop in healthy adults.*

perience, education, and general acculturation (Fig. 11-2). Measures such as verbal skills, general information, and judgment are held to be essential in the evaluation of crystallized intelligence (Birren and Schaie, 1977). Crystallized intelligence increases slowly in younger individuals, peaks between the ages of 41 and 60 years, and continues to develop in healthy adults. Perhaps the concept of crystallized intelligence explains why many individuals in middle adulthood are increasingly productive and creative in their professions and in their social and community activities.

Even though findings suggest that general intelligence in middle adulthood is dominated by crystallized abilities, fluid abilities are not greatly affected. While the decline in fluid in-

telligence is statistically significant, it may not be of great importance. The extent of intellectual development depends greatly upon the individual's motivation. Some researchers have hypothesized that intellectual decline occurs in individuals who are rigid or resistant to change (Stevens-Long, 1979).

PERSONALITY CHANGES

The process of aging has been thought to have a detrimental effect upon personality. Aging people are generally stereotyped as rigid, impatient, and focused on the past. However, studies have not supported this view; in fact, reports of studies suggest no systematic change in personality with aging. The ability to adapt to the process of aging and the demands of the environment seems related to the individual's self-concept and to the level of physical health. The healthy adult has been found to be flexible, resourceful, and optimistic (Butler and Lewis, 1977).

Erikson's View of Middle Adulthood

According to Erik Erikson (1963), the developmental task of middle adulthood is the attainment of *generativity versus stagnation*. Generativity, in Erikson's terms is a tremendous and sincere concern for oneself, as well as for the growth and development of children, peers, older adults, the community, and society. While this seems an awesome task, healthy people in middle adulthood possess the wisdom, strength, vigor, and experience to fulfill it. Generativity is accomplished through parenting, working in one's career, participating in community activities, or working cooperatively with peers, spouse, family members, and others to reach mutually determined goals (Lugo and Hershey, 1974). The experiences, personal and vicarious, of the generative adult culminate in mature judgment, greater insight into oneself, and a more discerning awareness of the needs of society. The mature adult has a well developed philosophy of life which serves as a basis for leadership, stability, and objectivity.

In those adults who do not develop generativity, stagnation and interpersonal impoverishment occur. Stagnation and self-absorption are seen in many adults who have been deprived, by choice or circumstances, of experiences that would have fostered psychosocial development in the previous phases of the life cycle. Such people tend to abhor the changes of aging, i.e., graying of hair, decreasing physical strength, etc. Often, the denial of physical changes takes the form of inappropriate dress, excessive makeup, and drastic changes in life-styles (Butler and Lewis, 1977). Inability to accept the physical changes and psychosocial challenges of the period can lead to a sense of lowered self-esteem and depression. Interestingly, such people continue to work and meet their needs yet continue to exhibit patterns of isolation, preoccupation with oneself, and lack of interest in others.

Peck's Expansion of Erikson's Theory

R. C. Peck proposed seven issues of adult development, four of which are relevant to middle adulthood. The first issue is *valuing wisdom versus valuing physical powers*. Successful aging depends upon greater use of mental abilities as physical stamina declines. One example is the professional athlete who resigns from an active player role to one of coaching or managing a team. Another task is *socializing versus sexualizing* in human relationships. Although sexual intimacy continues to be important, the adult needs to have interpersonal relationships that are characterized by understanding the needs and views of others. The third challenge, related to interruptions in relationships brought about by death, shifts in allegiances; or geographic distance, is to develop *cathectic flexibility versus cathectic impoverishment*. Cathectic flexibility is accomplished when individuals continue interactions with others by establishing new friends, renewing friendships, and developing relationships with extended family members. Closely related to this is the fourth task: *mental flexibility versus mental rigidity*. Individuals in middle adulthood must be open to new experiences and new interpretations. Holding on to traditional patterns of behavior and obsolete interpretations does not provide new vistas and can impede psychosocial and intellectual development.

Activity Theory

The activity theory is counter to the theory of disengagement, which was initially used to explain successful aging. While both theories were proposed as relevant to psychosocial adjustment in late adulthood, both have some applicability to middle adulthood. The disengagement theory holds that as a natural phenomenon of aging there is a strong tendency for one to withdraw from society and to become concerned with a few friends, family, and, eventually, only oneself (Cumming and Henry, 1961). In addition to the focus on self, changes in some of the external social factors—retirement, children launched from the home, and death of spouse—create a shrinkage in the social life space and, in turn, decrease the number of social roles held (Stevens-Long, 1979). One implication of this theory is that one's social life space is diminished in accordance with one's ego energy reserve. Thus, it was concluded that disengagement fostered a greater sense of psychological well-being because the individual no longer desires deep emotional involvement in activities and relationships, as was in middle adulthood. Further studies suggested that disengagement is simply one of several patterns of aging (Havighurst, Neugarten, and Tobin, 1968).

The activity theory maintains that those in late adulthood have essentially the same psychological and social needs as they had in middle adulthood, except for those needs created by normal physiological changes or alterations in health status (Havighurst, Neugarten, and Tobin, 1968). The activity theory holds that the aging adult desires to remain socially active, and that optimal aging occurs when one does so (Butler and Lewis, 1977).

Role Exit Theory

Although framed in a more sociological perspective, the role exit theory is essentially the same as the activity theory. According to it, as one ages a number of major social roles are terminated and there is an interruption of well-established patterns of social interactions (Blau; 1973). In role exit, or termination of roles (such as in cases of loss of spouse, retirement from work, children lauched from the home, and death of a close friend), one establishes new roles and new interests in order to maintain meaningful social interactions, and to prevent shrinkage of the social environment. It is felt by some that the role exit theory exaggerates the impact of social losses on many in late adulthood.

DEVELOPMENTAL TASKS OF MIDDLE ADULTHOOD

Middle adulthood presents new adaptive tasks that are different from those of young adulthood. Many tasks are accomplished without conscious awareness; the degree of accomplishment will affect the individual's adjustment to this phase of the life cycle. Mastery of the developmental tasks is equated with success, and assures that the individual will feel secure and will evidence continued growth. Lack of mastery tends to create frustration, depression, and anxiety in the individual and can impede one's ability to grow.

Havighurst's Developmental Tasks

Robert J. Havighurst (1974) identifies seven developmental tasks of middle adulthood. The first major task is reaching and maintaining satisfactory performance in one's career. During middle adulthood, people who have been employed most of their adult lives will attain the highest income and status of their careers. For others who enter the world of work later in their adult lives and for those individuals who enter new careers or new jobs, the task becomes that of achieving satisfaction in a job that interests them. Achieving social and civic responsibility is the second major task. Because individuals in middle adulthood are near the peak of their intellectual and physical functioning, participation in social and civic activities is expected.

For many women whose children have left the home, participation in organized social and civic activities provides opportunities for new experiences that are mentally stimulating and emotionally satisfying. Among working-class men and women, there is less participation in social and civic organizations; more participation has been observed in those organizations that are job-related, e.g., labor unions, or in groups such as neighborhood associations (Havighurst, 1974).

Another task is to accept and adjust to the physiological changes of middle adulthood. Age-related physical changes bring about a de-

cline in physical capacity that eventually necessitates adjustment in physical life-style and psychological adaptation.

For adults living with teenage children (e.g., biological or adoptive children, or extended family members) or aging adults (e.g., parents, in-laws, aunts, uncles, and significant others), the tasks of helping teenaged children to become happy and responsible adults, and of adjusting to aging parents are relevant.

Among those individuals who have been married for extended periods of time and especially those couples with children, the intervening years may have been focused on parenting and succeeding in one's career. Having accomplished these tasks by middle adulthood,

Figure 11-3. *Providing help to members of the younger and older generations is one of the goals of middle adulthood proposed by Joanne Sabol Stevenson.*

1. Developing socioeconomic consolidation

2. Evaluating one's occupation or career in light of a personal value system

3. Helping younger persons (e.g., offspring) to become integrated human beings

4. Enhancing or redeveloping intimacy with one's spouse or most significant other

5. Developing a few deep friendships

6. Helping aging persons (e.g., parents or in-laws) progress through the later years of life

7. Assuming responsible positions in occupational, social, and civic activities

8. Maintaining and improving the home or other forms of property

9. Using leisure time in satisfying and creative ways

10. Adjusting to biologic or personal changes

Figure 11-4. *Developmental tasks of Middlescence I. From Joanne Sabol Stevenson, Issues and crises during middlescence. New York: Appleton-Century-Crofts, 1977. (Adapted with permission.)*

the individuals are frequently ready to place more emphasis on the role of wife or husband. Middle adulthood, then, is a time for the task of relating to one's spouse as a person.

Finally, Havighurst identifies the task of developing adult leisure activities. For adults with lessened demands from growing children and career goals, this is a time not only for increased civic and social activities but for solely pleasurable activities.

Stevenson's Developmental Tasks

Joanne Sabol Stevenson (1977) proposes two sets of developmental tasks for those in middle adulthood. She subdivides middle adulthood into two periods: Middlescence I, or the "core years," from ages 30 to 50 years, and Middlescence II, from 50 to 70.

The major objective of the core years is to assume responsibility for one's personal growth and development, as well as for one's growth as part of organizational enterprises, e.g., industry, education, charitable organizations, religion, and the family. A second objective is to provide help to younger and older generations (Fig. 11-3). Seven of the ten developmental tasks of Middlescence I (Fig. 11-4) are parallel to the developmental tasks proposed by Havighurst. The three additional tasks are to develop socioeconomic consolidation, to develop a few deep friendships, and to maintain and improve one's standard of living. Stevenson's developmental tasks for what she calls Middlescence II are shown in Figure 11-5.

Social Roles and Role Transition

Individuals in middle adulthood assume numerous social roles: spouse, parent, child of aging parents, worker, friend, member of organizations, citizen. In each social role, there are expected associated behaviors established by the norms of society. Therefore, a social role is a pattern of learned and expected behavior. Changes in social roles occur throughout the life cycle; new roles are assumed and old roles are terminated. When there is a role transition, the individual is expected to learn the new behaviors appropriate for the new role. Through the process of socialization, adults learn to perform their social roles. Often, the adult prepares for a transition in role; this involves identifying the new norms and expectations that will be associated with the new social role once the transition is made. For example, many people in middle adulthood return to school to prepare for a career change. In some instances, the adult does not have an opportunity to prepare in advance for a new social role; an example is the sudden death of a spouse that leads one into widowhood. Adulthood involves some age-related change in social roles; among these are

1. Maintaining flexible views in occupational, civic, political, religious, and social positions

2. Keeping current on relevant scientific, political, and cultural changes

3. Developing mutually supportive (interdependent) relationships with grown offspring and other members of the younger generation

4. Reevaluating and enhancing the relationship with spouse or most significant other or adjusting to loss of them

5. Helping aged parents or other relatives progress through the last stage of life

6. Deriving satisfaction from increased availability of leisure time

7. Preparing for retirement and planning another career when feasible

8. Adapting to signals of accelerated aging processes

Figure 11-5. *Developmental tasks of Middlescence II. From Joanne Sabol Stevenson, Issues and crises during middlescence. New York: Appleton-Century-Crofts, 1977. (Adapted with permission.)*

parenting of adult children, grandparenthood, alteration in marital status, (e.g., widowhood), and assuming responsibility for aging parents.

Parenting of Adult Children

The parenting of adult children, whether they remain in the home or leave, should include encouraging autonomy and allowing independence. Family tasks and respnsibilities should be allocated among the family members on the basis of ability, interest, and availability (Duvall, 1971). Some parents are rigid and experience tremendous difficulty in allowing young adults to assume some tasks of the household. On the other hand, there are young adult children who are immature and perceive remaining

at home as an opportunity to delay assuming their adult responsibilities. These behaviors can impede growth and development of both the adult child and the parents.

After adult children leave the home, the family enters what is commonly referred to as the "empty nest" or "postparental" phase of the life cycle; both terms are quite misleading and frequently inaccurate. The "empty nest" label is often used to denote that after the children depart, parents, especially the mother, develop a profound sense of loss and loneliness. However, while this stage of the family cycle often constitutes a difficult transition, Neugarten (1970) found that, for many middle-aged women, satisfaction with life improved when the children left the home.

Divorce

Divorce among couples in middle adulthood is not as infrequent as it once was. Traditionally, married people have two major social roles: parent and spouse. All too often, the marriage is held together only by the parenting of children. In fact, for many couples, parenting is the only goal shared by the husband and wife. One study reported that while many middle-aged wives and husbands found satisfaction with their children, few found satisfaction with their spouses (Turner and Helms, 1979). Many couples realize their incompatibility in an earlier stage of the marriage but delay separation or divorce until the children have grown up and moved out of the home. On the other hand, some couples do not recognize their incompatibility or dissatisfaction with one another until the children leave the home. The couple then recognizes that their focus on parenting has led to an impoverished interpersonal relationship that cannot be repaired.

Another study reported that among couples married 18 years or more, marital problems were caused by the following: first, ill health

(29 percent); second, infidelity (26 percent); third, incompatibility (23 percent); fourth, sexual difficulties (15 percent); and fifth, financial problems (7 percent) (Turner and Helms, 1979).

Widowhood

Thanks to medical advances and increased longevity, the incidence of widowhood in middle adulthood is decreasing. In 1960, the percentage of widowed men and women between the ages of 55 and 65 years was 6.2 and 24.5 respectively; in 1971, the rates declined to 3.5 and 21.2, respectively (Ostfeld and Gibson, 1972). One reason for the higher rate of widowhood for women is that they tend to marry men older than themselves. Because of higher mortality rates, widowhood is higher among nonwhite women and men.

The reactions to the death of a spouse vary and are influenced by what sort of relationship the couple had, whether the death was sudden or anticipated, and by cultural traditions. Depression and grief are likely to occur in those persons whose relatioships were emotionally and physically fulfilling. In relationships that were less fulfilling, the widowed spouse may feel guilty because of anger toward the deceased and relief at the death. As one would expect, the emotional impact of a sudden death tends to be far greater than that of an anticipated one.

The impact of bereavement is described by Gorer (1965) as consisting of three stages of mourning: initial shock, intense grief, and gradual recovery. Shock is thought to last only a few days; generally, the individual is physically and emtionally exhausted. The stage of intense grief can last for months, and is characterized by episodes of crying and bewilderment. The third stage is characterized by acknowledgment of the spouse's death. Other reactions seen in the first month of bereavement include anorexia, weight loss, insomnia, fatigue, and difficulty in concentration and memory (see Figure

11-6). According to Kimmel (1974), the grief, physical exhaustion, and loneliness experienced by the widowed spouse may increase the incidence of a serious physical illness, a serious accident, or even suicide.

Widowhood represents the loss of a significant social role and a transition to a less well-defined one. While the transition is likely to be difficult, the widowed individual can eventually establish new social roles and new relationships.

Widowed individuals who were satisfied with their marriage are more receptive to the notion of remarrying. Studies suggest that people in the later phases of the life cycle who remarry after the death of a spouse tend to be happier than those who remain single (Woodruff and Birren, 1975).

Grandparenthood

Maintaining meaningful relationships with grandchildren can promote the growth and development of both the grandparent(s) and the grandchildren. Neugarten and Weinstein (1968) enumerated five styles of grandparenting: "formal," "fun seeker," "surrogate parent," "reservoir of family wisdom," and "distant." The "formal" grandparent essentially leaves parenting to the parents. He or she will occasionally baby-sit and provide other services for the grandchildren. The "fun-seeker" style is characterized by a desire for mutual enjoyment and entertainment, and time spent in mutual interests. The "surrogate parent" assumes parental responsibility for the grandchild either voluntarily or by request of the parent. Often, this style exists in extended families. The grandparents who perceive their role as teacher or advisor to the grandchildren are of the "reservoir of family wisdom" group. This style exists in many foreign cultures and is seen increasingly among American families. The "distant" grandparents' contacts with the grandchildren are infrequent and are usually limited to holidays and special family gatherings.

According to Neugarten and Weinstein (1968), the "fun seeker" and "distant" styles of grandparenting were the most common among grandparents under the age of 65 years. Two factors may account for the "distant" style of grandparenting among individuals in middle adulthood: first, geographical distances between families; and second, many grandparents

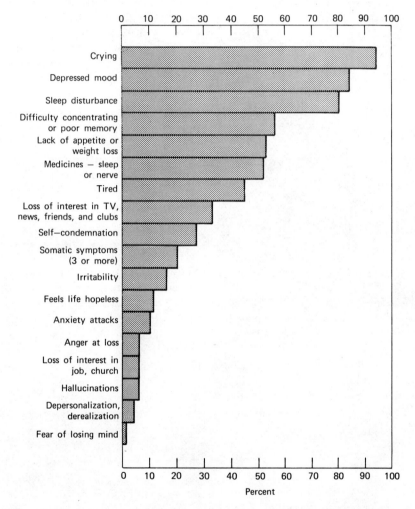

Figure 11-6. *Percentage of randomly selected recently widowed persons reporting various symptoms of bereavement (N = 109). Somatic symptoms include headaches, blurred vision, dyspnea, abdominal pain, constipation, urinary frequency, dysmenorrhea, and other body pains. (Reproduced with permission from Douglas C. Kimmel, Adulthood and aging. New York: John Wiley & Sons, 1974; adapted from Paula J. Clayton, James A. Halikes, and William L. Maurice, The bereavement of the widowed. Diseases of the Nervous System 1971, 32(9), 597–604.*

Figure 11-7. *For many, middle adulthood brings an increase in leisure time that can be devoted to such activities as physical exercise.*

are career oriented and are actively employed. Even though many grandparents are satisfied with their role as grandparent, not all are necessarily desirous of it.

Leisure

The persistence of the work ethic has affected the attitudes and values about the use of leisure time of many, especially those in middle adulthood. While earlier research found Americans to be more work-oriented than leisure-oriented, more recent writings note that leisure is becoming a way of life, and predict a leisure-dominated culture. McDaniels (1977) states that people, especially youth, increasingly want to be identified by their life-style and cultural taste, rather than by their occupations. For adults in middle adulthood who have not cultivated a use of leisure time, leisure is commonly seen as time free from work, but not all time spent away from the job is really free time; leisure consists of self-motivated activities, whether physical, intellectual, charitable, or creative (Fig. 11-7). For many adults, middle adulthood is a time of increased availability of money because of decreased financial responsibilities such as for care of children; increased availability of time, such as from earned vacation time; and increased opportunities for leisure activities through continuing education, adult education, and municipal and private enrichment programs. Leisure activities can foster a sense of self fulfillment, stimulate personal growth, and provide service to others.

Relationships With Aging Parents

Middle adulthood is the time when many individuals assume increasing responsibility for the physical, social, emotional, and financial well-being of their aging parents. The extent to which adult children assume responsibility for their aging parents depends upon several factors, including earlier child-parent relationship, cultural or ethnic influences, and individual level of growth and development. For most adults, the parent role has been defined partly in terms of assisting: the parent provides assistance to the child. As parents age, many adult children assume some responsibility for assisting the parents through the later years of their lives. Blenkner (1965) suggests the term "filial maturity" to explain this phenomenon.

Improved medical care, housing for the aged, and better pension or Social Security plans have increased the physical and financial independence of many aging parents. The majority of aging parents do not live with their children; by their own accounts, they prefer to live near, but not with, them (Burnside and Ebersole,

1979). The 1970 Census of households found less than 5 percent of aging parents living with adult children (Woodruff and Birren, 1975). In American society, independence and self-reliance are highly valued, and many aging parents will therefore resist being dependent on their children. It is not uncommon to hear adult children tell of an aging parent who lives independently, does all of the housework, drives an automobile, and is quite happy. Even so, for many aging parents the family is the nucleus of the social environment. This is especially true of the aging parent who is widowed, lives alone, and is without friends.

Some common psychosocial needs of aging parents—self-esteem, recognition, affection, belonging, and independence—can be fulfilled in a variety of ways. For example, one's self-esteem is maintained by emotional and financial independence and by respect and approval from others; recognition is maintained through assuming some responsibilities in the family and participating in community activities, the need for affection is fulfilled by feeling a true sense of belonging to the family.

Crises in Middle Adulthood

There are divergent views regarding the prevalence of crises in middle adulthood. Some hold that middle adulthood is characterized by a high incidence of marital dissatisfaction, alcholism, and psychosomatic illness. Conversely, others maintain that middle adulthood is not an especially stressful phase of the life cycle (Stevens-Long, 1979). Studies of adulthood tend to identify issues and conflicts that can lead to a disruption in the life cycle for some, but does not for the majority. Of course, there exists the potential for unanticipated crises such as the death of a child, death of a spouse, or divorce.

Middle adulthood is a time of introspection that brings an awareness of the relative imminence of one's death, and this in turn leads to

reflections on what life has been all about. During middle adulthood, one begins to view life in terms of time left to live, rather than time since birth. Thus, some adults who feel their life-styles are no longer fulfilling decide to make major life changes such as divorce or beginning a new career. Mid-life crises need not be viewed as negative; crises can be times of growth, of creative opportunity as well as turmoil.

While a positive view of middle adulthood has been presented, it is not a time of positive growth for all. Some adults are dissatisfied, unhappy, and do not adapt to this phase of the life cycle. In Erikson's terms, these people experience *stagnation*, and, perhaps, enter late adulthood in a state of despair.

COMMON HEALTH AND DEVELOPMENTAL PROBLEMS OF THE MIDDLE-AGED ADULT

The gradual age-related physiological changes of middle adulthood affect the functional and reserve capacities of many body systems; these changes increase the body's vulnerability to disease. In addition to age-related physiological changes, the stress of such events as change of occupation, change of residence, marriage, death of a spouse, or divorce may bring about alterations in health (Woodruff and Birren, 1975). It is important to bear in mind that positive sources of stress (e.g., a vacation) can cause the same results as negative experiences. (See the section on Stress in Ch. 10.)

Minority status and low socioeconomic standing are important influences on health status, and of course, the two are often found together. Those in low-income groups, especially those living in urban ghettos, are often exposed to a cluster of stressful influences, all of them detrimental to health: overcrowding, excessive noise, substandard or dilapidated housing, crime, malnutrition, etc. It has long been observed that those of lower socioeconomic standing are far less likely to seek preventive health care, waiting instead until a problem develops before consulting a doctor, dentist, etc. This in turn increases the incidence of morbidity, which almost certainly accelerates the process of aging. Adherence to inappropriate life-styles (among all socioeconomic strata) has also been cited as a leading cause of health problems, the prime examples being cigarette smoking, lack of exercise, and poor nutrition.

Obesity

Excessive eating and decreased physical activity are the two leading causes of obesity among middle-aged adults. A person is considered obese if he or she is 15 to 20 percent above the normal weight range for sex, age, and height. Some cultures increase the likelihood of obesity; the ethnic diets of Mexican-Americans, Puerto Ricans, and blacks have been implicated in obesity among these groups (Howard and Herbold, 1978). In our society, eating is an important social activity and becomes a part of many social events such as holiday celebrations, baby showers, weddings, and "tailgate parties" at sporting events. Obesity in middle adulthood is causally related to diabetes mellitus, hypertension, and osteoarthritis, and should any of these health problems already exist, obesity can intensify their effects.

Menopause

For the majority of women, menopause is gradual and the symptoms associated with it are mild; thus they are gradually adjusted to and at times even ignored. A small percentage of women have physical and psychological complaints.

Vasomotor instability (which causes "hot flushes") results from rapid changes in the diameter of the blood vessels and is thought to be caused by erratic swings of estrogen and progesterone. Hot flushes may last for a few seconds to an hour or more; most women accept and tolerate them as temporary discomforts that eventually disappear. If the hot flushes are severe enough, estrogen therapy may be considered. However, the dangers and disadvantages of exogenous estrogen should be discussed with the client.

Menstrual irregularities may take one of two forms: lighter, shorter periods with decreased flow, or heavier, longer periods. Heavy bleeding or intermenstrual bleeding is abnormal in middle-aged women. While most menstrual irregularities experienced by menopausal women are normal. some are not. Any woman with dysfunctional bleeding should be evaluated by a gynecologist to rule out fibroids, uterine and endocervical polyps, and cervical, uterine, and ovarian cancer, all of which are more common in women over 40 years old. Women over 40 should have a Pap smear at least every 2 years; those with dysfunctional bleeding or a family history of cancer of the reproductive organs should have Pap smears more frequently. Because of the increased incidence of cancer of the reproductive organs in women over 40 years some physicians recommend endometrial biopsy or the endometrial jet washer procedure on a yearly basis if there is irregular or intermenstrual bleeding.

Atrophic vaginitis (characterized by thinning mucosa, decreased vaginal lubrication and loss of pliability of vulvovaginal tissues) can result if endogenous estrogen levels are low. Women with atrophic vaginitis may have frequent vaginal infections, since the resistance to infection is decreased; dyspareunia is frequently the major complaint. Atrophic changes can be reversed with estrogen therapy but the risks and dangers of estrogen must be considered. Local estrogens (creams or lotions) are effective for relief of the symptoms associated with atrophic vaginitis; however, topical estrogens are absorbed to some degree into the general circulation, posing some risks.

Affective changes reported by some women include emotional lability, anxiety, headache, insomnia, depression, and increased fatigue. It is not yet known whether these symptoms are related to hormonal changes or to the psychological changes occurring during the middle years. Only about 10 percent of all women have severe, incapacitating menopausal symptoms (Martin, 1978).

Hypertension

Hypertension, a persistent elevation of the systolic pressure above 140 mm Hg and the diastolic above 90 mm Hg, tends to occur more frequently among older people in American society, and is primarily attributable to inelasticity of the arteries. An increase in blood pressure can be secondary to alterations of the cardiovascular, renal, adrenal, and neurologic systems, to stress, and to obesity. Blacks and females have higher rates of hypertension. A sustained elevation of blood pressure is significant because it can affect the heart, the kidneys, the brain, and the eyes. Hypertension is commonly referred to as the "silent killer," indicating very vague or absent symptoms. Some symptoms related to mild hypertension include anxiety, headache, insomnia, irritability, forgetfulness, fatigue, and occasional sensations of palpitations (Luckmann and Sorensen, 1980).

Osteoarthritis

Commonly referred to as "wear-and-tear arthritis," osteoarthritis is a degenerative change in the joints that occurs with aging. The spine and the weight-bearing joints of the lower extremities are commonly affected. It is not uncommon for people who have used crutches, manually operated wheelchairs, or walkers for extended periods of time to develop osteoarthritis of the joints of the upper extremities. Similarly, those who have engaged in occupations requiring lifting of heavy objects may develop degenerative changes of the upper extremities.

Most people over the age of 45 have some form of osteoarthritis. While the cause is unknown, some predisposing factors include aging, joint trauma, and obesity. The incidence is greater among women and may first become evident at the time of the menopause. Among women with osteoarthritis prior to menopause, its onset tends to increase the symptoms markedly (Luckmann and Sorensen, 1980). An aching pain, intensified by activity and exercise, is quite characteristic. Muscle spasms, night pain, and morning stiffness may also occur. Another manifestation is the development of nodules (Heberden's nodes) that cause disfigurement of the terminal interphalangeal joints. While the changes in the intraphalangeal joints are not disabling, osteoarthritic changes of the joints and associated ligament damage of the lower extremities can limit mobility. The hip is less commonly affected than the hands or knees, but involvement of the hip can be especially disabling (Luckmann and Sorensen, 1980). The pain can interfere with walking up and down stairs, sitting in and getting out of a bathtub, and arising from a chair. There is no cure for osteoarthritis. However, drug therapy for symptomatic relief has been effective, and surgical procedures have successfully restored functioning in the hips and knees.

Visual Changes

Presbyopia (discussed earlier) and glaucoma are the two major eye alterations that occur in middle adulthood. Presbyopia is not considered a severe visual impairment and is correctable with reading glasses or bifocals. Often, the need for reading glasses or bifocals is stressful for the individual; in these instances, the individual requires emotional support and an opportunity to discuss his/her feelings.

It is estimated that as many as two out of every 100 persons over the age of 40 have glaucoma. It generally develops between the ages of 40 and 65 years, and is caused by increased intraocular pressure, which can lead to irreversible alteration of the optic nerve, and to blindness. Family history, eye injury, and diseases of the eye are associated with the occurrence of acute or chronic glaucoma. Although acute glaucoma is less common, prompt medical attention is imperative to prevent severely impaired vision or blindness. Acute glaucoma begins suddenly; the symptoms include nausea, vomiting, pain in the affected eye, redness of the eye, and cloudy vision.

Because of its gradual onset and the absence of early symptoms, chronic glaucoma is often called the "sneak thief of vision." One of the earliest manifestations is a gradual loss of peripheral vision. Reports of walking into objects and not seeing objects or people on the affected side are common. Other symptoms of chronic glaucoma are severe headaches, blurred vision, tearing of the eye, halos around sources of light, and dull pain in the eye. There is no cure for glaucoma, but it can be controlled by drug therapy and surgical intervention. Because glaucoma is the second leading cause of blindness and is preventable, adults need to know its symptoms.

Cancer

Cancer is the second leading cause of death (after heart diseease) in the United States. Smoking, drinking, and suntanning are some of the practices most associated with the development of cancer. Cancers of the lung and esophagus are strongly linked to smoking and drinking. Adults must be acquainted with the seven warning signals of cancer (Persistent cough or hoarseness, change in wart or mole, abnormal bleeding or discharge, thickening or lump in breast or other body site, change in bladder or bowel functioning, a sore that heals very slowly, and indigestion or difficulty in swallowing), and must

seek medical attention immediately if any of these appear.

ANTICIPATORY GUIDANCE

Nutrition

Primary consideration must be given to dietary habits. Generally, there is less demand for calories as one ages because of decreasing basal energy requirements and lessened physical activity. For healthy adults of 50 to 60 years, the daily recommended caloric intake is 2,400 calories for men and 1,800 for women. After age 60, depending on life-style, the caloric intake may need to be reduced (Howard and Herbold, 1978). The diet needs to include recommended servings from each of the four food groups: milk, meat, bread, fruits and vegetables. In selecting foods, individuals in middle adulthood need additional guidelines.

Carbohydrates. Contrary to popular belief, the healthy adult may consume a variety of carbohydrate foods, while whole grain breads and cereals should be encouraged to increase the intake of fiber.

Fats. Saturated fat from animal sources should be decreased or eliminated. Lean meats, poultry, and fish contain lower proportions of fat.

Protein. Protein requirements remain the same throughout adulthood, although any impairment in digestion or absorption will increase protein requirements. Proteins are essential for maintaining and repairing body tissues and for increasing the body's resistance. Complete protein foods (meat, fish, poultry, eggs, and dairy products) contain all eight essential amino acids in approximately the correct proportions, whereas incomplete protein foods (vegetables, legumes, and grains) contain all eight essential amino acids but are low in one or more of them. Animal proteins are complete proteins that are well used in the body. Daily intake of protein foods should be encouraged (Howard and Herbold, 1978).

Iron. Diets low in protein foods tend to be also low in iron and can result in iron-deficiency anemia. Dark leafy vegetables, eggs, legumes, meats, poultry, and iron-fortified cereals and breads are important in the daily diet.

Calcium. Even though osteoporosis affects approximately 25 percent of adults in middle and late adulthood, low dietary calcium intake has not been established as the cause. Large doses of calcium, vitamin D, or fluoride have been used but their value has not been established (Luckmann and Sorensen, 1980). Some adults will have a history of increasing lactose intolerance (i.e., intolerance of milk and milk products) and are encouraged to include other calcium food sources such as dark green vegetables, nuts, and beans.

Vitamins. Vitamins C and E are frequently used by aging adults who believe that they delay the process of aging; however, no evidence to date supports that notion (Howard and Herbold, 1978). Adults who use mineral oil as a laxative must be advised to discontinue the use and to consume foods high in roughage to aid in elimination; mineral oil prevents the absorption of vitamins A, D, E, and K (Howard and Herbold, 1978).

The health care provider must consider the special nutritional needs of each individual bearing in mind that some dietary preferences may have ethnic origins. On the other hand, it must also be remembered that many individuals from a given ethnic group have not retained "cultural" dietary patterns, but have acquired "assimilated" dietary patterns that are more "typically American" than ethnic.

Changes in dietary patterns tend to create frustrations for many adults. Therefore, the individual will often need continuous support and encouragement. Groups such as T.O.P.S. (Take Off Pounds Sensibly), Weight Watchers, and community groups provide a tremendous amount of support for persons attempting to change their dietary habits and lose weight.

Physical Exercise

While some medical research suggests that aging is accelerated by inactivity, physiologists and biologists are reluctant to state that exercises alter the rate of aging. However, there is substantial evidence that physical activity reduces the incidence of alterations in health. Physical exercise and activity are believed to promote physical and emotional health and are extremely valuable for adults in middle adulthood (Fig. 11-8). Physical exercise of the leg muscles seems to facilitate the blood flow to the heart, thereby indirectly benefiting all body systems (Troll, 1975). Physical exercises strengthen the heart muscle, improve circulation, and improve respiratory efficiency. The adult in middle adulthood should be encouraged to include some type of active physical exercise into his or her daily schedule. The type of exercise will vary; generally, after ages 40 to 45 years, activities requiring speed, strength, and endurance should be replaced with those activities that require skill and coordination. Group activities may be more stimulating than solitary ones. Some excellent physical activities for adults include swimming, walking, golfing, bowling, jogging, and biking. People with health alterations such as pulmonary emphysema should seek medical advice before participating in physical activity. Physical activity has the potential for relieving symptoms of tension such as muscular contractions, headache, and neckache.

Smoking

"Warning: the Surgeon General has determined that cigarette smoking is dangerous to your health" is the message printed on cigarette packages. Yet, statistics show an increase in smoking among both sexes. The success of the campaign

Figure 11-8. *Physical exercise promotes physical and mental health and is extremely important for those in middle adulthood.*

to stimulate people to stop smoking has been slow; however, efforts must continue to affect this change. Stopping smoking benefits the individual immediately. The cilia, which are paralyzed by carbon particles, resume their function and clean the tracheobronchial tree; the smoker's cough decreases, and the lung capacity (maximum breathing capacity) improves. The individual must be stimulated and helped to stop smoking. All means available must be utilized; in addition, those assisting the individual need to display an attitude of understanding the difficulty of breaking this hazardous habit. Both the American Cancer Society and the American Lung Association have innumerable pamphlets and behavior modification programs to provide guidance and assistance to individuals who are striving to discontinue smoking. Health care providers have the responsibility to provide information about the effects of smoking to individuals. Early morning cough, increasing shortness of breath, and easy fatigability are early signs of alterations in the respiratory system.

Smoking habits can be used as a predictive factor in lung cancer; with one-half pack of cigarettes per day there is one and three tenths times the average risk; individuals smoking one pack per day have two times the risk, and two packs per day advances the risk to three times the average risk of developing lung cancer (American Cancer Society, 1980).

Dental Care

During middle adulthood, there is an increase in the incidence of periodontal disease, which begins in childhood, its prevalence and severity increasing with age. While prevention has been noted as the only method of control of periodontal disease, adults must continue good oral hygiene to delay the progression of the condition.

Dental plaque is believed to be the cause of periodontal disease, and control measures focus on its removal. Dental plaque is a colorless, transparent soft mass of proliferating bacteria with leukocytes, macrophages, and epithelial cells in a sticky matrix that adheres to the teeth. It can only be detached by mechanical removal. Therefore, rinsing the mouth as the sole measure is ineffective. Toothbrushing, flossing, and irrigation with pressurized water are the most effective methods for removing plaque. Brushing of the teeth after meals is highly recommended. Carbohydrates increase the acidity in the mouth; the longer carbohydrates remain in the mouth, the longer it takes for the pH to return to normal levels. Flossing or the use of water irrigation should be performed at least once a day.

Gingivitis, inflammation of the gums, is the early form of periodontal disease. It is manifested by bleeding of the gums when they are subjected to only slight trauma, as from normal brushing of the teeth. Unless controlled, the inflammation forms pockets around the teeth. These pockets gradually deepen and eventually the underlying tissue is destroyed and the gingiva separate from the teeth. Periodonitis is the advanced form and leads to the loss of teeth; it is characterized by the inflammation extending into the periodontal pockets and destroying the supporting structures of the teeth (Luckmann and Sorensen, 1980). Because of the seriousness of periodontal disease, all adults should be assessed to determine their needs for dental education and dental health care.

Drug Abuse and Alcoholism

Drug abuse among middle-aged adults is increasing as it is among those of all ages in American society. Drug *misuse* denotes an improper or indiscriminate use of common substances in ways that can cause acute or chronic toxicity. The misuse of substances is caused more by lack of knowleldge than any other cause. For example, chronic misuse of common aspirin has led to kidney damage and gastrointestinal bleeding. Drug *abuse*, on the other hand, denotes that the individual intentionally uses substances in excess continually or periodically. This behavior will also have damaging effects and usually leads to physical and psychological dependence.

Physical dependence is manifested by the individual's need for the substance in order to function "normally." Withdrawal symptoms are usually experienced when the individual discontinues the use of the substance. A craving for the sensations the substance produces constitutes psychological dependence. Mild psychological dependence on such substances as coffee and cigarettes leads to a feeling of uneasiness when the affected person is deprived of the substance. However, if desired, the individual can usually do without the substance. Severe psychological dependence is reflected among individuals who compulsively use a substance. Repeated office visits to physicians for medication and frequent visits to the emergency rooms because of "severe pain" are behaviors observed among individuals who are psychologically dependent on specific drugs.

Alcoholism is one of the greatest public health problems. It exists at every level of society and in both sexes. Excessive intake of alcohol leads to physical dependence. Often, alcoholism is not seen as a problem by the alcoholic individual. In part, this may be related to how the individual's peers define alcoholism. The National Council on Alcoholism issued the following criteria as a basis for the diagnosis of alcoholism: (1) alcoholic "blackouts," i.e., loss of memory for events during a period of drinking; (2) consuming a fifth or more of whiskey or its equivalent in beer or wine a day; (3) blood alcohol level above 150 mg/100 ml without observable evidence of intoxication; (4) gross tremors, hallucinations, or convulsions; and (5) continued drinking despite medical advice or family or job problems caused by drinking (National Council on Alcoholism, 1972).

Histories of drinking to relieve anger, depression, or fatigue; frequent automobile accidents; and frequent comments about the stopping of drinking may be indicative of alcohol abuse or alcoholism. Alcholism tends to cause malnutrition, obesity, liver dysfunction (cirrhosis),

and, eventually, changes in personality, e.g., apathy, listlessness, and short attention span (Butler and Lewis, 1977). During middle adulthood, many adults (especially widowed women) begin to abuse alcohol because of loneliness, boredom, or grief. Individuals taking mild tranquilizers, e.g., diazepan (Valium) or chlordiazepoxide (Librium), must understand that the alcohol and these drugs may potentiate each other. Activities to prevent loneliness, boredom, and uselessness are advisable. Alcoholics Anonymous (A.A.) has had impressive results in helping people to achieve and maintain sobriety.

Because alcoholism is a complex disease, the individual frequently requires a variety of serv-

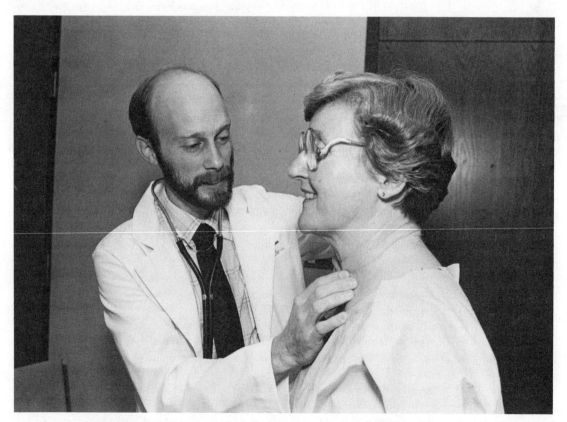

Figure 11-9. *The health care professional working with the middle-aged adult must expect to see a wide variety of values among clients.*

ices to control the compulsive and abusive use of alcohol. Since alcoholism tends to affect the total family, efforts must be directed toward including other family members in the plan of treatment. Preventive measures are the best control of this disease.

The Climacteric

Menopause marks a major change in the functioning of one aspect of a woman's body, and the nurse should address this event accordingly. Asking the client how she feels about approaching menopause is one way to begin. Fears and misconceptions about menopause can be explored and corrected.

For the small percentage of women who have severe symptoms associated with menopause it is important that the nurse acknowledge these symptoms, assist the client in locating competent medical treatment, and reassure the client that the symptoms decrease with time.

Since a small percentage of men may also experience symptoms such as mood swings, insomnia, lassitude, decreased sexual potency and vasomotor instability, the nurse should also discuss these symptoms with male clients. It is helpful to inform the client that most men report that symptoms disappear in two to three years (Hull and Ruebsatt, 1975).

HINTS ON WORKING WITH THE MIDDLE-AGED ADULT

The nurse or other health care worker working with the middle-aged client should allow the client to make his or her own decisions to as great as extent as possible, and encourage the client to take responsibility for those decisions. As is true of people of all ages, the middle-aged client must be treated as an individual; the health care professional must expect and accept a wide variety of values, perceptions, and aspirations,

which may be influenced solely by the client's personal background, or perhaps by cultural or ethnic factors (Fig. 11-9). The individual's strengths and insights should be acknowledged whenever possible. Finally, as always, the privacy and dignity of the middle-aged client must be respected and upheld by all health care workers, whether in taking a history or performing direct physical care.

SUMMARY

The healthy person functions at almost peak efficiency in the initial years of middle adulthood. In the absence of disease, and with good nutrition and physical activity balanced with rest, biophysiological functioning can be maintained during middle adulthood.

Intelligence remains essentially the same in middle adulthood, or increases slightly. Ability level is one factor that influences adult intellectual capacity; individuals who tested higher in earlier years tend to maintain that ability longer, or to improve it.

In Erikson's terms, the chief developmental task of middle adulthood is to achieve generativity, a sense of responsibility for others. Failure to achieve generativity results in what Erikson calls stagnation. To this task, R. C. Peck added those of valuing wisdom versus valuing physical powers, socializing versus sexualizing in human relationships, cathectic flexibility versus cathectic impoverishment, and mental flexibility versus mental rigidity.

The anticipated events of middle adulthood—launching children from the home, age-related physiological changes (e.g., menopause), grandparenthood, caring for one's aging parents—are those least likely to cause undue stress. Unanticipated events such as the death of a spouse or child, loss of job, or divorce may precipitate a crisis in middle adulthood.

Common health problems of the period include obesity, hypertension, osteoarthritis, changes in vision, and cancer. The two major preventive measures indicated for middle-aged adults are proper nutrition and a balanced regimen of rest and exercise. If these are observed, the years of middle adulthood can be among the most productive in the life span.

REFERENCES

Aiken, L. *Later life*. Philadelphia: W. B. Saunders, 1978.

Binstock, R. H., and Shanas, E. *The handbook of aging and the social sciences*. New York: Van Nostrand Reinhold, 1976.

Birren, J. E., and Schaie, K. W. (Eds.), *Handbook of the psychology of aging*. New York: Van Nostrand Reinhold, 1977.

Blau, Z. S. *Old age in a changing society*. New York: New Viewpoints, 1973.

Blenkner, M., Social work and family relationships in later life with some thoughts on filial maturity. In Shanas E., and Streib, G. F. (Eds.), *social structure and the family*. Englewood Cliffs, N. J.: Prentice-Hall, 1965.

Bullough, B. and Bullough, V. L. *Poverty, ethnic identity and health care*. New York: Appleton-Century-Crofts, 1972.

Burnside, I. M. *Nursing and the aged*. New York: McGraw-Hill, 1976.

Burnside, I. M., and Ebersole, P. *Psychosocial Caring Throughout the life span*. New York: McGraw-Hill, 1979.

Butler, R. N., and Lewis, M. I. *Aging and mental health*, St. Louis: C. V. Mosby, 1977.

Chew, P. *The inner world of the middle-aged man*. New York: McMillan 1976.

Cumming, E. and Henry, W. E. *Growing old: The process of disengagement*. New York: Basic Books, 1961.

Delury, G. E. *The world almanac and book of facts 1980*. New York: Newspaper Enterprise Association, Inc. 1981.

Diekelmann, N. *Primary health care of the well adult*. New York: McGraw-Hill, 1977.

Duvall, E. M. *Family development*. Philadelphia: J. B. Lippincott, 1971.

Erikson, E. *Childhood and society* (2nd Ed.). New York: Norton, 1963.

Finch, C., and Hayflick, L., *Handbook of the biology of aging*. New York: Van Nostrand, 1977.

Gerney, S. and Cox, C. *After forty: How women can achieve fulfillment*. New York: Dial, 1973.

Gorer, G. *Death, grief and mourning in contemporary Britain*. London: Cresset, 1965.

Graber, E. A., and Banber, H. R. K. The case for and against estrogen. *The American Journal of Nursing*, 1975, 75(10), 1766–1771.

Groer, M. E., and Shekleton, M. E. *Basic pathophysiology: A conceptual approach*. St. Louis: C. V. Mosby, 1979.

Havighurst, R. J. *Developmental tasks and education*. New York: David McKay, 1974.

Havighurst, R. J., Neugarten, B., and Tobin, S. S. Disengagement and patterns of aging. In Neugarten, B. (Ed.), *Middle age and aging*. Chicago: University of Chicago Press, 1968.

Hershey, D. *Life span and factors affecting it: Aging themes in gerontology*. Chicago, Ill.: Charles C. Thomas, 1974.

Holmes, T., and Rohe, R. The social readjustment rating scale. *Journal of Psychosomatic research*, 1967, 11, 213–218.

Howard, R. B., and Herbold, N. H. *Nutrition in clinical case*. New York: McGraw-Hill, 1978.

Hull, R., and Ruebsatt, H. J. *The male climacteric*. New York: Hawthorn Books, 1975.

Johnson, L. Living Sensibly. *The American Journal of Nursing*, 1975, 75(6), 1012–1016.

Jones, D., Dunbar, C. F., and Jirovec, M. M. *Medical surgical nursing*. New York: McGraw-Hill, 1978.

Kimmel, D. C. Adult development: Challenges for counseling. *Personnel and Guidance Journal*, November 1976, 103–105.

Kimmel, D. C. *Adulthood and aging*. New York: John Wiley & Sons, 1974.

Kohn, R. R. *Aging*. Kalamazoo, Mich.: The Upjohn Company, 1973.

Luckmann, J., and Sorensen, K. C. *Medical-surgical nursing: A psycholphysiologic approach*. Philadelphia: W. B. Saunders, 1980.

Lugo, J. O., and Hershey, G. L. *Human development: A multidisciplinary approach to the psychology of individual growth*. New York: MacMillan, 1974.

Maas, H. S., and Kuypers, J. H. *From thirty to seventy*. San Francisco: Jossey-Bass Publishers, 1974.

Martin, L. L. *Health care of women*. Philadelphia: J. B. Lippincott, 1978.

McDaniels, C. Leisure and career development at mid-life: A rationale *Vocational Guidance Quarterly*, 1977, 25(4), 344–350.

Mims, F. H., and Swenson, M. *Sexuality: A nursing perspective*. New York: McGraw-Hill, 1980.

National Council on Alcoholism, Criteria Committee. Criteria for the diagnosis of alcoholism. *American Journal of Psychiatry*, 1972, 129, 127–135.

Neugarten, B. L. A New look at the "crises" of middle age. *Geriatric Focus*, 1970, (9), 1–9.

Neugarten, B. L., and Weinstein,. K. The changing American grandparent. In Neugarten, B. (Ed.), *Middle age and aging*. Chicago: University of Chicago Press, 1968.

Ostfeld, A. M., and Gibson, D. C. *Epidemiology of aging*. Maryland: DHEW, 1972.

Peck, R. Psychological developments in the second half of life. In *Middle Age and Aging*, Chicago: University of Chicago Press, 1968.

Pruck, V. N. The mid stage woman. *The American Journal of Nursing*, 1975, 75(6), 1019–1021.

Pruck, V. N. *Quality of life: The middle years*. Acton, Mass.: Publishing Sciences Group, 1974.

Rainwater, L. *Behind ghetto walls: Black family life in a federal slum*. Chicago: Aldine, 1970.

Reinhart, A. M., and Quinn, M. D. (Eds.), *Current practice in gerontological nursing*. St. Louis: C. V. Mosby, 1979.

Rodman, M. J., and Smith, D. W. *Pharmacology and drug therapy in nursing*. Philadelphia: J. B. Lippincott, 1979.

Seyle, H. *The stress of life*. New York: McGraw-Hill, 1956.

Sheehy, G. *Passages: Crises of adult life*. New York: Dutton, 1976.

Smith, D. W., Bierman, E. L., and Robinson, N. M. *The biologic ages of man: From conception through old age* (2nd Ed.). Philadelphia: W. B. Saunders, 1978.

Stevens-Long, J. *Adult life developmental processes*. California: Mayfield Publishing Company, 1979.

Stevenson, J. S. *Issues and crises during middlescence*. New York: Appleton-Century-Crofts, 1977.

Troll, L. E. *Early and Middle adulthood*. Calif.: Brooks/Cole Publishing Company, 1975.

Troll, L. E., Israel, J. and Israel, K. (Eds.), *Looking ahead*. Englewood Clifs, N.J.: Prentice-Hall, 1977.

Turner, J. S., and Helms, D. B. *Contemporary adulthood*. Philadelphia: W. B. Saunders, 1979.

U.S. Department of Health, Education, and Welfare. *The breast cancer digest*. Bethesda, Md.: National Cancer Institute, 1979.

———. *Facts of life and death*. Bethesda, Md.: DHEW, 1978.

———. *Health status of minorities and low-income groups*. Bethesda, Md.: 1979.

———. *Women in midlife—security and fulfillment, Part I*. Washington, D.C.: U.S. Government Printing Office, 1978.

Vander, A. J., Sherman, J. H. and Luciano, D. S. *Human physiology*. New York: McGraw-Hill, 1980.

Vontress, C. E. Counseling middle-aged and aging cultural minorities. *Personnel and Guidance Journal*, November 1976, 132–135.

Woodruff, D. S., and Birren, J. E. *Aging scientific perspectives and social issues*. New York: D. Van Nostrand, 1975.

12

The Young Older Adult

AMIE MODIGH

The major changes in the young elderly (ages 65 through 80) are psychosocial. The physiological changes are much more subtle until after age 80, when psychosocial changes are more subtle. A knowledge of the normal changes of early older adulthood will enable the health care provider to help her clients of this age experience the period as "A Good Age," to use Comfort's (1976) phrase.

Studies and writings about old age are older than the Bible. Before 2000 B.C., Egyptian papyrus rolls bore such titles as *Book for Transforming An Old Man Into a Youth*, which recommended the taking of two emetics each month to ensure longevity. It also describes skeletal and cardiovascular changes in such

twentieth-century terminology "porotic changes in the bone" and "tortuous calcific vessels." The modern study of old age was first referred to as "gerontology" by Metchnikoff, a Russian zoologist working in Paris, in 1903. Six years later, Austrian-born Ignatius Leo Nascher, in New York, coined the word *geriatrics* to describe the study of the clinical aspects of aging (Freeman, 1979).

There was, however, little evidence of interest in these fields in the United States until more recently. The first issue of the *Journal of Gerontology* was published in 1946. The National Institute on Aging was pocket-vetoed in 1972, then reconsidered and finally signed into effect in 1974. The first chair in Geriatric Medicine

Figure 12-1. *The attainment of age 65 allows many the option to continue working, perhaps in a second career, to travel, spend more time with friends, to become involved in community affairs, and so on. These men are volunteers in a neighborhood improvement association. (Courtesy Durham Senior Citizen Council.)*

was established at Cornell Medical Center in 1978. Since 1970 there has been a surge of interest in all aspects of aging. This is evidenced by accelerated research, publications, and by a small but encouraging increase in the number of health care professionals entering the field of gerontology.

The attainment of age 65 has an emotional impact stronger than that of most other ages. To some, the age is synonymous with freedom to pursue a second career or hobby, opportunity to travel, visit friends, and spend more time with the family; such people envision reaching

age 65 as a pleasant goal, a stepping-stone to a new, exciting, and challenging part of life. Many find great joy and satisfaction in the opportunity to "be" instead of "do" all the time (Fig. 12-1).

Many benefits begin at age 65. Those who have retirement pay usually begin receiving it at this age. Social Security benefits, Medicare, eligibility for membership in certain senior citizen clubs, and lower taxes with double personal exemptions are among other benefits. Many businesses and organizations offer free services and reduced rates for trips, tours, and theater and concert tickets for senior citizens. A large number of universities offer free tuition to people 65 and over.

These benefits are, however, not something to which everyone looks forward. To some people, the 65th year is a dreaded birthday. The gold watch that may be presented at a testimonial dinner may be followed by good-byes from colleagues and friends who somehow never call for that golf date they suggested. These retirees would rather continue collecting a paycheck than a social security or retirement check, not only because the amount in the former was larger, but also because of the negative connotations of the latter. The thought of hearing the title "senior citizen" applied to themselves is often quite distasteful.

Our society has encouraged the negative connotations of "age 65." Since shortly after the Industrial Revolution, 65 has marked the end of most working careers, and until the 1970s many people were forced to retire at age 65. Working is synonymous with being useful in American culture; consequently, nonworking people may see themselves as falling into the "useless" category. It is easy to understand why

so many retired people question their self-worth and suffer from decreased self-esteem.

Society's retirement laws have, in essence, defined old age as beginning at 65. This definition is significant because attitudes and expectations flow from it. No doubt people act and feel old because of these expectations. It will, indeed, be interesting to follow sociological and psychological studies in states where the retirement age has been delayed to age 70. There are, of course, many occupations in which individuals do not retire at 65. Fishermen and farmers often experience a gradual retirement brought about by a gradual decline in strength and health. Self-employed business people, certain physicians, lawyers, clergy, and people in political positions retire at a later age, and often by choice.

It is interesting to examine the paradox of a society that contributes to the negative connotations of aging when, in fact, American society reaps many benefits from its elderly. They offer great service to society through voluntary work in churches, hospitals, and many other organizations. The elderly people who are allowed to work perform as well or better than

Figure 12-2. *Many older people stay abreast of political and civic matters. This man is reading a petition to be presented to a local housing authority. (Courtesy Durham Senior Citizens Council.)*

younger workers. In addition, they have fewer accidents, less absenteeism, and fewer job turnovers. The elderly are also very well informed on public issues and take an active part in politics as well as vote in larger numbers than other age groups (Fig. 12-2). People 65 and over are far more law abiding than any other age group except very young children; as Palmore (1977) points out, this age group is sent to prison at one-tenth the rate of the rest of the population.

What determines whether one will view one's own aging positively or negatively? The answer often lies in how one viewed the attainment of other milestones in the life span; the individual who dreaded his thirtieth birthday (and there are many who did) because he felt he would then be "over the hill" is very likely to be in the group viewing 65 in a negative fashion. On the other hand, the person who has looked forward to each new stage of growth and development as a new challenge or an opportunity for new activities and growth will most certainly do the same at 65.

SOME GENERAL FACTS ABOUT THE YOUNG OLDER ADULT

Great individual differences continue to exist among 65- to 80-year-olds, as they did earlier in life. Because a human being is the sum total of his existence, the longer he lives, the more of an individual he will become. This age group is therefore less homogeneous than earlier age groups. It is important for the reader to keep this in mind as general developmental changes are discussed. There are as many variations in a group of 65-year-olds as there are people 65 years old.

Everyone knows someone who is "75 years young"—the grandmother who is mistaken for the mother, or the physically and mentally active individual who loves to try new ideas and listen to new thoughts, who can converse on a variety of topics and is up-to-date with current events. Similarly, everyone knows someone who is "40 years old"—the mother who is mistaken for the grandmother, or the individual who is so rigid that he or she does not want to hear a new idea, much less try a new activity. These examples point out the fact that the individual personality, not age alone, plays an increasingly important part in continued growth and development.

Everyone involved in working with the older population needs to be aware of ageism, a word coined by Robert Butler in 1968. He defines this as "a process of systematic stereotyping and discrimination against people because they are old, just as racism and sexism accomplish this with skin color and gender" (Butler, 1975, p. 12). Ageism is a powerful force in American society, manifested in a wide range of stereotypes, myths, cartoons, jokes, and labels.

The elderly have done much to improve conditions for themselves in the United States and to help decrease ageism. One of the more active political organizations is the Gray Panthers, started in 1970 by Maggie E. Kuhn. She felt that many retirement clubs were "play pens for the old" and stated that the Gray Panthers' goal was to be actively involved on a legislative level in the promotion of freedom and justice for the elderly (Butler, 1975, p. 34). As one example, this organization has been very instrumental in changing forced retirement.

Cowgill and Homes (1972) have done extensive research on cultural variations in views of aging; one outstanding generalization states "the role and status of the aged varies systematically with the degree of modernization of society." Modernization tends to decrease the relative status of the aged. Since the United States is highly modernized, the status of our elderly should be very low. Generally speaking, this is true. The number of American medical curricula that include Gerontology is less than ten. While many schools of nursing include some gerontological content, this is not re-

Figure 12-3. *It is ironic that the United States, which benefits from an almost unparalleled level of medical technology, also harbors a uniquely negative view of aging. The casting aside of the aged, which necessitates the assertion that "we are somebody," would be unthinkable in a great many other cultures, including many that are regarded as "underdeveloped." (Courtesy Durham Senior Citizens Council.)*

flected in any state examination as of this writing. Most libraries have a limited amount of reference material in this field. Most of what little research has been done on aging has concentrated on the 5 percent of elderly who are in institutions. Age discrimination is blatant in employment.

The American attitude toward aging stands in stark contrast to that found in such countries as Thailand and Cambodia, where, despite underdevelopment and a much lower life expectancy, the aged are a valued part of the population (Fig. 12-3). Also, Cowgill and Homes's generalization notwithstanding, the highly modernized Japanese retain their veneration of the aged; their deeply ingrained cultural and religious beliefs (which include ancestor worship) are undoubtedly partly responsible. The Scandinavian countries also manifest an enlightened view of the aged; for example, Sweden does not mandatorily retire those over 65, provides free health care, and offers the elderly great discounts in travel, entertainment, and housing. Norway also provides free health care and even contributes to funeral expenses. Many of the same features are found in Denmark, which Meador (1980) characterizes as a place where old age is "a time to enjoy."

A notable exception to the rule of poor treatment of the elderly in the United States is found among blacks, especially poor blacks. Unfortunately, as more blacks move into the American mainstream, such traditions as respect for the elderly may also be left behind (Murphy, 1980).

Composition of the 65- to 80-Year-Old Age Group

One of the reasons for increased interest in aging is that it has been described as a social problem. Yet, we say with pride that one of our great medical achievements is that people now live longer. Why then is aging a problem? According to Atchley (1977) society is not prepared to deal with the very large and rapidly growing aging population. The speed of growth is impressive. In 1900, the United States had 3 million people aged 65 and over. By 1970, there were 20 million people in this category. The projected number for the year 2000 is 35 million; this means almost one out of every six Americans will be 65 or older by 2000.

In 1978, life expectancy for women was 77 and for men 68 (*Statistical Abstract of the United States*, 1978). The *Statistical Abstract of the United States* (1978) showed, in 1977, 8.2 million white males, 0.8 million black males, 11.8 million white females, and 1.1 million black females. The percentage of the total population aged 65 and over is as follows: ages 65 to 69: male, 3.5 percent; female, 4.2 percent; ages 70 to 74: male, 2.5 percent; female, 3.2 percent; ages 75 to 79: male, 1.5 percent; female, 2.2 percent (*Statistical Abstract of the United States*, 1978).

The elderly are far more educated than they were in the 1950s. Statistical figures for 1977 showed 37 percent of people 65 and over had finished high school and 8 percent had four or more years of college. These figures have already increased and will continue to do so dramatically. There are two reasons for this: college attendance is very high among the younger generation, and older people are returning to universities (*Facts About Older Americans*, 1978).

Although Florida has the highest percentage of retirees in proportion to the total state population (17.1 percent), California and New York have a larger number, over 2 million each. Florida has 1.5 million and Ohio, Illinois, Texas, and Pennsylvania each have over one million (*Facts About Older Americans*, 1978).

Health data for the 65 to 80 year group reveals that while acute conditions decrease with age, chronic conditions increase; 20 percent of people 17 and under have one or more chronic conditions. At 65 and over, the percentage increases to 85 percent. Thirty-nine percent are limited in some activity. Eighteen percent have some decrease in mobility, 6 percent have difficulty getting around alone, 7 percent need mechanical aid, and 5 percent were homebound in 1972 (*Facts About Older Americans*, 1978).

It is important to realize that the vast majority of the young elderly still manage to live outside of an institutional setting in spite of the above facts. Only 1.3 million, or less than 5 percent of people over 65, live in America's 18,000 nursing homes (National Nursing Home Survey Report for National Center for Health Statistics, 1977).

Cardiac conditions are the most common health problems, followed by cancer, arthritis, visual impairment, and emotional disturbances. The death rate per 10,000 for all forms of heart disease in 1976 was 2,906.9 in males and 2,039.3 in females; the death rate for cancer was 1,324.0 and 741.0 respectively (*Statistical Abstract of the United States*, 1978).

PHYSIOLOGICAL CHANGES

Subtle outward variations and organ changes occur in early old age. The graying of hair will generally have started in the fourth or fifth decade; it continues and may lighten to white. The balding process in men has generally halted by now and what hair is left will most likely remain through this period. The pubic hair becomes sparser, along with the axillary hair. The color in these two areas begins to change to a lighter shade in the late seventies.

Some women find, to their despair, that hair growth increases on the upper lip and chin. Men may experience hair growth in nostrils and external ear canals. Both the nose and the ears tend to enlarge externally; this is more pronounced in men.

The most striking visible skin change is the increase in wrinkles and excess skin folds, primarily in the face and neck. This may be noticed to a lesser degree on arms, legs, and the abdomen. Individual variations are very pronounced; they depend on a variety of factors: genetics, nutrition, loss of subcutaneous fat, decrease in oil-secreting glands, and decrease in the number of blood vessels that supply the skin. The wrinkling process is greatly speeded up by long exposures to wind, water, and the sun. The sun is the greatest offender, and people who have spent a great deal of time outdoors, such as fishermen and farmers, tend to show wrinkles earlier. Another group prone to early wrinkles are the dedicated sunbathers who spend hours letting the most damaging rays of the sun beat on as large a part of their bodies as the law will allow. It is a fascinating paradox that wrinkles in women's faces are associated with unattractive aging in American society, yet no other nation spends so much time getting a suntan.

The extra folds on arms, legs, and abdomen are dependent on the relationship between muscles and fat. Muscles will remain intact, if used, well into the seventh decade. Thus, the walker, jogger, gardener, etc. will display considerably less extra skin folds than his rocking-chair counterpart.

Physical stature will generally decrease by the end of this period of growth and development. The calcium in the skeletal system gradually seeps out, leaving the medullary cavity of bone wider and the compact tissue thinner. Maxi-

mum bone loss is said to occur between ages 55 and 65. Degree of bone loss varies greatly among individuals but tends to be more pronounced in women. These changes lead to certain degree of "shrinkage." Reduced bone mass can actually shorten the spine by as much as one-half to two inches. This can also result in a stooped posture of varying degrees, to which calcification and ossification of ligaments and sclerotic changes in the tendons further contribute.

Lean body mass decreases as the individual ages. Muscles and tissues are replaced by adipose tissue, which weighs less. The mean weight for men continues to increase until approximately age 50; women peak approximately ten years later (Bierman, 1976).

There is an increase in cross-linkage of collagen fibers, which results in decreased elasticity of muscle fibers. This, plus decreased subcutaneous fat, leads to flabby or wrinkled skin.

Theories of Aging

In many of his writings, Wilson (1974), offers convincing evidence that life is programmed from beginning to end. This is similar to what Hayflick (1975) refers to as the genetic theory, which focuses on alterations in DNA and RNA. Cape (1978) presents still other theories: the toxic theory suggests that an accumulation of toxic by-products of metabolism is responsible for the aging process. The wear and tear theory simply proposes the analogy of the wearing out of a machine. The third is an elaborate error (or relational) theory involving the cellular-extracellular level and stating that aging is initiated by random errors causing self-destructive disorders. Other theories include the auto-immune theory, which suggests that the body produces antibodies that destroy normal body cells. One or more of these theories may prove to be correct. In the meantime, we know that we age basically because cells either stop reproducing and die, or gradually lose part of their function.

Skin

Normal changes in skin and hair have minimal functional significance in the young elderly, but they are the most obvious outward changes of aging.

The epidermis, consisting mostly of dead cells that are replaced continually, function as protection and contain the pigment melanin, which determines the skin color. These cells become less hydrated and less supple, therefore, more fragile. In some areas the epidermis thickens markedly. The dermal layer contains nerves, blood vessels, hair folliciles, and sweat and oil glands. There is a decrease in blood supply as well as a decrease in the number and function of sweat and oil glands. The dermal layer undergoes a loss of hydration because of the decreased mucopolysaccharides (the water-binding component of connective tissue). Dryness of hair in this age group is one result of decreased hydration to the hair follicles. Of interest is that the graying of the hair is the result of the melanocytes at the base of the hair follicles becoming less effective in producing the normal pigments (Finch and Hayflick, 1977).

The third layer of skin, the subcutaneous layer, contains a layer of fat for insulation and support. The fat gradually reduces in mass, usually starting around the sixth or seventh decade. The support to the skin capillaries decreases, making them very susceptible to breakage from the smallest injury. Thus, one frequently observes small, ecchymotic lesions on the exterior surfaces of the forearms. This is most common in women in their late seventies (Cape, 1978). All of these changes lead to dry, wrinkled, scaly, and less elastic skin. The degree and age at which this occurs varies greatly.

The brown spots that frequently appear on the backs of the hands, arms, and occasionally on the face, which are falsely referred to as "liver spots," are benign, pigmented areas totally unrelated to the liver and probably due to a decrease in the ability of the melanocytes to produce the pigment melanin in an even distribution (Finch and Hayflick, 1977). The fingernails and toenails become thick and hard because of decreased vascular supply to the nailbed. This is generally more pronounced in the toenails.

Vision

Over 80 percent of people retain good vision throughout life, 15 percent have poor vision, and 5 percent become blind (Cape, 1978). By far the most common change in vision is presbyopia, farsightedness caused by aging. This occurs gradually, starting around age 40. The cause is decreased flexibility of the lens with a concomitant decrease in its ability to accommodate to nearby objects. Many people will need reading glasses to compensate for the change. It occurs at a greater rate and severity in females for unknown reasons. Cataracts occur to some degree in 95 percent of people over 65 (Duke-Elder, 1969). However, only a small percentage of affected people suffer progression of the cataract to the point of needing lens removal or of underoing a significant decrease in vision.

The cornea tends to flatten, causing astigmatism. There is a decreased blood supply to the retina, which in turn means that the rods and cones situated on the retina need more intense light in order to register visual information. The pupil becomes smaller. The iris, being a muscle, becomes more rigid, and it therefore takes longer for the pupil to constrict or dilate in response to changes in the amount of light. The lens becomes more yellow; this further decreases the amount of light entering the eye and, eventually, the retina. This so called yellow-filter effect makes it harder to distinguish less intense colors such as purples, blues, and greens. Yellows and reds are more intense and can be seen and distinguished with more ease.

The sclera also becomes less elastic and tends

to yellow from fat deposits. An insignificant fatty invasion of corneal margin, resulting in a gray ring (*arcus senilis*) around the iris, will sometimes cause worry in the elderly person, who may associate this with high cholesterol; there is no such relationship, and arcus senilis is a common phenomenon in 40 percent of people over 65 (Cape, 1978).

Vitreous "floaters" are normal opacities occurring as a result of normal aging changes. It is important to explain this to the elderly person as it can be a frightening experience. It is normal for small, moving black dots to appear and move as the eye moves. If flashes of light occur, this needs further attention, as it could indicate a retinal tear.

The outside structure of the eye undergoes some skin changes like the rest of the skin. The decreased orbital fat may give the eye a sunken appearance late in early old age. Decreased muscle tone and less elastic tissue may cause a degree of ptosis (drooping of the eyelid); for the same reason, varying degrees of ectropian (turning outward of the lower eyelid) or entropion (turning inward of the lower eyelid) may occur. Unless these changes interfere with vision or cause corneal irritation, corrective surgery is unnecessary.

Hearing

Hearing decreases gradually, starting at around age 40. Fifty-five percent of people over 65 experience presbycusis (hearing loss due to aging) (Saxon and Etten, 1978). Presbycusis is progressive but may reach a plateau and not progress for a long period of time. High-pitched sounds (higher than normal speech frequency) are lost first. As hearing deficit progresses, it involves the middle frequencies, which contain most of the speech sounds. Males are affected earlier and in larger numbers than females. Exposure to loud noises (over 80 decibels) over a period of time will produce permanent hearing loss.

Conductive hearing loss occurs in the external or middle ear. Cerumen (ear wax) thickens as one grows older and can actually occlude the ear canals. *Sensorinueral hearing loss* involves the inner ear in one of two ways. Either physical sound waves are prevented from turning into electrical signals, possibly because of insufficient blood supply to the cochlea, or electrical signals may be prevented from being transmitted to the brain through the auditory nerve. Loss of neurons in the auditory nerve may be the cause. Both types of hearing loss affect both ears; however, there is usually a considerable difference in the degree of loss in the two ears.

Smell and Taste

There is a gradual and small decrease in the sensory receptors located in the mucosa of the nostrils, causing a decrease in smell. Two-thirds of taste sensation are dependent on the ability to smell (Butler and Lewis, 1973), which is one reason for the decreased sense of taste in old age. Also, the tastebuds decrease in number and function. Well-papillated tongues provide a more acute sense of taste. Since tastebuds are also located in the upper gum, those who wear upper dentures, which occlude those tastebuds, will complain of decreased taste. It is usually well into the seventh decade before any significant change in taste occurs. Hermel, Schonwetter and Samueloff (1970) found no difference in sensitivity to salty taste but discovered decreased sensitivity to sweet, sour, and bitter tastes. This may be one reason why so many older people increase their sugar intake. Rovee, Cohen, and Shlapack (1975) feel that when taste and smell are reduced to the point of practical significance it is because of illness rather than to normal aging.

Touch

Skin receptors are slower to respond to all forms of touch—pain, pressure, cold, and heat—and there is an increase in the threshold of the stimulus needed to elicit a response. There is still debate as to whether this is caused by a decrease in the number of receptors or a slowing down of afferent nerve transmission (Reichel, 1978).

Digestive System

The greatest change in the digestive system occurs in the mouth. The salivary glands produce less saliva, which causes dryness of the mouth (xerostomia). Teeth are often lost because of poor oral hygiene earlier in life. However, wear and tear also makes the teeth subject to erosion and trauma. Shrinkage of the bone and gum also contributes to the loss of teeth. The rest of the digestive system remains remarkably intact throughout life. In fact, there has been no concrete evidence that there is any decrease in absorption of nutrients as a result of normal aging. There is a slight decrease in the strength of esophageal peristaltic waves, causing food to enter the stomach a bit more slowly. There is a decrease in the production of enzymes, ptyalin, and amylase, and in the secretion of gastric acids. The motor activity of the stomach slows down hunger contractions and delays gastric emptying. There are no significant changes in the small intestine, colon, or rectum; constipation is not a normal physiological change. A small increase of water absorption in the large intestine and a thinning of muscle layers can be contributing factors; however, constipation is caused more by poor food and fluid intake, lack of exercise, and, in the United States particularly, overuse of laxatives.

The pancreas produces less amylase and lipase; which may be why many older people do not tolerate fat as well as they did earlier in life. Calcium absorption is decreased, more so in the absence of adequate vitamin D.

Circulatory System

The changes in the circulatory system have a profound effect on all cells, organs, and systems of the body. The cardiac index (the amount of blood pumped by the heart per square meter of body surface) decreases by 35 to 45 percent between ages 30 and 90; this is caused primarily by changes in the arterial walls, namely a fragmentation of the molecule elastin and an increase in cross-links of collagen. Thus, the arteries become dilated and stiff, and are less sensitive to baroreceptors. It is still debated whether arteriosclerosis is a normal physiological change of aging or a disease (Finch and Hayflick, 1977). In any case, changes occur gradually, starting in the fourth decade. The heart compensates by beating faster and harder. This, in turn, leads to an increase in the size of the left ventricle and an increase in systolic blood pressure.

Other physiological changes include an increase in fatty tissues of the heart and a decrease in elasticity of cardiac valves. In spite of these gradual changes, people with "aged but not diseased hearts" can tolerate activity "adequate for all normal purposes, including quite rigorous activity" (Cape, 1978).

Respiratory System

Maximum breathing capacity decreases by 50 percent between ages 30 and 90. Great individual variations exist depending on life-style, where one lives, and exposure to pollution.

However, forced expiratory volume and vital capacity normally decrease because the lungs become less elastic. They contain collagen and consequently experience an increase in cross-linking. The muscles used for inhaling and exhaling weaken, leading to an accumulation of carbon dioxide. The bony structure of the rib cage accumulates calcium, resulting in a stiff, somewhat expanded position, which prevents complete expirations. There is a decrease in the number of alveoli, where gas exchange takes place, which is further aggravated by the decrease in blood supply to the lungs, and especially to the capillaries surrounding the alveoli.

In the absence of disease, these changes should have little effect on daily activities.

Genitourinary System

The glomerular filtration rate decreases between the ages of 30 and 90 by 45 to 50 percent. This is caused by a decrease in the number of nephrons and a decrease in blood supply to the kidneys. The kidneys maintain a normal fluid and electrolyte balance throughout life unless they are upset by disease. Total body water decreases considerably, although extracellular fluid remains constant; the decrease in intracellular fluid is probably the result of tissue death as a result of aging (Reichel, 1978).

The ureters, bladder, and urethra are composed of muscular tissue; they therefore undergo similar changes to those in other muscles, namely, a decrease in elasticity and strength. The most significant changes take place in the bladder. As the muscle tone decreases, the bladder capacity decreases from approximately 500–600 cc to 200–300 cc. Furthermore, the receptors in the bladder wall that trigger impulses indicating a need to empty the bladder decrease in sensitivity. At a younger age these impulses are felt when the bladder is less than half full; as one ages, the time between the recognition of the signal and the actual need to empty the bladder decreases greatly. There is also an increase in residual urine because of decreased muscle contraction.

The male has the added risk of developing an obstruction in the lower genitourinary tract because of an enlarged prostate gland. If the prostate enlarges, it may occlude the urethra and produce dysuria, frequency, or even anuria.

Reproductive System

The female usually experiences menopause some time between 40 and 55; thus, the young elderly will have already adjusted to this development. The ovaries, fallopian tubes, and uterus continue to shrink, and the ligaments supporting these structures lose their elasticity, which may result in cystocele (a herniation of the bladder through the vaginal wall) or rectocele (herniation of part of the rectum into the vagina). This is more likely to occur in women whose vaginal tissues have been repeatedly stretched by childbirth. Vaginal contractures cause the vaginal canal to continue to become shorter and narrower. The elasticity of the vaginal walls decreases, and the pH of the glandular secretion increases, becoming closer to neutral and sometimes even to alkaline, making infection more likely. The amount of secretions is less, sometimes to the point of not producing adequate lubrication for pleasurable sexual intercourse. Although the breasts decrease in size and firmness, the nipples do not lose their sexual responsiveness.

The fatty layers of labia majora decrease, which may expose the clitoris not only to external irritation but to bacterial invasion via the urinary meatus.

The reproductive ability of the male is not altered to any significant degree until extreme old age. The testosterone level declines, resulting in decreasing firmness and size of the testicles. The scrotum may show atrophy of the walls. There is a decreased production of sperm. More time and stimulation are required to produce penile erection, ejaculation is delayed, and the refractory period is greatly increased, i.e., the ability to attain a second erection may require from several hours to days.

Both men and women experience a decrease in the frequency and explosiveness of orgasm but there is no physical reason why sexual activity cannot continue to be a meaningful and pleasurable experience among the young elderly.

Endocrine System

Generally speaking, there are minimal significant changes in the endocrine system as one grows older. The pituitary gland undergoes minor changes and continues to function quite well in the absence of disease.

The thyroid gland becomes less active, resulting in a lower basal metabolic rate. Studies by Gregerman and Bierman (1974) demonstrated that "the thyroid of even very elderly persons appears to function adequately and maintains its reserve capacity." However, later studies showed hypothyroidism to be quite common among the elderly (Brocklehurst and Hanley, 1978). The pancreas functions adequately. It appears that insulin is produced at a slower rate in the 60 to 70 age group. Cape (1978) suggests caution in diagnosing elderly people as diabetic, since higher levels of blood glucose may be a normal physiological change in aging.

Musculoskeletal System

Since muscles are composed of nonmitotic (nonreplaceable) cells, it is vital to make every attempt to maintain them in good condition. Muscles must be used or they will atrophy. Even with use, there is a gradual change in muscles in the young elderly. The number of muscle fibers decreases and the lost muscle tissue is replaced by fatty tissue, which leads to a decrease in strength. The muscles of the arms and legs become thin and flabby. The age at which this occurs varies greatly and is directly related to the degree of activity.

Calcium is lost from the bones, leading gradually to osteoporosis. This occurs more frequently in women. Vertebral atrophy occurs in both sexes but the bone is less mineralized in females; this difference is believed to be related to estrogen loss in females, which is drastic after menopause (Talbert, 1977). A higher incidence of fractures exists in females in this and older age groups. A progressive loss of about 40 percent of bone occurs in women between the ages of 40 and 80. In men this process does not start until between age 55 and 65 and progresses at a much slower rate. After age 80 the process

continues at the same rate in both sexes (Finch and Hayflick, 1977). Joint movement is hampered as the tendons and ligaments stiffen.

Nervous System

The number of neurons decreases beginning at age 30; the decrease is not significant, but the availability of nutrients in the form of blood and oxygen is vital for the survival and normal functioning of the neurons. Fortunately, there is little decrease in blood flow to the brain in the healthy individual in early old age. Ordy (1975) showed that cerebral blood flows at 106 ml/per 100 gm of brain tissue per minute at age 6; this decreases to 63 ml at age 21 but only to 58 ml at age 71. Kokmen et al. (1977), in a study of 51 healthy people between 61 and 84 years of age, found that the aging process had minimal effect on most neurological functions.

According to Bromley (1974) the brain shows a decline in abilities after age 20. This decline is more rapid after age 75. The least decline is seen in vocabulary, retention of information, and comprehension.

PSYCHOSOCIAL DEVELOPMENT

There are numerous examples of people who have proven that age alone does not decrease intelligence, creativity, learning, and the ability to produce—people like Winston Churchill, Grandma Moses, Artur Rubinstein, Duke Ellington, Leo Tolstoy, and George Burns. On the other hand, it must also be admitted that not everyone makes a satisfactory psychosocial adjustment to old age.

The last of Erikson's famous "eight stages of man" is *ego-integrity versus despair* (Erikson, 1950). Erikson states, "each individual, to become a mature adult, must to a sufficient degree develop all the ego qualities mentioned, so that a wise Indian, a true gentleman and a mature peasant share and recognize in one another the final stage of integrity. . . . Ego-integrity implies an emotional integration which permits participation by fellowship as well as acceptance of the responsibility of leadership" (Erikson, 1950 p. 269). If one has successfully reached this integrity one does not fear death and deals fruitfully with one's days of living. Despair, on the other hand, is a feeling that time does not permit the start of a new life, and a bitter waiting for the end takes over.

Neugarten (1972) points out that no theory of personality adequately explains the impact of aging on personality. One of her conclusions is that the inner life of the individual has a greater impact on the quality of the aging process than does the relationship to the outer world.

Language and Speech

One's vocabulary continues to increase as one ages; the activities in which one is involved obviously determine the extent to which this occurs. Toward the end of early old age there may be some changes in the speech. The voice tends to take on a higher pitch, and the force may start to decrease. This can have a negative psychological effect; men tend to become quite disturbed if they notice a rise in the pitch of the voice, and those engaged in public speaking may notice to their despair that the voice is less forceful than it once was.

Normally there is little or no change in comprehension. It is often said that intelligence is developed during childhood, peaks in early adulthood, and then decreases, but there is much evidence that this is a fallacy. Woodruff and Birren (1975) found that 65- to 78-year-old test subjects scored as high as an adolescent group in both numerical and verbal skills. Interestingly, the 65- to 70-year-olds showed a significant drop in spaital visualization and abstract reasoning but 71- to 78-year-olds showed

a substantial increase in both areas. Of course, individual differences exist among all age groups.

Adjusting to New Roles

As was suggested earlier, how well one tackles the developmental tasks of aging depends to a large extent on how well he or she handled previous developmental tasks. Other factors also influence the success of adjustment; state of health is obviously an important determinant.

To retire *to* something rather than *from* something implies some kind of active involvement. People who have planned something to do make a more favorable adjustment to new tasks and new roles.

Kasschau (1974) found that people who participated in retirement preparation programs had positive results even if they did not see them as helpful at the time. These programs are reaching only a small amount of the work force. There is a need for more such programs, and more research to determine what their content should be.

A loss of social roles occurs with retirement, but the self-concept, the cognitive element of oneself, remains stable throughout later life. Thus, the retired banker, lawyer, brick layer, and carpenter tend to see themselves as such after retirement even though they are not functioning in those roles. However, the self-esteem, one's *feelings* about oneself, is much more vulnerable, and how well that is maintained depends to a large extent on how useful the individual feels (Fig. 12-4). Social interaction is very important for both sexes, as it replaces interaction on the job. In fact, Busse and Pfeiffer (1969) suggest that the retired person pursue a "social career."

People who never married apparently have an easier time adjusting to retirement. Gubrium (1975) found that these people did not experience loneliness in older age; he speculates that they made a satisfactory adjustment to living

Figure 12-4. *One's sense of self-esteem depends on how useful one feels; these board members meet with a representative of a housing complex to make decisions concerning elderly residents. (Courtesy Durham Senior Citizens Council.)*

alone long ago, probably by choice. While living a solitary life certainly means missing a lot of "the good things in life," it means one may escape some of the bad, such as the deaths of one's spouse and close friends. Never-married older women who by choice had careers instead of families tend to have much contact with extended family, but this has not been found to be true of never-married men (Gubrium, 1975).

Divorced older people adjust poorly to aging and have the highest mortality rate, those with the second highest being the widowed. The ideal, obviously, is the older couple. *Facts About Older Americans* (1978) showed that in 1977, 77 percent of men and 39 percent of women over 65 were married.

Economic Status

Adjusting to a new economic status is a necessity for the vast majority of older people. Almost 12 percent of couples live below poverty level; 18.7 percent had incomes of $15,000 or more. (*Facts About Older Americans*, 1978). Among the group considered heads of family, 8 percent of the males and 14 percent of the females live below the poverty level. Comparable figures for sole survivors of families are 25.9 percent of men and 31.5 percent of women. The majority of the 65 to 80 age group remain financially independent, and 18 percent of men and 8.1 percent of women remain in the work force.

Most people in this age group live on fixed incomes, which means that as the cost of living goes up they must change their life-styles in order to survive. This often means moving to a smaller house or apartment, moving closer to stores to decrease transportation costs, and, perhaps selling one's car and using public transportation. For some, this new economic status may severely curtail the opportunity for travel, thereby cutting down on social and family interaction. The largest part of older people's money is spent on housing (32 percent), followed by food, transportation, and medical care (25, 15, and 10 percent, respectively) (*Facts About Older Americans*, 1978).

Leisure

To derive pleasure from leisure time may seem like an easy task; however, for someone with no hobbies or interests outside of work it can become a major obstacle in adjusting to retirement. Using one's leisure time is important; if it is managed successfully, it can provide as much for one's self-esteem as a job role (Havighurst, 1969). Some older people are too hampered by ethnic or religious backgrounds to permit free exploration of leisure roles without guilt. For example the elderly population who come from German and certain Scandinavian backgrounds take a dim view of leisure and play activity; it is seen as useless and nonproductive. The majority of people do adjust, however, to leisure. The most popular leisure activity in the United States is watching TV, followed by visiting friends, reading, gardening, walking, and engaging in handicrafts (Atchley, 1977).

Family Adjustments

The majority of older people are married, and since at least one party will have worked before retirement, an adjustment must be made in the relationship. Whereas for years the couple spent short periods of time together in the evenings, weekends, and maybe vacations, often involved with children, now they are faced with 24-hour togetherness. Among today's 65 to 80 year olds, a large number of the women were home much of the time. These women often see themselves as "the boss of the house," and the husband is likely to feel unwanted and useless. This is a common cause for depression among men; they may be resentful but they blame themselves for not having anything to offer. Good communication and, again, preplanning can do much to produce a renewed and pleasurable relationship with one's spouse.

A much more difficult developmental task is coping with widowhood. Women face a greater adjustment in terms of identity in addition to the grief, since they have been looked upon as "the wife of ———" for a long time. This is particularly true of foreign-born women. A change may be forthcoming as a result of the women's movement.

Loneliness and poverty are common problems for widows. The woman who adjusts well usually gets involved with some type of work or leisure activity. However, if most of her friends are married, she will, after experiencing their initial concern, find herself left out of their circle. This may be because of a denial on the part of the friends of the possibility of widowhood for themselves, or because the widow might be seen as a threat by her friends, particularly if she is attractive and personable.

According to Lopata (1973), middle-class women do better as widows because their life has included a larger variety of friends and activities to which they can turn. The working-class widow has fewer involvements and personal resources, and finds herself more isolated.

Only 14 percent of males over 65 are widowers; most attain that status after age 75. Much less research has been carried out thus far on widowers, although it is known that they are more likely to remarry. They tend not to have a confidante other than their wives. (Widows often find that need met in friends of either sex.) Widowhood is not easy for either sex but the widower is generally far better off financially.

Forty-five thousand Americans aged 65 and over got married in 1970; for five percent of these people it was the first marriage (Atchley, 1977). McKain (1969) has done extensive studies in "retirement marriages." Important factors for a successful adjustment to this are a long period of acquaintance before the wedding, approval of the partners' children (and, to a lesser degree, that of friends), and an ability to adjust to life changes in the past.

The relationship with children should be an adult-to-adult one. This is sometimes difficult as parents continue to think of their children as a "child." The need for the parents to give advice and the reluctance of the child to accept it is an obstacle that must be overcome by both. However, the vast majority of families are able to work out a satisfactory adult relationship. This is especially true when the parents can maintain their own autonomy and independence.

Deaths of Significant Others

Coping with the deaths of one's parents is another task for 65- to 80-year-olds, especially for the younger sector of this age group. When this happens it brings into focus very clearly one's own mortality. Also, the loss of friends occurs with increasing frequency, and it is indeed true that those in this age group tend to read the obituary column very frequently. Even the healthy individual will experience yet another loss, that of physical strength and ability. In order to successfully deal with many of these developmental tasks it is vital for the individual to gain insight into his or her own feelings about death. Kalish (1976) found that death is less feared and denied in older people because they are constantly reminded of it by the loss of loved ones and declining physical stamina; he also found, not surprisingly, that strongly religious people show a minimal fear of death. However, Kalish found that confirmed atheists also had little fear of death. These who were not sure of their religious feelings showed the greatest fear of death.

People who believe that "death is the key to the door of life" feel that death gives meaning to life (Kübler-Ross, 1975). Abraham Maslow aptly said "If you're reconciled to death or even if you are pretty well assured that you will have a good death, a dignified one, then every single moment of every single day is transformed because the pervasive undercurrent—the fear of death—is removed" (Maslow, 1969).

Sexuality

It is important to remember that sexuality includes much more than just sexual intercourse; it involves love, affection, touching, kissing, and cuddling. In fact, couples who received much satisfaction from sexual feeelings and activities adjust to the changes that affect sexuality very satisfactorily. The couples who expressed their sexuality through sexual intercourse alone have a harder time dealing with these normal changes.

None of the physiological change that occur in early old age necessitates much change in sexual expression. Although Masters and Johnson (1966) and Pfeiffer (1974) found a gradual decline in sexual behavior, Kaluger and Kaluger (1979) state that research has demonstrated that the decrease in interest is not physically based but socially, culturally, and psychologically imposed.

Sex was essentially a taboo subject in America until the Kinsey report became widely read in the late 1940s. The research of Masters and Johnson in the 1960's further influenced the lifting *of this taboo*. Most schools in America today have some form of sexual education incorporated into the curriculum; this was the norm in grammar schools throughout Europe for most of the twentieth century. However, the subject of sexuality among the elderly has taken longer to emerge. Many children and young adults tend to think that their parents have passed the age at which sexuality is a significant part of life. Sexual interest among older peole is often the theme of cartoons and jokes. Richman (1977) and Davis (1977), in studies of such humor, found that over two-thirds of the jokes were negative, emphasizing physical decline and decreased attractiveness.

The Duke University Center for the Study of Aging and Human Development conducted a longitudinal study and found that 80 percent of 68-year-old men had an active interest in sex. Seventy-five percent of the same men 10 years later had the same active interest. Twenty-five percent of 78-year-old men were regularly sexually active. Sexual interest and activity in women was found to decline even less. The major reason for decline of activity in women was lack of a socially sanctioned partner (Solnick, 1978).

The elderly will benefit from increased awareness and acceptance of sexuality as a normal need. They will feel freer to express love and affection as well as to seek counseling when it is needed. Perhaps the near future will witness what Ruth Weg (1978) calls "a move away from the goal-oriented passion of youth to the per-son-oriented intimacy of the mature and later years" p. 63.

COGNITIVE DEVELOPMENT

Studies at the Duke University Center for the Study of Aging and Human Development showed that the ability to learn at age 80 is about the same as at age 12; the only difference found was in speed; the older person requires more time to learn (Solnick, 1978).

A 12-year study conducted by the National Institute of Health and the Philadelphia Geriatric Center showed "no significant decline on test performances" and saw "significant increases in ability" for the vocabulary and picture arrangement subtest of the Wechsler Adult Intelligence Scale (Birren et al., 1968). It can be concluded that in normal aging individuals there is no decline in the ability to continue learning. Physical tasks can be performed and learned at about the same rate also.

In the absence of dementia or arteriosclerotic changes, short-term, recent, and remote memory remain intact throughout early old age. There are, of course, great individual differences. As with physical function, the more the memory is exercised, the better it is maintained. The more intelligent the individual, the less susceptible he or she is to memory loss. Older people tend to retain things that they hear better than things they see; the best recall results from a combination of both.

When there is a decline in memory, recent and short-term memory are affected first. It is well known that many people have a hard time recalling a certain name especially in times of stress. While this phenomenon is shrugged off when it happens to someone in his forties or fifties, its occurrence in older people may prompt others to think, "Oh, well, what do you expect at that age." The older person is very acutely aware of this, which further increases his stress and makes it even harder to remember the name or item. Comfort (1976) suggests that some elderly people have better recall of the past because the memories are more pleasant than their present experiences. Cognitive development in the elderly will probably continue to improve in the future, one reason being that continuous efforts are being made to offer social and intellectual stimulation through such policies as free tuition at many universities.

Neither disengagement theory nor the activity theory (both discussed in Chapter 11) adequately explains the psychosocial aspect of aging. A third theory, called the continuity theory, denies that old age is a separate, unique state. It is a "complex interrelationship of biological and physical changes, personality, situational opportunities and actual experiences" (Vander Zyl, 1979). Maas and Kuyper (1974), who conducted a long longitudinal study that supports the continuity theory, found that older people who had similar life-styles or personalities were also alike in young adulthood.

COMMON HEALTH AND DEVELOPMENTAL PROBLEMS OF THE YOUNG OLDER ADULT

Although the normal physiological changes of early old age have minimal negative influence, as has been stressed, when disease strikes, it is harder for the body to cope. Smith and Bierman (1978) found that the most common causes of death in people over 65 are cardiovascular disease (50 percent), hypertension (16 percent), cancer (15 percent), respiratory diseases (7 percent), accidents (2 percent) and others (7 percent). Incidentally, these are the same diseases causing death in the 45 to 64 age group, except that in that group cardiovascular disease causes 39 percent of deaths and cancer, 24 percent. The incidence of suicide among white males in-

creases from age 45 on; it is interesting to note that, in the same age group, the suicide rate among white females and nonwhites decreases.

Skin Disorders

Skin problems are more resistant to treatment and cause more discomfort than in younger age groups. It is common to see severe reactions to insect bites, including those from mosquitoes. This may result in cellulitis, requiring soaks and aggressive treatment with antibiotics. The use of nonaerosol insect repellants should be encouraged. Tinea (fungus) infections are fairly common, affecting the skin between the toes. This can be a special problem for the diabetic person, who is more susceptible to foot infection. Such preparations as haloprogin (Halotex), clotrimazole (Lotrimin), and miconazole nitrate (Mica Tin) are helpful.

Herpes zoster has an acute and chronic phase. Visible and painful lesions of the skin, mucous membranes, or eyes mark the rather short-lived acute phase, followed by the chronic and often severe nerve involvement, which causes intense pain. Those 70 and over are five times more likely to develop this disease than middle-aged people (Hope-Simpson, 1967). This viral disease, which may occur in people who had chicken pox (the causative virus of which is structurally similar to that of herpes zoster) in their childhood, has as its worst complication so-called *postherpetic neuralgia*. This may last on and off for months, and in some people for the rest of their lives. Therefore the pain must be controlled during the acute phase; narcotics may accomplish this. The systemic use of steroids has been reported to decrease both inflammation and pain during the acute phase. The use of chlorprothixene (Taractan) has shown some encouraging results in the relief of pain (Faber and Burks, 1974).

By far the most common skin complaint is itching. As older skin dehydrates and loses its natural oils, it dries, flakes, and itches. Referred to as *senile dermatitis,* ᵗhis condition is seen with increasing frequency in later stages of older adulthood. It is worse during the winter, aggravated by the cold air outside and the dry heat inside. An inexpensive cream such as Eucerin, that replaces the skin's natural oils can be of great help.

Changes in skin pigmentation need close observation as they can vary from simple seborrheic keratotic lesions to actinic, basal cell, or squamous cell carcinomas. These should be evaluated by a dermatologist to determine proper treatment.

Decubitus ulcers (bed sores) are a constant threat to anyone who is immobile. They are caused by ischemia, cell death, and necrosis arising from pressure. The greater the pressure, the faster the ulcer will develop. The normal skin changes of older adulthood further increase susceptibility. By far the best treatment is prevention. A small erythema over a bony surface is the first warning sign. Moderate ulcers are treated with air, heat lamps, antibiotic ointments, and debriding agents, to mention just a few treatments. Systemic treatment with vitamin C and antibiotics are other methods used with varying success. Severe ulcers may require surgical intervention.

Eye Disorders

Acute *glaucoma* is a serious emergency but fortunately it is not very common in this age group. Much more common is open-angle glaucoma, which develops very gradually. Its symptoms may be so mild that the sight is damaged before the disorder is detected. It is highly recommended that everyone have a simple yearly tonometry test to detect changes in intraoccular pressure; to fail to do so would be particularly tragic, since the disease is treatable with mydriatics.

Cataracts are the most common chronic eye problem. Simply defined, this is any opacification of the lens. Duke-Elder (1969) found that 95 percent of people over 65 have some degree of cataract, but he emphasizes that the percentage of those whose cataracts cause visual disability is small. The success rate of surgical intervention is high. It is vital to have a yearly ophthalmological exam; if cataract is detected the client must be followed to determine if and when surgery is indicated.

Degeneration of the retina resulting from sclerosis of the choroidal vessels is a tragic condition for which there is no treatment. It affects the macular portion of the retina and tends to be progressive. Gradually, all central vision is lost; however, peripheral vision remains intact barring other complications.

Ear Disorders

There are few common disorders involving the ear aside from gradual, chronic hearing loss, which has already been discussed. Vestibular disorders are fairly common, however, and can cause unpleasant dizziness. Meclizine hydrochloride (Antivert) and vasodilators may offer some relief.

Nose Disorders

Chronic *rhinorrhea* (discharge of nasal mucus) is common. Postnasal drip is a frequently heard complaint; this results in repeated clearing of the throat. Dryness of the nostrils is also common. These are discomforts more than illnesses, and can be adjusted to fairly easily. A little petroleum jelly can offer relief of dryness of the mucous membranes of the nostrils, and antihistamines are sometimes helpful in decreasing rhinorrhea.

Mouth Disorders

Dental caries present few problems for older people; however, gum disease is common. The gingival margin recedes and gradually exposes the anatomical crown. As this continues, a gradual loss of teeth results. The rate of loss is highly individualized. The process can be greatly reduced with good oral hygiene; if partial or complete dentures are used, regular checkups are needed to assure proper function. Bones and gums shrink, necessitating adjustments of the dentures. *Leukoplakia* (white patches on the mucous membranes of the gums or tongue), commonly found in the upper limit of early old age, must be treated immediately, since this is a precancerous lesion.

Gastrointestinal Disorders

The gastrointestinal system remains remarkably intact, although the most frequently heard complaints of older people concern this sytem. Sklar (1978), in a study of more than 300 people aged 65 and over, found that 56 percent of gastrointestinal complaints were of functional origin, 10 percent originated from neoplasms, 9 percent from peptic ulcers, and the remainder represented a variety of conditions. The functional complaints are often related to preoccupation with elimination, poor eating habits, and a fear of disease or of growing old. The symptoms, whether functional or organic, are similar. All complaints warrant a thorough history-taking and physical examination.

Chronic constipation is a frequent medical problem; the single biggest cause is habitual use of cathartics. Lack of exercise and a lack of fiber and roughage coupled with decreased fluid intake contribute greatly. Treatment of constipation requires the reeducation and cooperation of the client; the earlier the problem is identified, the higher the success rate.

Chronic diarrhea may also result from overuse of laxatives. A word of caution is important here: a complaint of diarrhea necessitates a digital rectal exam to rule out fecal impaction. If impaction is present, the "diarrhea" may actually be liquid stool leaking around the fecal bolus. The impaction needs to be removed, and

a retraining program started. Chronic diarrhea can also result from diverticulosis of the colon, present in about 25 percent of people over 65. This can easily be diagnosed with a barium enema and sigmoidoscopy. Treatment with a diet high in roughage and fiber, and psyllium hydrophilic mucilloid (Metamucil), can effectively relieve the symptoms.

Cancer of the stomach, pancreas, colon, and rectum may occur; it is important that they receive early recognition. It is interesting to note that Sklar's study population showed 14 percent of unsuspected and asymptomatic malignancies in the gastrointestinal system on autopsy (Sklar, 1978).

Peptic and gastric ulcers can generally be managed with cimetidine (Tagamet). Antacids, taken before meals, are often used in conjunction with this medication. Attention needs to be paid to which antacid is used because some are more effective, and some are very high in salt content. Periodic evaluation of blood in stool and complete blood counts are important. Surgical intervention will depend on the client's general condition, type of ulcer, and the client's coagulability.

Gallbladder disease occurs with increasing frequency with age; 30 percent of people over 70 have gallstones (Spiro, 1970). Conservative treatment is indicated unless obstruction is present; stones can sometimes be dissolved with medications (Reichel, 1978).

Hiatus hernia occurs in approximately 67 percent of people 60 and over (Sklar, 1978). It is probably caused by relaxation of ligament and some atrophy of the muscles in the esophageal hiatus. Increased intra-abdominal pressure from obesity further contributes, causing the sliding hernia most frequently observed. The symptoms of heartburn and "sour burps" are caused by reflux of gastric content into the esophagus. This in turn leads to esophagitis with bleeding and severe pain. A barium swallow will enable a diagnosis of the condition. Treatment consists of a bland diet, exclusion of acid drinks, coffee, and alcohol, and using antacids before and after meals. It is helpful if the individual remains upright at least one hour following each meal. Weight reduction is extremely helpful. The avoidance of constrictive clothing, and elevation of the head of the bed are also advised.

Nutritional Problems

Many studies have demonstrated that increased longevity is related to a decreased caloric intake. The number of calories needed decreases by early old age, but a decrease alone does not assure good nutrition; as always, the quality of the diet is vital. A person's nutritional status can be considered good if he would not benefit by a restriction or an increase in the intake of a particular nutrient.

Loss of appetite stemming from decreased activity is common among young older adults. Decreased social stimulation, depression, and anxiety also lead to decreased appetite. In a small percentage of those whose intake is inadequate, problems in chewing are the cause. Mobility and transportation problems leading to an inability to get to stores or an inability to prepare food are additional practical causes; finally, eating snack foods is another cause of nutritional inadequacy.

Painful red fissures and inflammation of the tongue are caused by vitamin B deficiency. Deficiency anemia causes the tongue to become pale and atrophic, showing a reduction of tastebuds. Poor skin turgor may be caused by lack of fluid and fat intake. Breaking of skin around the mouth and increased brittleness of the fingernails have been associated with decreased niacin, riboflavin, and pyridoxine intake. Opaqueness of the bulbar conjunctiva may be associated with lack of vitamin A. Lack of vitamin C will result in slowed wound healing, and vitamin D deficiency will speed the development of osteomalacia.

Circulatory and Cardiac Problems

Ischemic heart diseases are very common in early old age. They primarily affect the coronary blood vessels and are caused by arteriosclerotic changes. One of the ischemic heart diseases, *angina* (paroxysmal chest pains caused by decreased blood supply to the myocardium), is precipitated by physical or emotional exertion and is usually relieved in a few minutes by rest or, almost immediately, by nitroglycerin. Overeating or sudden exposure to heat or cold may also bring on an attack, and a nightmare can bring on a nocturnal episode of angina. The pain may be directly over the heart or may radiate down the left arm or up to the neck and jaw.

When the pain is not relieved by rest or nitroglycerin there is usually a more advanced form of ischemic heart disease present. It may be difficult to distinguish the pain of angina from that of a true myocardial infarction. A "silent MI" may occur in older people, since they do not present the typical picture of clutching the chest, turning blue, and passing out; rather, the pain may be described as "a little pressing," "heartburn like," and the victim will exhibit weakness, shortness of breath, vomiting, diaphoresis, and mental symptoms such as confusion. Kannel, McGee, and Gordon (1976) found that over 25 percent of MIs go unrecognized. This author has seen numerous old MIs detected on routine electrocardiograms before admission to a retirement or nursing home. Many of these residents were not aware that they had ever had a heart attack.

Hypertension, which may lead to strokes and cardiac problems, is common in older people, as well as in younger age groups; it affects over 35 million Americans of all ages (Daniels and Gifford, 1980). Dizziness and headache may be the first symptoms. Clients who were hypertensive earlier in life may become normotensive and even hypotensive as they get older, and the body may therefore respond more strongly to antihypertensive agents. Thus, these clients should regularly have their blood pressures and other diagnostic tests performed.

Orthostatic hypotension (a sudden drop in blood pressure) may happen when changing from a lying or sitting to a standing position. This may also occur after a warm bath or after a large meal. Many drugs, including antihypertensive drugs, may cause this. The symptoms are light-headedness or dizziness, which may result in falls. Medical attention to adjustment of medications is needed. Teaching the client to change position slowly is often all that is necessary.

A great deal of controversy exists regarding when to treat high blood pressure. Many British physicians feel very strongly that an increased blood pressure as one ages is the body's normal compensatory mechanism. They further feel that keeping the blood pressure low with medication leads to confusion. Brocklehurst and Hanley (1978) advocate no treatment for blood pressure of up to 200/100 except in the presence of other indications, such as left ventricular hypertrophy. Generally, American physicians will treat much earlier; for example, Harris (1978) recommends treatment for blood pressures above 170/100 regardless of age.

Congestive heart failure stemming from a variety of causes is seen frequently. Shortness of breath, edema, and enlargement of the heart result, and require intensive medical attention. Proper adjustment of medication and activities is vital.

Respiratory Disorders

Older adults are subject to a variety of pulmonary problems. Decreased pulmonary function makes chronic obstructive pulmonary disease (COPD) very common. COPD is a combination of chronic bronchitis, emphysema, and, sometimes, asthma. Basically, air is trapped in the alveoli and there is always an excess of carbon dioxide, expirations being forced and incomplete. The most pronounced clinical signs are shortness of breath and chronic cough. The degree of shortness of breath indicates the severity of the disease, which may be so mild that dyspnea is noticeable only after running or walking, or so severe that even talking when sitting down produces severe symptoms. It is extremely important to make every effort to eliminate exposure to all forms of pollution, such as smoking. Prevention and early treatment of upper respiratory infections are necessary. A gradually increasing exercise program including teaching of proper breathing technique can be helpful; inhalation therapy and bronchodilators are additional aids in some cases.

Severe COPD often leads to malnutrition. Affected people experience air hunger and frequently breathe in gulps with the mouth open; air is thus swallowed. This is exaggerated during such exertion as eating, with the result that a sensation of being full is experienced after only a few bites. Consequently, these people do better with frequent small meals than with three large ones. A sufficient fluid intake is vital in order to keep secretions moist enough to be expectorated.

The incidence of tuberculosis increases with age. After age 65, it is five times more common in men than it is in women. It often has its origin in an old lesion, dormant for many years, spurred to activity by debilitation of some kind, such as disease or malnutrition. Tuberculosis is more resistant to treatment in the elderly than in younger age groups.

Genitourinary System Disorders

The female is particularly vulnerable to urinary tract infection; changes in hormone production and the proximity of the urinary meatus to the vagina and anus contribute to this. Other factors are the increase in residual urine present in the bladder. Several studies have shown urinary tract infection to be much more prevalent in females up to age 70, after which it appears to be equally common in both sexes. Brocklehurst and Hanley (1978) found that men who have had prostatectomies had more frequent urinary tract infections, the reason being that prostatic secretions contain an antibacterial substance that is lost when the gland is resected.

Benign prostatic hypertrophy occurs very frequently in men. It may cause dysuria, nocturia, dribbling, hesitancy, and, in extreme cases, anuria. The severity of these symptoms determines the treatment. According to Jaffe (1978), 70 percent of all obstructions can be removed by transurethral resection (TUR). This causes minimal risks, since neither general anesthesia nor significant blood loss is involved. Men are not rendered impotent by a TUR, contrary to an old wives' tale. Cancer of the urinary tract occurs more frequently in males; the prostate gland and bladder are the most common sites.

Urinary tract infection is very common in the small percentage of bedridden people among young older adults. Contributing factors are increased residual urine, incontinence, and catheterization. Extreme strictness in cleanliness, aseptic technique, and closed urinary drainage systems will help prevent urinary tract infections. A high fluid intake with acidifying agents and vitamin C plus a diet high in acid ash will further cut down on urinary tract infections. Checking laboratory results for blood urea nitrogen (BUN), creatinine, and bacteria level is important.

Endocrine System Disorders

By far the most common disease involving the endocrine system is *diabetes mellitus*, caused usually by the failure (for unknown reasons) of the islands of Langerhans of the pancreas to secrete enough (or any) insulin. Since insulin is vital in the metabolism of sugar, the result of this lack or decrease is elevated blood sugar levels. The elderly obese person is particularly susceptible to diabetes. Weight control with a special diet is sufficient for many affected people. Some people develop a mild form of diabetes that responds well to moderate diet and oral hypoglycemic drugs. A certain percentage will need daily insulin and a more carefully controlled diet. Symptoms are often mild in the young elderly and may include polyuria, polydipsia, weakness, increased perspiration, and slower wound healing. Whereas many people who developed severe diabetes earlier in life will have died from complications before reaching this age bracket, as improved control of diabetes advances, the number who survive to a "ripe old age" increases.

Among older diabetics, complications such as diabetic retinopathy, vascular insufficiency, diabetic peripheral neuropathy, and gangrene (or amputations necessitated by previous gangrene) are common. The severely diabetic client needs extensive care for any complication. Those who develop the disease late in life need to be taught to understand the disease and how to control it.

There is much controversy as to when to treat diabetes in the elderly. Ronald Cape (1978) and many others found that those in this age group consistently have higher blood sugar levels but are asymptomatic; apparently they produce adequate insulin but at a slower rate (Haunz, 1979). Clinical signs, including glycosuria, are important determinants in diagnosis.

Musculoskeletal Disorders

Muscle cramps are a frequent complaint. They tend to occur primarily in the calf muscles, although the foot, thigh, hip, and hand may also be affected. The exact cause of cramps is unknown; peripheral vascular insufficiency can contribute but is by no means always the cause. Unfortunately, cramps usually occur at bedtime, and relief is obtained by getting up and walking. In more persistent cases, a warm bath or the application of warm towels to the legs before bedtime may offer relief. Vasodilators and antispasmodics may also be used.

Osteoarthritis is common; 15 percent of men and 25 percent of women over 60 have this noninflammatory disorder of moveable joints (Grob, 1978); which is characterized by gradual deterioration of articular cartilage and the formation of new bone at the joint surfaces. The disease affects the weight-bearing joints, knees, hips, lumbar spine, and the terminal interphalangeal joints of the fingers. The main symptom is joint pain, often unilateral at first, which is relieved by rest. The pain is not described as intense but as constantly aching when the joint is moved, often accompanied by crepitation. Clients should be reassured that osteoarthritis rarely produces disability. Rest, good body mechanics, the use of a bed board and local supports such as canes or walkers, decreased excess body weight, heat, massage, and mild exercise can do much to make it easier to live with the condition. Drugs other than analgesics for pain are not considered helpful.

Only 2 percent of people over age 65 develop acute *rheumatoid arthritis*; most of those in that age group who suffer from rheumatoid arthritis have the chronic form. Rheumatoid arthritis is a systemic disease of the connective tissue. Symptoms such as joint inflammation and severe pain are common. In its chronic and progressive form, it often leads to skeletal deformaties.

Rest is extremely important during the acute phase. Physical therapy and drugs play a major part in alleviating the disease in the chronic phase. Aspirin is still considered one of the most effective drugs for those who can tolerate it. Other, newer drugs that have shown promising results include ibuprofen (Motrin), tolmetin sodium (Tolectin), and sulindac (Clinoril).

Osteoporosis is a disease causing rarefaction of bone that occurs with increasing frequency in men over 55 and in women over 45. Barzel (1978) states that it is four times more common in women and whites, especially Northern Europeans. Approximately 11 million people in the United States have osteoporosis, which is among the diseases of which it is unknown whether the disease is an extreme sign of normal aging or a truly pathological process. Osteoporosis is characterized by a decrease in bone quantity; resorption is most pronounced in the trabecular bones, although the cortices of long bones are also affected.

Vertebral bodies are most frequently affected because they consist mostly of trabecular bone. Thus, spinal osteoporosis results in, first, a gradual loss of height stemming from a change in the upper thoracic area, followed by painful fractures of the vertebrae in the lower thoracic and lumbar regions, leading to further decrease in height. These fractures are treated with complete bed rest and analgesics. Wrist fractures should be casted until complete union is seen. The use of metal pins and the encouragement of early ambulation is the treatment for hip fractures caused by osteoporosis. There is still considerable controversy over drug treatment of osteoporosis. Estrogen therapy once thought to prevent or slow down the process in postmenopausal women, has not proven effective, and is now suspected of contributing to the risk of cancer.

Since there is a definite low calcium level in people with osteoporosis it is generally believed that an increased intake of calcium may serve as a prophylactic agent. However, one must also be assured of a high intake of vitamin D in order for calcium to be properly used by the body.

Fractures, aside from those caused by bone diseases, are common in the elderly population, the most common type being Colles's fracture, which often occurs because of using the hands to brace a fall. The second most common is the hip fracture; treatment for this is surgical intervention with reduction, internal fixation, or use of metal end prosthesis. These procedures allow for early ambulation, which is crucial to prevent complications in this age group.

Nervous System Disorders

The most significant nervous disorders involve the vascular system. Cerebral arteriosclerosis is responsible for a variety of nervous system disorders, from mild *transient ischemic attack* (TIA) to a crippling or fatal stroke. A TIA is a focal loss of cerebral function that lasts for a few minutes; symptoms will vary depending on what part of the brain is involved. Complete recovery usually occurs within 24 hours. Almost half of those who have TIAs frequently will have a severe stroke within three years (Brocklehurst and Hanley, 1978). Stenosis of the carotid artery is a common cause of cerebrovascular accident (CVA) (stroke); endarterectomies are done with increasing frequency and success. The conservative treatment involves the use of anticoagulants. CVAs are responsible for 200,000 deaths annually in the United States; 75 percent occur in people over 70 (Saxon and Etten, 1978). Unfortunately, those who do not die have small chance for complete recovery of neurological function. The *Statistical Abstract of the United States* (1978) showed over 3 million stroke victims in the United States. The symptoms and severity of the disease depend on the part and extent of the brain to which the blood supply is cut off. A CVA may result from a cerebral thrombosis (most common in the elderly) or an embolism with a very sudden onset. Initial treatment consists of supportive measures and positioning to prevent contractures; passive range of motion for involved joints, and active range of motion for uninvolved joints must be done at least once a day. A high-protein diet, adequate fluid intake, frequent turning, and deep breathing exercises must be included early in the client's regimen. Active physical therapy is started between ten days to two weeks after the stroke. The degree of recovery obviously depends on the severity of the stroke. However, the degree of the client's motivation and the effectiveness of the health care professional in planning the rehabilitation program are of extreme significance.

Anemias

Iron-deficiency anemia is extremely common in the elderly and has a variety of causes, including slow-bleeding peptic ulcer, gastric ulcer, hiatus hernia, polyps of the small intestines, diverticulosis, and cancer of the colon. Iron-deficiency anemia is present in over half of Americans over 65; women are more commonly affected. Vague complaints of tiredness should always be followed up by a complete blood count and cell indices in order to rule out this very common and treatable anemia. Simple replacement therapy with iron for six to twelve months is the treatment of choice. If the client does not respond to this treatment and shows a low reticular cell count, a further search for infections or neoplastic disease is in order (Howe, 1979).

Pernicious anemia, caused by a lack of intrinsic factor, which in turn leads to malabsorption of vitamin B_{12}, must be treated in order to prevent irreversible brain damage. The onset is insidious, and when it occurs in the elderly the symptoms are often attributed erroneously to senility. The anemia progresses slowly. Pallor, shortness of breath, and chest pain are often the presenting symptoms. Neurological symptoms such as paresthesia may be followed by severe gait disturbances. *Macrocytic anemia* (folic-acid–deficiency anemia) calls for an evaluation of serum vitamin B_{12} level. Treatment consists of one vitamin B_{12} injection every four to five weeks.

Drugs and Drug Interactions

The absorption, transportation, metabolism, and excretion of drugs are slowed down in the elderly. Three times as many drugs are prescribed in the United States for clients 65 and over as for younger people (Cape, 1978). The most commonly prescribed drugs are hypnotics, psychotropics, diuretics, cardiac drugs, antihypertensives, and laxatives. Commonly used over-the-counter drugs include aspirin, acetaminophen (Tylenol), antacids, laxatives, and alcohol. The number of drugs used is frightening; the possibilities for adverse drug interaction are mind-boggling.

One of the difficulties in dealing with drug problems is that the client frequently assumes that over-the-counter drugs must be safe or they would not be sold over the counter, and that a doctor prescribing drugs has taken into account the possibility of interactions with other drugs that the client takes without the doctor's knowledge. It is vital to ask clients what they are taking, and to emphasize that "medication" means *any* medications, even aspirin or a laxative.

Alcohol, in combination with any sedative, will have a potentiating effect and will depress the central nervous system. Many of the over-the-counter antacids are very high in sodium and can lead to drastic results in people with cardiovascular problems. Mineral oil, frequently used as a laxative in the older population, prevents absorption of vitamin D, which is necessary for calcium absorption; this may increase the incidence and severity of osteoporosis. Iron interacts with antacids in such a way that iron absorption is decreased.

ANTICIPATORY GUIDANCE

Skin

Daily baths are not only unnecessary but further contribute to the drying of skin. If baths are preferred, oils should be added to the water, and a simple mosturizing cream can be applied after the bath, but clients must be advised that the addition of oils increases the problem of slipperiness in the bath.

Showers are more beneficial than baths since the skin is not soaked in water. The stream has a massaging, stimulating effect on the skin and improves decreased circulation.

The client should avoid direct sunshine, especially during the hottest part of the day. The sun further dries the skin and predisposes the skin to cancer. A light hat to protect against direct sunlight on the head and face is a wise precaution against sunstroke. Sunscreen creams should be applied to exposed skin while in hot sun. The need for some sunshine must be emphasized in relation to the need for Vitamin D.

Wash-and-wear clothes must be washed before they are first worn. They contain a product similar to formaldehyde to which many people are sensitive. Wearing wool directly on the skin tends to cause or increase pruritis. Soft cotton worn next to the skin when wool is also to be worn will maintain warmth and decrease itching. The ingredients of fabric softners must be checked carefully, many are allergenic.

The heat in most houses during the winter tends to be dry; this further aggravates dry and itchy skin. It is helpful to use a humidifier or to place a pot of water near the heat outlet, particularly in the bedroom, since increased humidity will benefit dry mucous membranes while the client is sleeping.

For people who are partially or completely immobile for any length of time, frequent changes of position are vital. Sheets should not be washed in harsh detergents. The use of sheepskin is most helpful. The use of an air mattress or waterbed, along with frequent gentle massage with oil or cream will help to keep skin intact. All of these considerations apply even more when the client is sitting, when most of the body's weight is borne by the ischial tuberosities. Areas most susceptible to decubitus ulcers are the buttocks, coccyx, hips, knees, ankles, heels, ears, shoulders, and elbows.

Proper care of the nails is important. Decreased vision, stiff joints, and obesity can make cutting one's tonenails (which harden in old age) a dangerous activity. The help of a podiatrist or a family member may be needed. The nails must be cut straight across to prevent ingrown toenails. Soaking the feet for 15 to 20 minutes in warm water will help soften the nails. Instruct the client to avoid long soaks, and explain that thorough drying between toes is important because excess moisture contributes to fungus infections. Proper scissors and good lighting are important. A magnifying glass can also be helpful.

Sensory Changes

In addition to yearly eye examinations, there are many practical considerations in compensating for decrease in vision. Adequate lighting, aside from its obvious necessity, is a vital safety factor (Fig. 12-5). Since the pupil takes longer to adjust to light changes, waiting a few moments before proceeding is a sensible safety factor when entering a dark area. The same is true when first going out in bright sunshine from a darker area. Some elderly people who enjoy the atmosphere of a candlelit restaurant carry a small flashlight in order to read the menu with greater ease.

Glaring light can be difficult and even painful. Sunglasses will do much to relieve this. Toward the end of this age span some people may find night driving and the glare of headlights especially difficult. Planning activities so that the driving is done in the day is obviously the best solution.

Labels on medication bottles usually have very small print. If difficulty in reading labels is not sufficiently helped by the use of a magnifying glass, rewriting directions in larger print and using color-coding to label medication bottles may be effective. If the latter solution is used, bear in mind the yellow-filtering effect and favor reds and yellows in devising a color code. Those with exceptionally poor vision can see white letters on dark backgrounds better than black-on-white. Because of slowed accommodation, great care should be used when walking down stairs or when one is likely to encounter an unexpected obstacle, such as a pet.

National and state library services for the blind have an enormous variety of "talking" books and magazines available. This service is free and offers hours of entertainment and stimulation for the blind and those whose vision is poor. The person who has decreased vision generally receives more attention and empathy than the hard-of-hearing person. This is often felt by the individual, and he may tend to withdraw into himself.

What may be misconstrued as paranoic thoughts are common in people with decreased hearing, since they are understandably likely to misinterpret voices and other sounds. The hard of hearing may seem inattentive, may show strained facial expressions, or may answer inappropriately, and may be misclassified as senile. Hearing loss is usually gradual. Teaching lip reading or sign language should be considered. Because the ability to hear consonants is lost earlier, substituting words with more vowels is helpful. Extraneous noises such as background music decrease the ability to hear.

In communicating with someone who is hard of hearing, place yourself directly in front of the person, speak slowly, enunciate well, and move your lips slowly. If one of the client's ears is better, try to direct the voice to that ear, but not at the expense of eye contact. Do not shout; in extreme difficulties, write. Make every effort to be patient. It is taxing to have long conversations with someone hard of hearing, but remember that it is taxing for the deaf person too. In group situations take time to explain in a word or two what the conversation topic is. This will do much to allay the deaf persons fear that he or she is the topic. A touch, a squeeze, a smile can convey more than a thousand words. It is important, however, to remember to let the deaf person see you before you touch him in order to avoid frightening him.

Many who are even slightly hard of hearing avoid radio and television, since they feel they

Figure 12-5. *This man has maintained a high degree of manual dexterity; that and an adequate light supply permit him to weave a complex pattern. (Courtesy Durham Senior Citizens Council.)*

disturb others in the household with the volume necessary for them to hear. Earphones may be obtained for these instruments.

Some hearing deficits do not improve with the use of hearing aids. On the other hand, it is a fallacy that people with "nerve deafness" cannot use a hearing aid (Harford, 1979). It is absolutely vital for anyone who is hard of hearing to be seen by an ear, nose, and throat specialist, and to have a complete audiometry test in order to determine if the person will benefit from a hearing aid, and if so, what type of hearing aid would work best. The kind and degree of hearing loss will determine the most suitable type to be used.

The nurse will encounter people who are hard of hearing who could benefit from a hearing aid, but who refuse to be tested because of vanity. This appears to be as common a phenomenon in men as in women. Patience and understanding, and gradually offering examples of the added enjoyment the individual can experience with full hearing is often a successful approach.

Thickened cerumen (ear wax) and, in men, hair growth in the ears can occlude the auditory canal and contribute to decrease in hearing. The client should be encouraged never to use cotton swabs; all they usually accomplish is to push the wax closer to the eardrum, making the removal more difficult. Ear checks, once or twice a year, are wise. If cerumen is present, the use of carbamide peroxide (Debrox) to soften the wax is advisable. Sometimes that alone will remove the wax, but it may have to be followed with a gentle irrigation with a syringe, warm water, hexachlorophine solution (pHisoHex), or half-strength hydrogen peroxide. The health care professional can do this easily in a clinic or at home. If the wax cannot easily be removed this way, referral to a physician is needed.

There is controversy regarding how much of a decrease in smell and taste the young older adult experiences. Both of these senses need assessment when planning for nutritional needs. Attractive looking food as well as pleasant odors stimulate the salivary glands and the appetite. Spices and flavor additives can be used to advantage even in restricted diets. Alcohol in small doses also serves as an appetite stimulant. Since there is a definite decrease in salivary gland function, moist food is essential. Peanut butter may be difficult to swallow; however, if the older person enjoys it, try adding jello and serving with milk or tea with lemon. Lemon or lemon juice is an excellent additive to many foods as it has a stimulating effect on salivary production. After radiation therapy of the mouth, neck, or throat, when the mouth will be extremely dry, drinking a glass of water with freshly squeezed lemon before eating may prove very beneficial. Sucking lemon drops or other hard candy will also aid the dry mouth, as will chewing sugarless gum.

Elderly people living along are particularly vulnerable to poor nutrition. The Meals-on-Wheels program guarantees at least one nutritious meal daily. Many churches, schools, or local organizations serve free or inexpensive meals and some even offer transportation. More and more elderly day-care centers are available. They usually offer meals and recreation. Some people who cook and eat alone at home have found that planning their meal around a favorite TV program has increased their enjoyment greatly.

Adequate nutrition for the elderly differs little from that for younger people; daily requirements include four or more servings of bread and cereals, one or more serving of citrus fruit, one or more servings of dark yellow or green vegetables, two or more serving of meat and meat products (including fish, eggs, and nuts), and two or more servings of dairy products. The sensation of thirst is diminished as one ages, but the body still needs about two quarts of water a day in order to maintain all body functions and homeostasis.

As has been mentioned, the older person's total caloric requirement is reduced, and a decreased fat intake is suggested. Protein, preferably high-quality protein from animal sources, should continue to make up about 25 percent of the total caloric intake. This is vital for the growth and maintenance of all tissues of the body. Vitamin and mineral requirement remain the same. Added calcium is important for bone preservation. Foods high in fiber and roughage are a key to the prevention of constipation.

Gastrointestinal Changes

Even though very little "wearing out" takes place in the gastrointestinal tract, older people frequently have gastrointestinal complaints. The decrease in certain enzymes may indeed slow the digestion of certain foods. The individual is probably the best judge of which foods may cause discomfort, excess gas, etc., and should avoid those foods.

Constipated clients should be warned against the overuse of laxatives. Reeducate the client regarding how the bowel functions; use pictures if necessary. An increased fluid intake is necessary to maintain softness of fecal material, while roughage, obtained from fruits, vegetables, and bran cereals will provide the necessary bulk. Exercises to strengthen the abdominal muscles are helpful. The consistency of the stool will indicate more about the precise nature of the individual problem. If the consistency is fairly soft, the problem may be lack of muscle tone. A laxative may be necessary until tone is rebuilt. If, on the other hand, the stool is dry and hard, a stool softener will be more efficient. If a laxative is used, it may be possible to gradually decrease it to the point of cessation. However, the longer someone has been a habitual user of a laxative, the more difficult it is to relearn. Clients who were accustomed to using both kinds of medications may be able to gradually reach the point at which only a stool soft-

ener is needed. Of the two drugs, the stool softener is far less damaging.

Exercise

The role of exercise is indeed important (Fig. 12-6). Exercise is best when it is part of one's life-style, but, Comfort states, "the problem is it needs to be lifelong" (1978). However, there is evidence that an exercise program can be instituted at a late age with much success. Sensible exercise has a positive effect on all systems, especially the cardiovascular, respiratory, and musculoskeletal system.

One of the best exercises and one that can continue the longest is walking. Brisk walking produces rhythmic activity of large muscle masses, improves breathing, and decreases blood pressure. Swimming is another excellent form of exercise and can be done by people who have difficulty walking. Jogging, bicycling, calisthenics, and dancing are other appropriate forms of exercise. The most important considerations are that the exercise is tailored to the individual, that it is done with regularity 4 to 5 times per week, and that it is an enjoyable activity.

It is vital for the individual who is not used to exercise to start slowly. As beneficial as proper exercise is, improper exercise can be as damaging. The client must start by having a medical examination.

Figure 12-6. *Older adults can continue to excel in athletics, especially if they have maintained an exercise regimen throughout their lives. This 71-year-old man is a local tennis champion. (Courtesy Durham Herald.)*

Sleep

Those who exercise regularly, in addition to enjoying the above-mentioned benefits, also sleep better. Complaints about sleep are frequent among the elderly; age brings some changes in sleep patterns (Kahn and Fisher, 1968; Butler and Lewis, 1973). The elderly tend to take longer to fall asleep, they do not sleep as deeply, and have more periods of waking up. They also tend to take naps during the day. Older people tend to develop individualized patterns of sleep to fit their needs just as younger people do. The use of sleeping pills, which may interfere with these patterns, is of questionable value. Sleeping pills that contain barbiturates should definitely be avoided since they tend to accumulate and cause intellectual impairment, slurring of speech, and unsteady gait. If a sleeping medication is temporarily necessary, one that is quickly eliminated by the body should be used.

Several alternatives to using sleeping pills can be suggested. Physical activity, taking fewer daytime naps, a warm bath, relaxing music, reading, and watching TV may all make falling asleep easier. Going to bed shortly after a big meal is not wise; however, having a small bedtime snack with warm milk is relaxing to many. In most cases a nightcap, glass of beer or wine, friendly conversation, or, occasionally, a mild antihistamine such as diphenhydramine hydrochloride (Benadryl) can be substituted for most sleeping pills.

A word of caution regarding sleep is in order: a drastic change in sleep pattern can be a signal of problems. If a client starts waking up much earlier than usual, or feels tired but is unable to return to sleep, be alert for other signs of depression.

Cardiovascular and Respiratory Changes

The elderly person's response to normal cardiovascular and respiratory changes should be motivated by common sense, i.e., if he senses that he needs a bit more time to carry out routine activities, or that he needs to rest more often, he should follow those instincts. Decreased circulation may cause cold feet and hands. The client should avoid constricting clothing on arms and legs, and should add layers of clothes before getting to the point of feeling cold. Nicotine further constricts blood vessels and should be discouraged if poor circulation is present. If total abstinence is impossible, the client should at least be encouraged to cut down, and to smoke only mildest filtered brands. Pipe and cigars are generally less harmful since they are rarely inhaled. Because of decreased respiratory function, hypoxia is felt at a lower altitude. If the client is at a higher altitude than he is accustomed to, walking, jogging, or bicycling may cause shortness of breath sooner.

Changes in Elimination

Decreased bladder size and sensitivity will necessitate more frequent urination. Bathrooms should be made readily available to the client. To know that these changes are normal will decrease the anxiety and make it easier for the individual to find his or her routine and learn some practical adjustments. It is wise to attempt emptying the bladder just before going out and just before going to bed even if the urge is not present. A decreased fluid intake before taking a trip or going to sleep is advisable. However, it must be understood that an adequate fluid intake is vital, and if less fluid is taken in at one time of day it must be increased at another time (to total at least two quarts a day). The elderly person should know which stores have public bathrooms. When outside the home, the client should plan to stop every two to three hours.

The male needs to understand the possible changes in the prostate gland. He needs to know that symptoms such as frequency, decreased urinary stream, difficulty in starting the stream, dribbling, and nocturia should be promptly brought to the attention of a urologist. A simple transurethral resection (TUR) can relieve these symptoms in the vast majority of people.

Reproductive and Endocrine System Changes

The health care professional must not only be knowledgeable about sex but be an understanding counselor. Therefore, if the professional is uncomfortable dealing with the subject of sex, at least know someone to whom the client may be referred.

Many couples have ceased their sex life because of dyspareunia (painful intercourse). With proper counseling and open communication between partners this is an easy problem to solve. Decreased lubrication of the vaginal walls must be overcome; prolonging foreplay is sometimes all that is necessary. A large number of couples have been helped by using lubricant creams. The male takes longer to obtain an erection, and he too can therefore benefit from prolonged foreplay.

Many misconceptions regarding sex in relation to heart strain exist. A client who has recovered from a myocardial infarction can safely have intercourse by the time he is permitted to walk up a flight of stairs. Masturbation can be engaged in earlier than that. Aphrodisiacs should be discouraged. The so-called "Spanish fly" and amyl nitrate are both dangerous. The first has a corrosive action that can lead to tissue destruction, and the second is a potent vasodilator that can cause coronary occlusion. Alcohol is not an aphrodisiac. A small amount may relax inhibitions, but too large an amount can preclude one from having intercourse.

Nervous System Changes

Normal neurological changes have little practical significance until later in old age. A decrease in reaction time may necessitate avoiding the operation of machinery or a car, although many people can continue to drive in early old age. People in this age group should be encouraged to remain socially and intellectually active.

Practical Implications of Psychosocial and Socioeconomic Changes

The health care professional needs to be aware of all available community resources and books that can be helpful in planning retirement. Or-

ganizations such as the American Association of Retired Persons (AARP) have chapters in most large and medium-sized communities; they offer lectures on such pertinent topics as financial management, insurance, and income tax preparation. AARP's bimonthly magazine, *Modern Maturity*, offers interesting and valuable information on a variety of topics. Providing tours to various parts of this and other countries at reduced rates for members is one of AARP's many services. Other organizations the young elderly should know about are the Gray Panthers and the Retired Professional Action Group (RPAG), founded by Public Citizens Incorporated, which was in turn founded by Ralph Nader. The National Caucus on the Black Aged (NCBA) and the International Senior Citizens Association (ISCA), founded originally in Denmark, with headquarters now in Los Angeles, offer much helpful information and activities.

The Foster Grandparents program gives low-income older adults an opportunity to work with neglected, physically or mentally retarded children. The Retired Senior Volunteers Program (RSVP) grew as a result of this project. Participants are paid if they are below a certain income level, and transportation is usually provided. Vista, the Peace Corps, Project Green Thumb, and the Service Corps of Retired Executives (SCORE), which offers its members' services as counselors and advisors to companies, are examples of groups that benefit from the services of retirees. Through the local State Council on Aging, the health care professional can learn what programs are available in her locality.

The Comprehensive Service Amendments to the Older Americans Act of 1973 required each state to establish a network of Area Agencies on Aging in order to receive federal funds. These agencies have grown, and local power and control have increased. They are effective advocates of older people in many areas.

Most retired people live on a fixed income and experience a decline in health at some time during the aging process. Whenever the cost of living in the United States escalates, health care, especially for the lower-income population, tends to suffer.

HINTS ON WORKING WITH THE YOUNG OLDER ADULT

Respect for the older person is crucial. The health care professional working with those in this age group should address them as "Mr. ———," "Miss ———," etc., in accordance with each person's preference. Nicknames such as "Gramps," "Honey," or "Grandma" must never be used except when requested.

Guard against stereotyping the older person. For example, not all older people love children; find out how the older person feels about children before suggesting participation in the Foster Grandparent program. "Old people are more religious" is another myth. No statement that begins "All older people are . . . " is true.

of adult children, and change in economic status. In Erikson's terms, the chief developmental task of the period is the achievement of ego-integrity versus despair. With regard to the withdrawal from the world of work, those most likely to succeed in adjusting are those who have made some plans for retirement, that is, those who are retiring to something rather than from something. Not surprisingly, those whose lives revolved around their jobs have the greatest difficulty in making the adjustment.

Common health problems of the period include gastrointestinal disorders (mostly of functional origin), heart problems, hypertension, chronic obstructive pulmonary disease, urinary tract infections in women, prostatic hypertrophy in men, diabetes mellitus, stroke (commonly arising from stenosis of the carotid artery), and anemia.

As always, a preventive approach to health care is essential. Provided that the young older adult follows a regime of good nutrition and adequate rest and exercise, and in the absence of disease, these years can be among the most vigorous in the life span.

SUMMARY

The major changes of early older adulthood (65 to 80 years) are psychosocial; physiological changes are more subtle until after age 80. The most striking physiological change of the period is the appearance of wrinkles and skin folds; other common changes are thinning of head and pubic hair, and the growth of hair on the upper lip and chin in women, and in the nostrils and ears in men. A majority of those in this age group will experience some impairment of vision and hearing as a result of aging, but only a small percentage are significantly hampered by these developments.

The psychosocial changes that commonly affect the young older adult include adjustment to retirement, widowhood, the independence

REFERENCES

Aaron, J. E., and Gallagher, J. C. Seasonal variation of histological osteomalacia in femoral neck fracture. *Lancet*, 1974, *2*, 84–85.

Ahammer, I. M. Social-learning theory as a framework for the study of adult personality development. In Baltes, P. B., and Warner-Schaie, K. (Eds.), *Life span developmental psychology, personality and socialization*. New York: Academic Press, 1973.

Aging. Denver radio program expands to other states. July/August, 1978, 8–10.

Atchley, R. C. *The social forces in later life*. Belmont, Calif.: Wadsworth, 1977.

Atchley, R. C. Dimensions of widowhood in later life. *Gerontologist*, 1975, *15*, 176–178.

Barzel, U. S. Common metabolic disorders of the skeleton in aging. In Reichel W. (Ed.). *Clinical aspects of aging*. Baltimore: Williams & Wilkins, 1978.

Bierman, E. L. Obesity, carbohydrates and lipid interaction

in the elderly. In Winnick, M. (Eds.), *Nutrition and aging*. New York: John Wiley & Sons, 1976.

Birren, J. E., Butler, R. N., Greenhouse, S. W., Sokoloff L., and Yarrow, M. R. *Human Aging*. Washington, D.C.: Government Printing Office, Public Health Publications No. 986, 1968.

Brocklehurst, J. C., and Hanley T., *Geriatric medicines for students*. Edinburgh: Churchill Livingstone, 1978.

Brocklehurst, J. C. (Ed.). *Textbook of geriatric medicine and gerontology* (2nd ed.). Edinburgh: Churchill Livingstone, 1978.

Bronley, D. B. The psychology of aging. In *The effects of aging on adult intelligence* (2nd Ed.). London: Penquin Books, 1974.

Busse, E., and Pfeiffer E. (Eds.), *Behavior and adaptation in late life*. Boston: Little Brown, 1969.

Butler, R., and Lewis M. I. *Aging and mental health*. St. Louis: C. V. Mosby, 1973.

Butler, R. C. Why Survive? In *Being old in America*. New York: Harper & Row, 1975.

Cape, R. *Aging: its complex management*. Hagerstown, Md.: Harper & Row, 1978.

Clark, M., and Anderson, B. *Culture and aging*. Springfield, Ill.: Charles C Thomas, 1967.

Comfort, A. *A good age*. London: Mitchell, Beazely, 1976.

Cooper, B. S., and Piro, Paula A. Age differences in medical care spending, fiscal year 1973. *Social Security bulletin*, 1974, *37*, 3–14.

Cowgill, D. O., and Homes, L. *Aging and modernization*. New York: Appleton-Century Crofts, 1972.

Cumming, E. and Henry, W. E. *Growing old: The process of disengagement*. New York: Basic Books, 1961.

Daniels, Linda, and Gifford, Ray W. Therapy for older adults who are hypertensive. *Geriatric Nursing*. May/June 1980, 37–39.

Davis, L. J. Attitudes toward old age and aging as shown by humor. *Gerontologist*, 1977, *17*(3), 220–226.

Devine, B. China's house of respect. *Journal of Gerontological Nursing*, 1980, *16*(6), 338–340.

deVries, Herbert A. Physiology of physical conditioning for the elderly. In Harris, R., and Frankel, L. J. (Eds.), *Guide to fitness after fifty*. New York: Plenum Press, 1977.

Duke-Elder, E. L. *A system of opthamology*, Vol. XI. St. Louis: C.V. Mosby, 1969.

Erikson, E. H. *Childhood and society* (2nd Ed.). New York: Norton, 1950.

Faber, G. A., and Burks, J. W. Chlorprothizene therapy for herpes zoster neuralgia. *Southern Medicine Journal*, 1974, *67*, 808.

Facts about older Americans, 1978. Washington, D.C.: United States Department of Health, Education & Welfare, Office of Human Development Services Administration on Aging, 1979.

Federal council on aging White House conference on aging chartbrook. Washington, D.C.: U.S. Government Printing Office, 1961.

Finch, C. E., and Hayflick, L. (Eds.). *Handbook of the biology of aging*. New York: Van Nostrand Reinhold, 1977.

Freeman, J. T. *Aging: Its history and literature*. New York: Human Sciences Press, 1979.

Geriatrics, 1980, *35*(1), 16.

Glick, I. O., Weiss, R. S., and Parkes, C. M. *The first year of bereavement*. New York: John Wiley & Sons, 1974.

Goldstein, Sidney. *Consumption patterns of the aged*. Philadelphia: University of Pennsylvania Press, 1960.

Gregerman, R. I., and Bierman, E. L. Aging and hormones. In Williams, R. H. (Ed.), *Textbook of endocrinology* (5th Ed.). Philadelphia: W.B. Saunders, 1974.

Grob, David. Prevalent joint diseases. In Reichel, William (Ed.), *Clinical aspects of aging*. Baltimore: Williams & Wilkins, 1978.

Gubrium, J. F. Being single in old age. *International Journal of Aging and Human Development*, 1975, *6*, 29–41.

Harford, E. R. Guidelines for hearing problems: Substituting management for myth. *Geriatrics*, December, 1979, 69–75.

Harris, R. Special problems of geriatric patients with heart disease. In Reichel, W. (Ed.), *Clinical aspects of aging*. Baltimore, Md.: Williams & Wilkins, 1978.

Haunz, E. A. Pitfalls in early diagnoses of diabetes. *Geriatrics*, October, 1979, 59–70.

Havighurst, R. J., Munnich, J. M. A., Neugarten, B. L., and Thomas, H. *Adjustment to retirement*. Assen, Netherlands: Van Gorcum, 1969.

Hayflick, L. Cell biology of aging. *Bioscience*, 1975, *25*, 629.

Hermel, J. S. S., and Samueloff, S. Taste sensation identification and aging in man. *Journal of Oral Medicine*, 1970, *25*, 39–42.

Homes, T., and Masuda, M. Life changes and illness, susceptibility. In Dohrenwent B. S., and Dohrenwent B. P. (Eds.), *Stressful life events: Their nature and effects*. New York: John Wiley & Sons, 1974.

Hope-Simpson, R. E.. Herpes zoster in the elderly. *Geriatrics*, 1967, *22*, 151.

Howe, R. B. Tips on diagnosing and treating anemia in the aging. *Geriatrics*, December, 1979, 29–36.

Human communication and its disorders. Bethesda, Md.: U.S. Government Printing Office, U.S. Dept. of Health, Education & Welfare, 1969.

Jaffe, J. W. Common lower urinary tract problems in older persons. In Reichel, W. (Ed.), *Clinical aspects of aging*. Baltimore, Md.: Williams & Wilkins, 1978.

Johnson, R. E., Pope, C. and Campbell, W. Reported use of nonprescription drugs in health maintenance. *American Journal of Pharmaceutical Medicine*, 1976, *33*, 1249–1254.

Jordheim, A. E. Old age in Norway—a time to look forward to. *Geriatric Nursing,* May/June, 1980, 46–47.

Kahn, E., and Fisher, C. The sleep characteristics of the normal aged male. *Journal of Nervous and Mental Disorders,* 1968, 477–494.

Kalish, R. A. Death and dying in a social context. In Binstock R., and Shanas E. (Eds.), *Handbook of aging and social sciences.* New York: Van Nostrand-Reinhold, 1976.

Kaluger, G., and Kaluger, M. F. *Human development: The span of life.* St. Louis: C.V. Mosby, 1979.

Kannel, W. B., McGee, D., and Gordon, T. A general cardiovascular risk profile: The Framingham study. *American Journal of Cardiology,* 1976, *38,* 46.

Kasschau, P. Re-evaluating the need for retirement preparation programs. *Industrial Gerontology,* 1974, *1,* 42–59.

Kennedy, C. E. *Human development: The adult years and aging.* New York: Macmillan, 1978.

Khairi, M. R. A., and Johnston, C. C. What we know and don't know about bone loss in the elderly. *Geriatrics,* 1978, 67–76.

Kokmen, E., Bossemeyer, R. W., Jr., Barney, J., and Williams, W. J. Neurological manifestations of aging. *Journal of Gerontology,* 1977, *32*(4), 411–419.

Kübler-Ross, E. *Death: The Final Stage of Growth.* Englewood Cliffs, New Jersey: Prentice-Hall, 1975.

Lidz, T. *The person: His development throughout the life cycle.* New York: Basic Books, 1978.

Lopala, H. Z. *Widowed in an American city.* Cambridge, Mass.: Schenkman, 1973.

Llinas, R. R. L. The cortex of the cerebellum. *Scientific American,* 1975, *232,* 56–71.

Maas, H., and Kuyper, J. *From thirty to seventy.* San Francisco: Jossey-Bass, 1974.

Maslow, A. Editorial. *Psychology Today,* 1969, *24,* 26–34.

Masters, W. H., and Johnson, V. E. *Human sexual response.* Boston: Little, Brown, 1966.

Match, S. K. Establishing telephone reassurance services. *The national council on aging,* 1972.

McKain, W. C., Jr. *1969 retirement marriage.* Storrs, Conn.: University of Connecticut, Agriculture Experiment Station, 1969.

Meador, R. Age in Denmark—a time to enjoy. *Geriatric Nursing,* May/June, 1980, 48–49.

Melmon, K. L. 1971 preventable drug reactions—causes and cures. *New England Journal of Medicine, 284,* 1361–1367, 1971.

Murphy, S. The black elderly today. *Modern Maturity,* June/July 1980, 48–51.

National nursing home survey report for national center for health statistics, 1977. *The National Health,* November, 1979.

Neugarten, B. L. Personality and the Aging Process. *Gerontologist,* December, 1972, 9–15.

Ordy, J. M. Nervous system, behavior, aging. In Ordy, J. M., and Brizzee, R. R. (Eds.), *Neurology of aging.* New York: Plenum Press, 1975.

Osaka, M. M. Aging and family among Japanese Americans: The role of ethnic tradition in the adjustment of old age. *The Gerontologist,* 1979, *19*(5), 448–454.

Palmore E. *Normal aging,* Vol. 1 and Vol. 2. Durham, N.C.: Duke University Press, 1970.

Palmore E. The joy of aging. *Duke alumini register,* May/June 1979.

Pfieffer, E. *Successful Aging.* Durham, N.C.: Duke University Center for the Study of Aging and Human Development, 1974.

Reichel, W. (Ed.). *Clinical aspects of aging.* Baltimore, Md.: William & Wilkins, 1978.

Richman, J. R. The foolishness and wisdom of age; Attitudes toward the elderly as reflected by jokes. *Gerontologist,* 1977, *17*(3), 210–219.

Rovee, C. K., Cohen, R. Y., and Shlpack, W. Life span stability in olfactory sensitivity. *Developmental psychology,* 1975, *11,* 311–318.

Saxon, S. V., and Etten, M. J. *Physical change and aging.* New York: The Tiresias Press, 1978.

Schiffman, S. Food recognition by the elderly. *Journal of Gerontology,* 1977, *32*(5), 586–592.

Schwartz, D. Self-medication for elderly out-patients. *American Journal of Nursing,* 1975, *18,* 1810.

Shomaker, D. M. Use and abuse of OTC medications by the elderly. *Journal of Gerontological Nursing,* January 1980, *1,* 21.

Sklar, M. Gastrointestinal Disease in the Aged. In Reichel, W. (Ed.), *Clinical aspects of aging.* Baltimore, Md.: Williams & Wilkins, 1978.

Smith, D. W., and Bierman, E. L. (Eds.), *The biological ages of man.* Philadelphia: W. B. Saunders, 1978.

Solnick, R. L. *Sexuality and aging.* Los Angeles: The University of Southern California Press, Ethel Percy Andrus Gerontology Center, 1978.

Spiro, H. M. *Clinical gastroenterology.* New York: Macmillan, 1970.

Statistical abstract of the United States, 1978 (99th ed.). Washington, D.C.: U.S. Dept. of Commerce, Bureau of the Census, 1978.

Sviland, M. A. P. A program of sexual liberation and growth in the elderly. In Solnick, R. L. (Ed.), *Sexuality and aging.* Los Angeles: University of Southern California Press, The Ethel Percy Andrus Gerontology Center, 1978.

Talbert, G. B. Aging of the reproductive system. In Finch, C., and Hayflick, L. (Eds.)., *Handbook of the biology of aging.* New York: Van Nostrand-Reinhold, 1977.

Toma, E. A. Aging of skeletal-dental systems and supporting tissues. In Finch, C., and Hayflick, L. (Eds.), *Handbook of the biology of aging.* New York: Van Nostrand-Reinhold, 1977.

United States Dept. of Health, Education & Welfare. *Human communication and its disorders.* Bethesda, Md.: U.S. Government Printing Office, 1969.

Vander Zyl, Psycho-social theories of aging activity, disengagement and continuity. *Journal of Gerontological Nursing*, May/June 1979, *15*, 45–47.

Wallace, D. J. The biology of aging: 1976 an overview. *Journal of the American Geriatrics Society*, 1977, *25*.

Weg, R. B. The physiology of sexuality in aging. In Solnick, R. (Ed.), *Sexuality and aging.* Los Angeles: The University of Southern California Press, The Ethel Percy Andrus Gerontology Center, 1978.

Weg. R. B. Nutrition and the later years. Los Angeles: The University of Southern California Press, The Ethel Percy Andrus gerontology center, 1978.

Wilson, E. The programmed theory of aging. In Rockstein, K. (Ed.), *Theoretical aspects of aging.* New York: Academic Press, 1974.

Woodruff, D. S., Birren, J. E. Age changes and cohort differences in personality. *Developmental Psychology*, 1975, *6*, 252–259.

Zborowski, M. *People in pain.* San Francisco: Jossey-Bass, 1973.

13

The Aged Adult

AMIE MODIGH

Ponce de Leon, the Spanish explorer, discovered Florida while searching for the Fountain of Youth in 1513. The search for ways to preserve youth continues; today sophisticated research is being conducted for the same purpose. Anthropologists are seeking to explain why inhabitants of certain places, such as Soviet Georgia and the Ecuadorian mountains, are more likely to live to as old as 150. Other researchers are exploring ways to repair DNA (the genetic material that controls life). There has been some startling evidence of the possibility of prolonging life in animals by manipulating DNA with ultraviolet light (Kent 1979). It is of interest that although there are more elderly people in the world than ever before, the maximum human life span has remained essentially the same (about 110 years) for 100,000 years (Cutler, 1980). If DNA repair can be effected in human beings, it may indeed be possible to prolong life.

There are two facets of aging. *Primary aging* consists of physiological and psychological changes, some of which are probably genetically determined. These changes can be slowed and compensated for with a variety of methods. However, *secondary aging* can be influenced more. Hans Selye defines secondary aging as resulting "from the sum of all the stress to which the body has been exposed during a life time and . . . is, in a sense, an accelerated version of normal aging" (Selye, 1956 p. 274). It is with regard to secondary aging that prevention, education, and guidance are so important.

The nurse should be not so concerned with adding years to life as with adding life to years.

People over 80 (with whom this chapter is concerned) are a fascinating group to study for several reasons. They comprise the fastest-growing age group; it is projected that by the year 2,000, people over 80 will make up 12 percent of the American population over 65 years of age. (The percentage in 1977 was 8.9 percent [1977 *Statistical Abstract*].) It is the least studied age group, and those who are healthy are especially neglected. Statistics usually refer to "65 and over," rarely to "80 and over." It is also the only age group of which it can be said that practically none of the researchers studying it has experienced being as old as the subjects.

Octogenarians living at the time of this writing have lived through almost unbelievable technological advances. When they were growing up, the fastest-moving person was the one with the fastest horse; today, people have been to the moon. To many elderly people, today's environment can be frightening and confusing.

POSITIVE AND NEGATIVE VIEWS OF AGING

There are no socioeconomic milestones such as retirement for aged adults. Thus, those people who have made a satisfactory adjustment to retirement view their eightieth birthday as little more than an occasion for a big party. Barring health problems, there is no reason why they should experience any negative changes.

A variety of studies have been conducted to determine the degree of satisfaction people have with their lives. The age group used most has been 65 to 70. Individuals who were satisfied with life before retirement generally continued to score high on life satisfaction tests after retirement (Atchley, 1978). Rosenberg (1970) stated that life satisfaction is related to the de-

gree of social interaction the individual maintains. It is safe to assume that many of the same factors influence life satisfaction in the older age groups as well. Thus, reasonably good health, meaningful and social activities, and a feeling of usefulness all contribute greatly to life satisfaction among aged adults.

The less fortunate persons who did not make a positive transition to retirement continue to experience negative feelings. These people tend to withdraw from social activities, express death wishes, and show a high incidence of suicide.

The attitude of children toward this age group varies little from what it is toward those 65 and over; they tend to view growing old negatively. Serock et al. (1978) feel that this could be changed. They propose that children in grammar school should be provided with factual information regarding all stages of life. They also feel that this would have a positive influence on viewing their own old age.

Not surprisingly, middle-aged people have a different view of old age. People in their fifties rarely view age 65 as "old" but frequently state that old age begins at 80. There must be some truth in Bernard Baruch's statement, "I'll never be an old man. To me, old age is always 15 years older than I am" (Comfort, 1976 p. 100).

GENERAL FACTS ABOUT AGED ADULTS

With each passing year more experiences are added to life, and aged adults have a greater stock of experiences than any younger age group. (The reader is again referred to Alex Comfort [1976] for diverse examples of successful, productive, and happy people aged 80 and over. Another applicable generalization is that people in this age group are more adaptable. They have had to adapt to a variety of events in life in order to make it as far as they have. Generally speaking, they have also experienced more losses than members of any other age group, losses brought about by death, financial setbacks, physical deterioration, decreased opportunities for socializing because of transportation or mobility problems, and so on.

There is little statistical data on the over-80 age group. The socioeconomic status of those in this group tends to decline in the vast majority of cases; this is particularly true for those living on a fixed income.

There is little change in where people in the 80-plus group live as compared to where they lived in earlier old age. Eighty-three percent remain in their own homes with a relative or friend, but 17 percent (as opposed to less than 5 percent in the 65 to 70 age group) live in nursing homes (Atchley, 1977).

The health data estimates 15 to 20 percent of people who are 85 and over have health problems of such magnitude that they must limit their daily activities. Chronic diseases are more common than acute diseases, cardiac problems being the most common health defect in the age group (Neugarten, 1975).

PHYSIOLOGICAL DEVELOPMENT

The *Encyclopedia Britannica* (1977) defines age as "the sequential or progressive change in an organism that leads to an increased risk of debility, disease and death," and human aging as "the sum of the changes that take place in the bodily structures, the behavior and attitudes, and the socioeconomic relationships of a person between his maturity and his death" (p. 109). Birren (1973) defines aging processes as "those which increase the susceptibility of individuals as they grow older to the factors which may cause death" (p. 23). Birren (1973) defines biological aging as a set of processes "such that the individual organism becomes increasingly likely to die the longer he lives even if he lives

in a constantly favorable environment." Thus, the final physiological development is death.

It is important to keep in mind that the aging process is not only a gradual downward process but one with plateaus. Visual change is a good example. Visual acuity decreases and glasses are adjusted accordingly, but it may then be several years before a stronger prescription is necessary. Thus, it is not unusual to hear a person in his eighties state, "I feel better than I ever have." It is also important to remember that all organs and systems are not affected in the same ways, nor does the aging process start or progress uniformly throughout the body. The physiological functions that are most important in the maintenance of homeostasis are adequate well into "advanced age" (Anderson, 1976).

General structural changes continue in the eighties. Almost everyone shows some decrease in height because of skeletal changes; this is more pronounced in the female. The spread of the pelvic bone starts to show in the vast majority of men, as do enlargement of the nose and external ear. There tends to be no further loss of hair on the scalp; although it becomes drier, while axillary and pubic hair continues to get sparser. The wrinkling process tends to slow down.

Skin

The continuing changes in the skin resulting from decreased blood supply, decreased fatty tissue, and decreased elasticity will gradually give the skin a thin and shiny appearance; this is more pronounced over bony areas such as the scalp and hands. Skin changes are more pronounced in exposed areas; in fact, it has been shown that, histologically, the skin of the buttocks remains "young." The receptors in the skin that signal changes in temperature and pain need more intense stimulation to produce a response.

Toenails show increased hardness as a result of decreased blood supply to the nailbed. The nails may also become brittle. The same process occurs to a lesser degree in fingernails.

Vision

By the eighth decade of life, nearly 90 percent of people have some degree of cataract (opacity of the lens), however, over 80 percent of people retain good vision throughout life. The rods and cones of the retina need increasingly more light to register visual information. Myopia tends to increase. In the presence of presbyopia or hyperopia this may lead to an ability to read even fine print without glasses. The individual who finds that he can do better without glasses will erroneously attribute the change to an improvement in sight; this is often referred to as "second sight." The sclera becomes increasingly yellow. The outside of the eye shows a similar appearance because of less orbital fat. *Ptosis* (drooping of the eyelids) is more frequent in this age group and may interfere with vision. As muscles and elasticity decrease, *ectropion* (a turning outward of the lower lid) and *entropion* (turning inward of the lower lid) occur with increasing frequency and may need attention.

Hearing

Sixty-six percent of people over 80 have significant hearing loss (Saxon, 1978). The eardrum thickens and becomes more rigid, resulting in decreased ability to transmit sounds. The accumulation of cerumen is accelerated; that, coupled with increased hair growth in men, further contributes to hearing loss, which is normally more prevalent in men.

Smell and Taste

Studies have demonstrated that there is a decrease in the ability to smell and taste; 30 percent of people over 80 have difficulty in identifying common substances by smell (Butler and Lewis, 1973). However, the remaining 70 percent can smell adequately. In the absence of disease, the decrease in taste buds and function are of little significance until very advanced age.

Digestive System

The decreased function of salivary glands leads to xerostomia, or dryness of the mouth. The loss of teeth is not an inevitable part of the aging process; there are, however, some factors that may contribute to loss of teeth, such as wear and tear of the teeth and resorption of the gum and underlying bony structure. The pulp of the tooth, where the vascular and blood supply are, shows an increase in fiber and a decrease in cellular nutrients. The loss of permeability in the dentin makes the tooth vulnerable to chips and fractures (Hudis, 1978).

Many of the changes observed in the 65 to 80 age group tend to plateau in the eighties. Although constipation is more prevalent in this age group, it is not believed to be caused by any of the normal physiological changes in the gastrointestinal system. There is some mucosal and muscular atrophy throughout the alimentary canal. However, no evidence points to any decrease in motor or secretory activity of the colon. The basal metabolic rate (BMR) is decreased by 20 percent in the eighth decade, and the number of calories needed to maintain normal body functions therefore decreases.

The gallbladder shows an increase in stone formation and infection in advanced age. Gallstones are present in 40 percent of people over 80, although no direct relationship between stone formation and the aging process has been established (Anderson, 1976; Sklar, 1978).

None of the changes in the gastrointestinal system interferes with normal living. Sklar (1978) states that "death in the aged is almost never attributed to a 'wearing out' of any of the digestive organs" (p. 174).

Circulatory System

The circulatory system shows an increase in the rigidity of the arterial walls. The cardiac index decreases by 35 to 45 percent between ages 30 and 90. A sharp drop occurs in the eighties, particularly among those who are sedentary. Thus, if the amount of blood pumped by the heart is decreased by almost half, all body systems will receive less nutrients and will experience decreased ability to rid themselves of waste products. Chughtai (1977) found while studying senile and nonsenile elderly that senility was far more prevalent in cases of significantly slowed circulation. He also found great individual variations in relation to age; thus, a 90-year-old with good circulation was found to show far less evidence of senility than a 60-year-old with poor circulation.

One of the body's compensatory mechanisms is the heart's increased effort to pump blood against the rigid arterial walls. This in turn leads to increased size in the left ventricle and increased blood pressure. Thus, a moderate increase in blood pressure can be viewed as a normal physiological change in aging.

Respiratory System

The 80-year-old's maximum breathing capacity is decreased to 40 percent of the capacity one has at age 20 (Cape, 1978). Decreased muscular strength and increased rib cage stiffness (from calcium deposits) are contributory factors. Asymptomatic emphysematous bullae occur along the anterior margins and apexes of the lungs. The ciliated columnar epithelium and mucous glands of the bronchi gradually atrophy, resulting in diminished self-cleansing properties of the lungs (Anderson, 1976). Thus, at advanced age, most people will experience some degree of what is called senile emphysema and mild chronic cough. In healthy older people these changes are not significant enough to alter daily activities, however, energetic exercise

without training is not possible. Hypoxia is felt at lower altitudes when maximum breathing capacity has decreased to this point. Thus, a healthy young person does not require additional oxygen at 10,000 feet but an octogenerian usually needs it before reaching 6,000 feet above sea level (Cape, 1978).

Urinary System

The glomerular filtration rate decreases so that by age 90 it is only 45 to 50 percent of what it was at age 30. Nephrons continue to decrease in number and the renal blood flow decreases by 53 percent by age 90 (Shock, 1968).

The bladder continues to shrink and may hold only 150 to 200 cc. Because the bladder is a muscle it decreases in strength, and emptying of the bladder is often inadequate. The time between sensation of fullness and necessity of emptying the bladder is shortened because of an increase in sensation threshold of the bladder receptors that send messages to the brain. Thus, when the message reaches the brain there is little time before emptying is necessary. The bladder sphincter also decreases in strength, making it harder to hold the urine in the bladder.

The prostate gland in males is one of the few tissues that increases in bulk with age; it will actually grow and add functioning glandular and stromal tissue. Because of its location at the base of the bladder surrounding the urethra, it is a frequent cause of trouble. As it enlarges, it occludes the urethra partially, spasmodically, and occasionally completely, leading to urinary hesitancy, diminution of force, dribbling, frequency, and nocturia. Only 10 percent of men evidence growth to the point of significant obstruction (Brosman, 1978).

Noctural frequency occurs in both sexes in advanced age and is associated with cortical inhibition over the sacral bladder center (Anderson, 1976). Milne (1980) found this in 50 percent of men and 70 percent of women over 80. A mild degree of urinary incontinence may occur because of the above-mentioned changes. However, severe urinary incontinence is usually associated with senile dementia. In the absence of disease, the body maintains its normal fluid and electrolyte balance amazingly well. However, when the balance has been disturbed, it requires much longer to return to normal than in the younger person.

Reproductive System

The changes in external and internal reproductive organs tend to plateau around 80 years of age, followed by gradual changes that continue at widely varying rates. Shrinkage of the external genitals with decreased elasticity of the vaginal walls and decreased glandular secretions are some of the changes occurring in the female. The vagina becomes less acidic, the ovaries and uterus continue to decrease in size, and the uterus becomes more fibrous.

The male experiences a continual decrease in testosterone level resulting in a decrease in the size and firmness of the testicles. The scrotum atrophies, and the penis requires more stimulation as well as longer refractory periods before becoming erect. Also, the force of ejaculation gradually diminishes.

The frequency and explosiveness of orgasm decrease in both male and female; however, there is no physiological reason why couples cannot continue a satisfactory sex life. (This is discussed further under Psychosocial Development).

Endocrine System

The exact role of the endocrine system in aging is unknown. The effects of hormonal changes on the male and female climacteric have received most attention in research studies. It is known that the thyroid gland's activity decreases, and this is believed to compensate for the increased responsiveness of the target organs. DeGroot and Stanbury (1975) warn that a slightly lower-than-normal thyroid level is not necessarily an indication of hypothyroidism.

There is a significant decrease in the size of the pituitary gland in the eighth decade. The weight decreases from 400 mg in adult males to 315 mg at age 80 (Asch and Greenblatt, 1978). However, there is no change in the secretory pattern. The adrenal gland, on the other hand, shows a slight increase in weight as well as some fibrosis and hyperplasia. Clinically it is rarely significant in the absence of disease. The pancreas continues to produce adequate amounts of insulin but at a slower rate.

Musculoskeletal System

By the eighth decade the majority of people show some atrophy and weakening of muscles. Again, this varies greatly among individuals. It is more pronounced in women and is directly related to the amount the muscles are used. Anderson (1976) found that the grip of the right hand of people age 60 was 44.0 kg for men and 32.4 kg for women; at age 89, these figures were 32.1 kg and 27.4 kg respectively. Muscle fibers continue to decrease in number and bulk. This is particularly noticeable in the small muscles of the hands, where the bones become very prominent. The muscles of the upper and lower extremities tend to take on a flabby appearance as the muscle tissues are replaced by fat. Collagen, which makes up 40 percent of body protein, undergoes deterioration; the fibers decrease in size and become irregularly shaped because of the development of cross-linkages between their molecules. Thus, the fibers thicken and become less elastic. Stiffness and a decreased range of motion in all joints may result. The bone continues to decrease in its total mass. By age 80, bone resorption in women will have resulted in approximately a 30-percent loss of the total adult bone mass; in men, the percentage is a much more favorable 10 percent.

Nervous System

Neurons continue to die gradually. The significance of this is still being debated as of this writing. The neurons do, however, require nutrients, since the nervous system will show rapid deterioration when cerebral blood flow is interrupted. It has been established that a decrease in cerebral blood flow is not caused by the aging process per se but by arteriosclerosis (Anderson, 1976).

Stimuli receptors generally increase their threshold. Heat, cold, pain, and pressure must be applied with greater intensity before the message is picked up and sent to the brain for a response.

The brain itself shows a decrease in size as age advances. The amount of the brain cells' water decreases from 92 percent at birth to 76 percent by age 90. There are other intricate changes in brain-cell chemistry, the significance of which is not known. As of this writing, research is being directed at the brain as the possible controlling mechanism of the aging process.

For obvious reasons, it has been difficult to study brain tissue in healthy adults. Most brain research has been done on monkeys and rats. A group from the University of California at Berkeley studied brain tissue in rats and found that the most significant reduction in the solidity of nerve and glia cells took place before the rat reached adulthood. The decrease in number after maturity until very old age was insignificant. This same group also studied the effect of environment on rats for 15 years. It was found that an inspiring environment had a stimulating effect on the cortex (Diamond, 1978). It can be assumed that an inspiring surrounding

that motivates and stimulates is likely to preserve human brain function as well.

PSYCHOSOCIAL DEVELOPMENT

The ability to adapt to one's environment, to learn, and to remember are indications of psychological age; habits, norms, and roles reflect social age. Vast individual variations exist in psychosocial development.

Bernice Neugarten (1978) states that "personality type is the pivotal factor in predicting which individual will age successfully and that adaptation is the key concept." It is felt that the way in which people age reflects how they have lived, barring catastrophic, physical, or psychosocial crises" (p. 45).

Neugarten (1978) described several types of personalities seen in older people. The "reorganizers" are the people who successfully replace old activities with new ones when they retire, becoming active in organizations, community, etc. As they grow older and less able to travel, they volunteer in hospitals, telephone services for shut-ins, and so on (Fig. 13-1). The "focused" limit themselves to one or two activities but devote a great deal of time and energy to them. Both of these types are very active and have high satisfaction levels. Neugarten includes the "disengaged" individual among these well-adjusted elderly but strongly emphasizes that the disengagement is voluntary and not imposed. The attitude of the disengaged (sometimes referred to as the "rocking-chair elderly") is typified by the statement, "I have waited for the day that I could sit in my backyard and do nothing since I was 40 and now I finally can." These people are content, and their disengagement is a matter of choice.

The less flexible group consists of the "holders-on." They are somewhat threatened by old age and do all they can to continue middle-age

Figure 13-1. *Finding new ways to make oneself useful is an important part of successful aging. These women volunteer their services as receptionists at the retirement home in which they live. (Courtesy Durham Senior Citizens Council.)*

activities. Usually successful and active, they remain satisfied. The "constricted" type defend themselves against aging by restricting their involvements with people and activities. Their activity level is low but for certain types of personality this affords a medium to high level of satisfaction.

Another personality type is termed "passive-dependent." One subdivision of this type is the "succorance-seeking" person, who remains fairly active and satisfied as long as there are one or two people to lean on to help meet emotional needs. Another passive-dependent type is the "apathetic" person, who has some things in common with the "rocking-chair" person except that his activity level is extremely low and he enjoys only medium to low life satisfaction.

Finally Neugarten refers to "unintegrated personalities who show a disorganized pattern of aging." Somehow they manage to live in the community with low activity and low life satisfaction. They display poor emotional control and their thought processes are deteriorated.

Language and Speech

It is not unusual for the vocabulary to continue to expand even in advanced age. There is an obvious relationship between the activities in which the individual engages and the increase in vocabulary. The voice gradually weakens because of senile bowing of the vocal cords and decreased elasticity of the laryngeal cartilages, resulting in a rise in pitch (Kaluger and Kaluger, 1979). Public speaking and singing is affected earlier than conversational speech, which eventually becomes forced and often characterized by increased pauses (Troll, 1971).

Successful aging depends to a large extent on how well the individual deals with developmental tasks. While "successful aging" can be defined in a variety of ways, for the purpose of this discussion, it suffices to define it as a subjective, inner satisfaction with life.

Coping with Widowhood

Statistics point out that by age 70 the majority of women are widows. The majority of men, however, do not become widowers until age 85. Mortality in general is higher among widows, and the suicide rate is higher among widowed people than among those who still have their spouses. However, a large number of people adjust to the loss and find life satisfying again. However, Gurin et al. (1969) found that the grieving process was unusually severe and prolonged among the very elderly. The widowed people also displayed a greater unhappiness and fear of death.

The importance of friends and family in offering support after a loss is well known. However, many elderly individuals face additional problems; to keep occupied and to be with friends is not always possible when widowhood occurs late in life. The surviving spouse may be financially "stranded" if a long illness preceded the death. The individual may be weak and limited in ability to travel. Atchley (1977) states that "most widowed people live alone and their major resource in coping with loneliness comes most often from friends rather than family" (p. 296). Many people have a decline in the number of friendships in old age, although a small minority actually have more friends; this is invariably related to high socioeconomic status and good health (Atchley, 1977). In addition to the threat of widowhood this age group faces frequent losses of friends and family members; one out of ten people over 80 has offspring who are 65 or over (Atchley, 1977). Siblings also play an important role in older people's lives; naturally, this is most true of never-married older people.

Most people in this age group who successfully developed an adult relationship with their children continue to enjoy this. In fact, Shanas et al. (1968) found in people over 80 that 98 percent of women and 72 percent of men either lived with or were very close to one of their children. Of course, many will also have adjusted (or not) to the role of grandparent by this time; 70 percent of older people have grandchildren. However, only a small percentage continue to be involved once the grandchildren are past a very young age. Since people tend to marry early, few grandparents older than 65 have young grandchildren. Troll (1971) found that "the valued grandparent" is an achieved role, not one that all grandparents attain. Thus, personal qualities and how they are viewed by the grandchild determine who is to be a valued grandparent.

Forty percent of older people in the United States are greatgrandparents. The small amount of research done finds this group more involved with great-grandchildren of all ages. It has also been found that children in a fourth-generation family are less prejudiced against older people (Atchley, 1977).

Dependence versus Independence

People who have always been independent will want to remain so until the very end. The fear of becoming dependent is widespread. The dependent role is so hard to accept because it means deferring to the benefactor and giving up the right to lead one's own life; that is what is expected in American society, of children, the poor, or any other dependent group. Anderson (1976) found that dependence, financial or physical, was the main cause of low moral among older people. Margaret Mead (1978), in interviewing many young mothers regarding what role they expected to play as grandparents, found that many expressed the intention to "get out of the way." Mead felt that many older people are similarly concerned with "not being a burden."

Bernice Neugarten (1975) speculates that as more people live to "old-old age" they will still live independently, but they will need support from both family and community. It appears that people who have learned earlier in life to adapt to temporary dependence are more able to do so again in old age.

Own Home versus Nursing or Retirement Home

There is little doubt that most people would choose to live in their own homes until the day they die; many people, indeed, do that. The efforts to increase community services so that more elderly can stay home will probably continue. The American Medical Association (AMA) has declared its support of more extensive home health care. In addition, two programs have realized significant savings. One study in Philadelphia by Blue Cross saved 4.6 million dollars over a three-year period by use of home health care. A similar study by Blue Cross in Connecticut reported over 4 million dollars in savings. As a result of these and other studies, the AMA has required "insurers to offer benefits for home health care" (*Geriatrics*, October 1979). Legislation to this end is already in effect in seven states (*Geriatrics*, October 1979). Studies have also shown that a lack of long-term care facilities is responsible for "wasting millions of dollars in Medicare funds each year" (*Geriatrics*, December 1979). The main reason for the expense is that acute care hospitals, which are far more expensive, are being used as housing for people who cannot get rooms in nursing homes; evidently, the 18,000 nursing homes in the United States are not adequate in quantity.

Regarding quality, less-than-favorable conditions still exist in many institutions, although they are less common. The standards that govern the eligibility of nursing homes for Medicare and Medicaid have led to much improvement. However, the drawback is a great many rules and regulations, which make for a too-rigid way of living for many people. Also, inspections of these facilities are by and large "paper inspections."

Several different types of nursing homes exist. There are skilled nursing facilities with 24-hour nursing service, supervised by a registered nurse and medical director. The intermediate care facility offers nursing care at a lower level, and supervision is not as comprehensive. Federal regulations require that these facilities and their administrators be licensed, and that they meet fire, health, and safety standards. Inspection of these homes is done by state health and welfare departments. Eighty-five percent of all nursing homes are operated for profit; many are good homes where the utmost of care and compassion are evident. However, there are still many that are sadly deficient and understaffed, and where "patient abuse is fairly common" (Brown, 1978).

Foster care facilities and homes for the aged do not provide medical care but may provide minimal nursing care. Sharing homes with friends is growing in popularity. Florida instituted a Share-A-Home Association in 1969; similar groups now exist in Georgia, North Carolina, Virginia, Ohio, and Kentucky. For an average cost of $400 per month, residents get a private room, three meals a day, laundry, and transportation to and from stores, church, etc. Rules are few and liberal but each person must be able to care for himself (Getze, 1980).

Another alternate living style is the planned retirement community. They first appeared in Florida and California but are now springing up throughout the country. They tend to be homogeneous with regard to residents' social class and ethnic background. The majority are still fairly expensive. Some of these facilities provide "life care," which means that residents buy their units and pay for monthly upkeep. However, if a resident uses up all his funds, the organization will continue to provide care. This type of arrangement allows individuals to remain in a homelike environment, and allows great opportunities for social interaction and the security of knowing that one's needs will be taken care of if needed.

Activities

Most older people establish meaningful activities to replace the ones they were engaged in while working. However, people in this age group often face restrictions on the type of activity they can carry on. Activity involving physical strength or driving a car may be curtailed. The individuals who conquer the developmental task of finding meaningful activities are the ones who have the most flexibility in terms of interests. For example, if one can no longer play golf but still enjoys taking walks and playing cards, one still has excellent opportunities for social interactions and satisfying activities.

Sexuality

The sexual needs of older people (discussed in the previous chapter) continue into later old age. How well they are met depends on how the individual adjusted to the physical changes of earlier old age, and, of course, on the availability of a partner. Again, it must be remembered that sexuality involves love, touch, cuddling, etc., in addition to or without intercourse. (See the further discussion under Anticipatory Guidance.)

Planning for Death

Although many older people will have come to grips with how they feel about death, they may need to review their thoughts and feelings with another person. Contrary to popular belief, most older people want to discuss death (but not to dwell constantly on the subject). Many want reassurance that they will not be alone when they die, or they may wish to discuss updating their will or giving away some possessions to children and other loved ones. It has been found that if people are allowed to discuss their thoughts on death it decreases their fear and enables them to use their energies for living.

A readjustment in one's thinking about death may occur if the individual makes a geographical move. Entering a retirement or nursing home makes the reality of "the final chapter" very real. A healthy conversation about death and wills (including living wills) is helpful to the individual. The living will is becoming increasingly popular; essentially, it is a statement of the desire not to be kept alive by "heroic measures" if recovery from physical or mental infirmity cannot reasonably be expected.

Life Review

Reflecting and reminiscing are healthy activities that help older people to "get their house in order." How well they accomplish this task is directly related to how content they are with their lives. Reminiscing can be a very pleasurable experience to those who have led a satisfying life. However, to those who have not, it can be painful. (This will be elaborated upon under Anticipatory Guidance.)

COGNITIVE DEVELOPMENT

For the normal and healthy elderly individual, general intellectual decline is a myth (Fig. 13-2). Baltes and Schaie (1974) tested four dimensions of intelligence: crystallized skills acquired through education and acculturation, cognitive flexibility (i.e., the ability to switch from one way of thinking to another), motor-visual flexibility (ability to switch from a familiar motor-visual task to an unfamiliar task and, visualization, involving tasks such as recognizing a pic-

Figure 13-2. *At 102 years of age, this mountain woman is still bright and alert. (Courtesy of Hugh Morton.)*

less well with tasks that involved learning a collection of numbers or memorizing a phrase or poem for which they saw no practical use. The old people saw no benefit in learning or trying to memorize "nonsense" items.

It is generally accepted that short-term memory declines with age but that long-term memory remains intact. However, both Clark (1968) and Raymond (1971) found short-term memory to have remained efficient with increasing age, although it may take longer for the individual to retrieve the information. It is certainly true that old people can often relate minute details of an event that took place in their childhood, and it is also true that they may not remember what they had for dinner yesterday. However, this is probably due to the differences in meaningfulness of the two events rather than to a function of memory per se. It is safe to say that there is some decline in all types of memory in extreme old age.

Conceptualization and problem solving are affected earlier than memory. Atchley (1975) raises an interesting point in relation to concept formation: primary and secondary schools of the 1920s and 1930s did not emphasize generalizations and inferential thinking, thus, the elderly have been less prepared to deal with abstract concepts. Consequently, they would not score high on tests that assess this type of intellectual ability.

Problem solving—using skills and perception to make choices—is particularly difficult if several items must be dealt with at the same time. The elderly tend to behave in a repetitive fashion, and if the nature and goal of a problem changes frequently, they are likely to have difficulty. If the problem and solution remains fairly constant, repetition can be used advantageously.

Response time increases with age. The increase in reaction time is negligent with simple tasks but becomes more significant with complicated ones. Botwinick (1973) concludes that

ture in an incomplete form. They found a decline only in motor-visual flexibility.

Schmavonian and Busse (1963), in attempting to measure degree of involvement in performing certain tasks, found that the meaningfulness of the task was extremely important. Hulicka and Weiss (1965) confirmed the idea while testing young people against 65- to 80-year-olds. The more meaningful the task the less difference between age groups in ability to carry out the tasks. Thus, they had no more difficulty than the younger age group in learning how to operate a new washing machine, lawnmower, typewriter, or other tasks for which they saw some practical use. They did

this is attributable to an increase in cautiousness in the elderly. He also found that when there was no time to be cautious, reaction time was faster.

COMMON HEALTH AND DEVELOPMENTAL PROBLEMS OF THE AGED ADULT

Geriatric medicine is defined by Sir Ferguson Anderson (1976) as "a branch of general medicine dealing with the health and with the clinical, social, preventive and remedial aspects of illness in the elderly" (p. 420). This definition refers to the many aspects of health with which the holistic approach is concerned. Holistic care has been frequently discussed and written about, yet it is simply a whole-person approach to health care. The physician who diagnoses and treats the elderly in the nonholistic manner is at a serious disadvantage; the patient's "medical problem may, indeed, not have its root in a physiological disease" (Portnoi, 1979, p. 22).

The emphasis in dealing with health problems of the very elderly should be on prevention. It is not sufficient to rely on the self-observation of symptoms by the elderly; services for prevention, education, examination, and counseling must be made available. Williams et al. (1973) found in a group of 300 people over 75 years of age that 77 percent were able to improve their health and performance of daily activities after being seen by a health care professional. Sir Ferguson Anderson (1976) suggests that a nurse is the most suitable professional to make the initial health and home assessment. He feels that less than 10 percent of aged clients would need medical referrals. The most common conditions in aged clients referred were cardiac failure, urinary symptoms, deafness, and failing eyesight. All were treatable if not curable, and all were certainly able to increase the quality of their lives.

Skin

The same skin conditions that occur in earlier old age may also be seen in the older elderly. The skin may, indeed, be even drier, and senile pruritis is a more common complaint. When the presenting symptom is pruritis, it is very important to rule out diabetes, chronic leukemia, renal disease, and internal malignancies, in addition to simple dry skin or psychosomatic cause.

Eye, Ear, Nose, and Throat

Eye diseases tend to plateau in the eighties except for macular degeneration. Entropion and ectropion occur more frequently in this age group and may need repair. Vestibular disorders still occur, often because of decreased blood supply or orthostatic hypotension. Thus, dizziness is a very disturbing problem for a fair number of elderly people. Relief can sometimes be obtained from meclizine hydrochloride (Antivert) or vasodilators. Deafness increases with age. In the 85 to 89 age group, 9.1 percent of men and 8.3 percent of women have significant deafness.

The nose and mouth are susceptible to the same problems that are seen in the younger elderly. There is an increase in occurrence of leukoplakia, especially in men. This should receive immediate professional attention since it is a precancerous lesion.

Gastrointestinal Disorders

This is the system from which the greatest number of complaints arise. The gastrointestinal system remains remarkably intact but functional symptoms are common. Nutritional problems continue to present the same menace as to the younger elderly. Malnutrition is a very real threat, especially in the uneducated, financially disadvantaged, and isolated older person. A common diagnosis in this age group is the so called "irritable bowel syndrome," which consists of periodic diarrhea and cramping. A history usually reveals long-standing symptoms often associated with stress. Diagnosis is made after other diseases of the gastrointestinal tract are carefully ruled out. Psyllium hydrophilc muciloid (Metamucil) and antiflatulents are sometimes used with varying success rates (Cooper, 1980). Abdominal cancers are fairly common, it is very difficult to differentiate gastric ulcer from gastric cancer.

Amberg and Zboralsko (1965) found gallstones in 39 percent of 1,057 patients over 80 years of age, sometimes leading to cancer. Anderson (1976) urges that cholecystectomy for the prevention of cancer of the gallbladder be restricted to those under 70. He and Reichel (1978) suggest surgery for people over 70 only in the presence of biliary obstruction. Reichel (1978) has evidence that conservative treatment consisting of reducing the infection and using medication to dissolve stones can be successful.

Constipation can become a problem to the point of causing bowel obstruction. Any elderly individual who is sedentary and complaining of abdominal pain, mild diarrhea, or other vague symptoms should be checked for impaction.

Circulatory and Cardiac Problems

A certain amount of arteriosclerotic heart disease is present in everyone by late old age. Atherosclerosis is also common, although not all cases are symptomatic. Some very encouraging studies show that it is possible not only to halt but to actually reverse the atherosclerotic process (Kent, 1979). Several studies have indicated that regression can be achieved with a low-fat, low-cholesterol diet and vigorous exercise;

these results were obtained in patients who had previously had heart attacks (Blankenhorn et al., 1974; Blankenhorn, 1978). It is hoped that the use of such a regimen might lower the incidence of death from heart disease; during the 1970s, half of all deaths in people 75 and over were caused by disease of the cardiovascular system (*Statistical Abstract of the United States, 1978*).

Cardiac arrhythmias are common, atrial fibrillation being the most frequently occurring type. Congestive heart failure also develops frequently after cardiac arrhythmias. Immediate medical attention is necessary and a careful regimen must be instituted. Rest is always required to some extent, although the dangers of complete bed rest are well-known. The use of a reclining chair is often a practical alternative.

The type of elevated blood pressure seen in this age group is benign, or essential. Anderson (1977) maintains that blood pressure of 215/110 in men and 230/110 in women over 80 years is not significant in the absence of disease. However, many physicians will disagree with this, and most American physicians will institute treatment for people in this age group who have considerably lower readings.

Respiratory Problems

The very elderly are vulnerable to a variety of respiratory problems. Because of the decrease in pulmonary function, they are extremely susceptible to bronchopneumonia. Chronic obstructive pulmonary disease is present in a high percentage of older people. Thus, a certain amount of coughing and expectoration of white mucus is common and is generally not treated with antibiotics. However, if the sputum changes to yellow and green, antibiotics are usually started at once; fever is not a reliable indicator of acute bronchial infection in the elderly but a change in sputum is. Early treatment is vital to prevent pneumonia.

The O Complex

Ronald Cape (1978), in his book, *Aging: Its Complex Management*, introduced a new way of viewing disease in the elderly. In agreement with the belief of other researchers that the endocrine and nervous systems are the major controlling systems in aging, Cape proposes that the brain is the main controlling factor in aging. Everything we do, feel, and experience goes through the brain. He further believes that the changes in the brain are caused not only by vascular insufficiency but by physical and chemical changes in the brain itself. According to Bromley (1974) the brain is at its optimal functioning level between ages 16 and 20, and a rapid decline occurs after age 75, the common manifestations in old age being loss of mobility or falling (failing motor system in the brain), confusion (failing visual, auditory and cognitive receptors in the brain), and incontinence (breakdown of the brain's control of micturition and defecation). Confusion is the most prominent sign.

Ample evidence exists to show that homeostatic breakdown is the fourth most common medical problem of old age. Cape (1978) states that "these are not diagnoses but clinical problems which stem from aspects of brain failure aggravated by a wide variety of conditions." The use of drugs to treat various specific conditions creates an added risk of adverse reaction. Thus, Cape adds to the list of other problems the problem of iatrogenic illness, and he refers to the five problems collectively as the "O Complex." The older the patient, the more commonly and severely these clinical symptoms appear. These five constitute an old-age syndrome which Cape and others feel should be regarded as the core of geriatric medicine.

Since there is strong evidence of the brain's dominant role in timing and masterminding the aging process, it is not surprising to find that failure of this organ has equally important clinical implications. The key to solving the problems of the O complex lies in the brain. Cerebral cellular metabolism is of greater importance to the very old than to anyone else. This being the case, the need for redoubling research effort in this area is obvious. A small gain in this field that would permit more elderly people to maintain their independence would pay enormous human and economic dividends. Recognition of the O complex means that useful explanations can be offered to patients. Understanding some of the background facts of a clinical problem may not cure it, but it can suggest to each individual possible methods of managing it and adopting rational countermeasures (Cape, 1978, p. 82).

Cape's positive outlook on aging and the many steps that health care professionals can take to increase the quality of life in old age, even when infirmity is present, is most inspiring.

Depression

Depression in the very elderly deserves special attention for several reasons. First, it is very common; over 20 percent of people over 75 suffer from significant depression (Weissman and Myers, 1978). Second, it often goes untreated in the elderly because they do not seek help, and the symptoms are quite different from those of depression in younger people. Rather than the guilt, delusions of persecution, and decreased self-esteem seen in younger age groups, one more frequently sees withdrawal, increased physical complaints, and a very pessimistic outlook on the future (Blazer, 1979). Among the elderly, depression is more common in men, whereas more women experience it at earlier stages. There is a high suicide rate among the depressed elderly; it is higher in white males over 80 than in any other group, as is depression. Atchley (1977) points out that suicide can be active or passive, an example of the latter being to stop taking medications or food.

In diagnosing depression, it is very important to rule out physical disorders along with thorough history taking and use of depression scales. It is important to differentiate between major depressive disorders and temporary dysphoric symptoms. Physical activity, social activity, counseling, and drug therapy are all part of successful treatment.

ANTICIPATORY GUIDANCE

The reader is reminded to review the previous chapter with regard to practical implications of physical and psychosocial changes. Many of the changes and actions required are similar for the older elderly; emphasis here will be placed on variations.

Skin

A larger number of people in their eighties or older are immobile. Thus, good skin care becomes even more important. Replacing natural oils by gently massaging with good, inexpensive creams and lotions should be part of a twice-daily routine for both men and women. Range-of-motion exercise can be accomplished at the same time. Special attention should be given to all boney areas. Obviously, if someone is bedridden or unable to do these things, it is important that someone else do the massaging and applying of creams. Also, a change of position, no less frequently than every two hours, is essential. In thin people of extremely old age, whose skin tends to be stretched and shiny, sleeping on sheepskin is helpful. If much time is spent in bed, an alternating pressure mattress or waterbed can be a great help in preventing decubitus ulcers. Proper nutrition and adequate fluid intake are both helpful in maintaining good skin turgor and texture.

Sensory Changes

Yearly ophthalmological examinations continue to be important. If the individual needs a cataract operation in advanced old age, it will take longer to adjust to the changes in vision. Thus, the nurse should spend a great deal of time preparing the individual, repeating what to expect until the information is thoroughly understood. She must be in close contact with the surgeon to be sure the right procedure is described. People with macular degeneration can be helped immensely by reeducation. They gradually lose their central vision but retain some peripheral vision, although a small percentage develop cataracts in addition and may thereby lose all vision. If the individual lives at home or with someone, orientation and practice sessions with a blindfold are extremely valuable. Keeping furniture and frequently used equipment in the same areas can help the individual to maintain independence and control of the environment.

For the fortunate 80 percent of people who maintain adequate vision, yearly examinations are still important. Elderly people must be made aware of the importance of proper lighting. A large study recently showed that of the people considered functionally blind, less than half proved to be so when examined in the hospital; it was concluded that by simply improving home lighting, the number of people considered visually disabled could be reduced from 520 to less than 300 per 100,000 of the people 76 years and over (Cullinan, 1979). In addition to making reading, cooking, and other activities more enjoyable, proper lighting decreases accident hazards.

The number of people with decreased hearing increases greatly, especially in men over 80. The older the person is, the more likely he or she is to be viewed as senile. Such people will sometimes respond inappropriately (verbally or nonverbally) to a question rather than suffering the embarrassment of having to ask that the question be repeated. Encourage the person to overcome the embarrassment and to ask to have things repeated if necessary.

The health care professional will need to be familiar with the different types of hearing aids in order to help the individual use it correctly. One of the reasons many older people have an unused hearing aid in a drawer is that they were not taught how to use it properly.

The American Association of Retired Persons (AARP) called on the federal government in February 1980 to continue support and increase research money to the National Institute of Aging (NIA). They also recommended that physicians' and audiologists' fees be covered by Medicare. It is indeed difficult to believe that these services have not been covered by Medicare previously. It is also interesting that this movement is another example of an attempt to improve the quality of life for the elderly, accomplished by the elderly themselves.

Regarding nutritional aspects, emphasis must be placed on the intake of quality calories (Fig. 13-3). The elderly are more able to eat frequent, small meals than three large ones a day. High-calorie snacks in liquid form are readily available at reasonable prices. Vitamin supplements may be necessary, particularly if a complete diet is not followed. The nurse needs to be aware of daily requirements and to suggest ways of

Figure 13-3. *Nutrition is as important in old age as throughout the life span. Older adults who have the opportunity to eat in the company of others often eat better than those who live alone. (Courtesy Durham Senior Citizens Council.)*

obtaining needed supplements within the realistic limits of the client's socioeconomic background. It is also worth noting that several studies of the late 1970s indicated that small amounts of alcoholic beverages not only serve as an appetite stimulant but have a beneficial effect on vasodilation, thus increasing circulation.

Elderly people should be encouraged to make a conscious effort to increase their fluid intake. A written record of how much fluid is drunk daily, kept by the person himself, will help in this effort, and also puts control in the individual's hands. Since the thirst sensation is decreased, creative measures must be used to ensure adequate intake, which aids most body systems.

A study of more than 200 older men and women over a five-year period showed that carefully prescribed exercises resulted in significant physiological improvement (deVries, 1979). Participants ranged from 52 to 88 years of age. In addition to cardiac and respiratory benefits, exercise firms the skin, decreases constipation, keeps muscles and joints firm and supple, and has a very positive effect on emotions. There are many suitable exercises that will accomplish different things. Walking, when possible, is one of the best and longest-tolerated exercises in later years. Swimming also maximizes rhythmic activities of large muscles. "Use favors function and disuse invites decay," according to Bortz (1980), who, along with other researchers, believes that supervised, planned exercise has great benefits for both the healthy and the ill individual.

The reader is referred to the previous chapter regarding practical implications of changes in reproductive, endocrine, and cardiovascular function. Other changes will be discussed here as they relate to the previously described O Complex.

Falling

Control of posture involves the musculoskeletal system. Some of the degenerative changes that take place in the spinal cord in old age tend to lead to a stooped position. Various sensory stimuli are also vital in maintaining an upright position. Seasickness is a good example of discord between visual and vestibular stimuli. A normal righting reflex stops most people from falling when they over- or understep a stair; this reflex does not work as well in the very old. These factors all predispose one to falls. According to Sheldon (1960) and many others, falls are almost twice as common in women. Falls occur so often in the elderly that they accept them as a normal event, and, unfortunately, they rarely attempt to find out the cause. The health care professional can be a vital source of help in this regard. Educating the elderly and encouraging a complete work-up is the first step. It is thought that over 80 percent of falls are preventable (Cape, 1978).

Sheldon (1960) and Clark (1968), in studying 500 and 450 people, respectively, note that almost half the falls were accidental. The biggest culprits were poorly lit staircases, low furniture such as stools, scatter rugs, pets, or loose items on the floor. These hazards are all avoidable! Making a home visit to work with the elderly to help him or her identify potential hazards and how to avoid them is much more helpful than just saying "don't do that." The more control the individual has, the more self-respect and self-esteem is maintained.

So-called "drop attacks" were the cause of 25 percent (Sheldon) and 16 percent (Clark) of the falls analyzed in the two studies. This is caused by a "sudden failure of antigravity muscles which are then temporarily paralyzed until some nervous pathway is reestablished" (Cape, 1978, p. 121). If the individual is helped up or even pushes with his feet against a wall, nerve connections are reestablished and he can again walk normally. Rarely does the individual remember falling. There is strong evidence from Sheldon's (1968) study that these falls are caused by a sudden movement of the head to one side that cuts off circulation in the vertebral arteries, which follow a serpentine course through the cervical spine if osteoarthritis is present. This theory is supported by the recollection of several people of having suddenly turned the head before falling (such as when crossing a street), and by the fact that wearing a simple neck collar seems to prevent attacks. True vertigo was responsible for 7 percent (Sheldon) and 4 percent (Clark) of the falls. Vertigo in the elderly is probably caused by degeneration of the cochlea and vestibule. Little but empathy and understanding can be offered. Those who suffer from vertigo should not be left alone while they are moving, if possible.

Postural hypotension is easily diagnosed; significant difference in blood pressure (20 mm Hg systolic and 10 mm diastolic) obtained while the client is lying and standing indicates its presence. The client can eliminate falls stemming from postural hypotension by getting from a lying to a sitting position slowly, then standing for two or three minutes before proceeding.

A variety of central nervous system lesions were responsible for 5 percent and 16 percent, respectively, of the falls studied by Sheldon and Clark. The last category, "miscellaneous," includes situations such as extending the head backward to get something from a high shelf, and weakness in the legs arising from lack of muscle use.

In addition to the injury that a fall can cause, it also leads to fear. As the individual loses confidence, reduce mobility and increased weakness results, and both factors increase the likelihood of falls. For this reason, it is important for the health care professional to encourage clients to find the underlying cause of their falls and then work on prevention.

It should be mentioned incidentally that, while it is true that the elderly are more prone to fracture, it is a fallacy that they take longer to heal. The elderly have just as much potential to heal as younger people, and large amounts of callus forms rapidly in fractured bones. Rehabilitation of muscles and joints and healing of injured tissues takes longer (Freehafer, 1978). Also, complications are more common. The use of mechanical devices such as joint prostheses makes it possible to start mobilization earlier, which in turn decreases complications.

Confusion

Confusion occurs frequently in the elderly. Cape (1978) distinguishes confusion from *dementia*, which is chronic confusion. The terms *confused* and *senile* are often popularly but unjustly applied to older people in the absence of scientific evidence. The following story is typical.

Mrs. X was the secretary of the local AARP chapter, charming, attractive, intelligent, and 81 years of age. One day she was taken to the emergency room with a cough and chills. She was admitted, an IV was started, side rails (which she did not want) were in place, and she was given a sleeping pill. The next morning she woke up with her bed wet, and she was crying. This author happened to know Mrs. X. and went immediately to see her. The write-up on her chart stated, "81-year-old white female, admitted last night with probable RLL [right lower lobe] pneumonia. Senile, incontinent, and combative." This woman had never been in a hospital and had never had a sleeping pill, side rails, or an IV. She vaguely remembered having tried to get out of bed to go the bathroom during the night.

A health care professional in this situation should have taken the time to explain, to repeat, and to realize that the hospital room is not like the client's own room, the bed is often high, and the bathroom not in the same familiar place. Nurses must realize that a sleeping pill dulls the sensations, that new places are frightening, and that too much new stimulation will result in confusion even in young people. Nighttime increases confusion greatly, especially in elderly people in strange environments.

Health care professionals need to be aware of

the difference between acute and chronic confusion. Acute confusion is characterized by a sudden onset, impaired memory, muddled thinking, disorientation, illness, and, possibly, hallucinations. Restlessness, insomnia, and anorexia are also common. The six most common causes of acute confusion are: pneumonia, cardiac failure, urinary infections, neoplasia, depression, and drugs. It is interesting to note how different age groups respond to deterioration of cerebral function. A child may experience convulsions, adults may experience chills, and the elderly, confusion. Thus, all of the above conditions need to be carefully ruled out when someone appears with acute confusion. The client should be treated as a normal person, and with as much kindness as possible.

Chronic confusion, on the other hand, has an insidious onset. All too frequently, underlying treatable conditions are ignored, often because the family (and, sometimes, the individual) attributes symptoms to "just getting old" or "getting senile." Personality changes occur but they are usually an exaggerated form of already-existing personality traits. Thus, the cheerful individual may become a little euphoric and the quiet individual more withdrawn. It is again important that the health care professional encourage the individual and the family to bring the client for a thorough medical examination. The chronic confusion may be what Cape (1978) refers to as the "iceberg phenomenon"; the confusion is the visible manifestation of a hidden underlying cause. What could be more tragic than finding out (for example) that someone has pernicious anemia after irreversible brain damage has taken place? If detected early, this disease is completely controllable with monthly injections of vitamin B_{12}. Psychiatrists encourage prompt recognition of the iceberg phenomenon since an early and aggressive attempt to deal with sensory deprivation, loneliness, and isolation can slow the onset of organic brain syndrome.

Obviously, some people do develop chronic dementias for which there is little hope beyond good institutional care. However, with knowledgeable health care professionals making the proper assessments, the number of elderly living out their lives in quality and dignity should increase dramatically.

Incontinence

Sheldon (1960) differentiates between "dribbling," which he found in 27.2 percent of men and 24 percent of women, and persistent incontinence, which he found in 23 percent of men and 20 percent of women in a 928-bed geriatric hospital in England. Incontinence is believed to be more common than figures show because so many people try to deny or conceal that they have this problem. This is most unfortunate because the sooner the cause is found the more likely it is that treatment will be successful. In most people with incontinence, something can be done to improve the situation. Ruling out and treating a medical condition should obviously be done first.

Once illness has been ruled out, rehabilitation programs should be started. There are helpful exercises such as squeezing the buttocks together and practicing starting and stopping the urinary stream; this will strengthen the sphincter muscle at the base of the bladder. Regular emptying of the bladder can be helpful if it conforms to the individual's own routine. Many institutions assist incontinent residents to the bathroom every two hours; that works for some people, but only if the bladder is filled to a certain capacity. A more successful approach, although initially more time-consuming, is to observe and keep a close record for 24 to 48 hours as to when the individual is incontinent, and then to take the client to the bathroom at the observed times of incontinence. This will lead to better results.

Ambulatory people have less problems with

incontinence. Much effort and ingenuity should go into helping these people so that they can lead normal lives. A variety of receptacles and equipment for incontinence are available; they can be used temporarily to allow the individual to attend social functions without fear of embarrassment. These include absorbent pads, plastic pants, rubber cups with leg bags, etc. The nurse should be familiar with these and encourage their use if necessary.

Fecal incontinence is less common in ambulatory people and also responds fairly well to a rehabilitation program built around the individual's normal habits. The use of suppositories or other laxatives to achieve complete evacuation once a day or every other day is usually fairly successful.

Homeostasis

The maintenance of the body's internal environment is controlled by the hypothalamus. It controls much of "body fluid and acid-base balance, blood pressure, temperature, appetite, thirst and reactions to stress of all kinds" (Cape, 1978, p. 158). Infection may be present even in the absence of elevated temperature; confusion may be the presenting symptom. Hypothermia, which is caused by faulty heat regulation, is a real threat to the very elderly. Exposure to low temperature may not always be felt by the elderly individual, and the use of drugs or alcohol will further delay the perception of impending severe hypothermia. Adequate heating in houses and provisions for emergency heat when the usual source is not functioning is important. Instruct the individual family to wear added clothing in cold weather and to avoid sitting still for extended periods. Advise the family to have body thermometers that register as low as 32°C (90°F). At this temperature, pallor, facial edema, and clouding of vision tend to occur along with decreased blood pressure and decreased heartbeat. If these symptoms are pres-

ent, immediate transfer to the hospital is necessary.

Stress, whether pleasant or unpleasant, has the same effects on body systems, including an increased work load for the heart, lungs, brain, muscles, and glands. Stressful stimuli activate the sympathetic nervous system and excite the suprarenal cortex, causing it to produce greater amounts of cortisol, which may contribute to longevity. Too much stress can cause death; however, it is important to realize that some stress is essential and beneficial. The nurse must assess the amount of stress the individual normally experiences and how he handles it. To simply tell the client not to engage in certain activities can destroy the individual's initiative and even the will to live. As Cape states, elderly people need "regular exposure to a variety of stress which should tax but not overwhelm them" (1978).

Iatrogenic Disorders

All ingested drugs undergo absorption, transportation, tissue and protein binding, metabolism, and excretion. All these processes are greatly slowed down in the blood stream of the aged adult. Therefore, generally speaking, the very elderly require considerably smaller dosages of a drug to obtain the desired effect. The occurrence of adverse drug reactions is higher in the older individual and in those who take several drugs.

The average resident of a nursing home receives as many as eight to ten drugs a day. Some elderly people receive that many at home in addition to an unknown amount of over-the-counter drugs. It is vital for the nurse to follow the research findings on drug reactions in the elderly closely. The use of a clinical pharmacist and a nutritionist on the health care team can obviously be of invaluable aid in this area, both for teaching the client and making accurate assessments of needs and treatments.

Caring for the Terminally Ill and Dying Patient

In the early 1900s, most deaths occurred at home. A family member or a significant other was often present. Even children and animals were part of the last stage of growth and development. Then, more and more deaths occurred in hospitals. The more technology advanced, the more people died in sterile intensive care units or in rooms with signs saying "No Visitors." With the advent of more realistic and honest talk about death, and that of the hospice movement, more people have chosen to die at home or at least in the presence of loved ones. As of this writing there are two professional journals on the subject of death, and courses in thanatology are offered at many colleges.

One of the people most responsible for the introduction of death as a topic to be discussed and studied is Elisabeth Kübler-Ross. Her first best-seller in 1969, *Death and Dying*, was followed by several other helpful books that deal with the emotional aspects of dying with regard to the family and the individual who is dying. In brief, Kübler-Ross strongly advocates honesty, support, help in avoiding pain, and a guarantee to the dying person that someone who cares will be there at the time of death if at all possible.

This author has had numerous requests from dying clients to be with them or to see to it that some other significant person is present at the final moments of life, and has observed that, when it is possible to honor such a request, the dying person invariably dies in peace, and usually with minimal apparent discomfort. Many times the dying person perceives a deceased loved one calling him, stretching out a hand, or just being there. Some have described a bright light or a peaceful scene just moments before dying. One resident whispered his wife's name, reached out his hand, smiled, and stopped breathing. Moody (1975) describes many similar situations, usually involving critically ill people who "came back" from the brink of death.

Several physiological changes occur in the dying person. There is a decrease in sensation, power of motion, and reflexes, starting in the legs and progressing to the arms. Any pressure, including from blankets, can be uncomfortable. The dying person needs frequent changes of position of the legs. Peripheral circulation fails, making the body cool to the touch, regardless of room temperature. Core body temperature increases so that the dying person is not usually aware of being cold; in fact, the restlessness sometimes observed in people near death is caused by sensations of heat that can be relieved with circulating air from an open window, fan, etc. Dying people often seek any source of light; therefore, shades should be left open during the day, and a light should be left on at night. Sit or stand near the person. Dying people are usually not in pain, but if pain medication is still indicated, IV is the appropriate route since peripheral circulation is decreased or absent. Remember that the last sense to leave is hearing; do not whisper! Reassuring the person that you are there and that you care can offer wonderful comfort.

Of course, no data exists regarding what it is like to die. However, the nurse would do well to recall the dying words of the famous British anatomist, obstetrician, and medical writer William Hunter, who whispered to his wife moments before his death in 1783, "If I had the strength to hold a pen, I would write how easy and pleasant a thing it is to die."

Practical Implication of Psychosocial Changes

To be most helpful to the elderly individual, it is vital to understand different personalities and cultural and ethnic factors. All these, plus past experiences, will influence the way the elderly person deals with the developmental tasks of this stage.

Widowhood in this age group disposes the individual to illness, depression, and suicide. One antidote is social activity; however, to find suitable activities involves an intelligent assessment and planning with, not for, the client (Fig. 13-4). During the acute grieving period, the individual goes through several stages. The first stage is usually one of shock or disbelief. This may be followed by a short period of denial. Then the individual begins to experience actual physical pain and intense sorrow. The physical complaints are often headaches, muscle aches, and pressure in the chest and abdomen, which tends to come in waves and is often accompanied by weeping. As the grieving person moves into a chronic form of grieving, he or she experiences guilt, anger, and hostility, and view the world and any would-be helper with great hostility. Gradually these feelings recede and the individual is again able to invest energies into other activities and people. The older the individual, the more likely he or she is to experience physical illnesses and complaints during the grieving process. Gramlich (1973) feels this is attributable to the fact that the elderly have gone through so many grieving processes previously. Therefore, a cumulative effect comes into play, and somatic complaints become more common.

It is very important to encourage the individual to go through the grieving process, to share guilt, hostility, and tears. Talking about the lost loved one is also healthy and therapeutic. Explain that the grieving process is normal and healthy. It must be experienced if the individual is to continue any kind of normal, satisfying living.

The length of a "normal" grieving process varies greatly, but six months might be considered average. However, shorter or longer periods are not abnormal. Periodic returns of grief at holidays, anniversaries, etc. are very common. Keeping this in mind can be helpful to both the health care professional and their clients when they experience periodic physical or emotional "downs." The cultural aspect of grieving must not be forgotten. Depending on degree of religious observance, it may be considered highly inappropriate for a widowed person, particularly a woman, to take any active part in social life for at least one year. In some European countries, the widow wears black for one year after the death of her spouse.

The loss of a close friend can be just as devastating to some individuals. A fair number of bereaved people actually express a fear of forming other close relationships because "it hurts too much" or "I can't go through that again." It has been this author's experience that those who can verbalize this do better physically and emotionally than those who deny their grief.

The family of the elderly person may need help from the nurse in planning where the person will live. The best situation is obviously one in which the whole family has planned and considered alternatives while everyone is able and rational. However, of the 17 percent of Americans 80 and over who live in nursing homes, very few had any part in the decision-making process. The routine assessment should include plans for such contingencies as the elderly person's getting sick, becoming the sole survivor of the family, or becoming unable to perform daily activities. Again, the emphasis should be on planning with and not for the individual, and on preplanning before there is a crisis. The elderly person ceases to see himself as a worthwhile individual when he loses all control of both activities and decision-making.

If a move to a nursing home is being considered, it is advisable that the elderly person see the facility before the move is made; a better adjustment and higher degree of satisfaction are more likely under such circumstances. The family and the individual need to be familiar with the Bill of Rights for Nursing Home Patients (Brown, 1977); this discusses personal liberty, the right to participate in medical decision making, the right to privacy, the right to control finances, the right to information, freedom of

Figure 13-4. *Activity is one of the most effective antidotes for depression. These men work on handicrafts at a day-care facility for the elderly. (Courtesy Durham Senior Citizen Council.)*

speech, religion, and assembly, and the right to protective measures during medical procedures. The elderly person and his family should also be made aware that there are over 100 legal services specializing in helping elderly people who are not treated fairly.

A routine assessment of the elderly person should include a question about what daily activities the client enjoys. This will give the health care professional some idea of the individual's interests. Eating, cooking, housework, and watching TV are among the most common activities for this age group.

Sexual activity and desire decrease to some extent in most aged individuals. If it does so

equally in both partners, there is no adjustment problem. Counseling becomes important when one partner is still very interested in sex and the other is avoiding it for whatever reason, resulting in frustration for both.

Wasow and Loeb (1979) interviewed 63 nursing home residents (27 males and 36 females) between the ages of 65 and 85. Eighty percent of the men and 75 percent of the women felt that older people should be allowed to have sex. Interestingly, when the question was changed to "should a person *your* age have sex?" 39 percent of the men and 53 percent of the women said "no." Lack of a partner was given as the major reason. The nurse needs to be aware, and to make it known to the client, that strong sexual needs are normal, but if desire has decreased or even ceased, that is also normal. If the nurse is not comfortable in talking about sex, she must at least know of someone to whom the individual or couple can be referred. Sexual needs may also be fulfilled through masturbation and homosexuality as well. Genevay (1978) offers some excellent suggestions for better understanding of both of these subjects as they relate to the elderly. She strongly advises that all health care professionals be aware of a variety of styles of sexual expression in order to become nonjudgmental and to be helpful to all aged people.

Carefully listening to elderly people reminisce can be enjoyable and an extremely valuable tool in learning how to best work with them. Their reminiscences will reveal likes, dislikes, and how they coped with developmental tasks in the past. Pleasant events, obviously, are not difficult to talk about. However, there may be some very painful experiences that still cause the client a great deal of anxiety. The client may need to be referred to a psychiatrist, especially if it is suspected that painful memories are responsible for somatic complaints such as headaches and "tight chest." Older people go through a "life review," and it is important to work through painful events as well as joyous ones in order to assure a peaceful aging. This also contributes to a decreased fear of death.

It is usually easy to get people to reminisce. If any prompting is required, it is helpful to ask about favorite foods, where the client grew up, how he celebrated holidays, what his mother did when he was sick or hurt, what he liked best about his job, and what he has enjoyed most since retirement.

Under normal conditions, the elderly person's memory and learning abilities are maintained in direct relationship to the amount of intellectual stimulation available. Memory loss is a frequent complaint, although, interestingly enough, studies have shown that people who

voiced these complaints often performed better on tests than those who claimed to have a good memory. Kahn (1975) and Zarit (1979) have drawn the conclusion from several studies and personal observation that poor memory arises from a self-fulfilling prophecy that loss of memory comes with aging. Depression is also a factor in loss of recent memory, although depressed elderly people often remember past events. Retraining programs of varying lengths and types have proven to be helpful in restoring memory function. The nurse working with elderly people should offer a great deal of reassurance that some forgetting is normal at any age, and should concentrate on relieving depression.

If memory loss is caused by organic brain syndrome, it is important for the individual and family to know that the syndrome is present and irreversible, and that its effects will come and go. However, the individual should still be encouraged to continue the activities he enjoys and to remain at home as long as it is safe. In fact, those who do live in a stimulating, familiar home environment tend to regress at a much slower rate.

HINTS ON WORKING WITH AGED ADULTS

It is safe to say generally of the very elderly that they need more time for most activities. If they are allowed adequate time, there are few things the elderly cannot do. However, as obvious as this would seem to be, time is often the thing least frequently given to them.

Conversations are often missed by the elderly because of the speed with which they are carried out. The elderly take a little longer to assimilate all stimuli, verbal or other. In conversations, they take longer to process the information and a little longer to produce the answer. The nurse

Question	General Areas of Assessment
"Tell me about your. . . ."	
1. Activities upon arising	Quantity and quality of sleep; elimination; degree of motivation to start new day; personal hygiene; medical problems; exercise habits.
2. Home	Number of occupants and relationship; type of plumbing (indoor/outdoor); stairs; safety factors; socioeconomic status.
3. Daily activities	Exercise; recreational activities; hobbies; friends; disabilities; TV and other sedentary habits; transportation; pets; religious activities; medical problems; napping patterns; emotional stability; reaction to crisis/stress.
4. Meals	Nutrition; likes and dislikes; impact of socioeconomic status on nutritional status; number of meals and snacks per day; alcohol consumption; fluid intake; medical problems associated with meals.
5. Bedtime and sleeping	Bedtime hour; sexual partner and activities; use of hypnotics; orientation; confusion.

Figure 13-5. *Modigh's Multipurpose Assessment Tool (MMAT).*

must bear this in mind when conducting even the briefest of conversations with an elderly person; if she asks "How are you?" and speeds away before the client can respond, she will certainly not convey sincere concern.

Physicians and others often make the mistake, in talking with the elderly client, of asking a question and, if no answer is given immediately, they ask another question. This further slows down the communication process and may even lead to incorrect answers. The client's anxiety level is raised and the communication process is further hindered. In extreme cases, such exchanges may cause a perfectly normal person to be mislabeled as senile or confused.

When conducting a health assessment of an elderly client, the nurse should not try to elicit a complete history in one sitting; better results can be obtained in several short sessions. Asking the client to describe a typical day is one excellent way of assessing many facets of the client's health, and is usually enjoyable for the client also (see Figure 13.5). The older the client—or if he is disabled in any way—the more important it becomes to take one's time

and conduct all conversations slowly. Blazer feels that misunderstandings that arise from poor communication patterns can lead to disordered behavior, and even goes as far as to say, "this country's malpractice crisis probably results more from poor communication between physicians and patients than from incompetent medical care" (Blazer, 1978). Conversations about sensitive subjects such as the client's illness or the loss of a loved one will provoke anxiety and will therefore require still more time. Impaired hearing and vision, as well as the caution with which elderly people approach new people and events, are other factors that necessitate taking more time for communication.

The elderly require more time for activities as well. A feeling of being rushed increases everyone's anxiety, but this is even more pronounced in the elderly. Again, this can lead to confusion and inappropriate behavior. Many a client has been mislabeled "confused" or "uncooperative" because of being told rapidly to do three or four different things at the same time. This occurs frequently in a stressful environment such as an emergency room, doctor's office, or treatment room.

Most very elderly people have a routine for their daily activities. Many for whom last-minute changes are difficult are often mislabeled "inflexible." For example, an elderly woman may be thrilled that her granddaughter is going to pick her up at 5 p.m. for Sunday dinner; when the granddaughter calls at 3 p.m. and says "I'm right near here; I'll pick you up in a few minutes, OK?" it's another story. The grandmother was planning to be ready and looking her best at 5 o'clock. The granddaughter can't understand why her grandmother cannot get ready in ten minutes, and an unnecessarily stressful situation results.

Given time, the elderly can be quite flexible, especially if they were flexible when they were younger. It is helpful to remind the family to give the elderly person as much notice as possible when changes in plans are necessary. Given time, courtesy, and respect, most people can adjust to a necessary change.

When the elderly are in an institution, it is particularly important to allow them time to carry out as many of their daily activities as possible. Although the nursing assistant or nurse could perform the task more quickly, the advantages of allowing the person to do for himself what he is able to do far outweighs the time factor. The resultant feeling of pride and control is worth a great deal toward maintaining self-esteem and self-respect, aside from the physical and emotional benefits of the activity itself.

The health care professional who is most successful in aiding the elderly client is the one who can best understand him. That means that, in addition to the general theoretical knowledge discussed here, an attempt to view the world through the client's eyes must be brought into play. Role-playing through a 24-hour experience of "being a resident" in a nursing home can serve as a marvelous teaching tool. By wearing slightly dirty eyeglasses and cotton in the ears, and by impairing one's own mobility for 24 hours, the nurse can get some idea of what the world is like for an elderly person.

Conveying respect and concern for the elderly client is vital. This includes calling people by the name and title they prefer. If the individual is in an institution, allow as much privacy and control as possible. Knocking on the door before entering may seem like an insignificant act, but it not only shows respect for the individual's privacy, but also gives the individual a sense of some control over his environment. This author recalls a 92-year-old resident who more often than not would say, "just a minute, please" in response to a knock on his door during rounds. One day the author asked him if this was an inconvenient time for him to be seen. He explained that he asked people to wait before entering his room simply in order to exercise his control of one of the few things he still *could* control.

SUMMARY

Of all people over 80 years of age in the United States, 17 percent live in nursing homes; the others continue to live in their own homes or with relatives or friends. The daily activities of 15 to 20 percent of those over 85 are limited to some degree by health problems, cardiac problems being the most common health deficit.

A decrease in height (caused by changes in the musculoskeletal system) is seen in almost all elderly people; it is generally more pronounced in women. Nearly 90 percent of those in the over-80 population have some degree of cataract, but over 80 percent retain good vision throughout life. A majority of people in the same age group suffer a significant hearing loss.

The level of general intellectual functioning need not decline among normal, healthy elderly people. Some decline in all types of memory is seen in very old age, but, as with nearly all aspects of growth and development, individuals vary greatly.

Common health problems of the aged adult include gastrointestinal complaints (usually of functional origin), essential hypertension, and bronchopneumonia. Arteriosclerotic heart disease is present to some degree in all elderly people. Cardiac arrhythmias (which frequently lead to congestive heart failure) are common. Ronald Cape has proposed that falling (or loss of mobility), confusion, incontinence, impaired homeostasis, and iatrogenic disorders comprise a syndrome of aging, which he refers to as the O Complex.

Aside from these health problems, anticipatory guidance should also address such matters as skin care, the selection of a hearing aid (if indicated), planning an exercise program, and coping with widowhood (and the loss of significant others) through planned social activity. If a move to a nursing home is being considered, it is best for the client to see the home first.

This age group provides a stimulating challenge to the health care professional. His or her role in working with the elderly is one of teaching, counselling, being an advocate as well as learning from so many fascinating people with a wealth of life's experiences. The nurse can, indeed, be instrumental in contributing to the fulfillment of the elderly's life satisfaction.

REFERENCES

Amberg, J. R., and Zboralsko, F. F. Gallstones after seventy. *Geriatrics*, 1963, *20*, 539.

Anderson, F. *Practical management of the elderly*. Philadelphia: J. B. Lippincott, 1976.

Asch, R. H., and Greenblatt, R. B. Geriatric Endocrinology. In Reichel, W. (Ed.), *Clinical aspects of aging*. Baltimore, Md.: Williams & Wilkins, 1978.

Atchley, R. C. *The social forces in later life*. Calif.: Belmont, Wadsworth, 1977.

Baltes, P. B., and Schaie, K. W. The myth of the twilight years. *Psychology Today*, March 1974.

Birren, J. E. Research on aging. A frontier of science and social gain. In Brantl V. M., and Brown, Sister M. R. (Eds.), *Readings in gerontology*. St. Louis: C. V. Mosby, 1973.

Blankenhorn, D. H. Reversibility of latent atherosclerosis studies by femoral angiography in humans. *Modern Concepts in Cardiovascular Disease*, 1978, 47, 79–84.

Blankenhorn, D. H., Brooks, S. H., and Selzer, R. H., et al. Assessment of atherosclerosis from angiographic images. *Proceedings of the Society of Experimental Biological Medicine*, 1974, *145*, 1298–1300.

Blazer, D. G. II. Techniques for communicating with your elderly patient. *Geriatrics*, November 1978, 79–84.

Blazer, D. G. II, and Friedman, S. W. Depression in late life. *Family Physician*, November 1979, 91–96.

Bortz, W. M. II. Effects of exercise on aging—effect of aging in exercise. *Journal of the American Geriatrics Society*, February 1980, *28*, 49–51.

Botwinick, J. *Cognitive processes in maturity and old age*. New York: Springer, 1973.

Botwinick, J. *Aging and Behavior*. New York: Springer, 1973.

Brody, H., and Vijayashankar, N. Anatomical changes in the nervous system. In Finch, C., and Hayflick, L., (Eds.), *The biology of aging*. New York: Van Nostrand-Reinhold, 1977.

Bromley, D. B. The psychology of aging. In *The effects of aging on adult intelligence*. (2nd ed.). London: Penquin, 1974.

Brosman, S. A. Benign prostatic hypertrophy: When should you consider prostatectomy for your patient? *Geriatrics*, April 1978, 25–34.

Brown, R. N. A bill of rights for nursing home patients. In *Focus-Aging*. Guilford, Conn.: Dushkin Publications Group, 1978.

Bullough E. S. Ageing of mammals. *Nature*, 1971, *229*, 608–610.

Butler, R., and Lewis, M. *Aging and mental health*. St. Louis: C. V. Mosby, 1973.

Cape, R. *Aging: Its complex management*. New York: Harper & Row, 1978.

Cermak, L. S. *Improving your memory*. New York: McGraw-Hill, 1975.

Chughtai, M. A., Cape, R. D. T., Harding, L. K., and Mayer, P. P. Mean cerebral transit time in demented and normal elderly persons. *Age-Ageing*, 1977, *6*, 248–252.

Clark, A. N. G. Factors in fracture of the female femur: Clinical study of the environmental, physical, medical and preventive aspects of this injury. *Gerontological Clinician*, 1968, *10*, 287–290.

Comfort, A. *A good age*. London: Mitchell Beazeley, 1976.

Cooper, H. L. Irritable bowel syndrome: Diagnoses by exclusion. *Geriatrics*, January 1980, 43–46.

Craik, F. I. M. Short-term memory and the aging process. In Talland, G. A. (Ed.), *Human aging and behavior*. New York: Academic Press, 1971.

Cullinan, T. R., Gould, E. S., Silver, J. H., et al. St. Bartholomew's Hospital Medical College, London: Visual Disability and Home Lighting. 1979, *Lancet 1*, 642–644.

Cutler, R. G. The evolution of longevity. *Geriatrics*, January 1980, 98–104.

DeGroot, L. J., and Stanbury, J. B. *The thyroid and its diseases*. New York: John Wiley & Sons, 1975.

deVries, H. A. Tips on prescribing exercise regimens for your old patient. *Geriatrics*, April 1979, 75–81.

Diamond, M. C. Aging and cell loss: Calling for an honest count. *Psychology Today*, September 1978.

Encyclopedia Britannica, Micropedia, Vol. VIII, 1977.

Fleiss, J., Gurland, B., and DesRoche, P. Distinctions between organic brain syndrome and functional psychiatric disorders, based on the geriatric mental state interview. *International Journal of Aging and Human Development*, 1976, 7, 323–330.

Freehafer, A. Injuries to the skeletal system of older persons. In Reichel, W., (Ed.), *Clinical aspects of aging*. Baltimore, Md.: Williams & Wilkins, 1978.

Genevay, B. Age kills us softly when we deny our sexual identity. In Solnick, R. (Ed.), *Sexuality and aging*, Los Angeles: University of Southern California Press, The Ethel Percy Andrus Gerontology Center, 1978.

Geriatrics, October 1979, (abstract), 25.

———. Abstract of report from KTTV, Fort Worth–Dallas. December, 1979, 21.

Getze, L. H. New idea: Share-a-home. *Modern Maturity*, February/March, 1980.

Gramlich, E. P. Recognition and management of grief in elderly patients. In Brantl, V. M. and Raymond, Sister M. (Eds.), *Readings in gerontology*, St. Louis: C. V. Mosby, 1973.

Gray, R. V. The psychological response. In *Dealing With Death and Dying*, Nursing 1976 Skillbook. Jenkintown, Pa.: Intermed Communications, 1977.

Gurin, G., Veroff, J., and Feld, S. *Americans view their mental health: A national interview study*. New York: Basic Books, 1969.

Harris, R., and Frankel, L. J. *Guide to fitness after fifty*. New York: Plenum, 1977.

Hudis, M. M. Dentistry in the elderly. In Reichel, W. (Ed.), *Clinical aspects of aging*, Baltimore, Md.: Williams & Wilkins, 1978.

Hulicka, I. M., and Weiss, R. L. Age differences in relation as a function of learning. *Journal of Consulting Psychology*, 1965, 28, 125–129.

Kahn, R. I., Zarit, S. H., and Miller, N. M., et al. Memory complaint and impairment in the aged. *Archives of General Psychiatry*, 1975, 32, 1569–1573.

Kaluger, G., and Kaluger, M. F. *Human development: The span of life*. St. Louis: C. V. Mosby, 1979.

Kent, S. Regression of Atherosclerosis. *Geriatrics*, December 1979, 78–85.

Kübler-Ross, E. *On death and dying*. New York: Macmillan, 1969.

Lerner, M. When, why and where people die. In *Focus-aging*. Guilford, Conn.: Dushkin Publications Group, Inc., 1978.

Lorayne, H., and Lucas, J. *The memory book*, New York: Ballantine, 1974.

Mead, M. Dealing with the aged, a new style of aging. In *Focus-Aging*. Guilford, Conn.: Dushkin Publishing Group, Inc., 1978.

Milne, J. S., Williamson, J., Maule, M. M., and Wallace, E. T. Urinary symptoms in old people. *Modern Geriatrics*, 1980, 2.

Moody, R. A. *Life after life*. New York: Bantam, 1975.

Neugarten, B. Developmental perspectives. In Brantl, V. and Brown, Sister M. R. (Eds.), *Readings in gerontology*, St. Louis: C. V. Mosby, 1973.

———. Personality and the aging process. *Focus-aging*. Guilford, Conn.: Dushkin Publishing Group, 1978.

———. The future and the young-old. *The Gerontologist*, 1975, 15(1), 4–9.

Pfeiffer, E. A short portable mental status questionnaire for assessment of organic brain deficit in the elderly patients. *Journal of American Geriatric Society*, 1975, 231, 433–441.

Portnoi, V. A. Abstract in *Geriatrics*, November 1979, 34, 22.

Raymond, B. J. Free recall among the aged. *Psychological Report*, 1971, 29, 1179–1182.

Reichel, W. *Clinical aspects of aging*. Baltimore, Md.: Williams & Wilkins, 1978.

Rosenberg, G. S. *The workers grow old*. San Francisco: Jossey-Bass, 1970.

Saxon, S. V., and Etten, M. J. *Physical changes and aging*. New York: The Tiresias Press, 1978.

Selye, H. *The stress of life*. New York: McGraw-Hill, 1956.

Serock, K., Seefeldt, C., Jantz, R. K., and Galper, A. As children see old folks. In *Focus-aging*, Harold Cox (Ed.), Guilford, Conn.: Dushkin Publishing Group, 1978.

Shanas, E., Townsend, P., Wedderburn, D., Frees, H., Milhoj, P., Stenhouweer, J. *Older people in three industrial societies*. New York: Atherton Press, 1968.

Sheldon, J. H. On the natural history of falls in old age. *Britain Medical Journal*, 1960, 2, 1685–1690.

Schmavonian, B. M., and Busse, E. W. The utilization of psychophysiological techniques in the study of the aged. In Williams, R. H., Tibbets, C., and Donohue, W. (Eds.), *Process of aging—social and psychological perspectives.* New York: Atherton Press, 1963.

Shock, N. W. Physiological theories of aging. In Rockstein, M. (Ed.), *Theoretical aspects of aging.* New York: Academic Press, 1974.

Sklar, M. Gastrointestinal disease in the aged. In Reichel, W. (Ed.), *Clinical aspects of aging.* Baltimore, Md.: Williams & Wilkins, 1978.

Smith, J. M., and Birren, J. E. Research on aging, a frontier of science and social gain. In Brantl, V. N., and Brown, Sister M. R. (Eds.), *Readings in gerontology.* St. Louis: C. V. Mosby, 1973.

Solnick, R. L. *Sexuality and aging.* Los Angeles, Calif.: University of Southern California Press, The Ethel Percy Andrus Gerontology Center, 1978.

Statistical abstract of the United States, 1978. (99th ed.). Washington, D.C.: U.S. Dept. of Commerce, Bureau of the Census, 1978.

Troll, L. E. The family of later life: A decade review. *Journal of marriage and the family,* 1971, *33,* 263–290.

Wasow, M., and Loeb, M. In *Geriatrics,* October 1979, 16.

Weissman, M. M., and Myers, J. K. Affective disorders in a U.S. community: The use of research diagnostic criteria in an epidermiology survey. *Archives of General Psychiatry,* 1978, *35,* 1304–1311.

Williams, E. I., Bennett, F. M., Nixon, J. V., Nicholson, M. R., and Gabert, J. D. Sociomedical study of patients over 75 in general practice. *Britain Medical Journal,* 1972, *2,* 445.

Zarit, S. H. Helping an aging patient to cope with memory problems. *Geriatrics,* April 1979, 82–90.

14

Loss, Death, and Grief

PATRICIA SHORT-TOMLINSON

The final stage of life is death. The trajectory of development ceases only at the moment of death; the nurse has the potential to play a powerful role in this conclusive phase of human development.

This chapter examines the concept of personal loss, which can be understood as any permanent loss of a significant person that requires adaptation through a grieving process. Emphasis will be placed on the family and the family's response to loss.

The experience of loss occurs naturally throughout the life cycle both for the individual and the family. Gains achieved at each stage of development are accompanied by certain losses or environmental benefits one must give up.

When a child starts to walk, he or she loses the comfort of being carried; the young adult beginning a first job loses a certain degree of dependency. Thus, life can be understood as the ebb and flow of gains and losses. The coping mechanisms the individual learns within the family, and the family's collective coping style, have much to do with the resilience and adaptability of the individual in the face of a major loss such as death.

For the dying, approaching death means coming to grips with the loss of life and leaving others behind; for the family or others it is dealing with being left and the meaning of the life they are losing or have lost. It is in this context that the grieving process occurs.

Grief is a somewhat predictable process from which new understanding and growth can arise. However, a family's acceptance of the inevitability of a member's death, and the resultant grief process, seldom synchronizes with the client's own perception. It is in dealing with this discrepancy that the nurse may intervene to allow both the family and client to deal healthily with death and the effects of loss. The crisis of death and loss and the process of grieving involves three perspectives: that of the dying person, that of the bereaved family, and that of the care-giving system.

By the very nature of the profession, nursing has always been linked with death, loss, and grief. Caring for the dying client, assisting with life-saving or life-prolonging measures, responding to the bereaved family immediately after a death, caring for a client with a potentially life-threatening disease, or helping a client with a decision about abortion are all common nursing actions. These activities involve sensitive physical care, emotional support, keen observation, teaching, counseling, and the coordination of other services and facilities.

However, although nursing provides a care-giving system, the nurse often fails to respond to the needs of the client or family for a number of reasons, including the organization of the health care system, societal norms, and myths, misconceptions, and insecurities held by the nurse herself. Because of these forces, in situations signaling the end of life, the nurse often copes in stereotypical ways not beneficial to the client or family. Some of these are to treat the event without emotional involvement, to overidentify with the client, or to handle the situation as a purely technical challenge.

Why are these common mistakes perpetuated? Lack of understanding and knowledge often lead to rigid and noncreative solutions in the face of clinical demands. One cannot give sensitive, therapeutic care in such critical clinical situations without understanding the significant dynamics that accompany loss.

GRIEF, MOURNING, AND BEREAVEMENT

Graves's psychodynamic interpretation of loss defines grief and mourning as follows: "Grief is the sequence of affective, cognitive and physiological states that follows directly after an irretrievable loss; mourning . . . is a complex and lengthy process that begins with denial of loss . . . and proceeds toward acceptance of the loss and integration of multiple intrapsychic shifts" (1978, p. 875). He also suggested that grief behavior in both humans and primates, as discussed by Bowlby (1969), is characterized by needing and seeking nurturing support from the immediate environment. This lends credence to the normative nature of the need for a strong support system during the acute grief phase. The term *bereavement* is most often used to describe the total psychological and physiological response to the loss of a significant relationship. *Mourning* is also defined as a general response to loss that enables one to give up attachment to a loved one with certain prescribed rituals or behaviors set in a cultural frame of reference, whereas *grief* may be understood as the acute psychological state, often accompanied by physiological symptoms, that follows personal loss. These physiological symptoms are usually similar to those of acute anxiety, and way include appetite and sleep disturbances, feelings of restlessness, inability to concentrate, and a sense of impending doom.

It is important here to digress and briefly consider some sociocultural implications in studying grief and mourning, because nurses participate in the structure and some of the rituals that characterize the modern death system.

Across cultures, social responses to death are characterized by features of the particular society in which the death occurs. Kastenbaum, in *Death, Society and Human Values*, introduced the concept of the "death system." This includes the people providing death-related services, places in which death and death-related behaviors take place, a structure or ritual for taking care of the dying, an inculturated warning and prediction system to prevent deaths by safeguards and effective interventions, a means of handling loss and integrating a death through the grieving process, and a means of disposing of the dead (Kastenbaum 1974).

Each society develops its own death system. For example, a society works out ways of handling responses to loss by developing "norms" of mourning. In turn, those norms are incorporated by families, and ultimately by the individuals raised in that family.

It is evident that what characterizes a society will affect the way in which the death system is manifested. Contemporary American society is no exception; although its complexity and heterogeneity make generalizations risky, for purposes of this discussion, the following cultural characteristics need to be considered when looking at individual and familial responses to loss and the relationship of the nursing role to that experience:

Overdenial. Our society controls death by institutionalizing it. Most deaths today occur in some form of hospital (Mauksh 1975) and funerals, by and large, are conducted by funeral "homes." Consequently, two phenomena occur: the family is separated from the handling of illness and death, reducing the impact and exposure to that part of life; second, handling of the dead and dying has been routinized by hospitals, thereby reducing the effect of the crisis of death by orderly procedures and organized efforts to hide the facts of death from patients and family.

High Expectations of Technology. The present purpose of the hospital, in contrast to its earlier function as a place for the dying, is to cure and restore. Because of this, the institution often views a death as a failure to fulfill its primary purpose (Mauksch 1975). This view causes a dilemma for the health profession. The medical-legal controversy surrounding prolonging life with technology underscores a professional eth-

ical burden. Often, the nurse or physican must make humane decisions that are inconsistent with institutional goals. Ultimately, patient cure is often equated with success while client death is perceived as failure. Furthermore, the expectation of consumers that nurses and physicians cannot make mistakes puts an awesome burden on the latter, which may lead to increased anxiety and poor decision making during critical periods in the care of the terminally ill (Fig. 14-1).

Another internal conflict affects the care of patients and families facing a death in a hospital setting. The physician is socially sanctioned as the "curer" and comforter, but is not prepared by education or tradition in the specific skills of supporting a client or family in the dying process. Until the early 1970s, it was expected that the physician would deal with a dying client or inform the family of the death. Although some nurses may have been relieved to avoid the responsibility, the result was a lack of involvement with the dying client or the client's family. This is beginning to change as nurses accept new responsibilities, which include assisting clients in dying peacefully. The nurse thanatologist is a new specialist in nursing who is emerging to meet the need of terminal care in hospitals.

Few Mourning Customs. In the past, ritualized mourning shared by the extended family and the community helped diffuse the grief. These rituals, exemplified by the Irish wake and other ethnic customs, allowed legitimate expressions of anger, guilt, loss, and a reintegration of residual infantile feelings re-awakened at the time of loss.

Now, mourning and the sense of loss are quite often not widely shared. This individualization and deritualization of bereavement make for more serious problems of adjustment. Thus, paradoxically, while death has become less disruptive for society, its consequences may be more severe for the individual.

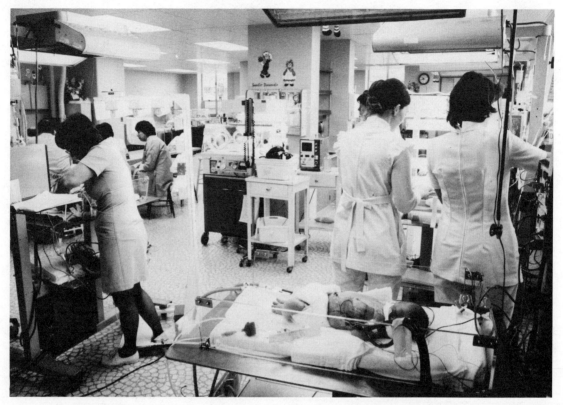

Figure 14-1. *A neonatal intensive care unit. The expectation that modern medical technology can restore health in every instance complicates both the role of the health care worker and the response to loss on the part of the individual and the family. (Courtesy University of Oregon Health Science Center.)*

Is our death system functional? Many believe it is not. For example, some of the technological advances since the early 1960s have produced a profound change in the nature of dying by greatly lengthening the average amount of time between the onset of a fatal illness and the end of life. The prolongation of life has increased the problems of personal dignity, control over one's fate, and isolation (Kübler-Ross 1975). Both the development of the hospice concept and other experiments in home care for the dying have resulted from the increased scrutiny and criticism of the dehumanizing effect of medical technology. The goal of the hospice movement is to provide an alternate place for

a person to spend his or her remaining time free of pain. The aim of such care is to make the body comfortable so that a person is free to prepare for death mentally and spiritually. Another goal is to enrich what people have left of their lives and to return control to the patient to take charge of important decisions.

Studies of home care for dying children and adults indicate economic and psychological advantages to the client and the family. Greater involvement, physical activities, and less loss of personal dignity are among the benefits (Martinson 1978; Kassokian et al. 1979).

It is clear that nurses play a central role in much of the American death system through their role in providing services, their relationship to the places or institutions in which death takes place, interventions to prevent death, and their primary care role in working with the grieving family. Despite the salience of the role of nursing, there is limited systematic application of theory-grounded practice in this clinical function.

THEORETICAL FOUNDATIONS OF CARE OF THE DYING AND THE BEREAVED

A theoretical framework is useful in examining complex phenomena in that it provides an organizational structure within which one can make predictions and justify interventions. This is particularly relevant to nursing. Biological theories of disease and health have always guided the practice of nursing; more recently, theories of stress, communication, learning, development, and interaction have provided the framework for an evolving theory of nursing.

Constructs from four theoretical concepts can be used to form the conceptual framework for clinical practice with the dying patient and the patient's family. These are attachment theory, developmental theory, crisis theory, and the stage theory of dying (Kübler-Ross 1975).

Attachment Theory

In order to understand the concept of loss as it relates to death of a significant person, some knowledge of the development of attachment is necessary. The attachment and loss concept is related to the idea of primary object attachment in the early mother-infant relationship first explored by Freud and later elaborated on by Bowlby and others (Bowlby 1969). In brief, early attachment should be considered in two ways: first, it provides the safety needed for survival at the most primitive level, and second, it sets the stage for later attachment patterns

played out in subsequent relationships. The early love relationship characterized by trust and dependency is transferred to other relationships as the child grows. By midchildhood the primary attachment has expanded to include other family members as well as peers, pets, toys, and persons deemed significant by the child.

During adolescence a somewhat significant change occurs with the primary attachment being gradually withdrawn from family relationships and transferred to someone outside the family group, quite often a person of the opposite sex during the sexual awakening. Thus, a secondary attachment is established. The effect of this is a reinvestment of trust and dependency in another while maintaining some of the earlier emotional attachment to the primary family. Attachment is inherently involved with care-giving and nurturance.

Loss of a significant other is one of the events most dreaded because it is linked to basic primary needs described above. Most people experience personal loss as a threat to basic security and survival. As a demonstration of this, the response to grief is often characterized by the type of autonomic nervous system reactions that are usually elicited when survival is threatened. Personal loss may be accompanied by feelings of extreme loneliness and a sense of abandonment because of the real and symbolic loss of a dependency object. Most often the loneliness following loss persists for some time because trust and dependency are not easily or quickly transferred to another relationship.

These ideas are central to an understanding of concepts of personal loss. Personal relationships increase throughout childhood and early adulthood, then stabilize and finally decrease during aging because of the loss of important friends through death. This curvilinear relationship has important implications in understanding individual response to death and the social network of the dying.

To summarize, primary attachment is that

first relationship experienced by the infant toward its caretaker, usually the mother. It is characterized by dependency and feelings of security when that person is around. Secondary attachment is the relationship to subsequent important people in a child's life but primarily that person on whom one becomes more dependent and feels most secure; quite often that person is a mate or another family member. Secondary attachment figures have a curvilinear relationship to age. Loss of a significant attachment figure is most often accompanied by feelings of severe loneliness and fears that one may not be able to survive alone. Physiological symptoms of acute anxiety are also often present.

Developmental Theory

The crisis of loss is experienced differently during different stages of life as perceptions and needs change. Developmental theory is based on the assumption of a maturation progression that is biologically determined and environmentally influenced. The idea that a readiness state must precede learning or developing a new ability is a central construct.

Two developmental perspectives on dying must be considered. The first perspective, related to one's own sense of mortality, changes as one ages. During childhood and youth, death is a concept that has very little personal meaning. Most energy in these years is spent on development. Up to the age of 7 a child has little cognitive ability to abstract the concept of death.

The 7-year-old engages in what Piaget calls "preoperational cognition," marked by magical thinking and a minimal ability to abstract. Since death or nothingness is an abstract notion, it is not surprising that a child of this age would have difficulty in seeing death as final.

However, this quality of "magical thinking," coupled with the egocentrism that is normal at this age, can have severe effects should a parent or sibling die. The child quite often feels more responsible and guilty for the death of a significant family member than does an older child.

The child's mode of thinking changes rather dramatically at about 7 when he moves to what Piaget calls the "concrete operational stage." This change involves a tremendous advance in a child's ability to abstract, to get outside of himself in the thinking process, and to use language to transmit ideas and information. This change allows the child to recognize the permanence of death. Thus, the older child not only understands his or her own mortality but will experience more acute grief at the death of a parent or sibling. After about age 9, and certainly from age 12 on, the child's ability to abstract and classify enables him to imagine many possibilities; should a death occur or be impending in the family, the child will undoubtedly want and need to talk about it.

In adulthood, the perception of one's relationship to time and death begins to change during midlife, becoming most prominent when parents age and die. This undoubtedly erodes one's own sense of immortality. It is hypothesized that, because of a recognition of death's increasing proximity, people during middle age often change their life-style to engage in activities put off during the busy years of child rearing. As further aging occurs, withdrawal from activities and friends may occur in an unconscious preparation for death.

The second developmental perspective is that held by society. In general, most people regard the death of a young person as untimely and tragic, and the death of an elderly person as less traumatic. A child's death, particularly, is viewed with special societal alarm, in part, because it respresents the loss of a life yet to be fulfilled (Fig. 14-2). The age of the dying member is also a significant factor in the response of the family. In American culture, children are greatly enough valued that death is almost universally held to be a tragedy, while death of the aged is often experienced with relief. Death of children carries with it other meanings; of the

Figure 14–2. *Mourning Parents, by Käthe Kollwitz. The death of child is viewed with special alarm in societies that value youth more than old age. (Courtesy Tourist Bureau, City of Cologne.)*

more-than-100,000 deaths of children under 14 occurring annually, the majority are caused by accidents and infections, for both of which parents tend to assume responsibility. As a result, parental guilt is one of the inherent features of most childhood deaths.

Crisis Theory

Crisis theory elaborates on developmental theory by focusing on certain developmental landmarks. Crisis theorists have speculated that there are predictable reactions to certain events that people experience throughout their lifetimes. Some of these predictable and universal events are called maturational, or developmental, crises. These are notable because they mark the transition points common in most lives such as adolescence, childbearing, and middle age.

A second type of crisis, proposed by the same theorists, are those known as situational crises. These, sometimes referred to as "nonnormative crises," simply refer to those unexpected events in life for which one is not prepared. It is hypothesized that these crises cause changes that lead to a disorganization of previous levels of function and an inability to use customary methods of problem solving, with an attendant potential for further growth and maturation. Those crises that entail personal loss include unexpected death, divorce, and parental abandonment. In this case, a distinction is made between death as it occurs in the last developmental stage and death as it occurs out of that normal sequence. The death of an aged person is less of a crisis because expectation and normative sequencing allow preparation. However, the effect of the loss of a family member in either case is a psychological crisis. Old roles have to be examined and reassigned. The change taxes psychological resources, and the loss may be perceived as a very real threat to life goals of individual family members.

Stress theory has been appreciably elaborated on by theorists interested in the reactions of individuals and families to the stress of loss and death. Hill developed the major concepts following his study of war separation and reunion. His ideas, commonly called the ABCX theory, have been further elaborated on by Lindeman (1965) and Lukton et al. (1974), but Hill's basic concepts have remained the same. Briefly, they are:

A (the event)—interacting with B (the crisis-meeting resources)—interacting with C (the definition of the event)—produces X (the crisis). The second and third determinants—the family and individual resources and definition of the event—lie within the family itself and must be seen in terms of the family's structures and values. The hardships of the event, which go to make up the first determinant, lie outside the family and are an attribute of the even itself (Hill, 1958, 141).

In a much simplified form, the basic implications of the propositions in this model for understanding the crisis of death are as follows:

1. The length of time and amount of preparation either an individual or family has correlates with the amount of impact the event (A) of death or terminal diagnosis has in producing a crisis.
2. The intrapersonal strength and interpersonal network (B) will determine the reaction to the death or terminal diagnosis.
3. The interpretation and meaning of death (C) will interact with the intra- and interpersonal resources to determine the degree of the crisis experienced by the individual or family.

More recently, Burr (1973) has strengthened Hill's original formulation with a deductive model for families. His theory enriches our understanding of the two salient features of crisis as it affects families, i.e., the family's vulnerability to loss and its regenerative power. A number of Burr's propositions are helpful in considering the family's resonse to death and loss.

Burr has hypothesized that the amount of change in the family system is positively correlated with the amount of crisis but is mediated by the following principles, which influence vulnerability and recovery: (1) The family's definition of the severity of the change (or loss) and the amount of time of anticipation influence family adaptability, integration, and cohesion of family members with regard to personal interaction and support; and (2) A feeling of guilt for the loss and an authoritarian family structure increase vulnerability to the stress event.

In this light one can expect that a sudden death could precipitate a crisis more easily than an expected death, and that an egalitarian family structure allows more reciprocal support.

Burr further hypothesizes that while vulnerability and regenerative power covary, certain situational factors affect recovery power but have only an indirect effect on vulnerability. For example, he concludes that the following increase the recovery potential in a family: extended family support, the degree to which family members care equally for each other, a consultative form of decision-making, and overall satisfaction, happiness, and stability in the marital situation.

In summary, understanding that both individuals and families go through a very disorganized and disruptive time during a death is important. The degree to which families may either weaken or recover and become stronger is related to individual adaptability, family structure, and an adequate support system.

Stage Theory of Dying

Closely linked to crisis theory is the so-called stage theory of death and dying. Most theorists agree that the grief process has a number of identifiable characteristics in the acceptance of one's own impending death or adjusting to the death of a significant other. While these stages are described differently by different experts, all agree in principle that one has to experience all phases to reach resolution. The process is called "grief work," and it is characterized by three stages: shock and denial, disorganization, and reorganization and resolution. These, in turn, are accompanied by somewhat typical behaviors. Kübler-Ross (1975) outlined five phases the dying person goes through; they have since been applied to the grief process of survivors. A brief description of these five phases follows.

The first stage is *denial*, a refractory period during which the bereaved person needs time to begin to incorporate the loss. It is an unconscious behavior precipitated by the need to maintain a particular relationship, marked by such statements as "No, I don't believe it" or "It can't be," or by a refusal to hear or understand what is being said.

The second stage, *anger*, represents a resistance to the idea of the loss. This is the most socially unacceptable phase, particularly for the survivor of death, and because of cultural norms dictating that one should not be angry with the dead or dying. Anger is unacceptable also in that the expression of hostility is often seen as "immature" behavior.

Bargaining characterizes the third stage; this represents an attempt to postpone the loss. Bargaining behavior is evidenced in such acts as seeking opinions or actually removing the family member from current medical care.

Depression often follows as the fourth stage; this accompanies facing the loss and the realization that the deceased is no longer available. Overwhelming feelings of loneliness, withdrawal, and periods of crying characterize this stage.

Finally, the fifth stage, *acceptance of the loss*, is reached. Affective response becomes more normalized, physiological reactions such as sleep and appetite disturbances abate, and social contact with others may resume.

The stage theory has been criticized for failing to account for distinctive differences in clients' age, sex, ethnicity, interpersonal relationships, and personality. Furthermore, many practitioners using Kübler-Ross's ideas focus on the order of the stages, failing to understand Kübler-Ross's own caution that an orderly progression does not always occur. The most serious criticism of Kübler-Ross's theory; however, is that it fails to draw a distinction between what does happen and what should happen (Kastenbaum 1977). If nurses assume that clients must reach the acceptance stage, or if higher value is placed on reaching acceptance than on dealing with the anger, there may be an impulse to push the patient along rather than allow the resolution to occur as it will, lending support and understanding as the emotional process unfolds.

Grief as a developmental process cannot be altered any more than any other developmental task. For instance, the child walks when he or she is developmentally able. The role of the parent is to create a safe environment and to be emotionally supportive. Similarly, the primary role of the nurse is to support the client in living through the grief process at his own pace. The inclination to push clients and families to an acceptance stage may be motivated by two factors. First, the family's attainment of the acceptance stage may be regarded as a "success," and this provides satisfaction to the nurse; second, clients' acceptance of death is far easier to deal with than either depression or anger.

It must be stressed that these categories are not phase-specific, i.e., one does not necessarily move from one stage to the next as if passing from one school grade to another. Rather, these stages should be seen as *primarily* characterized by the behaviors associated with them while behaviors of the other may also emerge or re-emerge.

The nurse may be asked how long it takes to adjust to a loss; estimates range from three months to a year or longer for the most acute response to loss. At least three things seem certain:

1. The amount of time needed to resolve grief is dependent upon the kind of support and intervention available.
2. The acute phase of grief occurs in the first 6 to 8 weeks. Later manifestations may reappear in the survivor, especially in the first year around significant "anniversary" times represented by any day or time holding special meaning.
3. "Grief work" may not be completed and a person may remain fixed in a behavior pattern such as is characterized by one of the phases.

It is of considerable consequence to understand the phases of the grief process because the behavior with which a nurse deals may be indicative of the stage of recovery or a failure to complete the grief work.

EFFECT OF FAMILY STRUCTURE ON RESPONSES TO DEATH

Certain properties of the nuclear family structure, which is predominant in American society, can be used to help predict the vulnerability of the family at the time of death. Both the general properties of families and their relationship to the community may help in this prediction.

Parad has described dimensions of contemporary family organization that potentially increase vulnerability to stress (1965). Organizationally, the young family is weak, quite often with as many or more dependents as there are adults. At a time of crisis the dependent members must rely on the adults for the necessary reorganization. With the growing preponderance of single-parent families, the burden of a crisis on such families is evident and should create concern. Unlike other organizations, a family has limited desire or capability in riddng itself of weak members and recruiting new ones.

The unequal age and sex distribution creates an awkward decision-making group in which authority is not well distributed and interpersonal relations are strained. This structure is also poorly suited as a work group because maintenance tasks are not evenly distributed. If a parent dies, these factors overburden an already wcak organization. Another factor is the increasing mobility in American society, which serves to weaken the family structure and increase vulnerability to stress and crisis.

Hill's notion of optimal support of a family during crisis is related to a positive correlation between agency and personal support. His idea suggests that, to the extent that both an agency and friends can offer support during a crisis, the family will derive the greatest benefit and will be the least vulnerable to a poor crisis resolution. This further suggests that when friends are not available, quite often an agency or professional will be relied upon.

In some instances, the family may fail to receive the kind of support it needs. Death may be uncomfortable for friends to deal with; not infrequently, when a terminal diagnosis has been made, visitors stop coming to the hospital or calling. If the health care team reacts similarly, the client quickly moves into a persecuted position in which feelings of isolation, meaninglessness, oppression, and inhumanness prevail. It is this state that the health care system is responsible for preventing. It should also be kept in mind that certain types of deaths are less normative and less "acceptable," and in such cases the family is given less support by friends and professionals. Some of these are deaths caused by homicide, child abuse, and suicide.

A number of other features of the family also determine its vulnerability to loss. The first of these is size. There is a greater expectation for a smaller number of people to handle the loss with less distribution of support. Since the early 1960s, the size of the nuclear family has shrunk from a relatively robust five to six persons to less than four. This figure has become all the more critical because of the isolation of the nuclear family from close relatives or other extensions of the family arising from greater mobility. Also, the family itself has changed from a productive unit to a consuming unit so that greater responsibility for economic maintenance of the family is vested in one or two members.

The functions of work and recreation, which are important parts of family life, are altered by a death. A change in the performance of these functions changes the family relationships and increases the vulnerability of the family. A death may cause increased economic stress and a decrease in leisure time.

It is also important to understand some of the basic conflicts survivors go through in order to increase sensitivity to the needs of those left behind after a death. Some survivors say that the grief syndrome persists for prolonged periods of time, and anger, loneliness, and a sense of loss are often overwhelming (McLaughlin 1978). Certain events and special days can trigger intense feelings. Support systems are difficult to maintain because many people may not want to deal with the survivor's depression after the first brief period of mourning. The task of everyday living and coping with minor problems may be overwhelming for a once-capable person. If there are children present, their anger at being left is often taken out on the surviving parent.

The parents of a newborn with a life-threatening condition must struggle with the issue of attachment. An ethical question nurses must face is whether to encourage parents to care for a child who is almost certain to die. One mother, having previously lost two infants to a rare metabolic disorder, was given no option of caring for them. When her third infant was born with the same syndrome and a similar prognosis, she asked to assume all his care in the hospital. In this case, the nursing staff were too worried about the outcome to consider the mother's need to nurture. As this mother proceeded with her caretaking task, she prepared herself for the infant's early death and provided a corrective experience for herself that enabled her to accept the deaths of her other infants.

When a family must attend to a client in terminal care, they often experience distressing feelings to which the nurse should be sensitive. They may feel helpless and dependent when others are in charge. Some families report feeling out of place, unsure of their role, and angry at being replaced by the nurse or physician. Many also experience a lack of empathy from nurses who do not acknowledge an awareness of the survivors' other responsibilities, which may include a job or obligations to other family members. Some feel isolated because their unique situation makes them feel different than others; interpersonal conflicts within the family brought about by the crisis, and the client's and hospital's demands may lead to disorganization and a further drain on their energies. Often, a simple

acknowledgment by the nurse indicating an awareness of these factors can decrease the feeling of isolation and make the family members feel supported. Greater sensitivity on the part of the nurse to the situation of the family members provides the environment in which the essential grief work can be done.

NURSING CARE OF THE GRIEVING

The nursing process originated in the scientific method and evolved into a systematic process of problem solving. The process consists of four major elements; assessment, planning, intervention, and evaluation, implemented within the context of a changing set of interactions with the client. The strategy of crisis intervention is to provide the individual with appropriate behavioral patterns that will enable him to deal effectively with a specific crisis (Lukton 1974), and the goal of the nursing process is to help the patient control his own life so that he can direct his energy toward achievements of his own choice. A combination of crisis intervention methods and the nursing process draws the best from both methods and provides a framework within which the nurse can work with the client who is dying and the client or family working through a death. Both methods rely on active interpersonal commitment, both imply a strongly supportive rather than interpretive role, both use observation and reports of client behavior as a main source of data from which to infer the degree of interference with normal functioning, both see the appropriate endpoint in intervention as being a healthy resolution of the problem.

Assessment

Using Hill's observations of family characteristics and the propositions from Burr's model, it is possible to speculate about the effects of

loss while predicting the likely type of resolution, given the existence of certain individual,
family, and situational characteristics. The data
base is the result of careful observation and sensitive interaction. Behavioral observation can
be of great value. Both children and adults often
express their grieving symbolically. Children
may reveal their feelings far more in play and
drawings than in other aspects of their behavior.
Grove (1978), in a brief narrative relating her
relationship with a 6-year-old boy dying of leukemia, describes a series of ship drawings done
during successive hospitalizations. The drawings changed only in size and color, from large
and bright to small and dark. Her understanding
of the child's perception of approaching death
came from her insightful analysis of the drawings.

Young children may also express their grief
at the loss of a parent similarly. One 7-year-old
child whose father died suddenly and tragically
drew a picture of flowers that were drooping
and flowing tears onto the ground (Fig. 14-3).

Adults as well may express their grief symbolically through their everyday behavior. A
woman who had just received a diagnosis of
inoperable lung cancer proceeded to buy a new
condominium and purchase expensive furniture, while she gaily planned for the future.
Another young woman in radiation therapy for
diffuse blastosarcoma began saving her money
to move to Ireland after her 21st birthday,
which was two years away. Both of these cases
represent denial behavior.

Similarly, acceptance can be shown symbolically. It is rather common to see people above
age 65 begin to make arrangements for the dis-
posal of accumulated wealth or material things,
not out of generosity but as a result of a decreased emotional investment in the objects.
This behavior represents symbolically what disengagement theorists and developmentalists
claim is the propensity of people to gradually
withdraw some of the emotional energy from
their work and social structures while turning

Figure 14–3. *Children are especially likely to express
their reactions to a death symbolically, as through drawings. This drawing was made by a 7-year-old whose father
died suddenly.*

inward in anticipation of their own eventual
needs. The increase in reminiscence during the
later decades in life during the normal aging
process should be understood as a life review
that is necessary as part of the preparation for
one's own death.

Intervention

Nursing intervention is influenced by three important factors: a culturally inculcated fear of
death, the attitude of the dying client, and the
personal style of the nurse. The fear of death
remains in most students throughout the course
of their nursing education despite the attempts
of nursing educators to make them more comfortable with the topic through courses on death

and dying (Hopping 1977; Benoliel 1974). It is important for students and faculty alike to know that changes in attitudes toward death, because they are embedded in our culture, are difficult. The goal in education in death and dying for nursing students may not be a change in attitude but an honest awareness of the distress and discomfort that dealing with death arouses and a greater sensitivity to the needs of the dying client and surviving family. This awareness and sensitivity, in turn, should be expected to change behavioral responses in the clinical setting.

The second dimension is the definition of the dying client used both within the culture at large and within the hospital. Glaser and Strauss (1965) developed two constructs, an awareness mode and a time context, within which a client is defined and which subsequently affect the behaviors displayed by the client and the staff in relation to his dying. The classifications of awareness are: (1) *closed awareness*, in which the client is not aware he is dying though everyone else is, (2) *suspected awareness*, in which the client suspects what everyone else knows and tries to confirm or refute his suspicions, (3) *mutual pretense awareness*, in which both client and staff know the client is dying but both avoid discussion, and (4) open awareness, in which both client and staff are aware that he is dying and this awareness is acted upon openly. The time context refers to the probability of death and the probable amount of time until death. The four categories, referred to by Glaser and Straussas "death trajectories," are: (1) certain death at a known time, (2) certain death at an unknown time, (3) uncertain death but a known time when certainty will be established, and (4) uncertain death and an unknown time at which questions will be resolved.

The interacting effects of these two dimensions create several possible cominations of staff-client interactions. For example regarding all of the death trajectories, only the open mode allows whatever time is left to be used for the psychological work of dying. The extent to which the nurse can help in this task is directly related to the degree to which she is willing to be open, to acknowledge and deal with the client's distress, and to confront her own attitudes about death.

The ability to function openly is related to the personal style of the nurse, the final dimension to be explored here. Investigators have found that the major fears of people who are dying do not include fear of death itself but fear of abandonment, dying alone, having friends and family disengage from them emotionally, fear of uncontrollable pain, and fear of becoming unable to make decisions about their own destiny (Deni 1978).

Keck and Walther (1977) found that nurses in their study spent the same amount of time with clients whether they were terminally ill or not, but could not identify ways in which emotional support was given although a certain percentage of the nurses' time was spent in nonphysical care activities. However, more time was spent with clients whose deaths were not certain to occur, in which cases intervention was seen as being potentially restorative.

The nurse's personal experience with death influences the sensitivity to the client and willingness to experience another's loss emotionally rather than relying on intellectualization or other defensive maneuvers. What behaviors should the nurse engage in to be most supportive? Quiet listening is far more successful than platitudes, and acknowledging feelings as legitimate, even when anger is directed at the hospital system, nurse, or doctor, is more beneficial than intellectualizing and explaining in biological terms what is happening.

The nurse's acceptance of death is crucial in working with clients facing certain death. The ability to "experience with" and a refusal to abandon the client through the difficult process of dying while lending hope for an eventual grief resolution to the family should be the goal of optimal nursing intervention.

Standards of Care

Kastenbaum (1977) reported on the conclusions and recommendations of the 1974 International Task Force on Hospital Care of the Dying established to define explicit standards of care for the terminally ill. Two lists of standards were developed: those that seemed to exist as unwritten, unofficial assumptions but which provided the implicit standards of worldwide practice in 1974, and recommended standards that should be the framework for a new approach to care of the dying. The recommended standards emerge from an underlying assumption that client, family, and staff all have legitimate needs and interests, whereas most care standards in use as of this writing reflect more concern with the needs of the staff and the institution.

Some of the implicit standards are:

1. The successful death is quiet and uneventful. The death slips by with as little notice as possible; nobody is disturbed.
2. Leave-taking behavior is at a minimum.
3. Strong emphasis is given to the body during the care-giving process. Little effort is wasted on the personality of the terminally ill individual.
4. The person dies at the right time, that is, after the full range of medical interventions has been tried but before the onset of an interminable period of lingering.
5. The patient expresses gratitude for the excellent care received.
6. Few people are on the scene. There is, in effect, no scene. Staff is not required to adjust to the presence of family and other visitors who might have their own needs that upset the equilibrium of routine.

Proposed standards include:

1. The terminally ill person's own preferences and life-style must be taken into account.
2. Permission of the client to engage in life-prolonging measures is essential.
3. Pain control is a goal of treatment.
4. A living will will be respected.
5. The client should have a sense of basic security and protection in his environment.
6. Opportunities should be provided for leave-takings with the people most important to the client.
7. Opportunities should be provided for experiencing the final moments in a way meaningful to the client.
8. Family should have the opportunity to discuss death, dying, and related emotional needs with the staff.
9. Family should have the opportunity for privacy with the dying person both while the latter is living and while he or she is newly dead.
10. Care-givers should be given adequate time to form and maintain personal relationships with the client.
11. A mutual support network should exist among the staff, encompassing both the technical and the psychosocial dimensions working with the terminally ill (Kastenbaum 1977, 389).

ANTICIPATORY GRIEF AND ANTICIPATORY GUIDANCE

Anticipatory grief and anticipatory guidance constitute another element of intervention. *Anticipatory grief* refers to the process of accomplishing part of the grief work before the actual bereavement. *Anticipatory guidance* refers to a teaching process that alerts the survivors to the feelings and experiences they will experience during bereavement; it is usually done after the death, when the fact of loss must finally be recognized.

The concept of anticipatory grief and guidance is related to crisis theory. Various authors have found that there is a causal relationship between the stress of bereavement and later emotional and physical problems in both children and adults (Stubblefield 1977). Lindeman and Caplan (1965) found that adequate management during the crisis of loss can prevent such problems. It is hypothesized that successful management of the crisis strengthens the personality and promotes more adaptive handling of future stresses (Parad 1969).

Preventive intervention consists of preparing the family for the aftermath of death and grief by providing information about the normative reactions and process of grief work they will experience. Alerting the family to special needs of individual family members helps them focus their responses to each other and understand new behaviors that may emerge under the stress of loss.

What is known of mourning from those who have experienced it forms the basis for anticipatory guidance to survivors. Unless they have experienced a loss like the one they are facing, the mourning process will be a new experience that will surprise, dismay, and sometimes frighten them.

Intrapersonal Changes and Coping

Many people in the midst of the early grief reaction question their sanity as they find their cognitive and behavioral responses altered. The type of behavior manifested is not nearly as predictable as the fact that behavior will change. For instance, a normally inactive person may heighten his level of activity as a means of coping with others and with the loneliness of loss, whereas a normally outgoing person may become introverted, using emotional energy to integrate the loss into the fabric of his life. Perhaps even more typical is unpredictable liability with fairly wide mood swings. These changes can be distressing but their effect can be modified through reassurance that they should be expected and are normal. These mood swings

are the result of a normal physiological response to psychological trauma.

Loss of self-esteem and self-confidence often plague the survivors as a result of the guilt that frequently accompanies death. Acute psychosomatic symptoms often accompany the loss. Changes in sleeping habits and loss of appetite with accompanying weight loss are typical. In fact, the degree of acute reactive depression can be measured by the extent to which these symptoms are present. The survivors should be encouraged to use pharmacologic support over a short period of time if these symptoms become too pervasive, since their energy is often needed during the day to cope with other's feelings and/ or a job. Clients often fear addiction, partly because it is hard for them to believe they will ever feel better again. The use of drugs to control the psychological pain is justified during this time.

Personal disorganization often occurs as result of the internal restructuring that precedes the integration of a loss. Patients should be encouraged to get back to work or to the established family routine as soon as possible. At the same time, they should be told to expect less of themselves in the performance of their everyday tasks; this may seem paradoxical, but the purpose is to create structure while realistically allowing for temporary functional limitations.

Judgment deficits almost always accompany high stress levels. Survivors should be cautioned to make as few decisions with long-range consequences (such as to move, leave a job, etc.) as possible in the few months following a loss.

Interpersonal Changes and Coping

Surviving families should be forewarned of the possibility of change in communication and intra-family problems during the grief state. That grieving greatly increases the potential for family disintegration has been demonstrated in studies of parents following the death of an infant, in families after the death of a parent, and in families in which a child dies after a lingering illness. In most cases, the family breakdown seems to be related to two major factors. The first is a breakdown of communication caused by a different pace of resolution of grief among the family members and a lack of mutual support. It is not always possible for grieving family members to nurture each other in the ways needed; thus, resentment and tension mount, and relationships deteriorate. Second, the illness preceding the death may have drained the family's economic resources; and the death itself may have caused the loss of a major wage-earner. The result is economic stress.

Once the family is made aware of the potential for a breakdown to occur, they should also be given instruction about seeking professional help. People often perceive turning to a mental health professional as representing a failure in handling their own lives. It should be explained, in such cases, that families or individuals who know when stress is likely to jeopardize their health, their work, and long-range relationships are acting positively and responsibly by preventing disturbances.

When the deceased is a parent, the family should understand that the remaining parent may feel anger and guilt. Children may also feel responsible for the death and may fear abandonment by the remaining parent.

Families should also be advised that taking on new attachments, be it the birth of a new baby to parents who have lost an infant or child, or remarriage for a person who has lost a spouse, should wait until the acute phases of grief are over, since healthy attachment is difficult to achieve while simultaneously detaching from a relationship that has just ended.

Anticipatory guidance should be directed toward assessing family supports, enlisting the help of outside agencies, and teaching clients about the grief process. One must always keep in mind the potential for long-range effects on family relationships if an adequate grief resolution is not facilitated. If guilt or anger are not expressed, accepted, and resolved the impact may play itself out in future family interactions. Clients should be reassured that their feelings are normal, and should be prepared for grief work to persist.

It is important that anticipatory guidance communicate:

1. The principles of the psychological dynamics that occur during the grief process in order to reassure the client and family that feelings of denial, anger, and depression are normal.
2. The need for use of interpersonal resources. Families and individuals can be helped to identify a significant support network and anticipate ways to moblize that network when needed.
3. That acute grief is temporary and that there is progression in the mourning process. Certain events can be predicted as benchmarks for progress. It is often extremely difficult for the person in the midst of his own despair to know when progress has occurred. progress will occur in small increments. To be able to focus on a small project, such as a small family outing, is a sign of the beginning of grief resolution, which people can be encouraged to watch for.
4. Hope and a reassurance that with time, adjustment will occur and psychological pain will abate.

CONCLUSION

Contemporary American culture tends to emphasize youth and to encourage alienation from the aged, and is more present- than future-oriented. Because of these attitudes, people tend to deny the eventuality of death and tend not to prepare for their own.

Death is the fate that awaits all people, yet most people give very little thought to their fate. Indeed, it is not easy for a healthy person to think seriously of dying. Yet to do so is one way to prepare emotionally for the feelings about death that are essential to developing a self-awareness. The nurse's doing so enables her to respond effectively to a client coping with death. Only recently has long-overdue concern been directed by professionals toward understanding death as the final stage of growth. Along with that has come an increasing awareness in nursing of the importance and responsibility in helping clients to die peacefully.

Attitudes toward the dying client and resources for clients and families must change. Despite increased knowledge about the complex interaction of physiological and psychological factors in health and illness, hospitals continue to attribute value to physical needs while ignoring emotional needs; this is no more clearly seen than in the part hospitals play in the death system. Hospitals have a codified procedure and a staff trained to respond most automatically to medical emergencies, yet they often fail to respond to the accompanying emotional trauma or emergency that accompanies life-threatening situations or death.

Nurses are in a position to make positive changes in the care of the dying and their families. Nursing theory, which integrates psychosocial and biological constructs, provides the basis for holistic care of the dying and the bereaved.

Although the nurse, by the very nature of her profession, is frequently in contact with the dying and the families of the dying, nurses often fail to respond to the needs of these clients; the reason for this lies in several factors, including the nature of the organization of the health care system, the nurse's own myths or fears about death, and the negative attitude toward death that is part of the American sociocultural heritage. Regarding this last factor, several related phenomena can be singled out, namely, the overdenial of death via institutionalization, the high expectations of technology (which lead to the perception of death as a failure on the part of the health care system), and the relative absence of mourning customs.

Kübler-Ross outlined five phases through which the dying person passes; they apply also to the survivors' struggle to integrate the loss of the deceased: denial, anger, bargaining, depression, and acceptance of the loss. Making the adjustment to a loss (which may or may not be marked by a strict progression through these stages) may take anywhere from three months to a year or more, the most acute phase of grief occurring in the first six to eight weeks. The changes brought about by the grieving process, in both individual behavior and interpersonal relationships, are not nearly as predictable as the fact that changes will occur. Anticipatory guidance should be directed toward assessing family supports, enlisting professional help if indicated, and teaching clients about the grieving process.

SUMMARY

Personal loss is any permanent loss that requires adaptation through a grieving process. For the dying, the approach of death means accepting the end of life; for the survivors, the impending death necessitates dealing with being left, and invariably brings about some reflections on one's own mortality and the meaning of life itself.

REFERENCES

Benoliel, J. Q. Anticipatory grief in physicians and nurses. In Schoenberg, B., et al. (Eds.), *Anticipatory grief.* New York: Columbia University Press, 1974.

Bowlby, *Attachment and loss,* Vols. I and II. New York: Basic Books, 1969.

Brantner, J. Positive approaches to dying. *Death Education,* 1977, *1*(3), 293–304.

Burr, W. R. Families under stress. In *Theory construction and the family*. New York: John Wiley & Sons, 1973.

Deni, L. Death and nursing care. *Journal of Nursing Care*, 1978, *11*(9), 20–23.

Epstein, C. *Nursing the dying patient: Learning processes for Interaction*. Reston, Va.: Reston, 1975.

Hill, R. Generic features of families under stress. In Parad, H. (Ed.), *Crisis intervention*. New York.: Family Service Association, 1965.

Hopping, B. L. Nursing students attitudes toward death. *Nursing Research*, 26, 443–447, 1977.

Glaser, B., and Straus, A. *Time for dying*. New York.: Aldine Press, 1965.

Groves, J. S. Differentiating grief, mourning and bereavement. *American Journal of Psychiatry*, 1978, *135*(7), 874–875.

Grove, S. I am a yellow ship. *American Journal of Nursing*, March 1978, 414.

Hall, J., and Weaver, B. (Eds.). *Nursing of families in crisis*. Philadelphia: J. B. Lippincott, 1974.

Kassokian, M. G. The cast and quality of dying—a comparison of home and hospital. *Nurse Practitioner*, Jan/Feb, 1979, *4*, 18–19.

Kastenbaum, R. *Death, society and human experience*. St. Louis: C. V. Mosby, 1977.

Keck, V. E., et al. Nurse encounters with dying and non-dying patients. *Nursing Research*, Nov/Dec, 1977, 26, 465–469.

Kübler-Ross E. *Death: The final stage of growth*. Englewood Cliffs, N. J.: Prentice-Hall, 1975.

Kutscher, A. H. *Death and bereavement*. Springfield, Ill.: CC Thomas, 1969.

Lascare, A. D. The dying child and the family. *Journal of Family Practice*, 1978, *6*(6), 1279–1286.

Lindeman, E. Symptomatology and management of acute grief. In Parad, H. (Ed.), *Crisis Intervention*. New York: Family Assoc. of America, 1965.

Lukton, R. et al. Crisis theory: review and critique. *Social Service Review*, 1974, *48*(3), 384–402.

Mauksch, H. O. The organizational context of dying. In Kübler-Ross, E. (Ed.), *Death, the final stage of growth*. Englewood Cliffs, N. J.: Prentice-Hall, 1975.

McLaughlin, M. F. Grief, who helps the loving? *American Journal of Nursing*, 1978, *78*(3), 422–423.

Martinson, I. *Home care for the dying child*. New York: Appleton-Century-Crofts, 1976.

Parad, H. (ed.). *Crisis intervention: Selected readings*. New York: Family Service Association of America, 1965.

Parad, H., and Caplan, G. A framework for studying families in crisis. In Parad, H. (Ed.), *Crisis Intervention*. New York: Family Service Assoc., 1965.

Reck, A. J. A contextual thanatology: A pragmatic approach to death and dying, *Death Education*, Fall, 1977, *1*, 315–323,

Sahler, O. (ed.). *The child and death*. St. Louis: C. V. Mosby, 1978.

Schoenberg, B., et al. *Loss and grief: Psychological management in medical practice*. New York: Columbia University Press, 1970.

Speck, R. Death, dying and bereavement. *Nursing Mirror*, 1978, *146*, 18.

Schoenberg, B., Carr, A., Peretz, D. and Kutscher, A. *Psychosocial aspects of terminal care*. New York: Columbia University Press, 1972.

Sherizen, S., and Lester, P. Dying in a hospital intensive care unit: The social significance for the family of the patient. *Omega: Journal of Death and Dying*, 1977, *8*(1), 29–40.

Stubblefield, K. S. A preventative program for bereaved families. *Social Work Health Care* Summer 1977, *2*, 379–389.

Williams C. C., et al. The intensive care unit: Social work interventions with the families of critically ill patients. *Social Work Health Care*, Summer 1977, *2*, 391–398.

National Center for Health Statistics Growth Grids

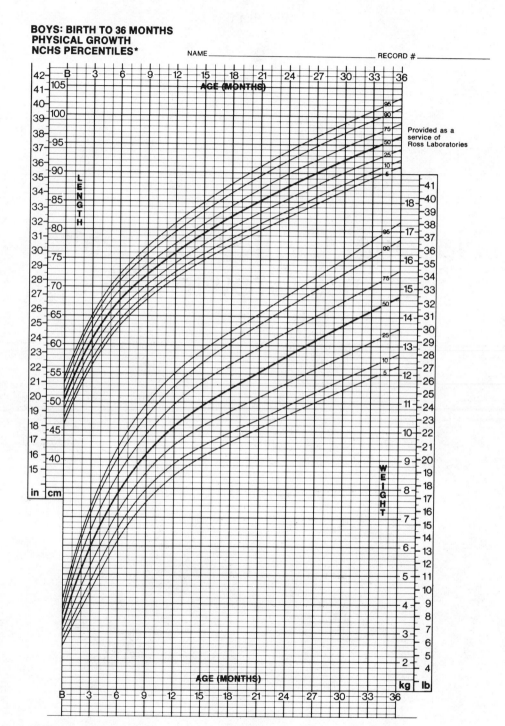

BOYS: BIRTH TO 36 MONTHS
PHYSICAL GROWTH
NCHS PERCENTILES*

NAME _____ RECORD # _____

Provided as a
service of
Ross Laboratories

AGE (MONTHS)

LENGTH

WEIGHT

AGE (MONTHS)

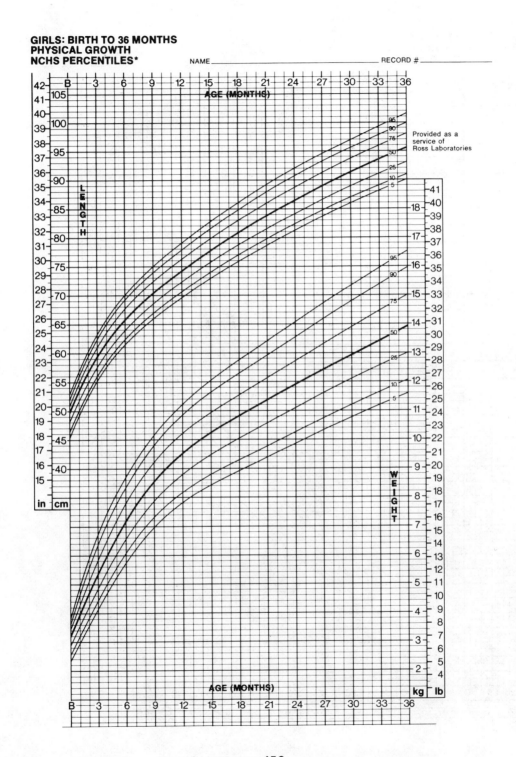

GIRLS: BIRTH TO 36 MONTHS
PHYSICAL GROWTH
NCHS PERCENTILES*

NAME _____ RECORD # _____

Provided as a
service of
Ross Laboratories

459

BOYS: PREPUBESCENT
PHYSICAL GROWTH
NCHS PERCENTILES*

NAME _____ RECORD # _____

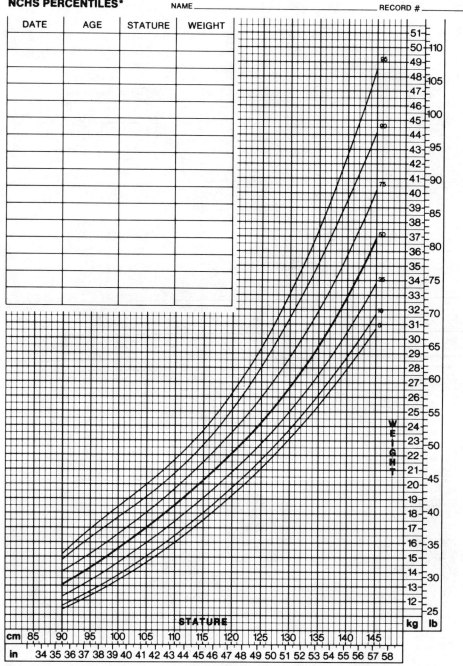

DATE	AGE	STATURE	WEIGHT

STATURE

cm 85 90 95 100 105 110 115 120 125 130 135 140 145 kg lb

in 34 35 36 37 38 39 40 41 42 43 44 45 46 47 48 49 50 51 52 53 54 55 56 57 58

GIRLS: PREPUBESCENT
PHYSICAL GROWTH
NCHS PERCENTILES*

NAME _____ RECORD # _____

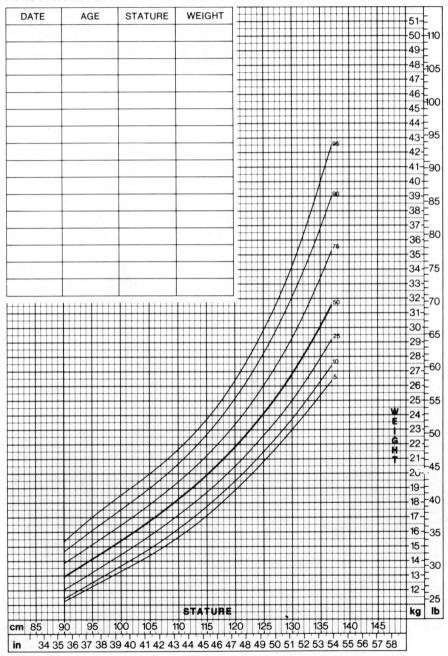

DATE	AGE	STATURE	WEIGHT

B

Denver Developmental Screening Test

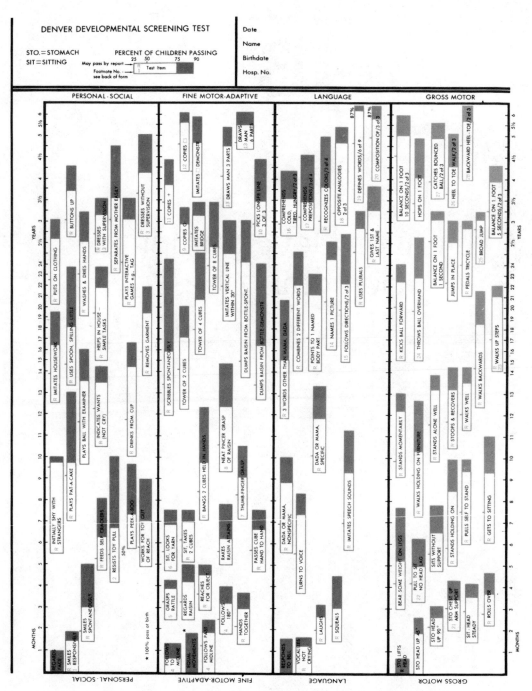

DENVER DEVELOPMENTAL SCREENING TEST

Reproduced with permission of William K. Frankenburg, University of Colorado Medical Center, 1969.

1. Try to get child to smile by smiling, talking or waving to him. Do not touch him.
2. When child is playing with toy, pull it away from him. Pass if he resists.
3. Child does not have to be able to tie shoes or button in the back.
4. Move yarn slowly in an arc from one side to the other, about 6" above child's face. Pass if eyes follow 90° to midline. (Past midline; 180°)
5. Pass if child grasps rattle when it is touched to the backs or tips of fingers.
6. Pass if child continues to look where yarn disappeared or tries to see where it went. Yarn should be dropped quickly from sight from tester's hand without arm movement.
7. Pass if child picks up raisin with any part of thumb and a finger.
8. Pass if child picks up raisin with the ends of thumb and index finger using an over hand approach.

9. Pass any enclosed form. Fail continuous round motions.
10. Which line is longer? (Not bigger.) Turn paper upside down and repeat. (3/3 or 5/6)
11. Pass any crossing lines.
12. Have child copy first. If failed, demonstrate

When giving items 9, 11 and 12, do not name the forms. Do not demonstrate 9 and 11.

13. When scoring, each pair (2 arms, 2 legs, etc.) counts as one part.
14. Point to picture and have child name it. (No credit is given for sounds only.)

15. Tell child to: Give block to Mommie; put block on table; put block on floor. Pass 2 of 3. (Do not help child by pointing, moving head or eyes.)
16. Ask child: What do you do when you are cold? ..hungry? ..tired? Pass 2 of 3.
17. Tell child to: Put block on table; under table; in front of chair, behind chair. Pass 3 of 4. (Do not help child by pointing, moving head or eyes.)
18. Ask child: If fire is hot, ice is ?; Mother is a woman, Dad is a ?; a horse is big, a mouse is ?. Pass 2 of 3.
19. Ask child: What is a ball? ..lake? ..desk? ..house? ..banana? ..curtain? ..ceiling? ..hedge? ..pavement? Pass if defined in terms of use, shape, what it is made of or general category (such as banana is fruit, not just yellow). Pass 6 of 9.
20. Ask child: What is a spoon made of? ..a shoe made of? ..a door made of? (No other objects may be substituted.) Pass 3 of 3.
21. When placed on stomach, child lifts chest off table with support of forearms and/or hands.
22. When child is on back, grasp his hands and pull him to sitting. Pass if head does not hang back.
23. Child may use wall or rail only, not person. May not crawl.
24. Child must throw ball overhand 3 feet to within arm's reach of tester.
25. Child must perform standing broad jump over width of test sheet. (8-1/2 inches)
26. Tell child to walk forward, 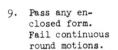 heel within 1 inch of toe. Tester may demonstrate. Child must walk 4 consecutive steps, 2 out of 3 trials.
27. Bounce ball to child who should stand 3 feet away from tester. Child must catch ball with hands, not arms, 2 out of 3 trials.
28. Tell child to walk backward, ⟵⊂∞∞⊃ toe within 1 inch of heel. Tester may demonstrate. Child must walk 4 consecutive steps, 2 out of 3 trials.

DATE AND BEHAVIORAL OBSERVATIONS (how child feels at time of test, relation to tester, attention span, verbal behavior, self-confidence, etc,):

C

Developmental Tasks Through the Life Span

Developmental Tasks of Infancy and Early Childhood

Learning to walk.

Learning to take solid foods.

Learning to talk.

Learning to control the elimination of body wastes.

Learning sex differences and sexual modesty.

Achieving physiological stabilty.

Forming simple concepts of social and physical reality.

Learning to relate emotionally to parents, siblings, and other people.

Learning to distinguish right and wrong and developing a conscience.

SOURCE: Reproduced with permission from Hurlock, Elizabeth, *Developmental Psychology*. New York: McGraw-Hill, 1959.

Developmental Tasks of Middle Childhood

Learning physical skills necessary for ordinary games.

Building wholesome attitudes toward oneself as a growing person.

Learning to get along with age-mates.

Learning an appropriate sex role.

Developing fundamental skills in reading, writing, and calculating.

Developing concepts necessary for everyday living.

Developing conscience, morality, and a scale of values.

Developing attitudes toward social groups and institutions.

Developmental Tasks of Adolescence

Accepting one's physique and accepting a masculine or feminine role.

New relations with age-mates of both sexes.
Emotional independence of parents and other adults.
Achieving assurance of economic independence.
Selecting and preparing for an occupation.
Developing intellectual skills and concepts necessary for civic competence.
Desiring and achieving scoially responsible behavior.
Preparing for marriage and family life.
Building conscious values in harmony with an adequate scientific world-picture.

Developmental Tasks of Early Adulthood
Selecting a mate.
Learning to live with a marriage partner.
Starting a family.
Rearing children.
Managing a home.
Getting started in an occupation.
Taking on civic responsibility.
Finding a congenial social group.

Developmental Tasks of Middle Age
Achieving adult civic and social responsibility.
Establishing and maintaining an economic standard of living.
Assisting teen-age children to become responsible and happy adults.
Developing adult leisure-time activities.
Relating to one's spouse.
Accepting and adjusting to the physiological changes of middle age.
Adjusting to aging parents.

Developmental Tasks of Later Maturity
Adjusting to decreasing physical strength and health.
Adjusting to retirement and reduced income.
Adjusting to death of spouse.
Establishing an explicit affiliation with one's age group.
Meeting social and civic obligations.
Establishing satisfactory physical living arrangements.

D

The Patient's Bill of Rights

These patients' rights policies and procedures ensure that, at least, each patient admitted to the facility:

1. Is fully informed, as evidenced by the patient's written acknowledgment, prior to or at the time of admission and during stay, of these rights and of all rules and regulations governing patient conduct and responsibilities.
2. Is fully informed, prior to or at the time of admission and during stay, of services available in the facility, and of related charges including any charges for services not covered under titles XVIII or XIX of the Social Security Act, or not covered by the facility's basic per diem rate.
3. Is fully informed, by a physician, of his medical condition unless medically contraindicated (as documented, by a physician, in his medical record) and is afforded the opportunity to participate in the planning of his medical treatment and to refuse to participate in experimental research.
4. Is transferred or discharged only for medical reasons, or for his welfare or that of other patients, or for nonpayment for his stay (except as prohibited by titles XVIII or XIX of the Social Security Act), and is given reasonable advance notice to ensure orderly transfer or discharge, and such actions are documented in his medical record.
5. Is encouraged and assisted, throughout his period of stay, to exercise his rights as a patient and as a citizen, and to this end may voice grievances and recommend changes in policies and services to facility staff and/ or to outside representatives of his choice, free from restraint, interference, coercion, discrimination, or reprisal.

SOURCE: DHEW, 1974

6. May manage his personal financial affairs or be given, at least quarterly, an accounting of financial transactions made on his behalf should the facility accept his written delegation of this responsibility for any period of time, in conformance with state law.

7. Is free from mental and physical abuse, and free from chemical and (except in emergencies) physical restraints except as authorized in writing by a physician for a specified and limited period of time, or when necessary to protect the patient from injury to himself or to others.

8. Is assured confidential treatment of his personal and medical records, and may approve or refuse their release to any individual outside the facility, except, in case of his transfer to another health care institution, or as required by law or third-party payment contract.

9. Is treated with consideration, respect, and full recognition of his dignity and individuality, including privacy in treatment and in care of his personal records.

10. Is not required to perform services for the facility that are not included for therapeutic purposes in his plan of care.

11. May associate and communicate privately with persons of his choice, and send and receive his personal mail unopened, unless medically contraindicated (as documented by his physician in his medical record).

12. May meet with, and participate in activities of social, religious, and community groups at his discretion, unless initially contraindicated (as documented by his physician in the medical record).

13. May retain and use his personal clothing and possessions as space permits, unless to do so would infringe upon rights of other patients, and unless medically contraindicated (as documented by his physician in his medical record).

14. If married, is assured privacy for visits by his/her spouse; if both are inpatients in the facility, they are permitted to share a room, unless medically contraindicated (as documented by the attending physician in the medical record).

E

Recommended Schedule for Active Immunization of Normal Infants and Children

Age	Immunizaion Recommended
2months	DTP,[a] TOPV[b]
4months	DTP, TOPV
6months	DTP[c]
1year	Tuberculin test[d]
15months	Measles, rubella, mumps[e]
18months	DTP, TOPV
4–6years	DTP, TOPV
14–16years	Td[f]—repeat every 10 years

SOURCE: Adapted from American Academy of pediatrics: Report of the Committee on Infectious Diseases. Ill., 1977. Copyright American Academy of Pediatrics, 1977.

[a] DTP—diphtheria and tetanus toxoids combined with pertussis vaccine.

[b] trivalent oral poliovirus vaccine. This recommendation is suitable for breast-fed as well as bottle-fed infants.

[c] A third dose of TOPV is optional but may be given in areas of high endemicity of poliomyelitis.

[d] Frequency of tuberculin testing depends on risk of exposure of the child and on the prevalence of tuberculosis in the population group. The initial test should be at or preceding the measles vaccine.

[e] May be given at 15 months as measles-rubella or measles-mumps-rubella combined vaccines.

[f] Td—combined tetanus and diphtheria toxoids (adult type) for those more than 6 years of age, in contrast to diphtheria and tetanus (DT) toxoids which contain a larger amount of diphtheria antigen.

F

Normal Physiological Parameters Through the Life Cycle

Table 1

Age	Vital Signs			Height		Weight		Blood Values		
	Pulse	Resp.	B/P	Cm	Inches	Kg	Pounds	Hgb (gm)	Hct	WBC/cu mm
Birth	140 ± 20	55 ± 25	$\frac{80}{46} \pm \frac{16}{16}$	50 ± 2	20 ± 1	3.4 ± 6	7.5 ± 1	20 (14–24)	53 (44–64)	19,000 (9–30)
14 days	135 ± 15	40 ±15						17 (15–20)	46 (42–60)	12,000 (5–21)
1 month	130 ± 20	35 ± 10	$\frac{80}{50} \pm \frac{20}{10}$	53 ± 2.5	21 ± 1	4.4 ± .8	10 ± 1.5	15 (11–17)	43 (35–49)	10,800 (5–19.5)
3 months				60 ± 2	23.5 ± 1	5.7 ± 8	12.5 ± 2	11 (10–13)	37 (31–41)	11,000 (5.5–18)
6 months	120 ± 20	31 ± 9	$\frac{90}{60} \pm \frac{28}{10}$	65.5 ± 3	26 ± 1	7.4 ± 1	16.5 ± 2.5	11.5 (10.5–14.5)	37 (30–40)	11,900 (6–17.5)
12 months	115 ± 20	30 ± 10	$\frac{96}{66} \pm \frac{30}{24}$	74.5 ± 3	29 ± 1.5	10 ± 1.5	22 ± 3	12.5 (11–15)	37 (33–42)	11,400 (6–17.5)
2 years	110 ± 20	25 ± 5	$\frac{98}{64} \pm \frac{26}{24}$	87 ± 4	34 ± 2	12.4 ± 2	27.5 ± 4	13 (12–15)	38 (33–42)	10,500 (6–17)
3 years	105 ± 15		$\frac{100}{66} \pm \frac{24}{22}$	96 ± 5.5	38 ± 2	14.5 ± 2	32 ± 5			9,000 (5.5–15.5)
4 years	100 ± 10	24 ± 4	$\frac{100}{66} \pm \frac{20}{20}$	103 ± 6	40.5 ± 2.5	16.5 ± 3	36.5 ± 5	13.5 (12.5–15)		
5 years	95 ± 15	22 ± 3	$\frac{100}{60} \pm \frac{14}{10}$	109 ± 6	43 ± 2.5	18.4 ± 3	40.5 ± 6		40 (31–43)	8,500 (5–14.5)
6 years	90 ± 15	21 ± 3	$\frac{100}{60} \pm \frac{16}{10}$	117 ± 7	46 ± 2.5	21.5 ± 4	47.5 ± 8			
8 years	85 ± 10	20 ± 3	$\frac{102}{60} \pm \frac{16}{10}$	129 ± 7.5	50.5 ± 3	27 ± 5	59 ± 11	14 (13–15.5)		8,000 (4.5–13.5)
10 years	80 ± 10	19 ± 3	$\frac{106}{60} \pm \frac{16}{10}$	139.5 ± 8	55 ± 3	32.5 ± 7	71 ± 14		42 (33–44)	

SOURCE: From C. S. Schuster, Normal physiological parameters through the life cycle, *The Nurse Practitioner: A Journal of Primary Nursing Care,* 2:25–28, Jan.–Feb., 1977.

Table 2

Age (Males)	Vital Signs			Height		Weight		Blood Values			Calories/24 hr		Protein
	Pulse	Resp.	B/P	Cm	Inches	Kg	Pounds	Hgb (gm)	Hct	WBC/cu mm	/Pound	Total	Gm/24 hr
12 years	69 ± 9	19 ± 2	$\frac{110}{64} \pm \frac{12}{16}$	150 ± 8	59 ± 3	39 ± 9	85 ± 12	13 (11–16)	38 (34–40)	8,000 (4.5–13.5)	30	2,400	48
14 years	65 ± 8	19 ± 3	$\frac{114}{68} \pm \frac{10}{14}$	163 ± 9	64 ± 4	49 ± 11	108 ± 20	14 (13–16)	41 (37–43)		31	2,800	54
16 years	63 ± 8	17 ± 3	$\frac{116}{70} \pm \frac{12}{14}$	172 ± 8	68 ± 4	59 ± 9	130 ± 20	15.5 (13–17)	45 (40–48)	7,800 (4.5–13)	31	3,000	58
18 years	61 ± 8	16 ± 3	$\frac{120}{72} \pm \frac{16}{14}$	174 ± 8	68.5 ± 4	63 ± 10	140 ± 20	16 (14–18)	47 (42–52)		25	3,200	62
18–22 years												3,000	60
23–50 years	70 ± 10	18 ± 2	$\frac{126}{74} \pm \frac{26}{16}$	175 ± 8	69 ± 4	68 ± 12	150 ± 25			7,500 (4.5–11.5)	18–25	2,700	56
51 + years												2,400	

Females	Vital Signs			Height		Weight		Blood Values			Calories/24 hr		Protein
	Pulse	Resp.	B/P	Cm	Inches	Kg	Pounds	Hgb (gm)	Hct	WBC/cu mm	/Pound	Total	Gm/24 hr
12 years	71 ± 9	19 ± 3	$\frac{110}{64} \pm \frac{10}{12}$	152 ± 8	60 ± 3	40 ± 10	88 ± 20	13 (11–16)	38 (34–40)	8,000 (4.5–13.5)	30	2,300	50
14 years	68 ± 8		$\frac{112}{66} \pm \frac{10}{12}$	160 ± 7	63 ± 3	50 ± 10	110 ± 18	13 (12–16)	39 (35–42)		24	2,400	51
16 years	66 ± 8	18 ± 3	$\frac{114}{70} \pm \frac{14}{12}$	162 ± 7	63.5 ± 3	53 ± 11	117 ± 17	13.5 (12–16)	40 (36–44)	7,800 (4.5–13.0)	21	2,400	53
18 years	65 ± 8		$\frac{120}{70} \pm \frac{16}{12}$			54 ± 11	120 ± 20				19	2,300	55
18–22 years												2,000	50
23–50 years		17 ± 3	$\frac{126}{74} \pm \frac{26}{16}$	163 ± 7	64 ± 3	60 ± 12	130 ± 25	14 (12–16)	42 (37–47)	7,500 (4.5–11.5)	15–20		
51 + years	70 ± 10											1,800	46
Pregnant							+15–30	12 (11–15)	36 (34–40)			+300	+30
Lactating							+2–5					+800	+20

475

Table 3

Age	/lb	/kg	total	gm	oz/lb	cc/kg	total	urine/24 hr	sleep/24 hr	Head Circum. (cm)
	Intake/24 hr							**Output**		
	Calories			*Protein*	*Water*					
Birth							45–90	15–60	22	
3 days	55	117				80–100	250–300			
14 days			kg × 120	kg × 2.2		125–150	400–500	40–400	16–22	34 ± 2.5
1 month		110				140–160	750–850	250–450	15–18	36.5 ± 2.5
3 months	50		kg × 110		2¼					40 ± 2.5
6 months		108		kg × 2.0		130–145	950–1,100	400–550	15–16	43 ± 3
12 months			1,100			120–135	1,100–1,300		13–15	46 ± 3
2 years	45	100	1,200	23	2	115–125	1,300–1,500	500–600	12–14	49 ± 3
3 years			1,300							50 ± 3
4 years			1,400			100–110	1,600–1,800	600–750		50.5 ± 3
5 years	41	90	1,600	30	1½	90–100	1,800–2,000			51 ± 2.5
6 years			1,800					650–1,000	11–12	51.5 ± 2.5
8 years	36	80	2,000	36		70–85	2,000–2,500			52.5 ± 2.5
10 years			2,200		1				9–11	53 ± 3
10–12 years						60–75				53.5 ± 3
12–14 years						50–60	2,200–2,700	700–1,500		54 ± 3
14–16 years						40–50			8–9	54.5 ± 3
16–18 years										55 ± 3
18–22 years	see Table 2								7–9	
23–50 years					¾	50	2,000–3,000			
51+ years								1,000–2,000	5–7	55.5 ± 3
Pregnant women										
Lactating women							3,000–4,500		9–10	

G

Points to Consider
in Selecting a Nursing Home

General Item	Factors To Consider
1. Location of Nursing Home	Is the home near to family, relatives and significant others? Rural or urban setting? Suitability of location for the individual. Are shopping centers and/or stores nearby?
2. License	Ask to see the Nursing Home's license. Date of renewal? Are they licensed for skilled and intermediate beds? Retirement Home? Does this home belong to any Long-Term Care Association? Is it a profit or nonprofit making home?
3. Fees and Expenditures	What is the monthly fee for room and board? What does it include? What does it not include? How often does fee change? What insurance policies can be used (Medicare, Medicaid, Blue Cross/ Blue Shield)? Is this a "guaranteed for life" home? Ask to see a copy of the contract (read the fine print). Does resident have to make the institution the benficiary of his will? What happens if one resigns?

General Item	Factors To Consider
4. Physical layout of Building and Rooms	Single or multistory building? Access to stairs, elevators, exits. Number of occupants per room? Who determines room selection and change of room and roommates? Are personal belongings allowed and encouraged? Any restrictions on visitors? What type of privacy is provided for spouses, loved ones? Are private telephones permitted? Is building equipped with adequate heating, lighting and air conditioning? Are rooms and general areas clean and well kept?
5. Administration	Make an appointment with the Administrator. What is his philosohpy of care for residents? General attitude. Explanation of "Bill of Rights". Any long range plans that might affect present residents? (e.g., expansion, move, etc).
6. Institutional Policies and Regulations	Read carefully the Handbook for residents. Is there a "Bill of Rights" for residents? Is leave of absence permitted? If so, for what length of time? Are alcoholic beverages allowed?What is Home's policy for gratuities?
7. Medical Care	Does the resident maintain his private physician? If not, what type of medical services does home offer? By Whom? Frequency of visits? What hospital is utilized when acute care is required? Emergency services within the home? Are Geriatric Nurse Practitioners and/or Physician Assistants employed by the home or any physicians? What other specialities are available? (e.g., ophtahalmologist).
8. Nursing Care	Make an appointment with Director of Nursing. What is the philosophy of Nursing Care? Number of RN's, LPN's and nursing assistants. Ratio of staff to residents per 24 hours (should be 3.0 or higher). What is the general morale or staff? Is continuing education provided for nursing staff? Are all residents out of bed daily, especially for meals? Does Nursing Care seem to be individualized for each resident? (Likes and dislikes included in care; title used for resident, etc). Observe a typical day for a resident (baths, medications, meals, rest periods, etc). Is reality orientation or some other form of mental stimulation provided in daily routine?
9. Allied Health Care	What other services are available? How often are they available during week? (e.g., physical therapy, occupational therapy, social worker, dietician, dentist, podiatrist). How does someone contract for their services? Fees?

General Item	Factors To Consider
10. Daily Activities in General	What time are meals served? Eat a meal there if possible. Evaluate taste and texture of food.
	Are snacks provided?
	May residents bring guests?
	What type of special diets are provided?
	Make an appointment with the dietician.
	Interact with several residents. Are they satisfied with care? Why or why not?
11. Recreational Opportunities	How many personnel work in recreation department?
	What type of programs are provided? Frequency? Ask to see a monthly schedule.
	What outdoor recreational activities are available?
	Are special field trips, parties arranged?
12. Transportation	Any public transportation in nearby vicinity?
	Does nursing home provide transportation for residents?
	Is there a fee for transportation?
	Can home's transportation be utilized for regular services (e.g., shopping, church services, etc).
Other Services	What type of housekeeping is provided for residents' rooms?
	Is there a difference for different levels of care?
	Is there a fee for services?
	Are any religious services provided in the home?
	Does a barber or hairdresser visit the home? Fee?

A Living Will

To My Family, My Physican, My Lawer and All Others Whom It May Concern

Death is as much a reality as birth, growth, maturity and old age—it is the one certainty of life. If the time comes when I can no longer take part in decisions for my own future, let this statement stand as a expression of my wishes and directions, while I am still of sound mind.

If at such a time the situation should arise in which there is no reasonable expectation of my recovery from extreme physical or mental disability, I direct that I be allowed to die and not be kept alive by medications, artificial means or "heroic measures". I do, however, ask that medication be mercifully administered to me to alleviate suffering even though this may shorted my remaining life.

This statement is made after careful consideration and is in accordance with my strong convictions and beliefs. I want the wishes and directions here expressed carried out to the extent permitted by law. Insofar as they are not legally enforceable, I hope that those to whom the Will is addressed will regard themselves as morally bound by these provisions.

Signed _____

Date _____

Witness _____

Witness _____

Copies of this request have been given to _____

_____

_____

_____

Glossary

accommodation (1) Adaptation or adjustment of the eye including changes in the ciliary muscle and lens in bringing light rays from various distances to focus upon the retina. (2) In Piaget's terms, changes made in the existing schemata to include new experiences.

acute otitis media Acute inflammation of the middle ear characterized by fluid accumulation.

adaptation The process of adjusting to new conditions.

amblyopia Dimness of vision.

ampulla The distal middle part of the oviduct, the usual site of fertilization of the ovum.

anaclitic depression A kind of depression seen in infants deprived of maternal contact, characterized by emotional withdrawal, arrested development, lack of appetite, and weight loss.

anal stage In Freudian theory, a stage in which gratification is derived from both elimination and retention of feces; occurs around age two.

animism The attribution of life and purposefulness to inanimate objects.

anticipatory grief The process of working through part of one's grief before the bereavement occurs.

anxiety A feeling of apprehension, uncertainty, or tension stemming from an imagined or unidentified threat.

Apgar scoring An evaluation of the newborn's basic life processes, made at approximately one minute after birth and again at five minutes after birth.

artificialism The belief that all objects and events have the sole purpose of satisfying human needs.

assimilation Perceiving and interpreting new information in terms of existing knowledge and understanding.

attachment An affectionate, reciprocal relationship between two people.

autonomy versus shame and doubt One of the developmental crises described by Erikson, typically occurring during toddlerhood; its successful resolution is characterized by a feeling of confidence and self-control.

basal metabolism rate (BMR) The amount of energy expended per unit of time under basal conditions; usually expressed as large calories per square meter of body surface per hour.

behavior modification The use of rewards and punishments to respectively reinforce and extinguish certain behaviors.

bereavement The psychological and physiological response to a significant loss.

blastocyst The cluster of cells that begins to differentiate itself into distinct parts during the germinal period of prenatal development.

body image The mental image one has of one's body.

caput succedaneum Swelling or fluid retention

in the soft tissue of the head resulting from trauma during delivery.

cardiac index   The volume per minute of cardiac output per square meter of body surface.

cataract   Partial or complete opacity of the crystalline lens or its capsule.

central visual acuity   Vision of the eye turned directly toward the object so that the image falls upon the macula lutea, providing the most acute vision.

centration   Inability to consider more than one characteristic of an object at a time.

cephalhematoma   An accumulation of blood between the periosteum and the flat cranial bones.

cephalocaudal development   Development that proceeds from head (cephalo) to tail (caudal).

cervix   The lower dilatable portion of the uterus.

chemical conjunctivitis   A condition commonly seen in the newborn after the instillation of silver nitrate drops.

child abuse   The physical or mental injury, sexual abuse, or maltreatment of a child under 18 years of age.

chorion   A protective outer sac that develops from tissue surrounding the embryo.

classical conditiong   A type of learning originated by Pavlov in which a subject responds to a previously neutral stimulus after it has been effectively paired with the unconditioned stimulus that originally produced the response.

climacteric   A transition period between the biological states of middle age and older age during which the body undergoes marked changes.

cognitive development   Development of the ability to understand and reason.

collective unconscious   Jung's term for the sum of all human memory.

conceptualization   The process of mentally conceiving ideas and forming concepts.

conditioning   Learning that occurs through association between stimuli and responses, as in classical or operant conditioning.

convergence   Inclination or direction toward a common point, center, or focus as the axes of vision falls upon the near point.

cooperative play   The highest form of social play, in which several children engage in a common activity.

coping   The use of available behaviors and defenses by the individual to decrease stress and deal with difficulties.

corpus   The body, or middle part, of the uterus.

critical growth periods   Periods during which specific tissues and organs display more rapid growth in utero and are more susceptible to environmental influences.

critical periods   Periods in development when the acquisition of certain behaviors is more easily facilitated.

croup   A viral infection in which relatively severe laryngeal spasms accompany a less serious inflammation.

crystallized intelligence   The use of reasoning to solve problems of everyday living; dependent on learning and experience.

Denver Development Screening Test (DDST)   A commonly used gross screening instrument based on predictable developmental patterns common to all individuals.

development   Physiological process by which the individual progresses from an undifferentiated state to a highly organized and functional capacity.

developmental assessment   A clinical estimate of developmental progress.

developmental crisis   A transient period of stress in an individual's life related to attempts to master certain developmental tasks of that period.

disengagement   The gradual withdrawal of an individual's emotional energy from his work and social structures while he turns inward in anticipation of his own needs.

dramatic play   A form of play involving pretending to be various animate and inanimate objects.

dysfunctional uterine bleeding   Irregular intermenstrual spotting or bleeding or excessive bleeding in a woman who has previously had normal menstrual flow.

dysmenorrhea Painful menstruation.

ectoderm The outside layer of the embryonic disc that gives rise to the skin and its appendages.

ego According to Freudian theory, the conscious component of the psyche that mediates the id and superego.

egocentrism The tendency of children to see things only from their own perspective.

ego-integrity versus despair The last developmental crisis described by Erikson, occurring in old age; its successful resolution is characterized by wisdom and the acceptance of one's impending death.

ejaculatory ducts Two ducts that release semen into the male urethra.

embryonic period The prenatal period from week 4 through week 8 during which organ growth and development takes place.

encopresis Involuntary defecation.

endometrium The inner vascular layer of the uterus which is shed during each menstrual period.

entoderm The inside layer of the embryonic disc which gives rise to structures such as the thyroid and thymus glands, the liver, and the biliary, intestinal, and respiratory tracts.

enuresis Involuntary discharge of urine, especially while asleep; bedwetting.

epididymis Oblong structure lying along the superior and posterior aspects of each testicle that stores immature spermatozoa.

fallopian tubes Tubular structures attached to the uterus at one end and opening into the peritoneal cavity. They are composed of four sections: infundibulum, ampulla, isthmus, and an interstitial section.

fetal period The prenatal period from week 9 through week 40 during which continued organ growth and differentiation take place.

fine motor behavior Movements of the fingers.

fluid intelligence The ability to perceive relationships, reason inductively, do abstract thinking, and form concepts; relatively independent of education and experience.

fontanel Six soft spots on the neonatal skull where the cranial bones have not yet ossified.

formal operations According to Piagetian theory, the stage at which a child becomes capable of hypothetical, deductive, and abstract thinking; occurs between 11 and 15 years.

fundus The upper, rounded top portion of the uterus.

gametogenesis The process through which immature male and female gametes mature and become ready for fertilization.

generativity versus stagnation In Erikson's terms, the developmental crisis of middle age; its successful resolution is characterized by involvement with the next generation.

genital stage A Freudian stage emerging at puberty in which the desire to achieve gratification via genital stimulation is increased; prerequisite to adult sexuality.

genotype The hereditary constitution of an organism resulting from its particular combination of genes.

geriatrics The branch of medical science concerned with aging and its diseases.

gerontology The scientific study of the phenomena and problems of aging.

grief A sequence of affective, cognitive, and physiological states following directly after an irretrievable loss.

gross motor behavior Activity requiring head, trunk, and extremity control.

growth The physiological process by which an organism assimilates or transforms essential, nonliving nutrients into living protoplasm.

habituation The process, resulting from prolonged stimulus presentation, by which one ceases to attend to the stimulation; learning.

hospice A facility that provides care for a person with a terminal illness; the aim of such a facility is to assist the person to remain free from pain and to prepare for death mentally and spiritually.

hydrocephalus An abnormal collection of cerebrospinal fluid in the ventricles of the brain.

hyperopia A refractive error in which the focus of parallel rays of light falls behind the retina; farsightedness.

id According to Freudian theory, the repository of all basic instincts; it is totally unconscious and operates on the pleasure principle.

industry versus inferiority The Eriksonian developmental crisis of school age; its successful resolution is characterized by a sense of competence and a good feeling about oneself.

initiative versus guilt In Erikson's terms, the developmental crisis of the preschool years; its successful resolution is characterized by eagerness to apply one's energy to new tasks and learning, as opposed to a sense of guilt and inadequacy.

intimacy versus isolation According to Eriksonian theory, the developmental crisis of young adulthood, the successful resolution of which is characterized by the ability to establish and maintain a loving relationship.

invention of new schemata A substage in Piaget's sensorimotor stage of cognitive development that occurs at approximately 18 to 24 months of age; characterized by true thought that enables the child to solve problems.

irreversibility A cognitive limitation of preoperational thought characterized by the inability to reverse a line of thought; for example, the 5-year-old who has a brother, asked if his brother has a brother, will say "no."

labor Series of processes through which cervical effacement (thinning out) and dilatation (opening) occur and during which the fetus descends through the birth canal and is delivered.

latency According to Freudian theory, a period of relative sexual quiescence, occurring around age 6.

libido Sexual drive.

linguistic theory Chomsky's contention that an infant is born with an innate ability to respond to and learn a language.

maturation The process of arriving at full development.

maturational crisis A predictable developmental period of disequillibrium.

menarche The onset of menstruation.

meningocele A deformity of the vertebral column characterized by a protruding sac encasing the meninges and cerebrospinal fluid.

menopause The time of cessation of the menstrual cycle and ability to conceive.

mesoblast A cellular layer of the embryonic disc that gives rise to muscle, skeleton, gastrointestinal tract, genetourinary tract, heart, circulatory and connective tissue.

milia Pinhead-size papules caused by sebaceous gland occlusion.

moro reflex The "startle" reflex seen in an infant, characterized by extension, encircling, and flexion of the arms in respnse to a sudden stimulus such as a sharp blow on the surface on which the infant is lying.

molding Temporary asymmetry of the fetal head resulting from overlapping of the bones of the skull during labor and delivery.

mourning A period of exhibition of grief over the death of a person.

myelin The tissue forming sheaths around nerve fibers.

myelination The process of accumulating myelin during the development, or repair, of nerves.

myometrium The middle layer of the uterus, composed of mostly smooth, highly contractile muscle.

myopia An optical defect, usually due to too great length of the anteriorposterior diameter of the globle, where by the focal image is formed in front of the retina; nearsightedness.

neonatal period   The first 28 days of life.

nursing bottle syndrome   Dental caries caused by prolonged use of a nursing bottle.

object permanence   In Piagetian theory, the ability to recognize that objects continue to exist when they are out of sight.

oral stage   The first stage in Freud's scheme of psychosexual development, in which the mouth is the center of gratification and exploration.

ossification   The conversion of tissue into bone.

otitis externa   Inflammation of the external ear.

otitis media   Inflammation of the middle ear.

ovaries   The female gonads; they are flat, ovoid or almond-shaped structures located in the upper pelvic cavity. They produce and release estrogen and progesterone.

parallel play   A type of play often seen in the young toddler, in which several children play alongside one another but do not play together.

person permanence   The ability to recognize that people continue to exist even when they are out of sight.

phallic stage   In a Freudian theory, a stage occurring in the preschool years, during which the child discovers that mainpulation of the genitals produces pleasurable sensations.

phenotype   The physiological traits that characterize a person; the visible manifestation of a genotype.

preconceptual period   According to Piaget, an early phase of the preoperational stage, occurring around age 2, during which the child is capable of creating mental images of objects or actions that are not present or are not being experienced.

premature baby   An infant born before the thirty-seventh week of gestation.

preoperational stage   According to Piaget, a pe-

riod of cognitive development occurring between ages 2 and 7, characterized by the ability to use mental symbols.

presbycusis   The loss of hearing that occurs with age.

presbyopia   Hyperopia and diminished power of accommodation of the eye caused by aging.

primary amenorrhea   Absence of menses by age 17.

primary circular reactions   A substage of Piaget's sensorimotor stage, occurring from 1 to 4 months, characterized by purposeful efforts to reproduce events (such as finger-to-mouth contact) that were previously attained by chance, and by the infant's preoccupation with his own body.

proximodistal development   Development that proceeds from near the central axis of the body toward the extremities.

puberty   The point at which one is sexually mature and capable of reproduction

pubescence   That period of the life span characterized by rapid physical growth, maturation of the reproductive organs, and appearance of primary and secondary sex characteristics.

realism   A cognitive limitation of toddlerhood, characterized by the confusion of physical realities with psychological events, such as the insistence that the events in a dream actually occurred.

reflexive stage   The first substage of Piaget's sensorimotor stage, characterized by reflexive responses; occurs during the first month of life.

regression   A defense mechanism characterized by a return to behaviors that are more appropriate to an earlier stage in life.

repression   A defense mechanism in which ideas and feelings are blocked from consciousness.

schema (pl., *Schemata*)   Piaget's term for a basic cognitive unit.

secondary amenorrhea   Absence of menses for

over 150 days in a woman who previously menstruated regularly.

secondary circular reactions A substage of Piaget's sensorimotor stage, occurring between 4 and 8 months, in which behavior becomes more purposeful, and in which the infant becomes more aware of his environment.

self-actualization Maslow's term signifying maximal utilization of one's potential for development.

sensorimotor stage Piaget's term for a period of cognitive developement lasting from birth to approximately 2 years of age, characterized by the combined use of the senses and body movements as a means of interacting with the environment.

separation anxiety A child's fear of separation from the significant other (usually the mother).

situational crisis A transient period of stress arising from an attempt to cope with a specific circumstance.

solitary play A type of play in which the child plays alone.

spina bifida A neural tube defect characterized by incomplete closure of the vertebral column and, sometimes, by protrusion of the meninges and spinal cord through the opening.

sublimation A defense mechanism characterized by the rechanneling of uncomfortable or unacceptable impulses or feelings into acceptable activities.

superego The Freudian term for that aspect of the psyche that functions as the conscience.

symbolic representation An act of reference in which a mental image is created to stand for something that is not actually present.

tertiary circular reactions A substage of Piaget's sensorimotor stage, occurring at 12 to 18 months of age, characterized by the seeking out of new experiences rather than duplicating those that have occurred previously.

testes The male gonads; they are ovoid and slightly flattened. Located in the scrotum, they produce spermatozoa and secrete testosterone.

trust versus mistrust The developmental crisis of Erikson's first psychosocial stage, during which an infant either does or does not develop basic trust in the environment.

uterus Muscular organ capable of growth and distention during pregnancy.

zygote A cell formed by the union of a sperm and an ovum.

Index